Geography of British Columbia

Geography of British Columbia

People and Landscapes in Transition

3rd Edition

Brett McGillivray

UBC Press · Vancouver · Toronto

21 20 19 18 17 16 15 14 13 12   5 4 3 2

Printed in Canada.

Library and Archives Canada Cataloguing in Publication

McGillivray, Brett, 1944-
Geography of British Columbia : people and landscapes in transition / Brett McGillivray. – 3rd ed.

Includes bibliographical references and index.
ISBN 978-0-7748-2078-3

1. British Columbia – Geography – Textbooks. I. Title.

FC3811.M33 2011   917.11   C2010-906812-2

---

e-book ISBN: 978-0-7748-2079-0 (pdf)

Canadä

UBC Press gratefully acknowledges the financial support for our publishing program of the Government of Canada (through the Canada Book Fund), the Canada Council for the Arts, and the British Columbia Arts Council.

UBC Press
The University of British Columbia
2029 West Mall
Vancouver, BC V6T 1Z2
www.ubcpress.ca

Contents

Illustrations, Figures, and Tables

ILLUSTRATIONS

FIGURES

TABLES

Preface

This text is the culmination of years of teaching the geography of British Columbia at Capilano College/UNIVERSITY. Several books have been particularly influential in its creation: Roderick Haig-Brown's *The Living Land* (1961), Mary Barker's *Natural Resources of British Columbia and the Yukon* (1977), and Charles Forward's *British Columbia: Its Resources and People* (1987). Each has done an admirable job in assessing the variety of landscapes and issues in British Columbia. Albert Farley's *Atlas of British Columbia: People, Environment, and Resources* (1979) is another important resource because it not only provides many useful maps but also contains much additional information about the province. I used *British Columbia: Its Resources and People* as a text for years because of my preference for a geography that examines themes and stresses an historical perspective. References to other works that have influenced this text can be found at the end of each chapter.

This edition of *Geography of British Columbia: People and Landscape in Transition,* like the previous ones, has come as a result of students and other instructors using the text and requesting updated and new information on the province. For example, information on climate change and its causes and consequences has been added, as has information on accreted terranes and a new section on volcanic activity and landscapes. British Columbia is a province in transition, and there have been many events and conditions that have altered and modified the landscape since the last edition. The treaty process continues to be a long, drawn-out affair, but it is essential to the economic, political, and social development of the province. The Tsawwassen Treaty, the first urban treaty, is now discussed in this edition, along with other agreements in their final stage of becoming treaties. In addition, First Nations have been asserting their influence over Crown lands, preventing activities that the provincial government has granted. The prime minister offered an apology, with compensation, to the Chinese who paid a head tax and to First Nations who attended residential schools. An apology was also extended to the Sikh community for the *Komagata Maru* incident. Because the apology was offered in an informal fashion, not in Parliament, however, it has sparked controversy.

Global uncertainty has escalated with conflicts in Iraq, Afghanistan, and other parts of the world, and pandemics such as H1N1 (swine flu) continue to alarm and threaten the public globally and, in turn, influence travel patterns and industries such as tourism. Perhaps the greatest economic threat has been the collapse of the US economy in late 2008, which set off a worldwide recession and political policies to buffer the economic downturn.

Since the publication of the last edition, the resource industries of British Columbia, which are so critical to the economic well-being of the province, have been subjected to major challenges. The world market prices of oil, natural gas, and coal have reached all-time highs, resulting in more exploration, new mines, and new debates about off-shore oil drilling. The world market price for many metals (e.g., copper, gold, silver, and molybdenum) has followed a pattern similar to that of energy resources with similar results: more exploration and mines. The recession of 2008, quite naturally, had a dampening effect on mineral prices, with one exception – gold.

The forest industry, which has been the economic backbone of British Columbia, did not experience the "good times" as did the mining and energy industries. It is an industry that has been mired in international conflict with the United States for a long time, although an agreement was signed in 2006. The agreement has not been favourable for British Columbia and has caused changes in the tenure situation. In addition, the forestry industry is not healthy economically – old plant and equipment, foreign competition, mountain pine beetle infestation, climate change, massive forest fires, and low world market prices have taken the competitive edge from mills. Furthermore, dependence on the US market for so many forest products has made wood commodities vulnerable, and the recession of 2008 has made the situation considerably worse by causing a great number of sawmills, planer mills, and pulp mills to close. All of these developments are discussed in the chapter on forestry.

The chapter on the fishing industry, which focuses on salmon, now has a more ominous tone. The condition of wild salmon has continued to deteriorate (with the exception of the unusually large sockeye salmon run in the Fraser river in 2010), and the value of farmed salmon now exceeds the commercial harvest value of all wild fish. The chapter examines the complexities of the issue and assesses policies to make the salmon fishery sustainable.

Electricity demand and production has likewise sparked new rounds of debate as to where and how future kilowatt hours will be generated. What role should Independent Power Producers (IPPs) play in relation to BC Hydro, and how controversial is the option of Site "C," the next megaproject on the Peace River to be developed? To answer these questions, concerns about the environment, including greenhouse gases and the need for green energy and conservation, must be considered.

Water is also increasingly an issue, not only in terms of how to make it safe (potable) but also in terms of how we treat it and return it to the system. Chapter 13 discusses water management and the vital need for source water protection as global warming changes discharge rates and creates demand for policy changes.

Tourism continues to be an important industry in British Columbia, but it too has been affected by many unpredictable world events such as pandemics, border-crossing hassles, changes to the value of the Canadian dollar, and recession. Conversely, tourism had a major economic boost with all the construction, tourism, and publicity generated by the 2010 Winter Olympics. The legacy of infrastructure will only enhance the tourism industry in the future.

The fluctuation in economic conditions for all of the resource industries in the province are reflected in demographic changes such as interprovincial migration and immigration generally, employment and unemployment, rural-to-urban migration, and the rates of growth or decline of communities. Single-resource communities continue to be the most vulnerable communities, and the pattern of urbanization clearly favours the core region of the province.

Despite new additions and emphases, this third edition of *Geography of British Columbia* remains selective, with an emphasis on the human side of geography, and designed for a one-term introductory course. I remain convinced that the topical or thematic approach, as opposed to a regional approach, is a better means of gaining the interest of students, especially since not all students have a background in geography or a familiarity with British Columbia. From my perspective, each theme is a story (containing many stories in turn), and these are stories that British Columbians as well as those interested in this province would benefit from knowing.

I also recognize that the study of British Columbia from a regional perspective is equally valid, and many instructors prefer this approach because it focuses much more on the human and physical features that make the regions within British Columbia unique. The thematic approach adopted here is not devoid of regional concerns, and all chapters, especially the first and last, pay considerable attention to the regional development of the province. The content of these chapters have been updated along with the statistics used in the tables and graphs. The stories of this province are only enhanced through this edition.

Acknowledgments

For this third edition I have a number of people to thank, including the librarians at Capilano University and my colleagues in the social sciences. My geography colleagues in particular have been most helpful and thanks goes to Charles Greenberg, Cheryl Schreader, Jeanne Mikta, Karen Ewing, Sheila Ross, and Diane Tanner. Bob Patrick at the University of Saskatchewan was particularly helpful with respect to the chapter on water. Wim Kok of Northern Lights College has provided many useful comments, particularly with respect to issues related to energy. Gilles Viaud from Thompson Rivers University has given me a considerable understanding of urban issues and I am grateful. The Geography Articulation Meeting held each spring has been a source of considerable information: the field trips at each of these events along with the mix of geographers provided insight and knowledge of the processes shaping the settlement and development of the various regions where the meetings were held. Thanks go to all those who organized the field trips and meetings. Past students have also had an influence on the content of this edition. My wife and partner, CarolAnn Glover, is again to be thanked for the time-consuming task of editing my many drafts of this third edition. Finally, I would like to thank Laraine Coates, production editor, along with the rest of the UBC Press staff, who assisted in putting this edition together. Any errors and omissions are the fault of the author.

Geography of British Columbia

British Columbia

A Region of Regions

USING GEOGRAPHY TO MAKE SENSE OF THE LANDSCAPE

The focus of geography is the landscape, both human and physical, and employing geographical knowledge gives meaning to the changes that are constantly shaping and modifying the landscape. Those who study the human landscape look at where people live, their activities, and how they have modified the landscape. Physical geographers, on the other hand, are interested in the many physical processes that influence the landscape. Of course, landscapes are frequently modified by a combination of physical and human processes, and common to both sides of the discipline is that geography always incorporates a spatial perspective. The two broad divisions in the field – the human and the physical – can be divided into a number of subfields. As Figure 1.1 shows, geography is associated with many other disciplines, but the spatial element keeps it distinct.

British Columbia is a large province, encompassing nearly 950,000 square kilometres, and is extremely varied from both a human and a physical perspective. Many nation-states are significantly smaller, and few have such a variety of landscapes. Nevertheless, fewer still have such a small population (4.46 million in 2009) in relation to area – a density of only 4.7 persons per square kilometre (Great Britain, four times smaller in size, had 264.9 persons per square kilometre in 2008 [CIA 2009]). It is a helpful exercise to compare the size of British Columbia to that of countries around the world and to locate the province in relation to those countries. A map of Canada reveals the size and spatial relationship of British Columbia to other provinces and territories within Canada. One theme of this text considers how British Columbia is unique within Canada.

One definition of geography is the study of "where things are and why they are where they are" (McCune 1970, 454). "Things" can be physical features, people, places, ideas (or human innovations), or anything in the landscape. "Where" questions concentrate on location as well as recognizing physical and human patterns and the distribution of various activities, people, and features of the landscape. Many of these questions can be answered simply by looking at a map, and students are encouraged to acquire a road map of British Columbia. Where is Barkerville? Where is the territory covered by Treaty 8? Where do earthquakes occur? Where are the sockeye spawning grounds? Where are Prince George and Prince Rupert? Knowledge of where things are is basic and essential geographical information. A useful beginning to test your knowledge of British Columbia is to draw a map of the province from memory and to place on it the features you consider important. This cognitive mapping exercise reflects individual landscape experiences (which can be shared with others) and demonstrates the importance of location. Using maps to answer "where" questions is the easiest aspect of geographical study.

Why are things where they are? "Why" questions are far more difficult than "where" questions and ultimately may verge on the metaphysical. Even so, students are encouraged to

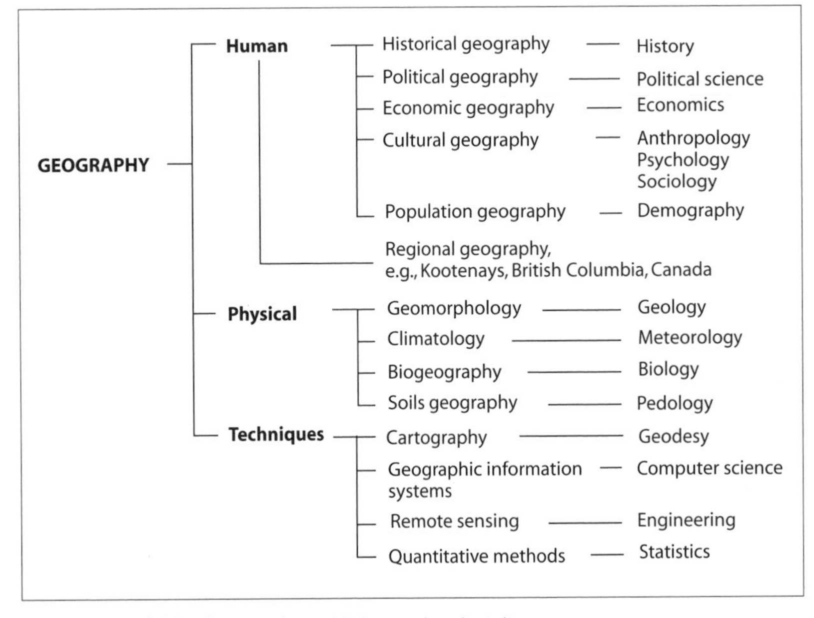

Figure 1.1 Subfields of geography and links to other disciplines

conduct research about and to analyze the various physical, economic, political, cultural, and historical factors involved in a specific location or locational patterns, whether it is the location of a type of vegetation, of a community, of a group of people, or of a resource. Why does sagebrush dominate the dry southern interior valleys? Why is Vancouver where it is and why has it grown so rapidly? Why did Barkerville become a ghost town? Why were the Japanese removed from the coast of British Columbia? Why did the Nechako River get dammed for hydroelectric power? Why is the Peace River region not part of Alberta? These questions are not easy and often require historical, physical, cultural, political, and economic assessments.

Other geographers add the "what" question: "what is the significance of these locational patterns?" (Renwick and Rubenstein 1995, 5). What influence do people have on the environment, and what influence does the environment have on people? Humans are constantly shaping and modifying the landscape to meet the demand for resources – clear-cutting forests, damming rivers, and building plants that pour emissions into the air and water – and these acts produce an environmental backlash to ecosystems and human health.

All of these questions – where, why, what – mean geography is a practical and pragmatic discipline, one that encourages an understanding of the surface of the earth on all geographic scales. Geography is a discipline that lends itself to being out of the classroom and in the environment, where one can read human and physical landscapes. The study of geography leads not only to an understanding of the processes responsible for settlement and development but also to an understanding of where people live and work and how they re-create and modify the surface of the earth. Geography allows us to recognize the range of physical characteristics responsible for mountain building and erosion; weather and climate patterns; and why some communities and regions are at considerable risk from floods, forest fires, or avalanches and how these risks can be reduced or eliminated. Geography develops the critical thinking skills necessary to unravel the complexity of spatial patterns, processes, and relationships, and these skills open up many career opportunities.

The geography of British Columbia is changing constantly. News headlines inform us of these changes and provoke where, why, and what questions: West Fraser to shut Kitimat mill; 535 jobs lost; Move will have huge impact on town, mayor says (29 October 2009) ... BC forest fires could cost $400-million (September 2009) ... Swine-flu vaccine shortage in British Columbia (October 2009) ... Pine beetle threatens Canada's boreal forest ... Mountain pine beetles are expected to wipe out 80 percent of BC's pine forest by 2013 (April 2008) ... British Columbia Utilities Commission rejects BC Hydro's long-term acquisition plan (August 2009) ... Vanished Fraser sockeye salmon run seen as disaster, mystery (August 2009) ... Prime Minister Harper offers full apology on behalf of Canadians for the Indian residential schools system (11 June 2008) ... Residential school compensation nearing completion (October 2008) ... First urban treaty in BC history takes effect today (3 April 2009) ... BC's forest jobs crisis (July 2008) ... Ottawa to announce $1 billion in forestry aid; BC expected to get up to $440 million of the package aimed at making the industry "greener" (June 2009) ... Recession slams tourist industry (August 2009). Common to such complex issues is the idea that human and physical processes are altering the landscape. The chapters that follow are intended to raise the where, why, and what questions and to provide the means to understand some of the processes influencing change.

Another theme throughout *Geography of British Columbia: People and Landscapes in Transition* is the importance of history, particularly how past decisions and actions have shaped the landscape of the present. From a European, colonial perspective, British Columbia has little history of settlement and development compared to eastern Canada or to many other nations in the world. Aboriginal people, however, have approximately 10,000 years of history in British Columbia, and anthropologists and archaeologists are still adding new evidence of their settlement patterns and use of resources. From the viewpoint of physical geography, changes to the landscape are often measured in several hundreds of millions of years, and the BC landscape is no exception.

The combination of physical processes in British Columbia has produced a spectacular variety of mountains, rivers, lakes, islands, fjords, forests, and minerals. Within this physical setting, First Nations people, and later non-Natives, settled, exploited, and altered the landscape, sometimes irreversibly. These physical-human interactions prompt us to look at the landscape another way, namely, from an environmental perspective.

Meshing with the idea of change over time in such a rugged and physically challenging province is the theme of movement over space. **Time-space convergence** (sometimes referred to as time-space compression, or collapse) refers to the changing technologies of movement that shrink time and space. Today, for example, a flight from London, England, to Vancouver, British Columbia, takes approximately nine hours. In 1790, a trip to the Pacific Northwest (there was no Vancouver until 1886) from London meant sailing around South America and took nearly seven months. By 1890, the voyage from London to Vancouver was reduced to three weeks with the introduction of steam-driven vessels and the building of the Canadian Pacific Railway across Canada. This change in transportation affecting the movement of goods and people had major implications, not only to settlement patterns but also to resource development. Railways (along with ship transport) promoted the movement of high bulk, low-value goods (e.g., wheat, lumber, and coal) over great distances with relatively low freight rates. With the advent of telegraph lines in 1890, movement of information between London and Vancouver was reduced to three days (Harris 1997). By comparison, current satellite systems provide instant global communications, with a consequent worldwide reorganization in the production of goods and services and spatial relationships generally.

Time-space convergence describes a shrinking world; however, the shrinking of time-space does not occur evenly. Geographic locations that are connected differ, sometimes greatly, from locations that are not connected. Freeways such as the Coquihalla give the advantage of rapid movement to southern British Columbia communities, while northern and coastal communities are significantly more isolated with only secondary roads or, in some cases, no roads. Similar comparisons occur with airports, railways, port facilities, pipelines, and communications infrastructure. These transportation developments play a significant role in settlement, development, and economic advantage throughout the province.

Closely associated with time-space convergence is **spatial diffusion**. Here, the focus is on the movement (or flow) from one location to another of goods, innovations and ideas, services, and people, and tracing where they move. For example, the spatial diffusion process is used to trace where new innovations in farm equipment, new seeds, or agricultural practices occur, and where they are adopted. The spatial diffusion process is also used to describe events, such as the waves of smallpox epidemics and their impact on First Nations, the evolution and pattern of salmon cannery location, or the spread of high speed Internet service. All of these movements were influenced by "carriers" and "barriers" (Gould 1969). Carriers are instrumental in the spread and adoption of innovations, goods acquisition, or the contraction of diseases; barriers prevent, or block, this movement. "Relocation diffusion" refers specifically to the movement of people from one place to another. Both barriers and carriers apply to relocation diffusion, but the terms "push" and "pull" factors are also applied. Push factors include the many factors that cause people to move, such as overpopulation, warfare, religious persecution, and a host of other political, economic, and social forces. Pull factors are the opposite, reflecting the various conditions that attract people to a new location. Both push and pull factors have been responsible for moving people to British Columbia.

Statistics are useful in assessing trends and patterns, and Table 1.1 indicates the increases in British Columbia's population along with the transformation from rural to urban environments. Isolation was a major factor in early non-Native settlement and development. However, time-space convergence increased accessibility and allowed greater opportunity for economic trade, development of resources, and settlement, particularly urban settlement. These changes affected the regions of British Columbia in vastly different ways and also resulted in a much more complex society. The conquering of distance also facilitated the global transition, especially in trade and investment, from the Atlantic to the Pacific, tying British Columbia closely to the Asia-Pacific region.

Population increase is determined by calculating natural increase (births minus deaths) and net migration (immigration minus emigration) and combining the two results. Figure 1.2 illustrates how important immigration

Table 1.1

Rural and urban population, 1871-2011

Year	Population	10-year change	Percentage Rural	Percentage Urban
1871	36,247		84.8	15.2
1871	36,247	–	84.8	15.2
1881	50,387	14,140	72.0	18.0
1891	98,173	47,786	57.5	42.5
1901	178,657	80,484	49.5	50.5
1911	392,480	213,823	48.1	51.9
1921	524,582	132,102	52.8	47.2
1931	694,263	69,681	56.9	43.1
1941	817,861	123,598	45.8	54.2
1951	1,165,210	347,349	47.2	52.8
1961	1,629,082	463,872	27.4	72.6
1971	2,184,621	555,539	24.3	75.7
1981	2,744,465	559,844	22.0	78.0
1991	3,282,061	537,596	19.6	80.4
2001	4,078,447	796,386	18.0	82.0
2011[a]	4,572,600	494,153	15.0	85.0

a Estimated from BC Stats (2009).
Sources: Statistics Canada (1983a), series A2-14; 1983c, series A67-69; 2001a, table 051-0001; 2001b, table 109-0200; BC Stats (2009).

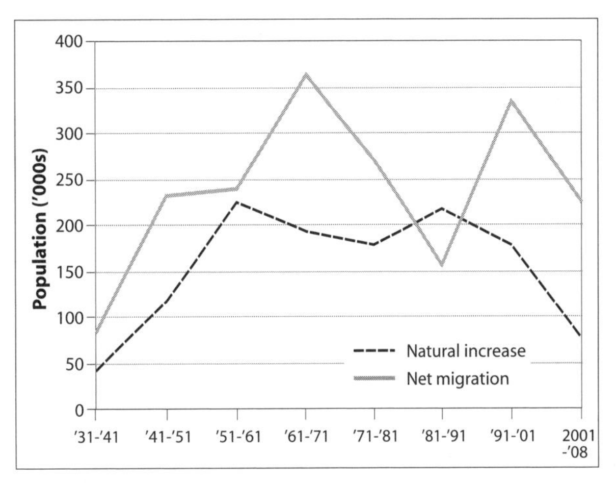

Figure 1.2 Natural population increase and net migration, 1931-2008
Sources: Statistics Canada (1983a, series A339-349; 2004, table 051-0013; 2010, table 051-0004).

has been to the overall population of the province. Immigration policies and economic conditions, however, influence these increases. The recession of the 1980s, for example, significantly reduced the number of people migrating to British Columbia. The recession that began in 2008 will likely have similar adverse effects on immigration.

TOPICAL OVERVIEW

Geography of British Columbia employs a primarily topical approach. Chapter 1 is an exception to this general direction in that it provides an introduction to geography and a regional overview of British Columbia's settlement and development. Chapters 2 to 6 introduce the basic physical and human landscapes and some of the processes that change these landscapes. Chapters 7 to 15 take an economic perspective as they concentrate on resources, resource management, and communities that depend on resources.

Most of this text concerns human geography, but the physical landscape is not ignored. Chapter 2 discusses the landscape in terms of the various physical processes that have changed and shaped it. These processes have created distinctive regional variations throughout British Columbia as well as marked contrasts to the rest of Canada.

Chapter 3 bridges the realms of physical and human geography with its focus on our human use systems and the risk from geophysical hazards. Chapters 4 to 6 present human geography themes: the historical geography of European settlement, the long history to present-day use of the landscape by First Nations, and, finally, the story of Asians in British Columbia. Each of these themes takes a historical perspective in examining the factors that influence the many changes to and regional variations in the landscape.

Resources, the theme of Chapters 7 to 15, have played and continue to play a large part in the economic well-being of this province as well as in attracting settlement. Chapter 7 begins with a discussion of resource management issues and the roles played by governments and corporations. Our use of, and dependence on, resources in this province have spatial and regional patterns. Each resource has its own unique history of development and influence on British Columbia. Consequently, forestry, fishing, metal mining, energy, agriculture, water, and tourism are examined in separate chapters. Single-resource communities, the theme of Chapter 15, illuminate the human link between resource development and the people who depend on it for a living.

Chapter 16, the final chapter, summarizes many of these themes and developments by reviewing 200 years of urbanization in this province. An urban view enhances the regional perspective of British Columbia because it shows the growth, and occasionally the decline, of communities throughout the province.

REGIONAL OVERVIEW

A **regional geography** approach is a means of assessing geographic areas that have common physical or human/cultural characteristics and can thus be distinguished from other regions (Gregory 2000, 687). From a regional perspective, British Columbia is a unique province within Canada for a variety of reasons. Physical characteristics set it apart from all other provinces. It has the youngest and highest mountains in the country and is often described as a vertical landscape. It also has the greatest amount of fresh water in Canada, an essential resource for the five species of Pacific salmon that also provides the potential for hydroelectric power. The highly indented coastline, "punctured by fjords," spans some 41,000 kilometres (Dearden 1987, 259). Weather and climate produce other distinctive patterns. The relatively mild, wet west coast, with the warmest winter temperatures in Canada, stands in contrast to the considerably colder and drier interior, with desert conditions in the southern river valleys. The interrelationship of climate, soils, and vegetation produces distinctive patterns from west to east in the province and also from south to north because of the eleven-degree span of latitude. Vertical change due to high mountain ranges produces regional variations similar to latitudinal differences.

Distinctive physical characteristics and a unique global location have influenced the human characteristics of the province. The pre-contact population of First Nations, particularly in coastal locations and along salmon-bearing rivers, was greater than anywhere else in Canada (Muckle 1998). Non-Native "discovery" and settlement was also unique in that it occurred from the west rather than from the east. British Columbia went through distinctive territorial struggles to become a British colony and further political struggles to establish its present boundaries. Its connection to the Pacific, and particularly to Asia, increased as transportation systems were developed. No other province has such a long history of immigration by Asians – first Chinese, and later Japanese and Sikhs. Nor did any other province gain the reputation of being so adamantly racist.

British Columbia has an abundant supply of resources, which have been the main attraction for the population and the reason for its rapid growth. Yet the physical characteristics of the province initially made resource extraction and export to distant markets difficult, and regionally differentiated patterns of settlement and development resulted (Robinson 1972). British Columbia is a region of regions and can be divided and subdivided, as it is throughout this text, on the basis of both physical and cultural characteristics.

River drainage systems, plateaux, mineral deposits, forests and vegetation, frost-free days, latitude, elevation, and precipitation are physical criteria by which the province can be divided into distinct regions. On the human or cultural side (**cultural regions**), features such as a common language or religion can demarcate regions, as can other political, economic, and social factors. For example, regional district and health board boundaries represent the organization of space based on political decisions, whereas fishing zones, tourist areas, forestry regions, newspaper circulation areas, policing jurisdictions, and school districts are regions derived more from economic and social functions. For many people who live in a particular area and have shared historical experiences, such as First Nations, there is a sense of place, or "nationalism," that comes with a connection to the land.

Traditional regional geography is a means of dividing British Columbia into parcels, or regions, for more critical examination of its characteristics and to make sense of its diversity. Critics of this approach point to the separateness of the resulting geographic areas; regions are not islands unto themselves but are linked in ways that are not captured by traditional regional geography. Moreover, the characterization of a region may be appropriate only at one point in time; consequently, regions need to be reconfigured as conditions change.

Today, regional geography is sometimes referred to as "reconstructed regional geography," thus distancing itself from traditional regional geography (Pudup 1988). This new direction takes into account the many complex relationships within any landscape, the many interactions linking adjacent regions, and even global conditions. An understanding of these relationships may be accomplished with a host of analytical tools, as well as by borrowing from other disciplines.

This new regional geography recognizes that regions can change over time and that regions may even overlap. For example, from a cultural and historical perspective, territorial boundaries or semi-nomadic regions divided Aboriginal peoples of the Pacific Northwest. These regional boundaries shifted because of warfare, scarcity of resources, and changes in technology. The greatest change of all, however, came with the arrival of Europeans, who reorganized the landscape into very different regions and placed First Nations on small parcels of land called reserves. The political boundaries of British Columbia have been drawn a number of times, but it is only since the 1990s that non-Natives have recognized the historical boundaries of First Nations.

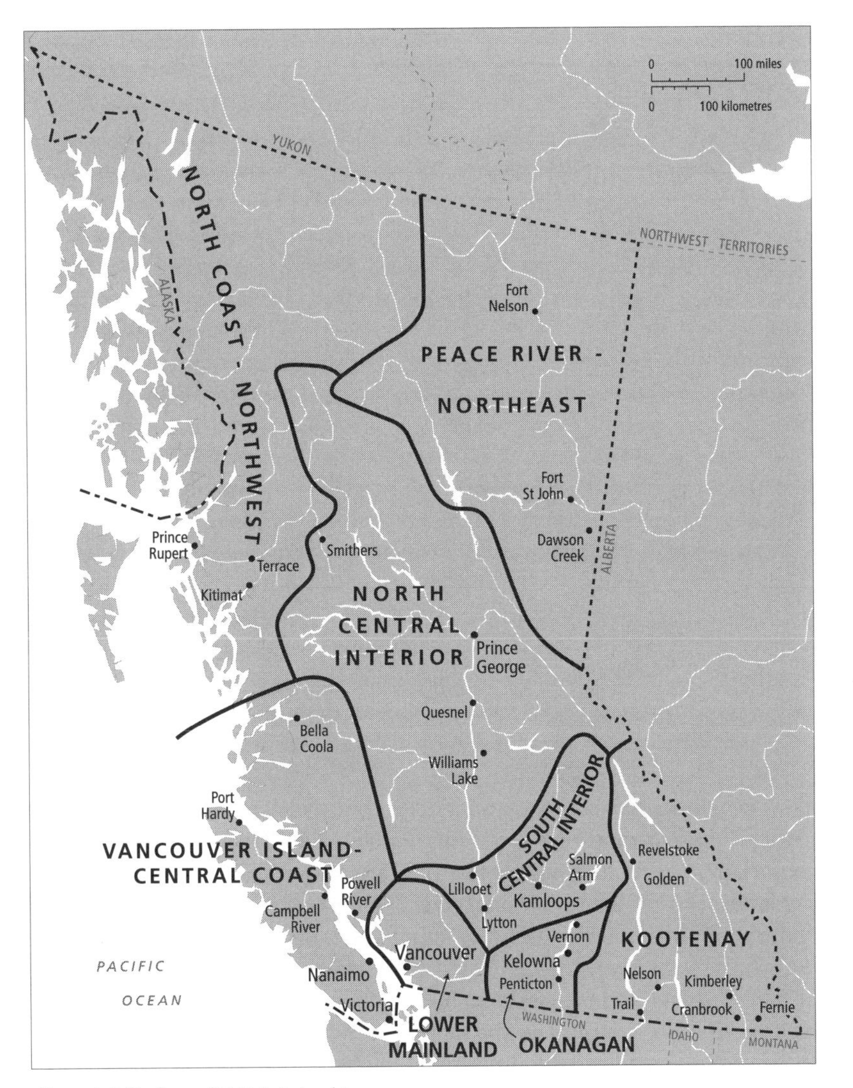

Figure 1.3 Regions of British Columbia

When dividing the province into meaningful regions, external regionalization must also be acknowledged. British Columbia was initially claimed by Spain before coming under the colonial control of Britain. Confederation broke the bonds of colonization and relieved the anxiety that this territory would be annexed by the United States; spatially, British Columbia then became one region of an independent Canada. More recently, the signing of the Free Trade Agreement (1989) and North American Free Trade Agreement (1994), along with increased trade and investment in the Asia-Pacific region, has placed British Columbia, and Canada, in a new, more global, regional economic alignment.

The organization of a regional geography of British Columbia must take all of these changing conditions into consideration; consequently, there are many regional approaches used to accomplish this goal. Figure 1.3 divides the province into eight regions, devised mainly by considering historical development in combination with census subdivisions. Table 1.2 provides the population change for these regions, spanning over 120 years. Maps, graphs, and statistics are some of the key tools of the geographic trade (the "techniques" shown in Figure 1.1). They

Table 1.2

Population by region, 1881-2008

Year	Vancouver Island/ central coast	Lower Mainland	Okanagan	Kootenay	South central interior	North central interior	North coast/ northwest	Peace River/ northeast
1881[a]	18,777	7,949	1,316	863	4,725	7,550	7,376	923
1891[a]	39,767	23,543	3,360	3,405	6,390	4,889	16,839	n/a.
1901[b]	54,629	53,641	7,704	32,733	14,563	5,123	9,270	948
1911[b]	84,786	183,108	21,240	51,993	24,103	9,011	16,595	1,644
1921	119,024	256,579	23,728	53,274	32,232	18,615	18,986	2,144
1931	133,591	379,858	30,919	63,327	37,621	23,236	18,689	7,013
1941	164,751	449,376	40,687	72,949	37,394	26,272	18,051	8,481
1951	233,250	649,238	62,530	93,256	50,363	41,324	20,854	14,395
1961	312,160	907,531	86,230	107,466	57,346	89,085	38,203	31,061
1971	415,254	1,256,425	130,498	125,643	106,993	133,906	63,080	45,155
1981	517,536	1,434,739	196,774	145,412	139,175	186,992	68,376	55,463
1991	611,654	1,829,537	240,291	141,480	142,628	190,141	67,975	58,355
2001	717,997	2,378,240	310,602	163,502	167,364	210,621	66,673	63,448
2003	719,914	2,440,750	315,690	164,012	167,721	209,075	65,461	63,957
2008	764,314	2,602,099	344,482	164,610	180,736	198,913	59,376	67,070

a Some approximations for regions as the province was divided into only five electoral areas.
b Some approximations for regions as the province was divided into only seven electoral areas.
Sources: Census of Canada (1951, table 6-84-88; 1971, table 8-92-96); BC Stats (2004b; 2008).

allow geographers to begin to understand the dynamics of the where, why, and what questions. For example, the construction of a map of each of the eight regions detailing such features as mountains, rivers, incorporated communities, Aboriginal title boundaries, and transportation systems, engenders familiarity with the vastness of the landscape, the physical and human features that distinguish each region, and the factors that integrate these separate regions with other parts of Canada and the world.

Monitoring regional population change, as Table 1.2 does, is another interesting exercise that compares and contrasts the regions with the greatest and least growth, as well as the fluctuations that occur within any of the eight regions. Graphing the absolute growth of each region is also useful as it gives a picture of the rate and trend of change and a sense of historical development. A comparison of just two regions – the Lower Mainland and Vancouver Island/central coast – provokes some interesting questions about rate of growth. Why did the Lower Mainland, a geographically smaller area, outstrip the Vancouver Island/central coast region so rapidly between 1901 and 1911? Bear in mind that the Lower Mainland includes the City of Vancouver, incorporated in 1886. Observe the slope of the graph lines between the years 1921 and 2008 in Figure 1.4, which shows the population increases of the two regions. What accounted for the different rates of population growth during this period, and why has the Lower Mainland population continued to grow at a more rapid rate?

The answers lie partly in the political decision to locate the Canadian Pacific Railway (CPR) terminus and an international port at Vancouver and in the consequent stimulation to economic growth and thus population. Employment opportunities related to obtaining and processing the primary resources of fish, forests, mines, and agriculture played a role in each region, and the port facility of Vancouver greatly widened the catchment area for exporting resources. The First World War, the opening of the Panama Canal, the Depression of the 1930s, and the Second World War were significant global events that affected each region. Following the war, technologies of

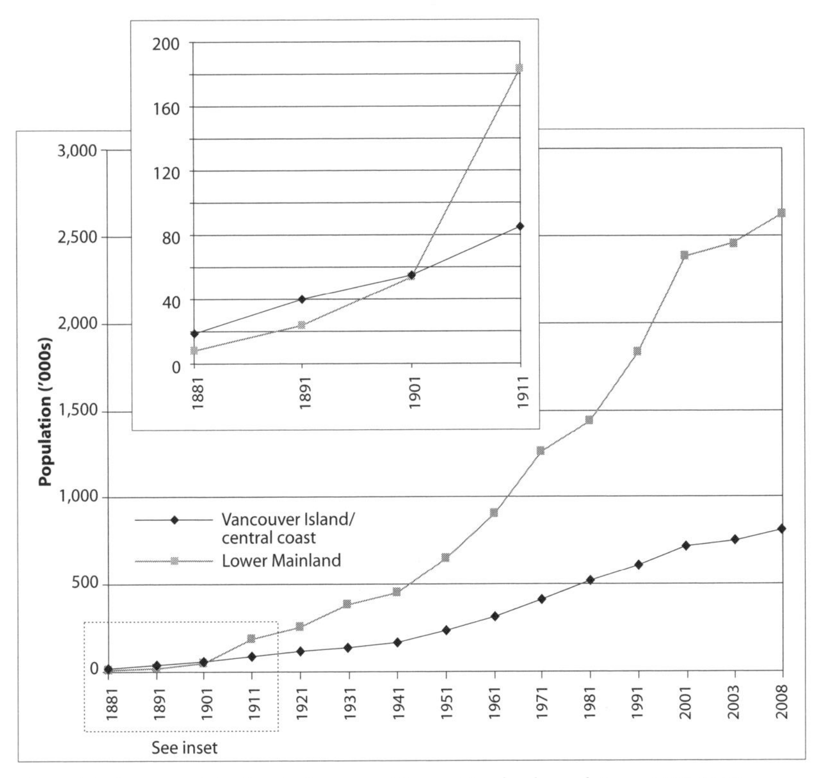

Figure 1.4 Vancouver Island/central coast and Lower Mainland populations, 1881-2008
Sources: Census of Canada (1951, table 6-84-88; 1971, table 8-92-96); BC Stats (2004b; 2008).

time-space convergence, in combination with more global means of producing goods and services, altered the way people made a living and reorganized the value of resources. The Lower Mainland region, with its greater connectivity – roads, rail, port, international airport, and conference centres – increasingly gained the greater proportion of the population.

A similar comparison could be made for any two regions and would require further analysis of the many factors influencing regional growth. Census figures can also reveal the ethnic composition of each region, and again these figures can form the basis of a host of social, ethnic, political, and economic questions. These lead in turn to geographic analysis of ethnic groups from First Nations and Asians to the British in British Columbia.

The development of communities provides further geographic knowledge about the settlement of British Columbia. Table 1.3 ranks by population the ten largest municipalities from BC Stat's estimates for 2011 and from Statistics Canada's community profiles. It is interesting to compare these populations to selected census years in the past (in which the ranking is also given). Many of our present-day communities did not exist before the twentieth century; others have changed boundaries; some were much more significant in the past; and others have lost population. What are the factors responsible for the growth or decline of communities, and how are communities connected to their regions and to other communities?

The regional comparisons in Figure 1.4 are reinforced in a comparison of the older centre of Victoria with Vancouver. The extremely rapid rise in Vancouver's population since its incorporation in 1886 illustrates the powerful influence of its status as a national railway and major port facility and, later, the focus of other railway, road, and highway systems and site of an international airport. Vancouver has also attracted the major financial institutions and head offices for many of the resource industries operating throughout the province and the Asia-Pacific region. As the provincial capital, Victoria has experienced the growth of government and services along with infrastructure developments that link it to Vancouver Island's resources and to the mainland, but these economic links are not nearly as extensive as those of Vancouver.

Other communities throughout British Columbia have changed their population ranking over time mainly through the expansion of transportation and resource development and processing. By the 1980s, however, the new urban growth dynamics of the tourism and retirement industries, along with technologies that shrink time and space, had affected some communities more than others. These communities, and their growth (or

Table 1.3

Population of municipalities for selected years, 1881-2011

Municipality	2011	2001	1996	1921	1901	1891	1881
Vancouver (CMA)	(1) 2,394,270[d]	(1) 2,073,681	(1) 1,891,465	(1) 163,220	(1) 27,010	(2) 13,685	–
Victoria (CMA)	(2) 358,672[d]	(2) 325,569	(2) 304,287	(2) 38,727	(2) 20,919	(1) 16,841	(1) 5,925
Kelowna (CMA)	(3) 191,114[d]	(4) 100,496	(4) 105,403	2,520	–	–	–
Abbotsford-Mission (CMA)[a]	(4) 180,945[d]	(3) 153,801	(3) 89,442	–	261	–	–
Kamloops (CA)	(5) 92,882[e]	(5) 80,655	(5) 76,394	(7) 4,501	(8) 1,594	–	–
Nanaimo (CA)	(6) 92,361[e]	(6) 76,185	(7) 70,130	(5) 6,304	(4) 6,130	(4) 4,595	(2) 1,645
Prince George (CA)	(7) 83,225[e]	(7) 75,567	(6) 75,150	2,053	–	–	–
Chilliwack (CA)	(8) 80,892[e]	(8) 65,672	(8) 60,186	(9) 3,161	277	–	–
Vernon (CA)	(9) 55,418[e]	(9) 34,957	(9) 31,817	(8) 3,685	802	–	–
Penticton (CA)	(10) 43,313[e]	(10) 32,339	(10) 30,987	–[b]	–	–	–
New Westminster	–[c]	–[c]	–[c]	–[c]	–[c]	(3) 6,641	(3) 1,500
North Vancouver	–[c]	–[c]	–[c]	(3) 7,652	–	–	–
Prince Rupert (CA)	13,392[e]	15,282	16,714	(4) 6,393	–	–	–
Nelson (C)	9,258[e]	9,703	9,585	(6) 5,230	(5) 5,273	–	–
Rossland (C)	3,278[e]	3,804	3,802	2.097	(3) 6,156	–	–
Fernie (C)	4,217[e]	4,812	4,877	2,802	(6) 1,640	–	–
Revelstoke(C)	7,230[e]	7,826	8,047	2,782	(7) 1,600	–	–
Trail (C)	7,237[e]	7,905	7,696	(10) 3,020	(9) 1,360	–	–
Greenwood (C)	625[e]	695	784	371	(10) 1,359	–	–

Note: CMA = Census metropolitan area; CA = Census area; C = City; DM = District municipality. The numbers in parenthesis indicate the municipality's size rank.

a Boundary change and incorporated as a city, 1995.
b Incorporated in 1909, but population less than 1,000.
c Included in Vancouver CMA.
d 2011 population projections.
e 2006 census population.

Sources: Census Canada (1951); BC Stats (2004a; 2004b; 2009).

decline), are intimately tied to the regions in which they are located.

In the sections that follow, the eight regions shown in Figure 1.3 are briefly described in terms of their distinctive physical characteristics and historical development.

Vancouver Island/Central Coast

This region combines Vancouver Island with the central coast, which extends from Powell River north to Bella Coola. Vancouver Island has a rugged spine of mountains, referred to as the Insular Mountains. The central coast is part of the Coast Mountains range, which has peaks reaching considerably higher elevations and includes one of the highest mountains in British Columbia, Mount Waddington (4,016 metres). These mountains run mainly north-south and influence weather and climate conditions significantly. The prevailing westerlies often result in torrential rains on the west side of Vancouver Island and on the mainland where it is exposed to the open Pacific. The Olympic Mountains of Washington State and the Insular Mountains of Vancouver Island provide a **rain shadow effect** on the east side of Vancouver Island and on the southern end of the central coast. The Pacific Ocean at these latitudes (approximately 48°30' to 52° north) is considerably warmer than the Atlantic Ocean on the east coast of Canada, providing this region with the mildest winter climates in the country. Precipitation variations due to the rain shadow also influence vegetation and, in particular, the growth of Douglas fir in the drier areas. There are no large rivers on either Vancouver Island or the mainland, but the many small rivers and streams are important for fish habitat and some hydroelectric power.

Historically, the Vancouver Island/central coast region has been home for many First Nations. The peoples in this region experienced the longest exposure to non-Natives, however, and diseases took a huge toll. It was here that the Spanish and British squared off in the 1780s over territorial claims for colonization and the valuable sea otter trade. By the early 1800s, British Columbia was embroiled in an overland fur trade struggle between the aggressive North West Company and the Hudson's Bay Company. In 1821, the merger of these companies resolved the dispute and left the Hudson's Bay Company with a monopoly over the territory.

Non-Native settlement was sparse and temporary until a series of political and economic events occurred in the mid-1800s. The Oregon Treaty of 1846 annexed the British territory and forts south of the forty-ninth parallel, and as a consequence, Victoria was established at the south end of Vancouver Island. In 1858, gold was discovered on the lower reaches of the Fraser River, triggering an avalanche of miners seeking their fortune. Victoria became the main port of entry for much of this activity. Further discoveries of gold in the Cariboo region enhanced Victoria's position as the area's main administrative and service centre. The two separate colonies of Vancouver Island and the mainland were amalgamated in 1866 with Victoria as the capital, a position it has maintained since British Columbia joined Confederation in 1871.

The location of the capital attracted most settlers in the region to the southern end of Vancouver Island. The rest of the island was opened up in response to resource discovery, transportation developments, and technological change. Coal was an important resource found before the discovery of gold. The discovery of other minerals, such as iron ore on Texada Island and copper at the north end of Vancouver Island, created more jobs, but most of this activity did not occur until after the 1960s. Salmon fishing and canneries sprang up along the island and mainland coasts, and farming settlements were established mainly in the southeast of the island. By far the most important industry throughout the region was forestry. Large lumber mills such as the one in Chemainus were in operation by the 1880s, followed by pulp mills in the communities of Ocean Falls, Port Alice, and Powell River in the early 1900s.

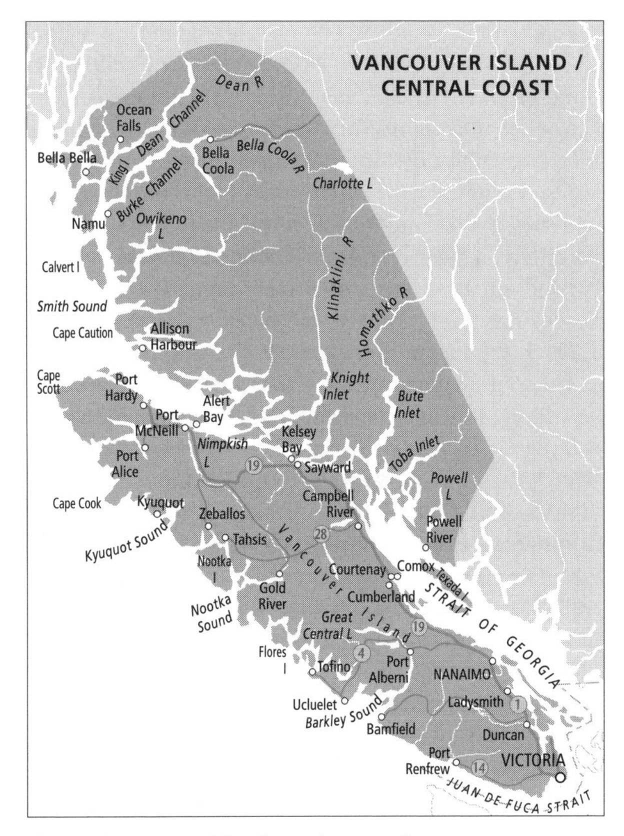

Figure 1.5 Vancouver Island/central coast region

Historically, railways were the most important means of land transportation on Vancouver Island. The Esquimalt and Nanaimo Railway (E&N) opened in 1886 and was the most important line, as it came with a provincial land grant to over one-quarter of the island. Roads appeared first on the southeast side of Vancouver Island and eventually linked the southern end with the northern as well as sending tentacles across to the few communities on the west side of the island (Wood 1979). Ocean-going transport provides the main link between the mainland coastal communities (e.g., Bella Bella, Bella Coola, Ocean Falls, and Powell River) and Vancouver Island. The highly indented and rugged mainland coastline hindered road development in the early days of settlement. Only Bella Coola was linked by road to the rest of the province.

Today, more than 750,000 people live in this region. The forest industry, which has been the economic backbone for much of Vancouver Island's history, has faced economic uncertainty for the past ten years, resulting in many sawmill and pulp-and-paper mill closures and increased unemployment. Farming, including farm-gate wineries, a wide range of recreation and tourism options, commercial fishing, and fish farming (although controversial) are other areas of employment. The Canadian Forces Base in Comox, which employs approximately 1,500, is significant to the Comox-Courtenay region.

The southeastern portion of the island has experienced some unique economic dynamics. It has attracted a huge number of retirees because of the mild winters with relatively little snow to shovel and year-round recreational activities. This part of the island also attracts the greatest number of tourists in the region because direct ferry links with the Lower Mainland make it easily accessible. As well, the southern end of Vancouver Island is intimately linked to the Lower Mainland region in the provision of administrative and service functions for western Canada and the Asia-Pacific region, making it part of the core, or heartland, of the province, as opposed to the periphery, or hinterland.

Lower Mainland

The climate of the Lower Mainland is similar to that of the southeast coast of Vancouver Island although it has higher precipitation, including more snow in winter, the higher the elevation. The Fraser River, the largest river system in the province and most significant to the salmon fishing industry, is an important physical feature of this region.

Historically, salmon and other resources of the water and land attracted many First Nations to the region. For non-Natives, gold was the main attraction following its discovery on the Fraser in 1858. Agricultural settlements soon followed, but securing these rich agricultural lands from the threat of floods has not been easy. Before the establishment of Vancouver, major sawmills on Burrard Inlet exported lumber, and canneries operated at the mouth of the Fraser River.

Transportation has been a major factor in the growth and development of this region. The completion of the CPR at Port Moody and its extension to Vancouver in 1886

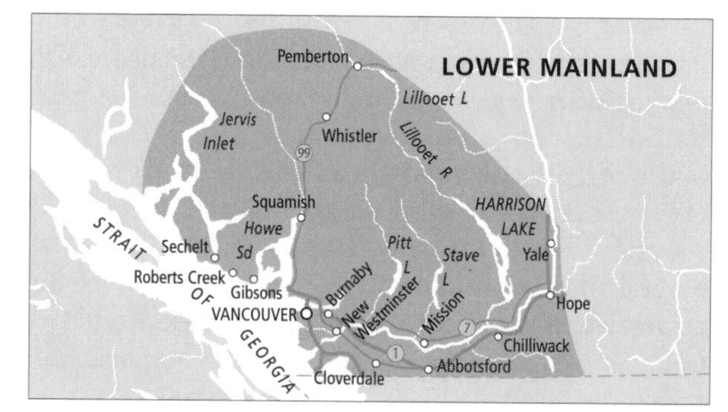

Figure 1.6 Lower Mainland region

was the catalyst for the rapid growth observed in Table 1.3. Vancouver, with its national railway and international port, was the main centre for this relatively small geographic region and was largely responsible for the growth of the adjacent Fraser Valley to the east, the Squamish-Whistler-Pemberton corridor to the north, and the Sunshine Coast to the northwest. The mountains and valleys framed the transportation links and settlement patterns for this region. The Sunshine Coast has a linear settlement pattern following the Strait of Georgia and is connected to Vancouver via ferry at Horseshoe Bay in West Vancouver. The Pacific Great Eastern Railway (renamed the British Columbia Railway, or BCR, in 1972 and taken over by CN in 2004) initially ran between Squamish and Quesnel (1921), and then was extended south to North Vancouver (1956) and north beyond Quesnel. This railway line has been an important transportation link to the ports at Squamish and North Vancouver. The Sea-to-Sky Highway is the main transportation system today as it winds its way beside Howe Sound to Squamish and then follows valleys leading past Whistler to Pemberton, Lillooet, and the interior of the province. Vancouver, Whistler, and all the communities between them, as well as the route that links them, have gained much attention and economic investment since the announcement that Vancouver/Whistler would host the 2010 Winter Olympics.

For the Fraser Valley, the river was originally the main transportation system. The construction of the CPR, and later the Canadian National Railway and British Columbia Electric Railway, made the region accessible. Road systems were built in the early 1900s, and eventually the construction of the Trans-Canada Highway and other highways linked Vancouver and the Fraser Valley to the rest of the province and south to the United States.

Favourable climate, superb natural features, highways, railways, port facilities, an international airport, and many commercial links to the rest of Canada, Asia, and the world make Vancouver a world city. This region has become the focus of the high-tech and film industries in the province along with tourism, international banking, finance, insurance, real estate, the head offices of resource-based industries, and most international immigration to the province. Forestry, fishing, and agriculture, the region's main industries historically, continue as important sources of employment. As well, many economic and administrative activities are shared with the southern end of Vancouver Island, making the combined region of the Lower Mainland and southern Vancouver Island the heartland of British Columbia. The Lower Mainland has 60 percent of the province's population, and this margin will increase in the future.

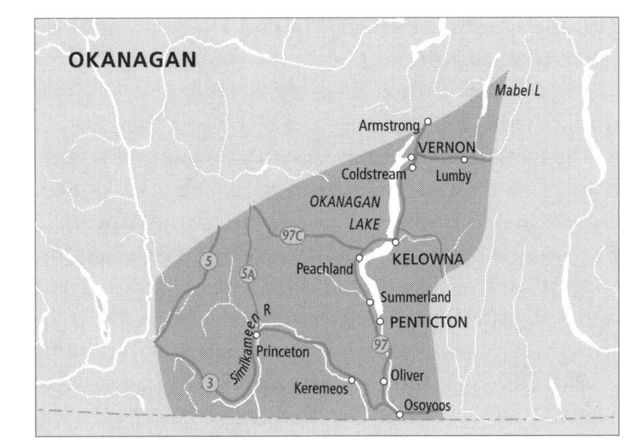

Figure 1.7 Okanagan region

Okanagan

The Okanagan Valley lies between the Cascade Mountains to the west and the Monashee Mountains to the east. There are several lakes in this valley, Okanagan Lake being the largest. The region's southern location between two large mountain chains results in a continental climate with typical temperature extremes of hot summers and relatively cold winters. The vegetation of this arid valley consists mainly of grasses and sagebrush with few trees; forests grow on the moister mountain slopes.

The region became the home for the Okanagan First Nations. Some non-Native settlement occurred with the fur trade, but much more took place as the region became recognized for its farming potential as a fruit-growing area. The Okanagan is one of the few places in Canada where apples and "soft fruit" such as peaches and cherries are produced, but irrigation is necessary in this dry belt. A number of communities evolved to serve the growing agricultural settlement. Vernon, Kelowna, and Penticton, all on Okanagan Lake, became the most prominent. Boats on Okanagan Lake served an important transportation function, and later railway lines were built. The Kettle Valley line (1915) provided the link between the Kootenays and the Lower Mainland, giving access to the southern Okanagan, while a branch line of the CPR linking Kamloops to Kelowna (1925) served the northern portion of the region.

Several changes occurred after the 1960s. Tourism, which had mainly been a summer activity, expanded into a year-round endeavour, with golf courses and ski runs. The dry climate and four distinct seasons, combined with relatively low land and housing prices and easy access to the Lower Mainland, made this a favourable location for retirement. These characteristics led to increased population, urban sprawl, and land use conflicts, especially over agricultural land, until the implementation of agricultural land reserves in 1972. The motors of fishing and pleasure boats carried Eurasian milfoil into the region, where it spread through the water systems, converting once sandy beaches to a mass of weeds and jeopardizing the growing tourism industry. Eradication programs in the 1970s made use of herbicides that provided some control but also raised concern about potential carcinogenic effects.

Other industries expanded in the region, including mining and forestry, both of which increased the employment and population base. With the signing of the Free Trade Agreement between Canada and the United States in 1989, agriculture changed rapidly. Fruit crops continued to dominate, and to compete with US producers with access to inexpensive labour, BC producers developed new varieties of apple trees requiring considerably less labour. One of the most significant changes was to the grape and wine industry. Forced to compete globally, it met the challenge by growing new grape varieties. New rules permitting the sale of wine from farms was a large part of the success of the industry, which now attracts many tourists.

The Okanagan has many physical assets and a fairly diversified economy, making it one of the rapid-growth regions in the province (Table 1.2). Within the Okanagan, Kelowna has become the most important service, administrative, and manufacturing centre, with a regional airport and a highway link to the coast via the Coquihalla Highway. Forest fires, however, have been particularly hazardous for this region. The firestorm of 2003 incinerated 238 homes (CBC 2003), and the firestorm of 2009 forced the evacuation of thousands of residents in the Kelowna area.

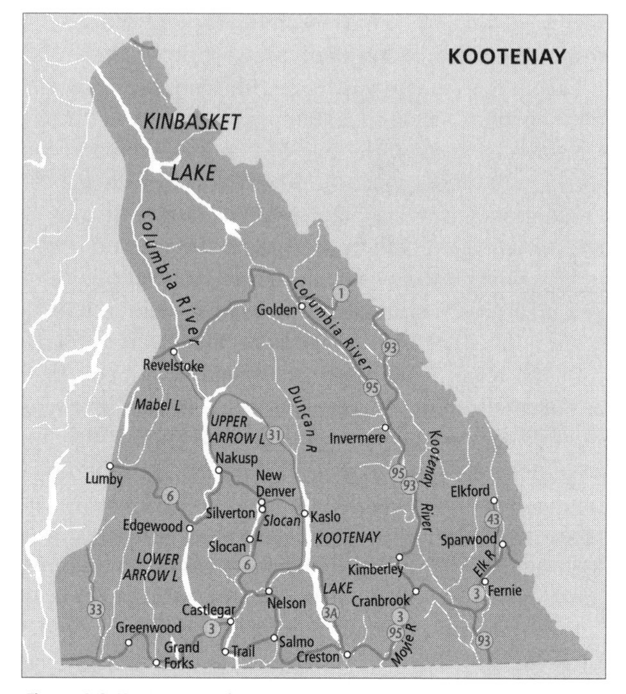

Figure 1.8 Kootenay region

Kootenay

Mountains, rivers, and valleys are the main physical features that define the Kootenay region. The rugged mountain chains run north-south: first the Monashees, then to the east of them the Selkirks, farther east the Purcells, and finally the Rockies. All the rivers and lakes in these valleys are part of the Columbia River system, which exits British Columbia at Trail as the Columbia flows into the United States. Climatically, this region is similar to the Okanagan, but slightly colder in winter, not as hot in summer, and with slightly more precipitation.

Several First Nations resided in this region: the Kootenai to the east, the Okanagan to the west, and the Shuswap Nation to the north.

Census Canada has traditionally divided the Kootenays into east and west, but the divisions share a valuable mineral resource base. Gold, silver, coal, copper, lead, and zinc were all discovered and became the lure to settlement and development following the fur trade era. With the Crow's Nest Pass Agreement of 1897, the CPR became the principal landowner, provider of rail transportation, and developer of the resources. The Agreement was a deal struck between the federal and provincial governments and the CPR to run a branch line from Lethbridge, Alberta, through the Crowsnest Pass to the mineral-rich Kootenays, ending initially at Nelson. Through the Agreement and their subsequent purchase of railway grants, the CPR acquired millions of acres of land, coal deposits, metal mines, a major smelter at Trail (Cominco), and West Kootenay Light and Power. The CPR exerted enormous control over this region. Other railway companies built lines and acquired land grants, but the CPR purchased most of these over time, consolidating its hold on the Kootenay economy. The population changes in this region (Table 1.2) hide the boom and bust cycles experienced by individual mines and smelters. The list of **ghost towns** in the Kootenays is sufficient to warrant several books and articles (Barlee 1970, 1978a, 1978b, 1984).

Historically, settlement patterns in the area have been influenced mainly by mineral exploitation, transportation developments, and agricultural opportunities. Other factors also led to settlement. For example, the Doukhobors migrated to the Kootenays between 1908 and 1913 in an attempt to escape religious persecution in Europe and political persecution in Saskatchewan. The area had good agricultural land for their communal lifestyle and appeared to be relatively isolated from government interference. In the early 1940s, many Japanese families, evacuated from the coast, were sent to communities and work camps throughout the Kootenays such as Greenwood, Sandon, New Denver, and Slocan City.

Forestry has been another resource activity in this region. The Kootenays experienced forestry expansion after 1961 with a pulp mill in West Kootenay at Castlegar and in East Kootenay at Skookumchuck. In the 1960s, the provincial government became involved in hydroelectric megaprojects. Through the "two rivers policy," the Peace River, in the northeast, and the Columbia River, in the Kootenays, were developed for hydroelectricity simultaneously. The dams also provided flood protection for cities in Washington State. Other dams, such as the Revelstoke Dam, were constructed later to fulfill increased electrical demand by British Columbians.

Currently, the Kootenay region remains dependent on resource development and export. Mining, forestry, and hydroelectric energy are vital to the economic well-being of the region, although both mining and forestry have seen setbacks. The closure of the Sullivan Mine (the mainstay of Kimberley from 1909 to December 2000), the slowdown in production at the Trail smelter, and fluctuating coal prices have reduced employment in mining throughout the region. Forestry has been hit by major downturns in the demand (and price) for softwood lumber and pulp and paper. Tourism has offered some diversification, as the region has many hot springs and lakes, and opportunities abound for skiing, hiking, and sightseeing. Investments in tourist infrastructure such as ski facilities (e.g., Panorama, Fernie, and Kimberley), golf courses, and casinos have increased tourist visitations. Nevertheless, being a considerable distance from major urban populations in Alberta, British Columbia, and the United States, the region is likely to show slow overall growth in the future.

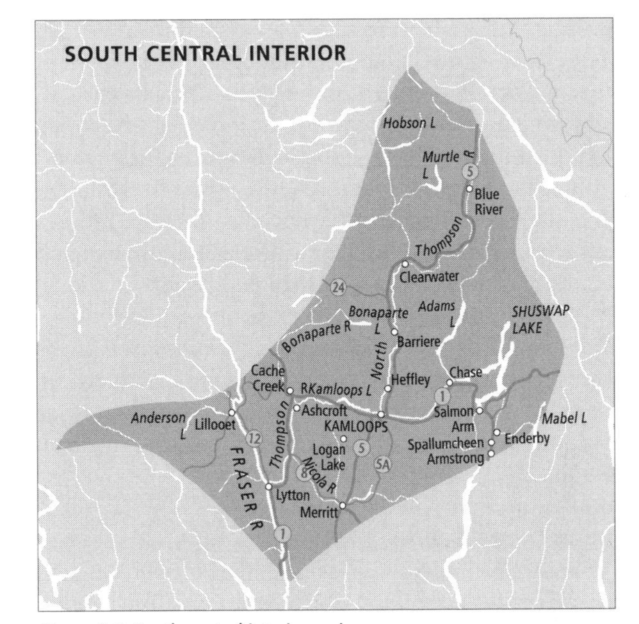

Figure 1.9 South central interior region

South Central Interior

The south central interior is largely identified by the Southern Interior Plateau. The region extends west of Lillooet to Revelstoke on the Columbia River, but it is the Thompson River system that mainly defines this region. The Thompson River valley is hot in summer and cold in winter, with precipitation occurring mainly on the surrounding mountains. The Secwepemc Nation of the Interior Salish have been the traditional users of the land.

Early European interest in this region was due to furs and gold. The Thompson River was an important "highway" for the fur trade, and Kamloops was established as a fur trade post in 1812. Small amounts of gold were discovered in the region in the mid-1850s but not in amounts significant enough to create a gold rush. The gold rush to the Cariboo in the 1860s, however, attracted cattle drives and cattle ranching to these interior grasslands.

When the CPR main line was built, Kamloops became the main supply and service centre for the region. This location, where the North and South Thompson Rivers meet, was enhanced by later transportation developments such as the extension of the Canadian National Railway (CNR) down the North Thompson and through Kamloops on its way to Vancouver. With the development of road systems, Kamloops was on the Trans-Canada Highway, and it is currently at the end of the Coquihalla Highway, which runs to the coast.

Farming and ranching were the main industries in the south central interior until the 1960s, when forestry, mining, tourism, and the retirement industry were added. Forestry has been the most important, with a pulp mill established in Kamloops in the mid-1960s and sawmills and other forest-product manufacturing throughout the region. Mining has also played a significant role, with a copper smelter at Kamloops (now closed) and mines through the Highland Valley from Ashcroft to Merritt. Tourism is increasing in importance largely because of the transportation routes (especially the Coquihalla Highway, which, after twenty-two years, no longer has tolls) that lead to Kamloops and branch north to Jasper and Edmonton and east to Banff and Calgary. Accessibility and many favourable features have also attracted retirees to this region. The south central interior is another region with a relatively rapid increase in population, as shown in Table 1.2, but it is dominated by one urban centre – Kamloops. This centre – which includes Thompson Rivers University and other institutions that perform medical, transportation, and other administration functions – has become an important service centre for the south central region.

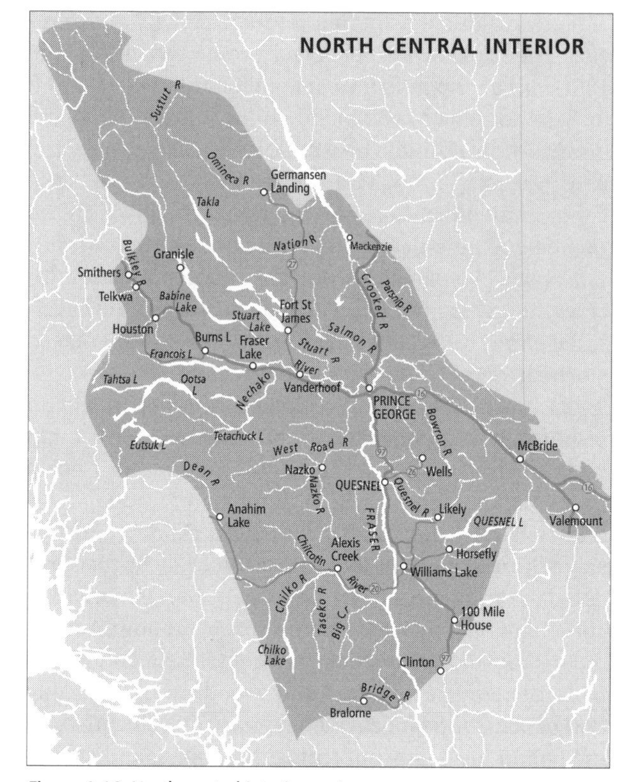

Figure 1.10 North central interior region

North Central Interior

The north central interior is one of the largest regions in the province and is defined by the Northern Interior Plateau. The northern half of the Fraser River, with its many tributaries, is a large part of this region, although the area extends westward to include Smithers and the Bulkley Valley. Temperature regimes are colder than are southern interior locations in winter and not as hot in summer. The mixing of Pacific and Arctic air masses results in increased precipitation, but this varies through the region in relation to the configuration of the mountain chains.

Historically, the area has been home to many First Nations, such as Sekani, Nat'oot'en, Wet'suwet'en, and Dakelh. The history of non-Native settlement began with the overland fur trade, which relied on the Fraser River for transportation. Forts were erected through the region in the early 1800s, but it was not until the Cariboo gold rush of the early 1860s that more permanent non-Native settlement ensued. A number of routes were used to gain access to the Barkerville area from the Lower Mainland until the Cariboo Road was constructed in the mid-1860s. Gold was also discovered in the northern part of the region, in the Omineca Mountains for example, but these finds were not sustainable. The development of Barkerville and other mining communities also attracted ranching to the southern part of the region, including the Chilcotin Plateau. By the end of the 1860s, however, the gold discoveries were minimal, and the miners were disappearing along with the communities. This large region remained relatively uninhabited until the CNR connected it to the port of Prince Rupert to the west in 1914. From the south, the Pacific Great Eastern Railway (PGE) was built in a series of stages: Squamish to Quesnel in 1921, Quesnel to Prince George in 1952, Prince George to the Peace River in 1958, and Prince George to Fort St. James in 1968. Prince George, like Kamloops, became a transportation hub of rail lines, and later highway systems, serving central and northern British Columbia.

Since the 1960s, as the population growth in Table 1.2 indicates, there has been considerable interest in this region. The mining of copper, molybdenum, and gold brought many workers into the area, but the main industry was forestry. The forests were largely untouched until the 1960s, when the increase in world demand for forest products saw a massive expansion in the industry throughout the north central interior. The manufacturing of plywood, lumber, and other forest products was integrated with pulp-and-paper mills built at Quesnel, Prince George, and in the new town of Mackenzie.

Today, some of the mines have closed, and the forest industry is in a serious downward cycle (e.g., all mills are presently closed at Mackenzie), reminding us that much of the region is extremely resource dependent. The mountain pine beetle infestation, which has devastated forests, has been another serious blow to this industry. Prince George has emerged as the largest centre. With its accessibility, the University of Northern British Columbia (UNBC), and its many services and administrative functions, the city has become an important service centre to central and northern British Columbia, performing a role similar to the one that Edmonton performs for Alberta.

North Coast/Northwest

The north coast/northwest is another large region, isolated for much of its area because of its rugged, mountainous landscape. The Coast Mountains, which extend the length of this region, are among the highest in British Columbia. The northwest corner of the region consists of the St. Elias range, where Fairweather Mountain (4,663 metres) is the highest peak in British Columbia. The coast is highly irregular and the northern coast makes up the Alaska Panhandle. The heavily forested Haida Gwaii (formerly Queen Charlotte Islands, composed of over 150 islands) is also part of this region. Climatically, much of the area is exposed to the Pacific and to the westerly flows of wind (that is, wind flowing from the west). Its northern location brings considerable exposure to the Aleutian low pressure system, which brings much rain in the summer and snow in the winter. The Skeena, Nass, and Stikine are the largest of the very significant river systems flowing to the Pacific.

The north coast has been home to high density populations of First Nations, whereas First Nations in the northwest have had much lower populations. A number of factors led to early European development of the north coast/northwest region. In the late 1700s, the sea otter trade aroused considerable interest. The Russians, who erected fur trade posts across Alaska and the Panhandle – the strip of Alaska that extends south along the coast – were the first to exploit these valuable furs. They laid territorial claim to Alaska, thus cutting off the northwest from the sea. Subsequently, Russia sold this territory to the Americans in 1867. Without access to the Pacific, the northwest portion of this region was left largely to overland fur trade interests.

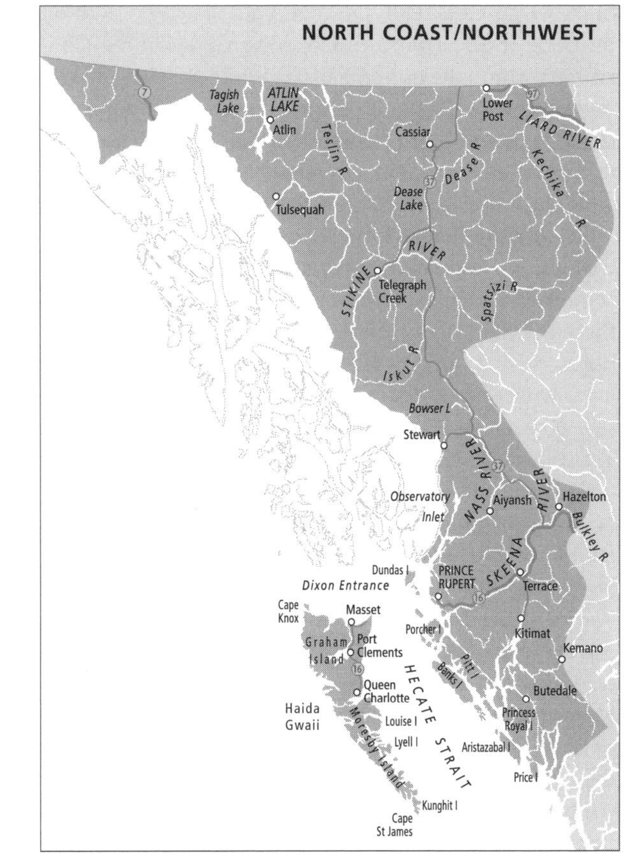

Figure 1.11 North coast/northwest region

Other resources encouraged temporary settlement. Salmon was valuable, for example, and canneries became evident by the late 1800s and early 1900s, especially at the mouth of the Skeena and the Nass and along the coast. Except for those at the mouth of the Skeena, most disappeared in the 1950s with improvements in fishing technology. Coal was discovered north of Prince Rupert and was mined for several years until the richer and more accessible coal mines of Nanaimo and other Vancouver Island locations were developed in the 1850s. Small amounts of gold were discovered on Haida Gwaii in 1850 and on the Skeena River in 1863, but no gold rush occurred in either case. In 1898, the Yukon gold rush opened up the northernmost portion of the region. The famous Chilkoot Pass, and the building of the White Pass and Yukon Railway in 1902, meant that most gold seekers passed through the northern tip of British Columbia before entering the Yukon. There was plenty of interest in this region in this period but little permanent settlement.

The completion of the CNR to Prince Rupert in 1914 was another important event, and it increased permanent settlement. Charles Hays, a prominent entrepreneur in the early history of Prince Rupert, was a major shareholder of the railway. He was an industrial waterfront landowner and an avid promoter of the town, but he never saw his vision materialize – he was a passenger on the Titanic in 1912 and, unfortunately, was not counted among the survivors. Prince Rupert was also the site of an early pulp mill, which encouraged growth as well. A copper smelter was built at Anyox, north of Prince Rupert, but lasted only until the Depression. The Second World War prompted construction of the Alaska Highway, giving some accessibility to the remote northwest portion of the region. One of the first areas outside the Lower Mainland to expand with the post-Second World War boom was Kitimat. This planned town was built to house the workers for a new aluminum plant in the early 1950s.

As in other regions, the mining and forest industries in the north coast/northwest region underwent major expansion beginning in the 1960s. The Stewart-Cassiar road linked the Alaska Highway through the northwest to Prince Rupert. A major copper mine near Stewart, a large open-pit mine for asbestos at Cassiar, and a number of small gold mines brought employment to these rather isolated locations. The development of Quintette and Bullmoose mines and the new town of Tumbler Ridge in the Peace River region – a coal mining region known as northeast coal – in the early 1980s resulted in the rail line being double tracked to Prince Rupert and a large coal port being built at nearby Ridley Island. With fluctuations in coal shipments, new investments in Ridley Island have seen the development of new bulk handling facilities (e.g., for petroleum products and wood pellets) and the ability to handle container shipping. A planned expansion of the aluminum smelter at Kitimat was cancelled in 1995 for environmental reasons, but aluminum production remains important to the region. The forest industry expanded in a number of directions. Much of the valuable timber from Haida Gwaii was harvested and barged south for the mills of the Lower Mainland. A pulp mill was opened at Kitimat in the 1960s, but it closed in 2010. Prince Rupert has emerged as the largest centre for the region, but the downturn in the forest industry resulted in the closure of a number of sawmills in the area and its pulp mill in 2001, putting into perspective the importance of the forest industry to this city and the region as a whole.

Northeast of Prince Rupert, the Nisga'a of the Nass River Valley were the first to enter into a modern peace treaty in British Columbia (signed in 2000), setting a precedent for future treaties throughout the province. The treaty resulted in a cash settlement, land ownership (i.e., their land base is no longer a reserve), a Native-owned forestry company, and a substantial share of the Nass River commercial salmon fishery.

There are few roads or rail lines through this area even today, and its growth is tied closely to resource development. The north coast/northwest region is slow growing, and recently, population has declined.

Peace River/Northeast

Most of the Peace River/northeast region does not fit the broad physical description of British Columbia as a mountainous, vertical landscape. This flat, sedimentary region east of the Rockies is physiographically similar to the prairies. The two major rivers, the Peace River in the south and Liard River in the north, are part of the Mackenzie River system, which drains into the Arctic Ocean. The region contains areas of permafrost, bog, and boreal spruce forests. Temperatures are cold in winter and surprisingly warm in summer, when the days are long, inducing **convection precipitation**.

North West Company fur traders were the first non-Natives to enter the region, and it is here that the earliest fur trade forts in British Columbia were erected. Discoveries of gold on the Peace River in the 1860s warranted the inclusion of this territory into British Columbia, but the finds were insufficient to sustain permanent settlement. The Yukon gold rush at the end of the nineteenth century initiated Treaty 8, covering the region north of Edmonton, the northwest corner of Saskatchewan, and the Peace River/northeast region of British Columbia.

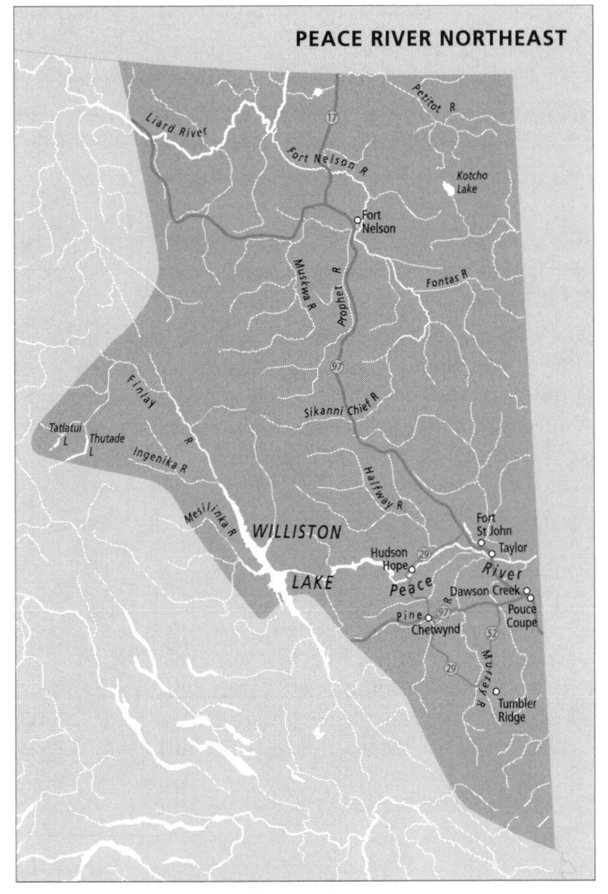

Figure 1.12 Peace River/northeast region

The Canadian government believed that this would be a route to the goldfields and that the treaty would be a means of avoiding conflict. The First Nations of the Peace River/northeast region were the only people to be included in a numbered treaty in the province.

Few agricultural settlers ventured this far north until the homesteads of the south and central Prairies were no longer available. The development of hardy, early maturing wheat facilitated agricultural homesteads by the 1920s and '30s. Problems of accessibility were improved when the rail line from Alberta was extended to Dawson Creek in 1930. The Alaska Highway was constructed during the war and Dawson Creek became Mile 0, helping to open up the region. The federal government made more farmland available at the end of the Second World War for returning servicemen. The PGE was finally extended to Fort Nelson by 1971. Wheat farming on the excellent soil and cattle rearing have been the main agricultural activities of the Peace River area.

The discovery and development of oil and natural gas in the 1950s encouraged investment, the building of pipelines, and the movement of considerably more people to this region. By the 1960s, the Peace River had become one target of the massive hydroelectric plan referred to as the two rivers policy. The plan involved constructing the W.A.C. Bennett Dam, which in turn created the largest reservoir, or artificial lake, in the province, Lake Williston. As well, transmission lines were built to connect the hydroelectric power source to southwestern British Columbia. All this attracted even more people to the Peace River. The energy crisis, beginning in the early 1970s, sparked another round of oil and gas exploration and development.

The sedimentary basin also contained coal, and in the early 1980s, northeast coal began to be developed in the area. The new town of Tumbler Ridge housed the miners, an electric rail line of the British Columbia Railway (formerly the PGE) was constructed, and millions of dollars were spent upgrading the CNR line from Prince George to Prince Rupert and, as discussed above, building a coal port at Ridley Island. The costs borne by both provincial and federal governments to export the coal were massive. Unfortunately, the world market demand and price for coal declined, and so did the contracts with Japanese buyers. The Quintette mine closed in 1999 and the Bullmoose in 2003, leaving Tumbler Ridge struggling to convert to a retirement, tourist, and recreation community. Fortunately, the price of coal rebounded, and a new coal mine, Wolverine, opened in 2006. The discovery of dinosaur fossils has also resulted in tourism related to paleontology.

Today, Fort St. John and Dawson Creek are the largest centres in the Peace River/northeast region. The forest industry employs many people in two pulp mills, sawmills, oriented strandboard plants, and a plywood mill. Agriculture has been difficult because the price of grain has been low consistently, and in 2003 the beef industry was hit by bovine spongiform encephalopathy (BSE, also called "mad cow disease"). BC Hydro has brought the controversial Site C Dam proposal, on the Peace River near Fort St. John, out of mothballs, which has many farmers, environmentalists, and others concerned about another flooding of the Peace River valley. Oil and natural gas particularly continue to be important, and exploration for coalbed methane is taking place. When the price of coal increased between 2003 and October 2009, three new coal mines opened in the region. Tourism has increased in response to a variety of recreational opportunities, the diversity of the landscape, and the discovery of dinosaur footprints at Tumbler Ridge and near the W.A.C. Bennett Dam. The Alaska Highway continues to be a major tourist attraction.

With all of this, however, comes the potential for conflict between the resource industries and those interested in preserving the wilderness. The Muskwa-Kechika Management Area Act, signed in 1998, attempts to accommodate all interests in the very large wilderness region from Lake Williston north to the Yukon border. The economy and growth of the Peace River/northeast region is tied to its diverse resource endowment. It is a large region, the population is relatively small, and growth, or decline, occurs in relationship to resource demand.

SUMMARY

Geography is a practical discipline that assesses the human and physical factors influencing the surface of the earth. The focus in this text is the landscapes of British Columbia. The province is divided into eight regions for the purpose of examination in the first and last chapters, but the remaining chapters employ a thematic, or topical, approach. The history of what has gone on before is an important theme for all chapters.

Time-space convergence and spatial diffusion are basic geographic concepts essential to understanding movement over space and through time. Statistics, graphs, and maps are common tools and techniques of the geographer, whereby the province and its regions can be more fully understood.

The regional perspective is important, not only in recognizing British Columbia's unique physical and human attributes, but also in assessing the more global and external forces shaping this province. Both external and internal factors influence growth and development, but not equally throughout the province. We have seen that some of the province's eight regions have grown considerably more than others.

Most important, *Geography of British Columbia* has been written with a minimum of geographic jargon in its attempt to tell the stories of human and physical influences shaping the landscapes of this province.

REFERENCES

Barlee, N.L. 1970. *Gold Creeks and Ghost Towns of Southern British Columbia.* Summerland, BC: Self-published.

–. 1978a. *The Best of Canada West.* Langley, BC: Stagecoach.

–. 1978b. *Similkameen: The Pictograph Country.* Summerland, BC: Self-published.

–. 1984. *West Kootenays, the Ghost Town Country.* Surrey, BC: Canada West Publishers.

Census of Canada. 1951. *British Columbia Population by Census Subdivisions, 1871-1951.* Table 6/6-84 to 6-88. Ottawa.

–. 1971. *British Columbia Population by Census Subdivisions, 1961 and 1971.* Table 8/8-92 to 8.96. Ottawa.

Dearden, P. 1987. "Marine-Based Recreation." In *British Columbia: Its Resources and People,* ed. C.N. Forward, 259-80. Western Geographical Series vol. 22. Victoria: University of Victoria.

Gould, P.R. 1969. *Spatial Diffusion.* Resource paper no. 4. Washington, DC: Association of American Geographers.

Gregory, D. 2000. "Regions and regional geography." In *The Dictionary of Human Geography,* ed. R.J. Johnston, D. Gregory, G. Pratt, and M. Watts, 687-90. 4th ed. Oxford: Blackwell.

Harris, C. 1997. *The Resettlement of British Columbia: Essays on Colonialism and Geographical Change.* Vancouver: UBC Press.

McCune, S. 1970. "Geography: Where? Why? So What?" *Journal of Geography* 7 (69): 454-57.

Muckle, R.J. 1998. *The First Nations of British Columbia.* Vancouver: UBC Press.

Pudup, M.B. 1988. "Arguments within Regional Geography." *Progress in Human Geography* 12: 369-90.

Renwick, W.H., and J.M. Rubenstein. 1995. *An Introduction to Geography: People, Places, and Environment.* Englewood Cliffs, NJ: Prentice Hall.

Robinson, J.L. 1972. "Areal Patterns and Regional Character." In *Studies in Canadian Geography: British Columbia,* 1-8. Toronto: University of Toronto Press.

Wood, C.J.B. 1979. "Settlement and Population." In *Vancouver Island: Land of Contrasts,* ed. C.N. Forward, 3-32. Western Geographical Series vol. 17. Victoria: University of Victoria.

INTERNET

BC Stats. 2004a. "British Columbia Municipal Census Populations, 1921-1971." www.bcstats.gov.bc.ca/data/pop/popstart.asp.

–. 2004b. "Municipal and Regional District Total Population Estimates, 1981-2004." www.bcstats.gov.bc.ca/data/pop/popstart.asp.

–. 2008. "Regional Population Estimates and Projections." www.bcstats.gov.bc.ca/data/pop/pop/dynamic/PopulationStatistics/SelectRegionType.asp?category=Census.

–. 2009. *British Columbia Population Projections, 2009 to 2036.* www.bcstats.gov.bc.ca/data/pop/pop/project/BCtab_Proj0906.pdf.

CBC. 2003. "BC Fire Talk Tape." Radio One. 14 November. www.cbc.ca/thecurrent/.

CIA. 2009. *World Fact Book.* www.cia.gov/library/publications/the-world-factbook/index.html.

Statistics Canada. 1983a. *Historical Statistics of Canada.* Catalogue 11-516-X1E. Population 1871-1971, series A2-14. www.statcan.gc.ca/.

–. 1983b. *Historical Statistics of Canada.* Catalogue 11-516-X1E. Natural increase and net migration 1931-1976, series A339-49. www.statcan.gc.ca/.

–. 1983c. *Historical Statistics of Canada.* Catalogue 11-516-X1E. Rural and urban population 1871-1971, series A67-69. www.statcan.gc.ca/.

–. 2001a. Population 1981-2001. CANSIM Table 051-0001. www.statcan.gc.ca/.

–. 2001b. Rural and urban population 1981-2001. CANSIM Table 109-0200. www.statcan.gc.ca/.

–. 2004. Natural population increase and net migration 1976-2001. CANSIM Table 051-0013. www.statcan.gc.ca/.

–. 2010. Components of population growth, Canada, provinces and territories, annual (persons). CANSIM Table 051-0004. www.statcan.gc.ca/.

2

Physical Processes and Human Implications

Much of the attraction, and beauty, of British Columbia comes from its great variety of physical features. Its many scenic landscapes of rugged mountains, large and turbulent rivers, diverse vegetation, variety of fauna, and often isolated settings – all changing from season to season – are both great assets and great challenges. The province has a rich resource base of minerals, energy, forests, fish, agriculture, and tourism. The physical setting has also been a powerful influence on settlement, the urban system, and transportation corridors, as well as posing a major risk to the human population.

This chapter begins the discussion of physical processes by drawing attention to the geologic time scale and the distinctive rock structures that make up the crust of the earth. The ability to assess the age of and distinguish among different rock types – even those billions of years old – gives a sense of the generalized geological makeup of British Columbia. The study of processes that create these landforms and transform the surface of the earth is known as **geomorphology**.

Another major focus in this chapter is the set of physical processes related to weather and climate. The spatial and seasonal patterns of temperature and precipitation vary greatly throughout the province, with broad distinctions between the coast and the interior and between southern and northern locations.

Soils and vegetation influence each other and are a product of many other physical processes, including variations in climate and geomorphology. The vegetation patterns of British Columbia are very distinct and obvious, ranging from huge coastal coniferous forests to southern interior valleys of cactus and sagebrush. These physical landscapes have major implications for human habitation of the province.

From a human perspective, significant changes occur within periods of ten to twenty years in terms of dress, popular music, political policies, and employment opportunities. From a physical perspective, it is necessary to think in terms of tens, or even hundreds, of millions of years when considering changes to the landscape (Table 2.1). The geological makeup of any region is very complex, and one means of unravelling this complexity is to assess the age of rocks. Life forms on the earth have been a major influence in dating rocks because their fossils remain as a benchmark of time. Phanerozoic rocks, containing fossils of complex life forms, date from 545 million years ago. As Table 2.1 indicates, there are further divisions within this time period. Of course there are rocks much older than these fossiliferous ones; they are known as Precambrian rocks and were created between 4.6 billion and 545 million years ago. This is a huge expanse of time and divisions here are made through carbon dating.

Table 2.1

Geologic time scale

Eon	Era	Period	Years before present
Phanerozoic	Cenozoic	Quaternary	
		Holocene	10,000
		Pleistocene	1,800,000
		Tertiary	
		Pliocene	5,800,000
		Miocene	23,800,000
		Oligocene	33,700,000
		Eocene	54,800,000
		Paleocene	65,000,000
	Mesozoic	Cretaceous	146,000,000
		Jurassic	208,000,000
		Triassic	248,000,000
	Paleozoic	Permian	280,000,000
		Carboniferous	360,000,000
		Devonian	408,000,000
		Silurian	438,000,000
		Ordovician	505,000,000
		Cambrian	545,000,000
Precambrian			
Proterozoic			2,500,000,000
Archean			3,800,000,000
Hadean			4,600,000,000

This province has undergone profound transformations, especially over the last 200 million years. Climatic changes have caused ocean levels to rise and fall and glaciers to scrape and modify the landscape; tectonic processes have been responsible for mountain building, volcanic activity, and the addition of islands to British Columbia. The landscape is by no means static.

THE ROCK CYCLE: FORMATION AND DESTRUCTION OF ROCKS

Igneous rocks are created when molten **magma** cools and hardens into a solid state. As the temperature drops, crystals form in the magma, like ice crystals forming in water. These crystals are called minerals and have different compositions. Some typical minerals found in igneous rocks are feldspar, quartz, hornblende, and olivine. Because hornblende and olivine contain iron, they are darker than feldspar and quartz. Igneous rocks containing mostly feldspar and quartz minerals are usually light in colour, an example being granite. Igneous rocks that have mostly iron-rich minerals are often dark grey or black; basalt is an example. Basalt is denser than granite since it contains larger proportions of iron and other heavy minerals. Granite is a typical continental rock, while the majority of oceanic rocks are basalt.

The size of the minerals in an igneous rock reflects the rate of cooling of the magma. If the magma cools relatively fast (over days, years, or even decades), the crystals are so small that a microscope is needed to see them. These quickly cooled rocks are found near or at the earth's surface, primarily as a result of volcanic activity, and are known as extrusive or volcanic igneous rocks. Basalt is an extrusive igneous rock and is commonly found in lava flows, dikes, and sills. Dikes and sills form as magma pushes into older rocks; sills form parallel to surrounding rock layers, while dikes cut across the rock structure (Figure 2.1). The cooling process of basalt frequently produces a pattern of cracks resulting in columns of four-, five-, or six-sided rock, creating what is known as columnar jointing. These distinctive columns can be seen along many of British Columbia's highways, an indication of the province's extensive volcanic history.

Magma that cools tens of kilometres below the earth's surface may require millions of years to solidify. This time allows the crystals to grow to a size where they can be seen by the human eye. These intrusive or plutonic rocks, of which granite is an example, are often found in huge plutonic masses known as batholiths (Figure 2.1); much of the Coast Mountains range of western British Columbia is batholithic. Although formed deep below the surface, granitic batholiths may be pushed upward during large-scale crustal movements. As surface weathering

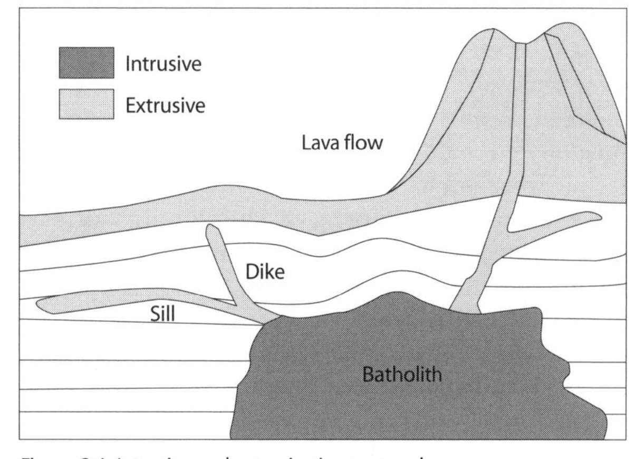

Figure 2.1 Intrusive and extrusive igneous rock

and erosion strip off the overlying rock formations, the batholiths are exposed.

Rocks at the surface are exposed to weathering processes caused by water, air, and vegetation growth and decay. **Weathering** may cause a rock to fragment into small, sedimentary pieces called **clastics**. Weathering by surface water and groundwater may also cause the minerals within rock to dissolve, making water "hard" and oceans salty. After weathering, sediments and dissolved minerals may be moved by streams, waves, wind, glaciers, and gravity to a new location, a process known as erosion. These sediments are eventually deposited, often in layers. The layering effect happens when different sizes and shapes of sediments are deposited, such as small, rounded particles on top of large, angular ones. This usually occurs when the agent of erosion changes, such as from wind to water. As sedimentary layers accumulate, the bottom ones are pushed deeper into the crust, causing compaction and heating. As well, groundwater can infiltrate the sediments and precipitate (or solidify) some of the dissolved minerals.

The precipitated minerals act as a cement to bond the sediments together. Compaction, heating, and cementing change the sediments into rock. Not surprisingly, rocks formed from unconsolidated materials bonded together are known as **sedimentary rocks**. Some common examples of clastic sedimentary rocks are sandstone (made from sand-sized sediments) and shale (made from clay and silt-sized sediments). Precipitated minerals may also form a second type of sedimentary rock called evaporatives. Rock salt, for example, forms in the shallow waters of warm oceans; as the water evaporates, the salts are precipitated. A third sedimentary rock type is formed through organic processes. As plant and animal products accumulate, their remains may eventually turn into rock. Coal is an example. As well, tiny organisms utilize minerals dissolved in lakes and oceans to form their shells and skeletons (similar to the way in which humans extract calcium from food to make bones). When the organisms die, their hard parts sink to the bottom of the water body and accumulate to form limestone.

Sedimentary rocks are located throughout British Columbia. Some are found in their original horizontal positions, but many are folded, tilted, and fractured. These deformations occur during times of mountain building, a prime example being the Rocky Mountains on the east side of the province.

The third category of rocks, after igneous and sedimentary, is called metamorphic. When rocks are exposed to extreme pressures and temperatures, or to chemical infusions from nearby magma bodies, the minerals within the rocks change in shape and chemistry. The original rock does not melt; it metamorphoses into a new rock. Limestone, for example, will metamorphose into marble. Metamorphic rocks are mostly associated with mountain building and igneous intrusions. In British Columbia, metamorphic rocks can be found in the Coast Mountains in conjunction with their huge batholithic cores.

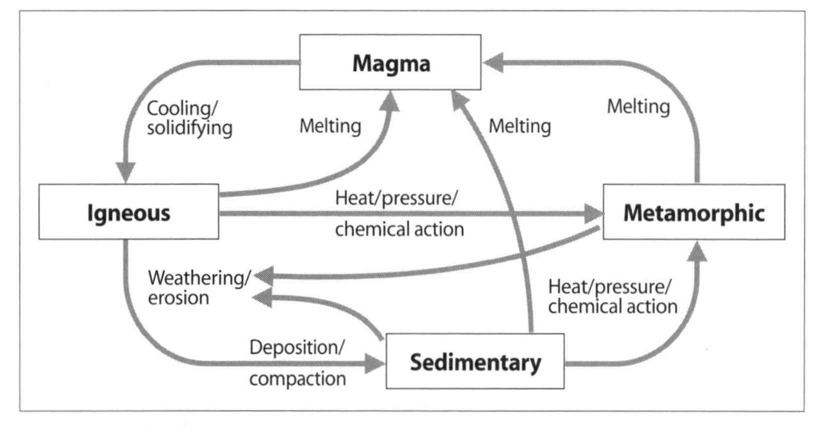

Figure 2.2 Rock cycle

The **rock cycle** reminds us that rocks are not static in geologic time, and that all can be related (Figure 2.2). Beginning with the molten state, igneous rocks are "born." Like all other rocks, however, they are subject to the processes of weathering and erosion. As particles are deposited and/or precipitated, igneous rocks can become sedimentary rocks. Both sedimentary and igneous rocks can be subject to intense heating, pressure, and chemical action, becoming metamorphic rock in the process. All three rock groups may go back to the molten form.

From this basic information about rocks, the generalized geology of British Columbia can be examined, recognizing that rocks and landforms were created during different eras, as Figure 2.3 demonstrates. Much of the Coast Mountains range and south central British Columbia is made up of intrusive igneous and is approximately 100 million years old. Consequently this region should be characterized by granitic types of rock. Of course, the actual geology of any region is much more complex, and those familiar with the Coast Mountains just north of Vancouver around Squamish and Whistler will recognize Mount Garibaldi, the Black Tusk, and obvious basalt columns as extrusive igneous formations. Mount Fissile, behind Blackcomb Mountain, has sedimentary rocks containing fossils of seashells at over 2,000 metres in elevation. Many mountain-building processes have been at work in the various regions of British Columbia. The generalized geological map gives an overview of the dominant rock type only.

The interior of the province and the north end of Haida Gwaii are a mix of flat-lying lava and some sedimentary rock. Much of the interior of the province was under water for considerable periods, and sedimentary rock should be found in these locations. About 40 to 50 million years ago, however, significant volcanic activity produced huge lava flows that covered the sedimentary levels. In areas like the spectacular Helmcken Falls on the Murtle River in Wells Grey Park, the much harder basalt rock, twenty to thirty metres thick, can be seen quite clearly over the top of the much softer sedimentary layers.

The third category, flat-lying or gently dipping sedimentary rocks formed some 65 million years ago, forms a plains-like landscape such as the Peace River/northeast region, which is structurally part of the Great Plains of North America. The mineral resources found in this region include oil, natural gas, and coal.

The Rockies are made up of folded sedimentary rock. Some 150 million years ago, there was an enormous amount of pressure on flat-lying sedimentary rock that folded it up to elevations in excess of 4,000 metres. Subsequent mountain-building episodes occurred, the last one approximately 65 million years ago. A drive through the Rockies not only gives the impression of a vertical landscape but also shows the folding of the sedimentary layers.

Most of Vancouver Island, the south end of Haida Gwaii, and much of the interior are formed of folded and faulted volcanic and sedimentary rock. These areas were covered by water in the last 500 million years and are characterized by sedimentary rock. Approximately 150 million years ago, volcanic activity and pressures were exerted to push these sedimentary and volcanic structures into very rugged landscapes.

The final category, found mainly in southeast British Columbia in the Kootenays, is metamorphic rock. These very hard rocks are the oldest rock structures in the province, dating back one billion or more years.

Recognizing the processes that formed the three basic rock structures, having some idea of the age of landforms, and being able to generalize about where these rocks and landforms can be found is the beginning in understanding the geomorphology of British Columbia. In the context of millions of years, mountains have been built up and worn down, and it is important to examine the factors that shape and change the physical landscape in more detail.

DEFORMATION OF THE EARTH'S SURFACE

Deformation is a broad term describing the rather complex geologic processes of mountain building and changes to the surface of the earth. The forces that fold, tilt, and fracture rocks are produced deep below the earth's surface and are explained by the theory of **plate tectonics**. Figure 2.4 shows a cross section of the earth, which is made of several components. At the centre of the earth is a solid inner core ringed by a liquid outer core. This core is ringed by the mantle, which is solid but behaves like a plastic substance because of the extreme pressures and temperature. The mantle is also composed of a number of layers, each with its own characteristics.

Close to the top of the mantle is the asthenosphere, which has temperatures and pressures enough to melt the mantle rocks partially. Because the asthenosphere is in a semimolten state, convection currents develop. The moving plumes of molten material put pressure on the solid rocks above, forcing them to break and shift position. The brittle rocks above the asthenosphere belong to a layer known as the lithosphere. The bottom part consists of rocks from the upper mantle, while the top part of the lithosphere contains ocean crust and continental crust. Each of these regions – core, mantle, and crust – has a different mineral composition. The core is composed mainly of iron; the mantle has large amounts of a rock called peridotite; the ocean crust is primarily basalt; and the continental crust contains mainly granite. Sedimentary and metamorphic rocks are also found throughout the crust. The lithosphere, which contains both mantle and crustal materials, is thinnest beneath the oceanic crust, where rising convection plumes from the asthenosphere can force the lithosphere to fracture and move. The broken parts of the lithosphere are known as plates, and most **volcanic** and **earthquake activity** occurs at their edges.

The theory of plate tectonics suggests that the earth's lithosphere consists of seven large plates, as well as many small ones. Several of the large plates are capped primarily by oceanic crust; others contain both ocean and continental crust. The main plates making up North and South America are shown in Figure 2.5.

Where magma finds its way to the surface of the earth and splits the lithosphere apart, it creates a divergent boundary or **rift zone** (Figure 2.6). The cooling magma forms new igneous rock in the rift zone. This rock, in turn, may be broken apart by new magma upwelling from below. Rocks formed and fractured in the rift zone make up what are known as the trailing edges of plates. Roughly half of the fractured rock is attached to the trailing edge of each plate and the other half to the second plate. The new rocks, although solid, are still very hot; they may have temperatures of several hundred degrees Celsius. The high temperatures cause the rocks to expand. As the rocks are forced away from the rift zone, they begin to cool, contracting and shrinking in volume. A look at the cross section of a rift zone in Figure 2.6 shows that the ocean becomes deeper on either side of the zone because the cooling crust contracts and thus compresses downward. The oceanic crust also becomes smoother the farther it is away from a rift zone because sediments from the continents are deposited in the ocean basins, level the irregularities of the igneous crust, and form a continental shelf.

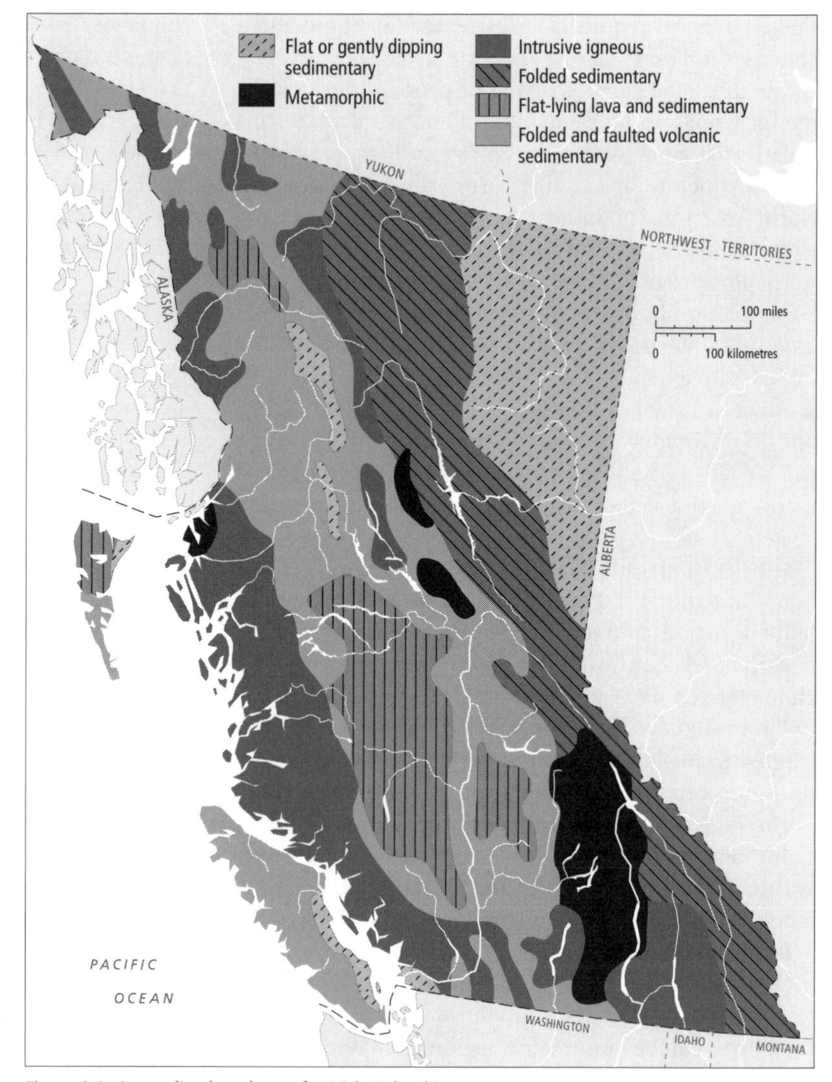

Figure 2.3 Generalized geology of British Columbia
Source: Modified from Ryder (1978).

Rift zones separate the plates, but in doing so, they create collisions in other parts of the globe. When two plates are forced together, the edges may fracture and buckle, forming mountain ranges. This process is known as **orogeny**. Prime examples of orogeny occur when the edge of a plate contains continental crust. The Andes on the west coast of South America were formed in this manner. The Pacific Northwest – comprising British Columbia, Washington, Oregon, and northern California – is another region where orogenic processes are at work. Besides colliding and sliding past each other, sometimes one of the plates, usually a denser, oceanic one, is forced downward and underneath the other. (Recall that oceanic rocks such as basalt are much denser, or heavier, than continental rocks such as granite.) This process is called **subduction**, and the major subduction zones for North and South America are shown in Figure 2.5. Continental rocks rarely subduct, which is why they are some of the oldest in the world – up to 3.96 billion years old. The subducting plate forms a trench in the ocean floor as much as five kilometres deep. As the plate subducts deeper, it warms. Eventually the temperature is high enough to allow some of the plate to melt. If enough magma is produced, it will flow upward to create plutonic batholiths within the crust and volcanic eruptions at the surface.

The new rocks created at a rift zone (the trailing edge of a plate) are gradually transported to a subduction zone (the leading edge of a plate), where they are eventually

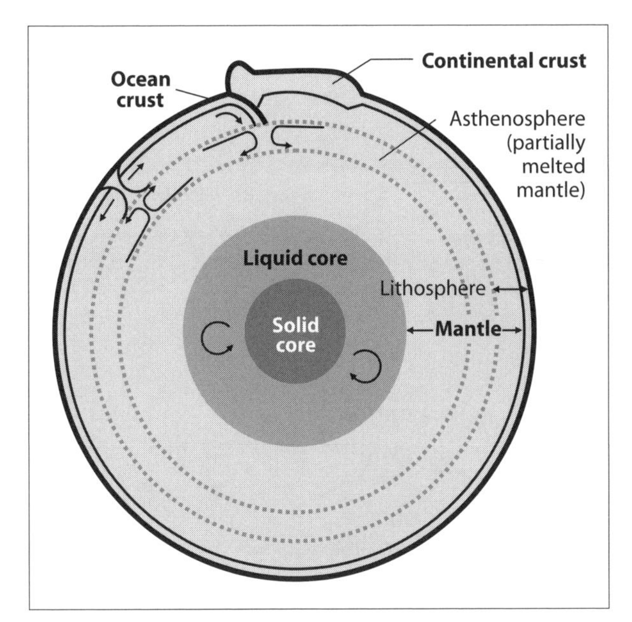

Figure 2.4 Cross section of Earth

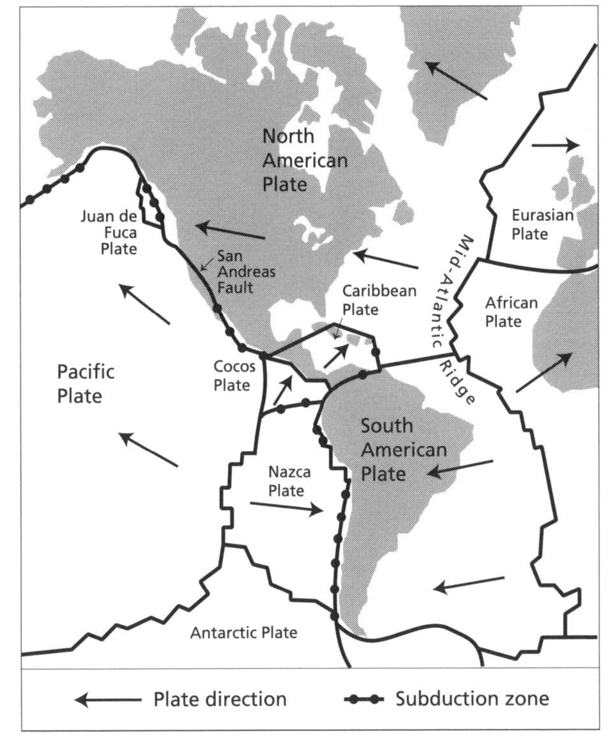

Figure 2.5 Major oceanic and continental plates

destroyed (returned to the magma state). On large plates, it may take an average of 250 million years to form, transport, and finally destroy the rocks. In this way, the plate acts as a giant conveyor belt. This analogy applies only to some parts of the world, however, as in other regions the plates collide and slip past each other to form **transform faults**.

The leading edge of the Juan de Fuca Plate, as can be observed in Figure 2.7, is in close proximity to Vancouver Island, where it is then subducted under the North American Plate. For the coast of British Columbia, this means that there is a relatively narrow continental shelf, and therefore little buildup of sediments. By comparison, the trailing edge of the North American Plate, shown in Figure 2.5, is in the middle of the Atlantic Ocean. Consequently, there is a very wide continental shelf, with a buildup of sediments off the Maritime provinces. Sedimentary buildup also provides potential for oil and natural gas deposits, and the wider the continental shelf the greater the potential.

Rift zones, subduction zones, and transform faults all occur along the west coast of North America. California is marked by the famous San Andreas fault, where the Pacific Plate moves in a northerly direction past the North American Plate. Each movement produces an earthquake, and one of the characteristics of the San Andreas fault is how it appears to "lock" in specific locations, such as San Francisco. Here, the pressure builds up until a major shift takes place. In 1906, this had catastrophic

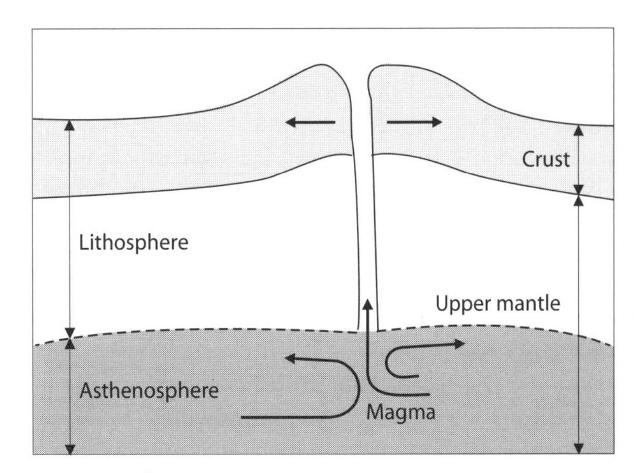

Figure 2.6 Rift zone

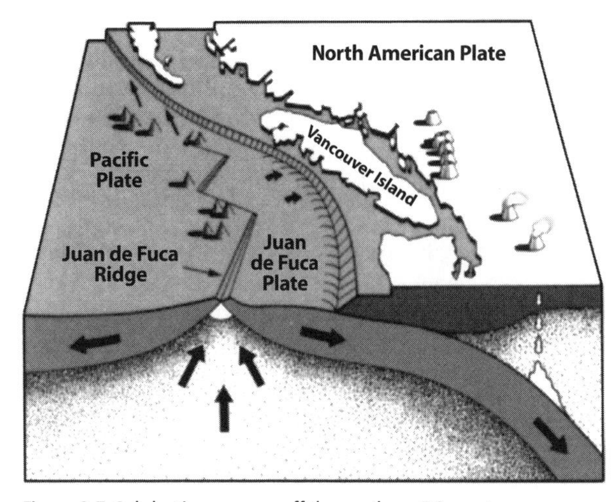

Figure 2.7 Subduction process off the southern BC coast
Source: Energy, Mines and Resources Canada (n.d., n.p.), courtesy of Pacific Geoscience Centre, Geological Survey of Canada.

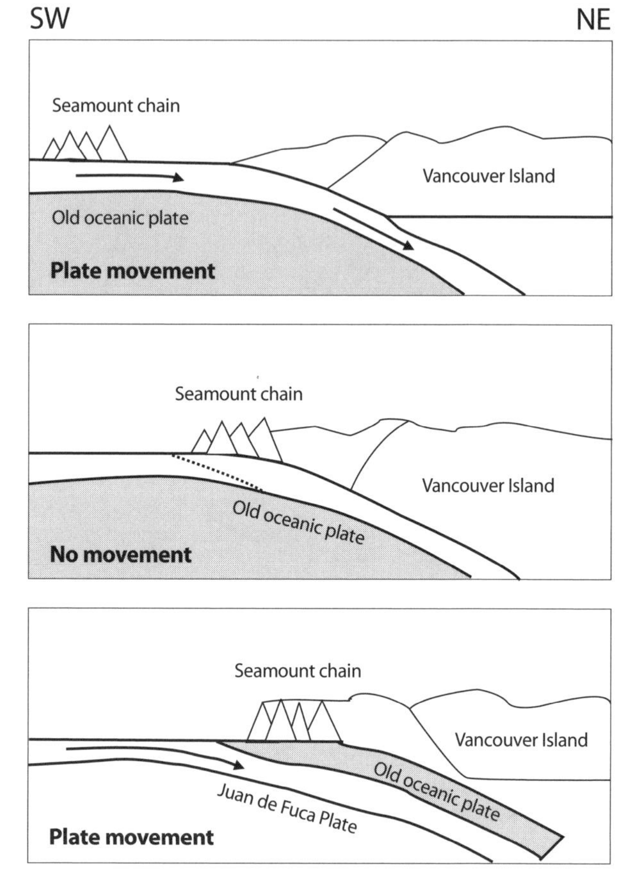

Figure 2.8 Accreted terrane process for southern Vancouver Island
Source: Modified from Monger (1990, 9).

consequences. From northern California to the end of Vancouver Island, a much smaller plate can be observed. This is the Juan de Fuca Plate, whose jagged rift zone is only 200 to 300 kilometres west of Vancouver Island (Figure 2.7). The subduction process also produces earthquakes, sometimes large ones, as the oceanic plate slides under the continental plate. Deep oceanic troughs, orogeny, batholiths, and evidence of volcanic activity (some of it quite recent, such as Mount St. Helens) are characteristic of the northwest Pacific region.

North of Vancouver Island, the Pacific Plate continues its northerly migration past the North American Plate with a transform fault in immediate proximity to Haida Gwaii. This movement, with its history of major earthquakes (e.g., magnitude 6.6 in 2009), continues all the way to Alaska, where another very active subduction zone occurs. The highest mountains in Canada – the Mount St. Elias range in southwestern Yukon and the northwest corner of British Columbia – attest to the mountain-building pressures of this collision. This region is also one of the most seismically active areas of Canada and the United States.

Subducting plates sometimes carry fragments of oceanic and continental crusts from other regions. These fragments, of differing compositions and ages, are known as **terranes**. Instead of being subducted along with the plate, the terranes are pushed up, or accreted, against the edge of the second plate. Figure 2.8 shows an example of a small terrane, the Seamount chain of mountains, being accreted on to the south end of Vancouver Island. The diagram shows not only the dynamics of mountain building and easterly migration of the Juan de Fuca Plate, but also that the southern end of Vancouver Island is composed of geologic structures formed in other regions at other geologic times.

Terranes are a big part of British Columbia. Geologic evidence suggests that the whole Cordilleran region, from the Rockies to the Insular Mountains, is a series of accreted terranes. Figure 2.9 organizes these terrane additions into five distinct belts: Foreland, Omineca,

Intermontane, Coast, and Insular. The Foreland Belt is represented by the Rocky Mountains, which form the eastern edge of the Cordilleran. The Omineca Belt includes the Omineca Mountains as well as the Purcell, Selkirk, Columbia, Monashee, and Cariboo ranges. The Interior Plateau region of the province makes up the Intermontane Belt, while the Coast Belt includes the rugged Coast and Cascade ranges. The Insular Belt is made up of the mountains on Vancouver Island, Haida Gwaii, and the Alaskan Panhandle. These dynamic tectonic forces have resulted in the highly complex geology of the province, with a very rugged landscape rich in minerals.

Piecing together the geologic history of British Columbia is a complex task, but it can be approximated with a basic understanding of the tectonic processes responsible for earthquakes and volcanic activity and for the terranes that accrete onto the continental crust. By meshing geologic time and recognizing rock formations, we can substantiate that "two hundred million years ago, there was no British Columbia – at least there was no British Columbia west of the present Rocky Mountains. The low shore of the continent sloped off into the sea perhaps where Calgary and Dawson Creek are today, and the continental shelf extended westward to the Okanagan-Quesnel-Cassiar areas" (Cannings and Cannings 1999, 11-12). Approximately 170 million years ago, the first set of terranes were added onto the craton – the ancient stable geologic formation of the Canadian Shield, including its perimeter of continental shelves. This is the Omineca Belt in Figure 2.9, and the collision of this terrane initiated the process of lifting, folding, and thrusting these sediments "at least 150 km eastward onto the edge of the old continent" to form the Rocky Mountains (Christopherson and Byrne 2009, 377).

By about 85 million years ago, more terranes were added and the force of this addition transformed the low-lying sediments into the lofty Rockies. The famous Burgess shale in Yoho National Park, with its fossils of plants and animals dated at 520 million years, results from the collision that uplifted the ocean floor of the craton. Adding to the complexity was volcanic activity in the interior of British Columbia at approximately 60 to 40 million years ago, due to a much larger subduction zone than presently exists. Massive amounts of lava flowed over existing sediments to form the Interior Plateau. The

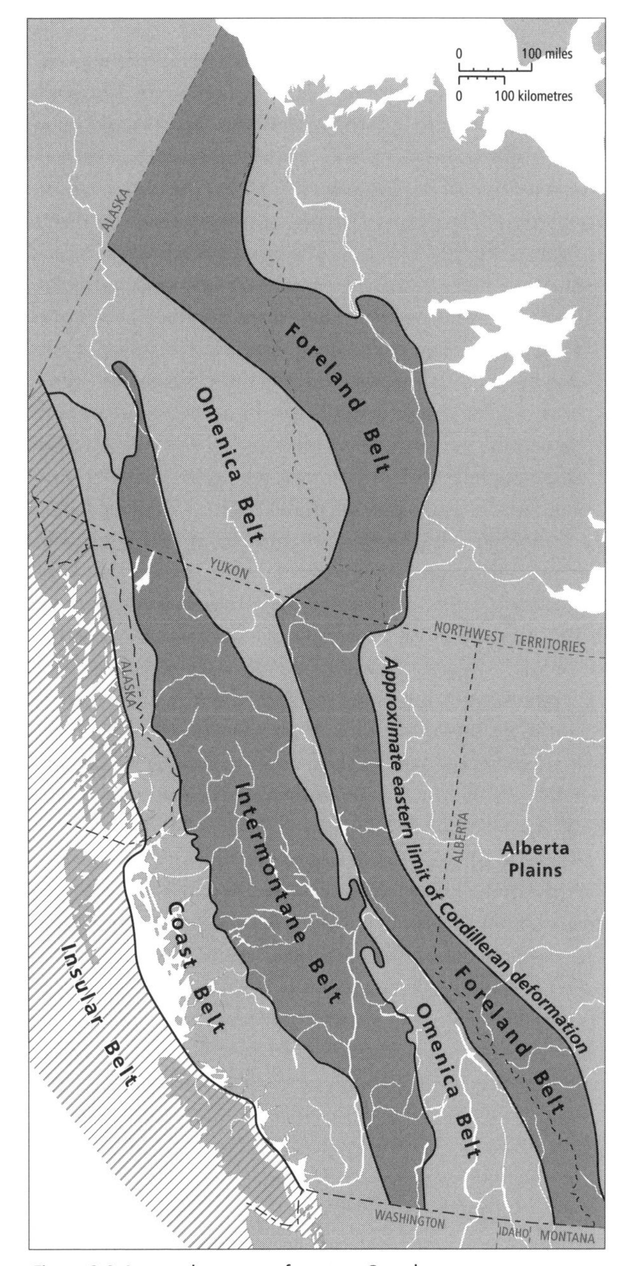

Figure 2.9 Accreted terranes of western Canada
Source: Modified from Natural Resources Canada (2007).

subduction zone migrated westward to its present location off the coast of Vancouver Island but, in the process, far more volcanic activity occurred, more terranes were added, and considerable compression pushed up the Coast Mountains. In the far northwest corner of the province, "the Yakutat Terrane is crunching into the Chugach Terrane. There some of North America's highest and most spectacular mountains rise virtually from the seacoast, and the force of the impact continues to push them up at the remarkable rate of 4 centimetres a year" (Cannings and Cannings 2004, 25).

Another piece of the puzzle that requires explanation, from the earlier discussion on the generalized geology of the province, is how metamorphic rock that is over a billion years old can be located in the Kootenays. Of course, over a billion years ago, this landform was not part of British Columbia; it was formed elsewhere in the world and, through tectonic forces, this terrane was moved and joined onto the continental plate. It should also be noted that the tectonic forces responsible for moving terranes also moved continental and oceanic plates. Neither British Columbia nor the continental plate were at the present latitude (or longitude) throughout geologic history. A fossil record of great tropical forests and dinosaurs, in a warmer climate, indicates a location much closer to the equator. In the future, further terranes could be added to British Columbia, and it is speculated that "Vancouver Island will eventually become part of the mainland" (Foster Learning 1997-2004).

Because of this complex geological journey, western North America, British Columbia in particular, remains an extremely volcanic landscape that has experienced eruptions in fairly recent times. Figure 2.10 outlines a number of volcanic belts. To the north, the Pacific Plate is actively moving northward along a transform fault, resulting in the Stikine Volcanic Belt, "the most active volcanic region in Canada, containing more than 100 volcanoes, 3 of which erupted in the last few hundred years" (Natural Resources Canada 2004a). The Anahim Volcanic Belt is a series of "hotspot" volcanoes similar to those creating the Hawaiian Islands. The youngest is the Nazko Cone in the Chilcotin region. A series of seismic tremors occurred in this location in 2007, leading to speculation of another volcanic eruption. The Wells Grey-Clearwater Volcanic Field is made up of layers of thick basalt. It is a region where "individual volcanoes have been active for at least the last 3 million years, during which time the region was covered by thick glacial ice at least twice, prior to the well known Fraser Glaciation (also known as the "Wisconsin Glaciation"). Volcanic eruptions underneath and through the thick blankets of glacial ice produced numerous unique glacial volcanoes and deposits" (Natural Resources Canada 2004a). The most southerly belt, Garibaldi Volcanic, is a product of the subduction zone activity. These statovolcanoes are "the most explosive young volcanoes in Canada" (Natural Resources Canada 2004a). Although Mount Meager erupted more than 2,000 years ago, its eruption was of similar magnitude to that of Mount St. Helens in 1980. Volcanic

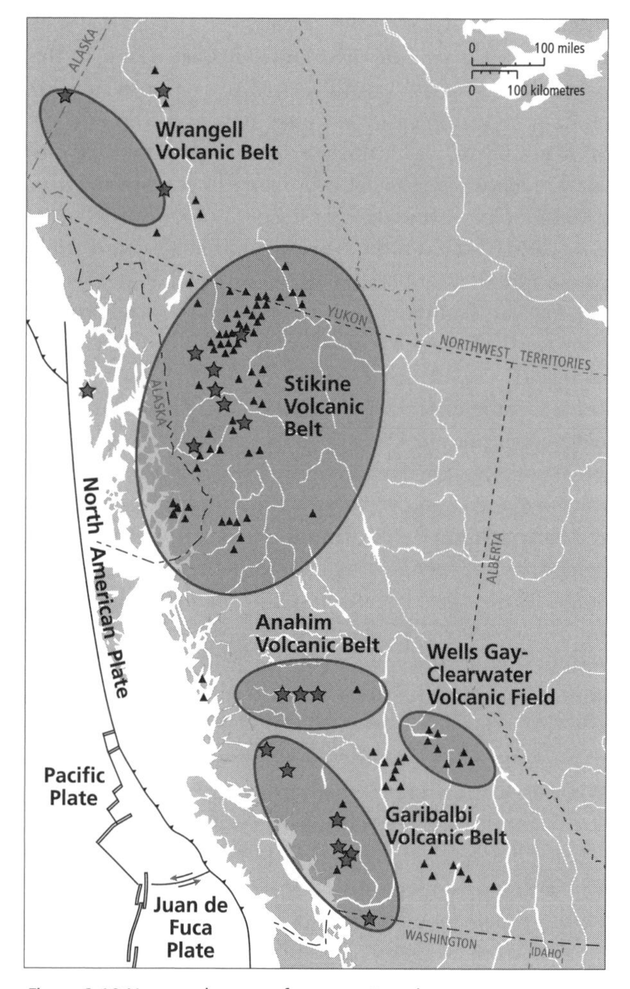

Figure 2.10 Young volcanoes of western Canada
Source: Modified from Natural Resources Canada (2004b).

activity is clearly not only a mountain-building process in this province, it is also a potential natural hazard.

Deformation of the earth's surface may occur as a result not only of tectonic factors but also of **isostasy**, a process of loading and unloading the surface of the earth with sediments and ice. As sediments erode from higher elevations such as the Rockies (which, keep in mind, are made up of sedimentary rock), these majestic mountains are gradually losing their lofty elevations. Erosion over the past 150 million years might lead one to expect that the "soft" sedimentary rocks would have been reduced to plains. The Rockies should be thought of instead as a floating raft piled high with cargo: as the cargo, or weight, is removed, the raft rises up. For every three metres of erosion to the Rockies, there is approximately two metres of uplift, or isostasy.

The last glacial age is the most recent and obvious example of isostasy. The glaciers built up to a depth of 1,500 to 2,000 metres, their enormous weight pushing the surface of the earth down perhaps several hundred metres. The level of the oceans of the world fell also to produce this volume of ice, but receded only tens of metres. As the ice melted – which happened in British Columbia only in the last 10,000 years – the ocean levels rose and, eventually, so did the land. This uplifting of the land, or isostatic rebound, takes thousands of years and still continues today. The lag in time between a relatively rapid rise in sea level and relatively slow rise in the level of the land has produced beach formations at different elevations around the world.

WEATHERING AND EROSION

Weathering describes a number of processes responsible for breaking down rock over time. **Erosion**, on the other hand, refers to the movement of rock materials via a number of agents – gravity, water, wind, and ice – working singly or in combination. Erosion continues the weathering process and helps to form sediments.

Mechanical weathering involves the physical destruction of rocks into smaller and smaller components. Freezing and thawing represent one common and powerful type of mechanical weathering in British Columbia. Rain or melting snow runs into cracks and fissures of rocks, where it is subject to freezing. The expansion of the resulting ice often fragments the rock particles.

Chemical weathering, as the name describes, refers to a chemical action that breaks rock down. The common substances of water, oxygen, and carbon dioxide, or some combination of these agents, can dissolve and chemically react with the minerals that make up rock. Sands, silts, and clays are often the product of chemical weathering, as is the red "oxidized" colour of soil.

The agents of erosion often work in combination. Gravity, for example, pulls broken rock and other weathered material downward, but when these materials are lubricated with water they are often carried away much more readily. Gravity is a powerful force and responsible for a host of erosion activities, referred to as mass wasting. These can include major landslides, such as the one near Hope in 1965, rockfalls, debris torrents, slumping, and soil creep. In all cases, the landscape is altered by particles moving down slope, sometimes rapidly and sometimes extremely slowly.

Flowing water, in the fluvial process, is an effective agent of erosion in British Columbia because of the many turbulent streams and rivers. Water carries particles ranging from boulders to clay particles, carving up the landscape and depositing the materials along stream beds where the stream slows, and eventually into the deltas of lakes or into the ocean. In the southern half of the province the Fraser and Columbia River systems dominate, and in the north the Skeena, Nass, and Stikine are the major rivers flowing to the Pacific. The northeastern portion of British Columbia is drained toward the Arctic by the Peace and Liard river systems (Figure 2.11). Where these rivers and streams flow through mountainous terrain they cut the land into V-shaped valleys. Where there is less relief, the river systems tend to meander and form broad river valleys.

The greatly indented coastline of British Columbia, along with its many islands, is subject to continual erosion from the actions of the ocean. Waves, in particular, are relentless in pounding and modifying the coastline. Currents and tides also assist in the erosion process, wearing down and carrying away materials. In other regions of the coast, where these materials are deposited, islands and coastlines are built up.

Wind plays a somewhat minor role in the erosion process. It picks up and disperses fine silts. In the dry interior of British Columbia, dust storms are not uncommon.

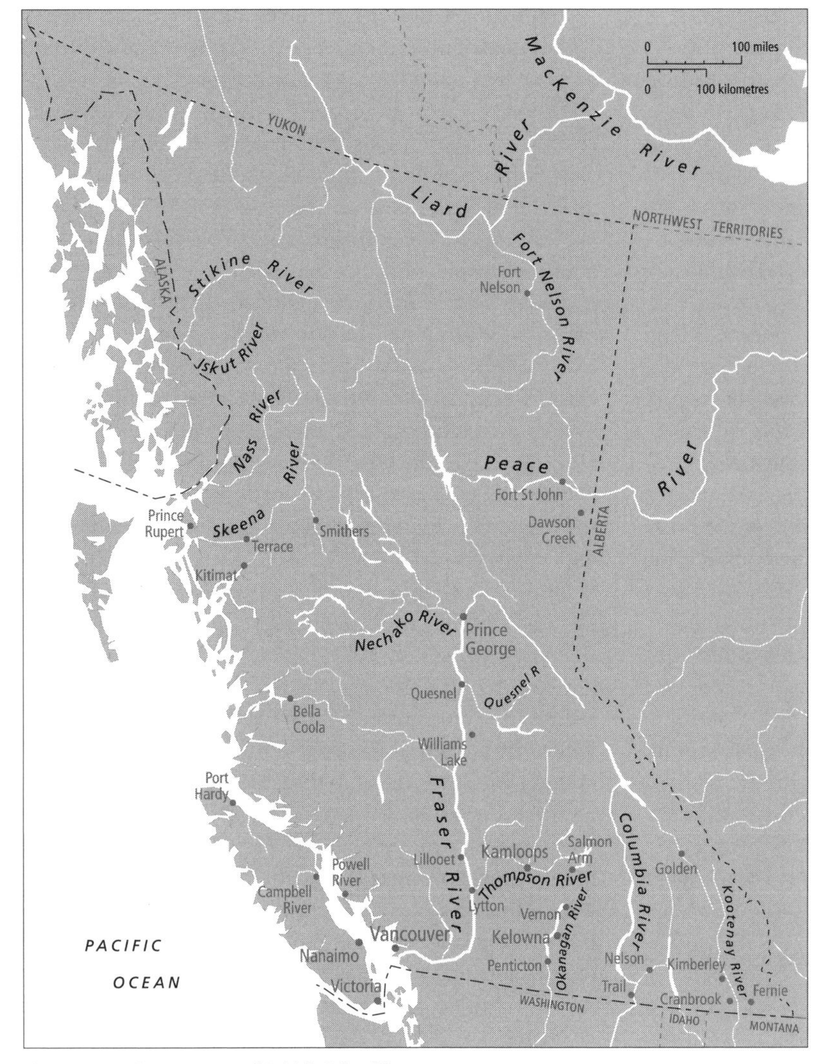

Figure 2.11 River systems of British Columbia

Wind-blown material may also act as an abrasive agent as it "sand blasts" landform structures.

The final factor, and a highly noticeable one because of its recent activity, is glaciation. There have been approximately twenty ice ages in the past 1.8 million years. The most recent began some 75,000 years ago, reaching its peak 18,000 years ago and covering most of British Columbia and Canada (Bone 2002, 48). With a cooling climate at the beginning of the glacial age, heavy snowfall in the high alpine areas of the mainland gradually compacted and transformed into icy glaciers. These moved down through stream and river valleys all the way out to the Strait of Georgia, where they coalesced with others flowing from the mountains of Vancouver Island.

These huge accumulations of ice, which reached 1,500 to 2,000 metres in height, carried a great deal of material with them and carved the landscape with their movement. Glacial erosion acted like a bulldozer. Glacial movement through existing river valleys scoured out the V-shaped valleys and turned them into U-shaped valleys. The process did not end at the coastline but continued along the ocean floor. Subsequent global warming melted the glaciers, raising sea levels, and despite isostatic rebound, left a coastal landscape of "drowned" glacial valleys called fjords – a significant asset for deep harbour ports and coastal navigation. In the Coast Mountains and other ranges, peaks under 2,000 metres have been rounded off through the movement of glaciers.

Moraines are another common landform in British Columbia. These are the large, linear piles of boulders, gravel, and sand that accumulated around the edges of glaciers. Erratics, large boulders that have been transported by glaciers, appear most obviously on flat landscapes such as the Fraser Valley.

Today, glaciers at high elevation in alpine areas are the remnants of what once covered the province.

WEATHER AND CLIMATE OF BRITISH COLUMBIA

Weather refers to the day-to-day atmospheric conditions that influence plans for any number of outdoor activities or travel. Climate is the longer-term effect of weather. It is revealed by the collection of weather statistics over months and years that enable climatologists to assess the averages, and extremes, of atmospheric conditions for any location or region. For example, the climate for the southwestern corner of British Columbia, in comparison to the rest of the province or to Canada as a whole, makes this region one of the most attractive in the country. Climatic conditions have been observed to have cycles, some of which are longer than others. El Niño conditions, for example, occur approximately every eight to ten years, resulting in much greater precipitation for coastal British Columbia. An ice age represents a much longer climatic cycle.

Figure 2.12 illustrates the global movement of air currents. The sun is the engine, or driving force, of weather and climate variations throughout the world. At the equator, a huge volume of air rises because of the intense surface heating. As the air rises thousands of metres, a low pressure zone is produced at the surface. This is typically characterized by precipitation, because the warm moist air rises, expands, and cools, causing condensation. High above the surface, the rising air diverges to the north and south poles. At approximately thirty degrees north and south of the equator these air masses descend, producing high pressure zones as they descend and compress. These subtropical high pressure belts are the driest regions of the world. In the northern hemisphere some of the descending air moves back along the surface in a southerly direction, attracted to the low pressure zone. Other portions of the air masses continue in a polar direction. The surface air mass collides with an air mass driven toward the equator by polar high pressure zones. Where these collisions take place a low pressure zone is produced, referred to as the polar front. Thousands of metres above the polar front are high velocity, easterly moving winds known as the jet stream.

The prevailing wind patterns shown in Figure 2.12 are influenced by flows of air between high and low pressure zones, by the rotation of the earth, and by the seasons of the year. During summer in the northern hemisphere the days are longer, the sun is higher in the sky, and the wind patterns shift northward. During the winter there is less sun and the wind patterns shift southward. British Columbia is largely influenced by winds known as the westerlies and by two pressure zones: the Aleutian low (producing cloudy and wet weather) to the north, and the Pacific high (producing clear and dry weather) to the south. The jet stream demarcates these two pressure zones and, as Figure 2.13 shows, moves in seasonal patterns.

There are other influences on the weather and climate of British Columbia. The relatively warm Pacific Ocean directly affects the coastal regions. Farther inland the land mass and mountains have the greatest influence. Land heats up more rapidly than water and to higher temperatures, and cools down more rapidly and to lower temperatures. As Figure 2.14 illustrates, the province has

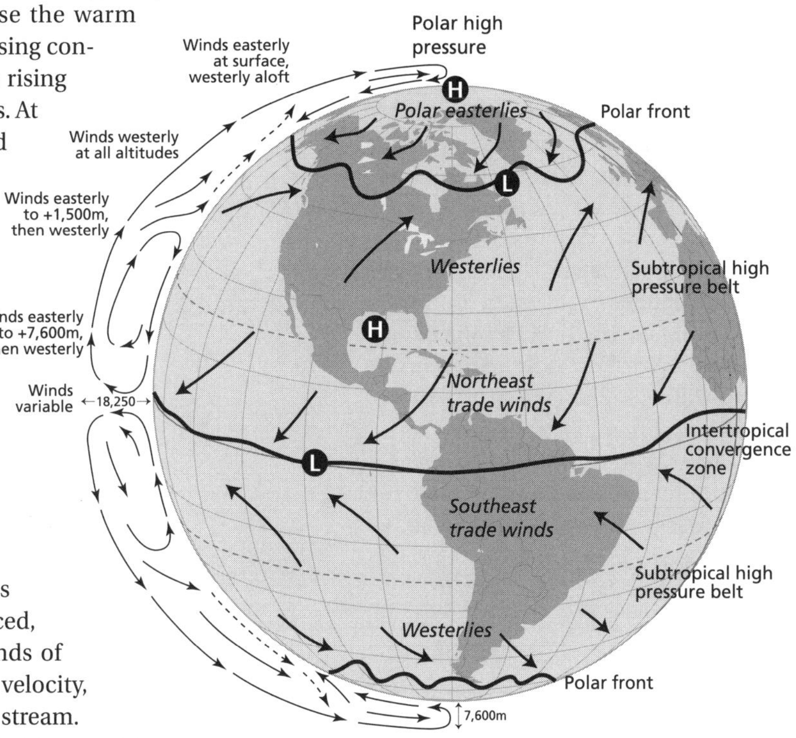

Figure 2.12 Global climate
Source: Modified from map by W. Heibert in Welsted, Everitt, and Stadel (1996, 32), with permission.

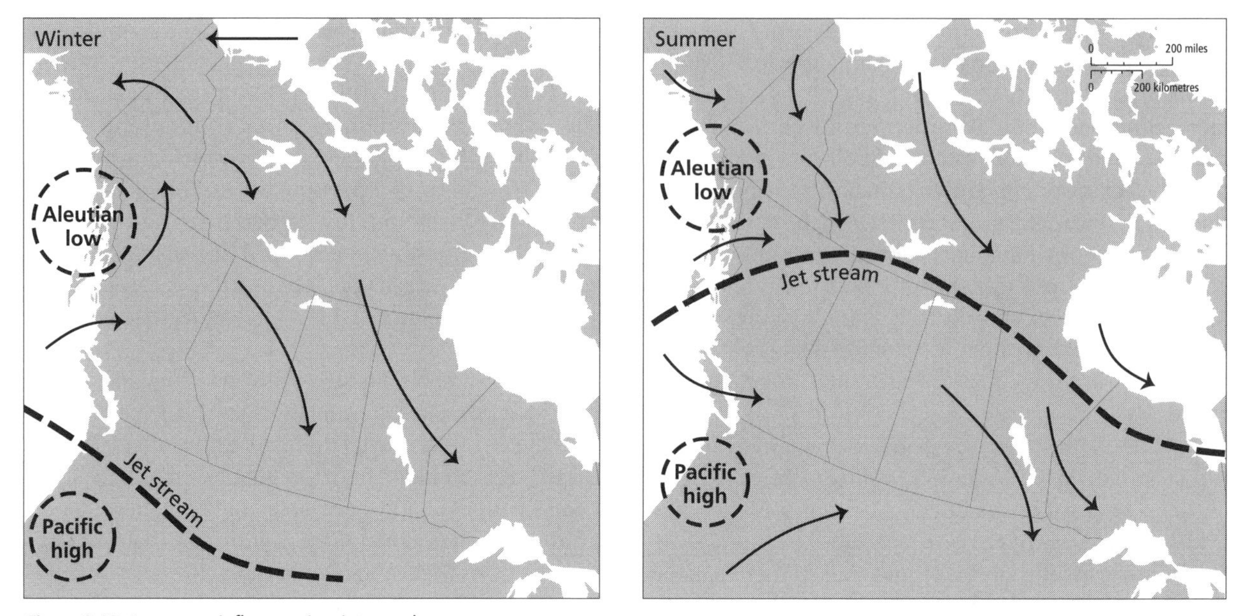

Figure 2.13 Jet stream influences in winter and summer

three climate regimes, or climate regions: Pacific, Cordilleran, and Boreal (Hare and Thomas 1974). Pacific Canada is the coastal region. Unique in Canada for its mild winters, it is often described as having a modified Mediterranean climate. The Cordilleran is defined by the many mountain chains between the Coast Mountains and the Rockies. The Boreal regime occurs in the plain-like northeast region of the province. Boreal also defines a vegetation of scrub forest typical of northern climates. Both Cordilleran and Boreal climate regimes are less subject to maritime influences than to continental air masses, which give them far greater extremes.

Prince Rupert, Vancouver, Fort Nelson, and Penticton can be found in Figure 2.14 and represent north and south coastal locations and north and south interior locations. Table 2.2 gives their mean monthly temperature and precipitation. Graphing the data would quickly show the

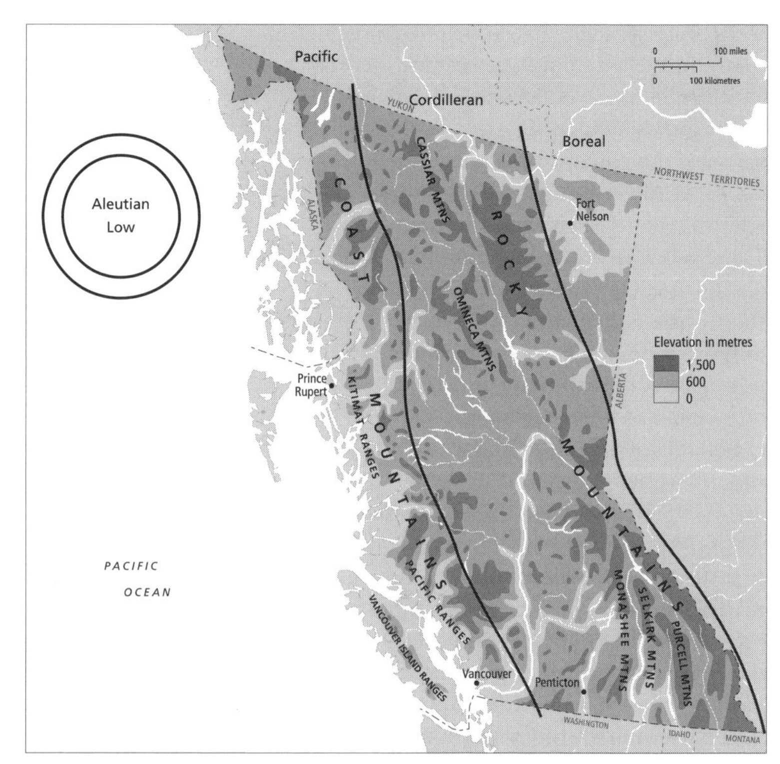

Figure 2.14 Climate regimes of British Columbia
Source: Modified from Hare and Thomas (1974, 13).

Table 2.2

Climate data for selected communities

	Vancouver 49° 11' N 123° 10' W Elevation: 4.3 m		Prince Rupert 54° 17' N 130° 26' W Elevation: 52 m		Penticton 49° 28' N 119° 30' W Elevation: 342 m		Fort Nelson 59° 50′ N 122° W Elevation: 364 m	
	Daily average (°C)	Precipitation (mm)	Daily average (°C)	Precipitation (mm)	Daily average (°C)	Precipitation (mm)	Daily average (°C)	Precipitation (mm)
January	3.3	154	1.3	257	-1.7	27	-21.2	20
February	4.8	123	2.5	204	0.7	23	-16.1	16
March	6.6	114	3.9	192	4.7	22	-7.7	14
April	9.2	84	6.0	179	9.0	27	2.9	18
May	12.5	68	8.7	140	13.6	37	10.0	48
June	15.2	55	11.1	124	17.4	39	14.9	69
July	17.5	40	13.1	114	20.4	28	16.8	85
August	17.6	39	13.5	155	20.1	31	14.9	69
September	14.6	54	11.3	244	14.9	25	9.1	40
October	10.1	113	7.9	379	8.7	20	0.6	30
November	6.0	181	4.1	304	3.1	27	-13.0	24
December	3.5	176	2.2	302	1.1	28	-19.9	18

Note: This table presents climate normals for the years 1971-2000.
Source: Environment Canada, National Climate Data and Information Archive (2010).

similarities and differences between these communities and the regions they represent.

Temperature differences between the coastal and interior locations are greatest in the winter months, when Vancouver and Prince Rupert have considerably milder temperatures than Penticton and Fort Nelson. Winter temperatures in the coastal region are mainly influenced by a relatively warm, large body of water – the Pacific Ocean – and the prevailing westerlies. In the summer, the ocean is slow to heat up and therefore provides a modifying influence. One of the most noticeable and dramatic influences on coastal temperatures occurs in the winter months. Frigid polar air, being dense and heavy, moves as a high pressure air mass from the Arctic down through the interior of British Columbia and funnels through the mountain passes to the coast, bringing freezing temperatures to Vancouver and Victoria.

As one travels inland, the moderating effect of the Pacific Ocean becomes less influential and the speed and intensity with which land can heat and cool becomes more influential. Consequently, temperatures in Fort Nelson and Penticton have greater variation than those in Prince Rupert and Vancouver.

Winter temperatures differ considerably between the two interior communities of Penticton and Fort Nelson, mainly because of latitude. Fort Nelson, just south of the sixtieth parallel, has only a few daylight hours in winter and, even then, the sun is at an extremely low angle. Penticton, just north of the forty-ninth parallel, gets a good deal more incoming solar radiation because of the increased number of sunlight hours in the winter. Fort Nelson is also much closer to the freezing Arctic high pressure air masses, which are not hindered by any physical barriers, while Penticton is protected by a number of mountain ranges that act as barriers to these Arctic air masses. Extremely cold winter temperatures (below –25°C) are relatively rare in Penticton.

In a vertical landscape such as British Columbia, elevation also influences temperature. Those participating in alpine activities will experience considerably cooler

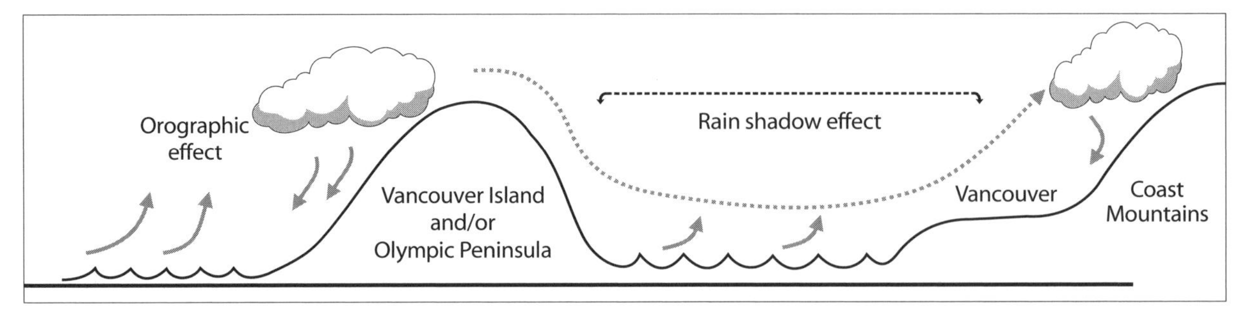

Figure 2.15 The rain shadow effect

temperatures than elsewhere year round. At very high elevations (over 3,000 metres) extremely cold temperatures can be experienced even in summer.

Considerable differences in precipitation can be seen throughout the year between Vancouver and Prince Rupert even though both have coastal locations. A number of factors are at play here. For the Vancouver area, the rain shadow effect has a significant role (Figure 2.15). The relatively warm prevailing westerlies move over the Pacific, absorbing moisture as it evaporates from the ocean. This air mass is forced to rise (a phenomenon known as the **orographic** effect) over either the Olympic Mountains of Washington State or the Insular Mountains of Vancouver Island, where the mass cools and contracts. This causes condensation, resulting in a great deal of precipitation on the western slopes – some of the highest precipitation in Canada. As the air mass descends the eastern slopes, it expands and warms and has the ability to absorb more moisture as it crosses the Strait of Georgia and Vancouver. The North Shore Mountains and other ranges of the Coast Mountains then force the air mass to repeat the orographic effect. The rain shadow region receives considerably less precipitation. It is worth noting that within Greater Vancouver significant differences in precipitation exist between south Delta and North Vancouver, just thirty-five kilometres away (Figure 2.16). Even though Prince Rupert has Haida Gwaii to intercept the westerlies, there is little rain shadow effect. Haida Gwaii does not have high mountains like Vancouver Island, and the distance from Haida Gwaii over the Hecate Strait to the mainland is considerably farther than across the Strait of Georgia. Consequently the air mass is still laden with moisture as it is forced to rise up the

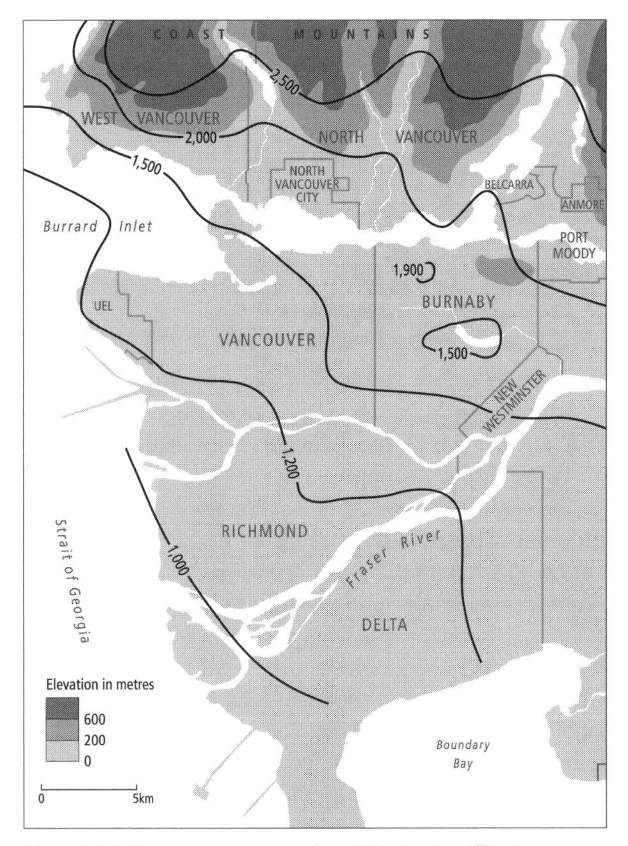

Figure 2.16 Vancouver area annual precipitation in millimetres
Sources: Modified from Wynn and Oke (1992, 25).

rugged Coast Mountains, which form the backdrop to Prince Rupert.

The Aleutian low brings rainy and turbulent weather, but its position is changeable, influenced by seasonal changes in the jet stream (see Figure 2.13). In the summer, Prince Rupert is still to the north of the jet stream wave and very much under the influence of the Aleutian low, whereas Vancouver is influenced by the Pacific high.

Precipitation patterns in the interior are very different from those of the coast, in both quantity and season. The greatest influence in the summer is from incoming solar radiation, which heats up the land rapidly, leading moisture to evaporate, rise, cool, and then condense into dark thundershowers. This convection process is particularly common on the prairie-like landscape of Fort Nelson. Penticton is nearly a desert. Its location in the southern Okanagan Valley is well inland, and the many mountains between the Pacific and Penticton wring out most of the moisture. This rugged topography ensures that most of the precipitation falls on the mountain peaks and relatively little reaches the dry valley bottom. What moisture does fall in the summer months is mainly due to convection.

During the winter, precipitation in the interior is predominantly in the form of snow. In northern, flat locations such as Fort Nelson, the surface has high reflectivity. Incoming solar radiation is therefore reflected back rather than absorbed and the atmosphere has little chance to build up moisture. It is difficult for Pacific flows of moisture to penetrate this far inland. Penticton, on the other hand, does have a marginal increase in precipitation during the winter months. The more southerly swing of the Aleutian low at this time of the year can, and does, bring more moisture. The Monashee Mountains to the east of Penticton accumulate the precipitation in the form of snow, which serves the skiing industry.

THE INFLUENCE OF CLIMATE CHANGE ON BRITISH COLUMBIA

An Inconvenient Truth (2006), Al Gore's film and accompanying book, may put to rest denials of the effect of greenhouse gases (GHGs). It has enhanced recognition that there are serious consequences to climate change, both globally and locally. Although land use practices – including agriculture, landfills, and forestry – are a factor, the key player in the global game of climate change is energy use because the principal emissions of GHGs come from burning fossil fuels (Fraser Basin Council 2006, 12). The Fraser Basin Council states that "this buildup of GHGs is contributing to rising average temperatures, changes in wind and precipitation patterns and increases in the frequency of severe weather events" (p. 12).

Ian Walker and Robin Sydneysmith (2008) outline the consequences of climate change for the various regions of British Columbia. Water shortages are going to be a major issue as temperatures rise, causing increased evapotranspiration and less precipitation, and glaciers will lose mass or, in many cases, disappear over the next century (p. 341). These developments will directly influence hydroelectric production, especially in the winter months, when there is increased demand. Natural hazards (e.g., firestorms, coastal flooding, high winds, avalanches, and hail) will increase in frequency, and agriculture will be hit by serious water shortages (drought) in some regions. Other regions, however, will increase the range of crops produced. While warmer waters place both ocean and freshwater fish in jeopardy, warmer weather and hotter, drier summers subject forests to pest infestations and have other adverse effects on existing coniferous forests (Wilson and Hebda 2008). "A billion or more pine trees are now dead in the interior of the province, the result of an insect attack of unprecedented proportions, made worse by warmer than average winter temperatures. Meanwhile, due to unusually dry conditions, forest fires burn with increasing intensity. Such fires result in uncontrolled, large pulses of GHG emissions into the atmosphere, which then increases the risk of future fires" (Parfitt 2010, 5-6).

Seventy-five percent of Canada's mammal and bird species, 70 percent of its freshwater fish, 60 percent of its evergreen trees, and thousands of other animals and plants make their home in British Columbia (Pojar 2010, 5). The spatial pattern of these plants and animals is influenced primarily by climate change, and Jim Pojar reminds us that "ecosystems do not migrate – species do" and "most species cannot disperse (move) quickly enough to keep pace with the projected changes" (p. 5). Vegetation change is already occurring and with it, animal habitats. To ensure that species will survive, we will need to study and understand new migration paths.

Understanding the consequences of climate change is an important step toward establishing solutions. And solutions do exist. The 1997 Kyoto Agreement was designed to reduce GHGs. As of 2007, it has been ratified by

174 nations, but the United States, which produces about one-quarter of the world's GHGs, has not signed the agreement. There are a number of options to reduce GHGs, including a major shift in energy use to green, renewable energy sources such as wind, solar, geothermal, and small hydro. The auto industry must produce more efficient, less emission-producing vehicles – for instance, hybrid vehicles or vehicles that use biodiesel and hydrogen. More funding for public transit will encourage people to use their cars less and, on an individual level, biking, walking, car pooling, and conserving energy are measures that will reduce each person's environmental footprint.

SOILS AND VEGETATION

Soils are the most fundamental element for growing plants, but fertile soils are in short supply in British Columbia: "Approximately two-thirds of the area of British Columbia consists of mountain slopes, rocky land, and water areas. Over the remaining third of the province, the parent materials of the soils have been largely formed by glacial drifts" (Dalichow 1972, 9).

Soils consist of mineral and rock fragments that have undergone varying degrees of physical and chemical weathering along with decaying organic matter, resulting in many combinations of minerals, water, and air. The organic matter, or vegetation, is greatly influenced by climate. As a consequence, two main soils are found throughout British Columbia, forest-related soils on the coast and semi-arid grassland soils in the central interior.

Soils have been influenced also by the bulldozer effect of glaciation where material (till) has accumulated. As a consequence, soils have a variety of textures and are continually changing over time.

A cross section, or profile, of a typical coastal soil reveals the characteristics in various layers, or horizons (Figure 2.17). Here, the top layer of the coniferous forest soil is made up of undecayed organic matter (O layer) and gives way to the A-horizon, where the organic matter decays and is added to weathered rock. These uppermost layers are the most subject to weathering and erosion. Below the A-horizon is the B-horizon, which accumulates minerals that have been filtered, or leached, out of the A-horizon. The C-horizon is the most stable layer and is often called the parent material (Lavkulich and Valentine 1978). Somewhat different characteristics make up the dry interior grasslands soils, where considerably less leaching throughout the horizon occurs. Figure 2.18 is a profile of soil typical of such areas as the Okanagan. Owing to glaciation, the parent material in much of British Columbia has been transported and mixed so that it rarely bears much resemblance to underlying bedrock.

Vegetation is tied closely to soils and climate and, as both vary greatly throughout the province, it will come as no surprise that there are also many vegetation zones. British Columbia is known for its forests and, in particular, the coniferous forests that dominate many regions. There are considerable differences between the types of

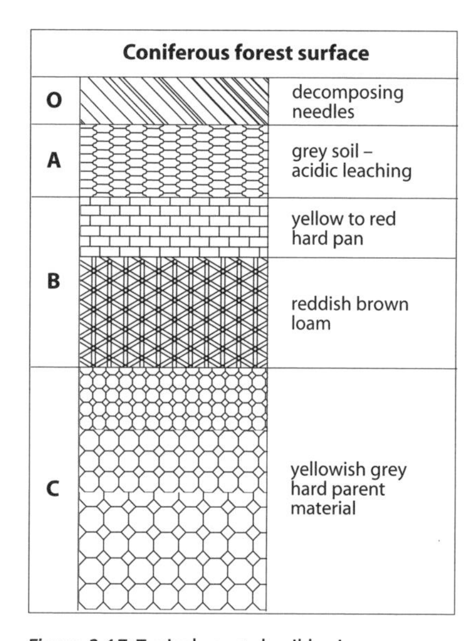

Figure 2.17 Typical coastal soil horizon
Sources: Modified from Valentine and Lavkulich (1978); Witherick et. al. (2001, 40-1, 204).

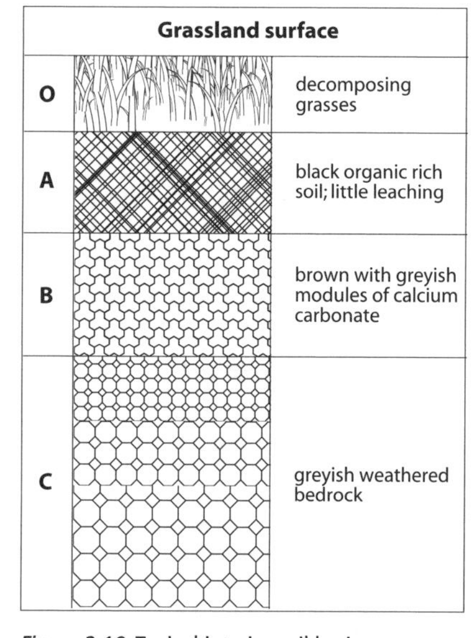

Figure 2.18 Typical interior soil horizon
Sources: Modified from Valentine and Lavkulich (1978); Witherick et. al. (2001, 40-1, 204).

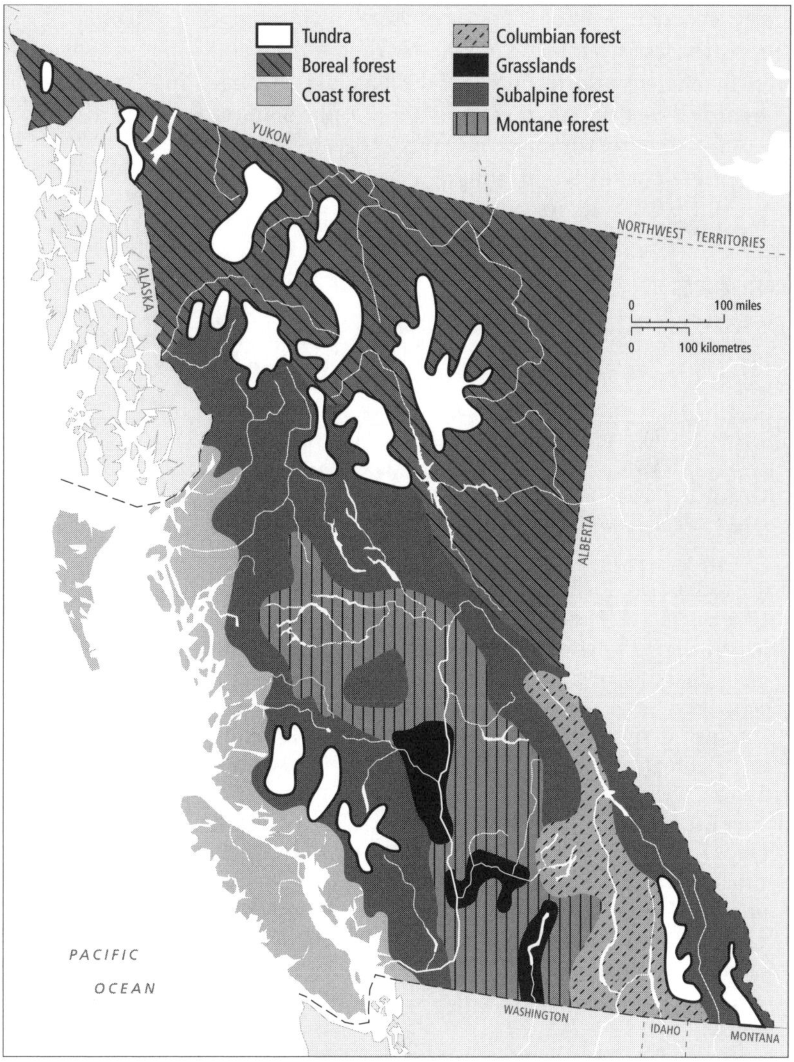

forests on the coast and in the interior. Forests also differ from north to south with latitude change, often mirroring the changes to forest species wrought by altitude in this mountainous province (Figure 2.19).

The coast forest is dominated by hemlock, cedar, and in the drier southern portions, Douglas fir. The trees here grow to an enormous size. Subalpine forest refers to vegetation affected by mountain elevation; species such as alpine fir, hemlock, and yellow cedar grow on coastal mountains, while lodgepole pine and spruces occupy interior mountain elevations. Even higher elevations, with extreme climate and scarce soil, feature tundra vegetation of moss and lichens. The montane forest covers much of the south and central interior. Because it is considerably drier than the coast and has much greater extremes in temperatures, this region is more prone to forest fires. To the south, forests of Ponderosa pine dominate, while varieties of spruce are found in the more northerly regions. Grasslands are also part of this landscape and occupy the Interior Plateau along with southern arid valleys. Here, bunch grass was the dominant vegetation until massive overgrazing occurred from the 1860s Cariboo gold rush on. Sagebrush, which cattle do not eat, has become the most common vegetation today. Columbian forest is found in the mountainous Kootenay region, which has considerably more moisture than the montane forest region. Forests of hemlock, cedar, and Douglas fir again appear, but not at nearly the size of those on the coast. Finally, to the north is the boreal forest, characterized primarily by spruce and aspen; however, their size and distribution is affected by latitude and permafrost.

Figure 2.19 Generalized vegetation patterns in British Columbia
Sources: Modified from Barker (1977, 79); Dalichow (1972, 12); and Jones and Annas (1978, 37).

SUMMARY

The physical processes of British Columbia have had, and continue to have, a powerful influence on shaping the landscape of the province. These processes cannot be ignored in examining the human development of the landscape. The historical use of the land by First Nations, and later by the first non-Natives to the region, was greatly influenced by the physical landscape. Over time, with the creation of new technologies, the physical environment has played less of a role. Yet our modern economy is still very much tied to the resources of the land, as our activities are to the daily weather conditions. And whenever there is a flood, earthquake, avalanche, or wind storm, the power of nature is very apparent.

REFERENCES

Barker, M.L. 1977. *Natural Resources of British Columbia and the Yukon*. Vancouver: Douglas, David and Charles.

Bone, R.M. 2002. *The Regional Geography of Canada*. 2nd ed. Don Mills, ON: Oxford University Press.

Cannings, S., and R. Cannings. 1999. *Geology of British Columbia: A Journey through Time*. Vancouver: Douglas and McIntyre.

–. 2004. *British Columbia: A Natural History*. 2nd ed. Vancouver: Greystone.

Christopherson, R., and M. Byrne. 2009. *Geosystems: An Introduction to Physical Geography*. Toronto: Pearson.

Dalichow, F. 1972. *Agricultural Geography of British Columbia*. Vancouver: Versatile.

Energy, Mines and Resources Canada. N.d. "Earthquakes in Southwest British Columbia." *Geofacts*. Pamphlet. Sidney, BC: Geological Survey of Canada, Pacific Geoscience Centre.

Fraser Basin Council. 2006. "Climate Change." In *2006 State of the Fraser Basin Report: Sustainability Snapshot*, 12-13. Vancouver: Fraser Basin Council, 2006.

Gore, A. 2006. *An Inconvenient Truth: The Planetary Emergency of Global Warming and What We Can Do about It*. Emmaus, PA: Rodale Press.

Hare, K., and J. Thomas. 1974. *Climate Canada*. Toronto: John Wiley.

Monger, J. 1990. "Continent-Ocean Interactions Built Vancouver's Foundations." *Geos* 4: 7-13.

Walker, I., and R. Sydneysmith, 2008. "British Columbia," Chapter 8 in *From Impacts to Adaptation: Canada in a Changing Climate 2007*, ed. D.S. Lemmen, F.J. Warren, J. Lacroix, and E. Bush, 329-86. Ottawa: Natural Resources Canada.

Welsted, J., J. Everitt, and C. Stadel. 1996. *The Geography of Manitoba: Its Land and Its People*. Winnipeg: University of Manitoba Press.

Witherick, M., S. Ross, and J. Small. 2001. *A Modern Dictionary of Geography*. 4th ed. London: Arnold.

Wynn, G., and T. Oke, eds. 1992. *Vancouver and Its Region*. Vancouver: UBC Press.

INTERNET

Environment Canada. 2010. "Canadian Climate Normals or Averages 1971-2000." National Climate Data and Information Archive. climate.weatheroffice.gc.ca/climate_normals/index_e.html.

Foster Learning. 1997-2004. "An Overview: The Shaping of Western Canada." OTS Heavy Oil Science Centre. www.lloydminsterheavyoil.com/geooverview.htm.

Jones, R.K., and R. Annas. 1978. "Vegetation." Sec. 1.4 in *The Soil Landscapes of British Columbia*, ed. K.W.G. Valentine, P.N. Sprout, T.E. Baker, and L.M. Lavkulich. BC Ministry of Environment: Soils. www.env.gov.bc.ca/soils/landscape/1.4vegetation.html.

Lavkulich, L.M., and K.W.G. Valentine. 1978. "Soil and Soil Processes." Sec. 2.2 in *The Soil Landscapes of British Columbia*, ed. K.W.G. Valentine, P.N. Sprout, T.E. Baker, and www.env.gov.bc.ca/soils/landscape/2.2soil.html.

Natural Resources Canada. 2004a. "Geological Survey of Canada: Volcanoes of Canada." gsc.nrcan.gc.ca/volcanoes/.

–. 2004b. "Young Volcanoes of Western Canada." gsc.nrcan.gc.ca/volcanoes/images/fig04_e.jpg.

–. 2007. "Cordilleran Geoscience: The Five Belt Framework of the Canadian Cordillera." gsc.nrcan.gc.ca/cordgeo/belts_e.php.

Parfitt, B. 2010. *Managing BC's Forests for a Cooler Planet: Carbon Storage, Sustainable Jobs and Conservation*. Canadian Centre for Policy Alternatives. www.policyalternatives.ca/sites/default/files/uploads/publications/reports/docs/ccpa_bc_managingforests.pdf.

Pojar, J. 2010. *A New Climate for Conservation Nature, Carbon and Climate Change in British Columbia*. West Coast Environmental Law. wcel.org/publications-search.

Ryder, J.M. 1978. "Geology, Landforms, and Surficial Materials." Sec 1.3 in *The Soil Landscapes of British Columbia*, ed. K.W.G. Valentine, P.N. Sprout, T.E. Baker, and L.M. Lavkulich. BC Ministry of Environment: Soils. www.env.gov.bc.ca/soils/landscape/1.3geology.html.

Valentine, K.W.G., and L.M. Lavkulich. 1978. "The Soil Orders of British Columbia." Sec. 2.4 in *The Soil Landscapes of British Columbia*, ed. K.W.G. Valentine, P.N. Sprout, T.E. Baker, and L.M. Lavkulich. BC Ministry of Environment: Soils. www.env.gov.bc.ca/soils/landscape/2.4order.html.

Wilson S.J., and R.J. Hebda. 2008. *Mitigating and Adapting to Climate Change through the Conservation of Nature*. The Land Trust Alliance of British Columbia. www.landtrustalliance.bc.ca/research.html.

Geophysical Hazards

Living with Risks

A geophysical hazard, or natural hazard, can be defined as an assessment of risk from the earth's forces. This implies two components: the jeopardy for people and their property, and the natural forces at work. In hazards research, unless humans or their property are involved in the extreme geophysical event, by definition there is no hazard.

The first component demands an examination of community locations, transportation systems, industrial processes, and human activities generally – in short, the process of decision making in relation to the physical environment. The decision-making process is crucial; conditions and technologies shaping our human landscape change over time, and so does our knowledge about the potential for disaster or catastrophe. Recognizing risks and developing measures to reduce or eliminate them requires decision making at various levels of society, from the individual through all levels of government. Furthermore, institutions such as universities, legal firms, banks, insurance companies, and engineering firms must be part of the solution. This province, with its rugged, mountainous landscape, tectonic activity, turbulent rivers, and exposure to many climate regimes, has many geophysical hazards. Unfortunately, lack of knowledge or simply ignoring risks has resulted in a great deal of death and destruction. And as the population grows, the risk of future disasters often increases. Land-use planning is an absolute necessity.

The second component involves understanding the physical environment, or forces of the earth, which can be placed into three categories:

- *Tectonic hazard* refers to spreading of the sea floor and the consequent collision, or subduction, of continental and oceanic plates. The risk here is from earthquake activity, volcanic eruption, and tsunamis.
- *Gravitational hazard* refers to the discharge down slope of surface material such as rock, earth, snow, and all manner of debris under the force of gravity. Snow avalanches, rock- and mudslides, debris flows, and debris torrents are all examples of gravitational hazards.
- *Climatic hazard* is evidenced in unusual weather conditions resulting from extreme temperatures, lack of moisture (drought), excess moisture (floods), lightning, hail, and violent winds (hurricanes, typhoons, and tornadoes).

Although categorized separately, hazards can overlap. An earthquake (tectonic hazard), for example, may trigger a landslide (gravitational hazard). A volcanic eruption, such as that of Mount St. Helens in 1980, may cause debris torrents, floods, hurricane-force winds, and may even affect the global climate because of the amount of ash and debris dumped into the atmosphere. The eruption of Mount St. Helens was a catastrophic event.

NATURAL HAZARDS MODEL

The following model separates the two components, or systems, in order to examine the dynamics of each in assessing risk and reduction of risk from natural hazards (Figure 3.1). The term "system" implies that complex processes are often involved, and the potential for conflict between the two is reflected by the term "versus" in the model. Each system requires further investigation.

The natural/physical system represents all the physical processes, outlined in the three categories above, that affect the landscape of British Columbia. Chapter 2 examined many of the processes responsible for geophysical hazards: tectonic activity, the gravitational activity of a vertical landscape, and the forces responsible for weather and climate variations throughout British Columbia. The concept of an extreme geophysical event recognizes that

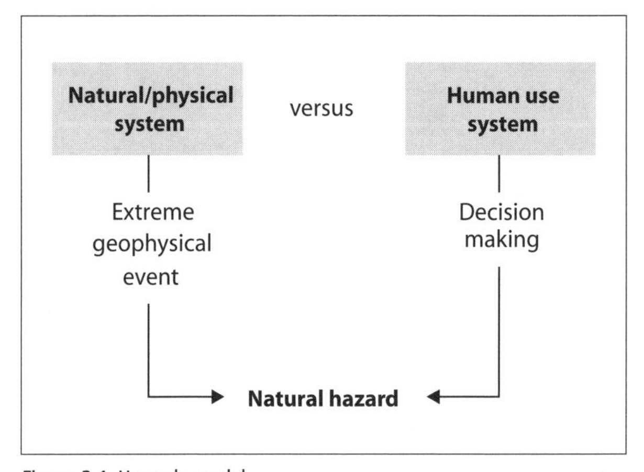

Figure 3.1 Hazards model

the forces of nature do have extremes, referred to as earthquakes, typhoons, floods, and so forth. They may be infrequent, but they are not unnatural. The challenge is to understand when, where, and why these physical processes occur.

Because people or property have to be involved for an event to be categorized as a hazard, the many earthquakes recorded off the west coast of Vancouver Island each year are not a hazard. Similarly, though avalanches occur throughout the mountainous terrain of the province in winter, only a few affect people and consequently fit into the hazard model.

Analyzing the human use system involves analyzing how location decisions are made. Where are homes, commercial structures, institutional buildings, industrial parks, railway lines, highway systems, and recreation facilities located? All have been governed by decision making. Housing subdivisions, for example, are frequently located on flat land as a result of political decisions governing zoning, density, and services available. Economics plays a role in decision making by developers, who in turn involve architects and financial institutions as they decide on the type of buildings and how many will be built in any location.

Decisions about location can be understood from a historical perspective. In British Columbia, river systems served as an effective means of transportation and irrigation, and river flood plains provided fertile soil for agriculture and flat land for housing. Subsequent decisions to build and expand on these initial locations have resulted in whole communities being situated on flood plains. Unfortunately, the decision-making process of the human use system often has little respect for the natural/physical system until the flood occurs.

When the two systems are in conflict during any hazard, it may help to ask which system is to "blame." In other words, when a flood causes loss of life or property damage, is it the fault of the extreme geophysical event or of the human use system in which decisions were made to build in a risky location? The literature on hazards suggests that the natural/physical system is passive (Burton, Kates, and White 1978; Whittow 1980), implying that it is absolutely necessary to gain as much knowledge about our physical environment as possible. In studying river systems, we come to understand the processes for water discharge, how a river creates a flood plain, and where the risk of locating is greatest. In other words, "flooding is a hazard only because humans have chosen to occupy flood-vulnerable areas" (de Loë 2000, 355).

The human use system is active and therefore must take the responsibility for damage and loss of life. Location decisions are usually made from an economic and political perspective, with little consideration of the physical environment. When the flood occurs or an avalanche hits the community, however, there is a tendency to view the extreme geophysical event as the cause of the catastrophe. In fact, these disasters are commonly referred to as "acts of God," and insurance companies normally do not insure against them. In this simplistic way, extreme geophysical events in nature are put into a realm that suggests we know nothing about physical processes and have no responsibility for our location decisions. Governments usually come to the rescue when a catastrophe occurs, compensating individuals for losses. The negative aspect of this process is that compensation is viewed as a right rather than a gift, and responsibility is rarely taken at the individual level. The importance of the hazards model is its ability to help us understand the processes operating in each system, so that decision making will seriously consider the natural/physical system and thus reduce or eliminate risks.

The story of Walhachin, on the South Thompson River between Cache Creek and Kamloops, illustrates the conflict between the two systems. In the early 1900s, American owners of these bench lands, or terraces, sold off ten-acre farms, mainly to British remittance men from England. (The men had commissions in the military and their families had money.) Land in the Thompson River Valley, which was known as Walhachin, was promoted as comparable to that of the Okanagan and the Wenatchee Valley in Washington; by the beginning of the twentieth century, both had excellent reputations for fruit farming. Immigration to Walhachin began by 1910, and the immigrant remittance men soon discovered that this was no Okanagan Valley. The area was a desert, and the struggle to obtain water for irrigation was never ending. In those days, it was not possible to pump it from the Thompson River. Over 200 kilometres of flumes were built to carry water from dammed-up streams. Even with these expensive efforts, there was rarely enough water.

The soil was alkaline and not conducive to tree fruit agriculture; to make matters worse, the region was subject to frosts severe enough to kill fruit trees. With the advent of the First World War in 1914, nearly all the men from Walhachin enlisted in the armed forces. Most never returned, and the community was abandoned by the 1920s (Riis 1973).

The town of Sandon, in the West Kootenay region, had a more disastrous story. It was a fairly typical mining town, hastily constructed following the discovery of rich veins of silver in the 1890s. The town was located in the Selkirk Mountains, where peaks can exceed 2,700 metres, six metres of snow in winter is not uncommon, and stream valleys are narrow. When the snow melts in the spring, Carpenter Creek, which winds through the community, becomes a raging torrent. Many of the trees surrounding the community were cut, leaving Sandon vulnerable to avalanches and mudslides.

The hazards were many for Sandon. The town was made of haphazard frame structures built too close together, and when fire broke out, it burned to the ground – the upper town in 1900 and the lower town in 1906. After the fire of 1900, the town was rebuilt, and much of Carpenter Creek was encased with a culvert shaped like an upside-down U, so that a road could be put over the top. Of course, this required constant maintenance so that debris from nearby logging would not build up at the front end of the encasement. Over time, the price of silver declined and so did the mining, and more and more people abandoned the town. Maintenance of the Carpenter Creek encasement was not kept up, and eventually most of the town was destroyed as the creek sought its own course.

Another mining community, Britannia Beach, just north of Vancouver on the way to Squamish, faced its share of natural hazards as well. Unlike Sandon, Britannia Beach was a planned company town. Though the main community was located on the shores of Howe Sound, a number of self-contained mining camps, including Jane Camp, were established in the mountains above the main town. The physical setting had steep, mountainous terrain and heavy rainfall in the winter months, with snow at higher elevations. Geologists had warned of the instability of the slopes above Jane Camp. Unfortunately, no action was taken and a major landslide occurred in 1915, claiming over fifty lives and destroying much of the camp.

The main community of Britannia Beach was unwisely situated on the flood plain of Britannia Creek. The accumulation of debris in the creek as a result of all the logging activity associated with the community and from the mining process caused water to pond behind debris dams. The dams broke with the rains of 1921, and the combined flooding and debris torrent washed many of the homes located on the flood plain away into Howe Sound. With little warning, great destruction occurred and many lives were lost.

These examples are just a few of the many tragic stories that colour the history of British Columbia. Common to each is how the human use system and the natural/physical system combine to create catastrophes. Had people or corporations made decisions based on the physical characteristics of the region, the risks could have been reduced or eliminated. Part of the calamity is that knowledge about the natural/physical system was available, but ignored.

MEASURING EXTREME GEOPHYSICAL EVENTS

There are a number of ways to measure extreme geophysical events, and these become the basis of risk assessment for any location. *Magnitude* measures the intensity of the event and thus describes the potential for damage. A common measure of magnitude is the Richter scale for earthquakes. The waves radiating out from an earthquake are calculated by a logarithmic calibration. A reading of 4.0 or less means some degree of shaking but usually little damage. A reading of 7.0 or higher indicates a catastrophic event. Among the climatic hazards, high winds are measured by various categories of speed. Hurricanes, for example, begin at 121 kilometres per hour and are feared for their destructive power. Floods, another climatic event, are measured by the rising volume of water, usually past a base level. All extreme geophysical events are measured by magnitude in some way.

Frequency refers to how often an event occurs in any location. Lightning-induced forest fires are fairly common in the interior of the province during July and August but much less so in coastal locations, and they are very rare in either location during the winter. Some river systems, such as the Fraser, have a high potential to flood each spring, while others are considerably less likely to do so. As can be observed from these examples, seasons influence the frequency of some extreme geophysical events. This is not the case for tectonic hazards.

Speed of onset describes how rapidly the event occurs and, consequently, how much warning is likely. Earthquakes occur with little or no warning, whereas snowmelt floods may take weeks to raise water levels enough to inundate areas.

Duration is the length of time the event lasts. Earthquakes may be over within seconds. When a flood occurs, it may take weeks before water levels recede. Wind storms are often over in hours.

Spatial pattern may be the most important descriptive measure of all because it assesses whether any pattern can be discerned on the landscape where extreme geophysical events occur. In other words, can it be mapped? Spatial patterns remind us that the various measures of assessment should not be isolated. It is of fundamental importance to recognize that most geophysical events have a spatial pattern for which risk can be estimated. Clearly, some locations are at greater risk than others for particular types of natural hazards. Earthquake risk in British Columbia, for example, is much higher in coastal locations than in the interior. If one requires zero risk of earthquake, moving to Saskatoon would accomplish this goal. Of course, this geographical location exposes one to the risk of frigid temperatures in the winter.

Combining all these types of measurement gives some ability to predict extreme geophysical events. We have not refined prediction to the point of being able to state the exact day and time that a particular magnitude of earthquake will occur, although research is being conducted toward that end. The accumulation of magnitude, frequency, and location measures over time gives us statistical data about how often floods, earthquakes, and other events have occurred, and at what intensity, in various locations. Statistical probability then allows us to assess the risk of any location. It is thus essential to the decision-making process when choices of location are made.

RESPONSES TO EXTREME GEOPHYSICAL EVENTS

Individuals are often unaware of the range of hazards and consequently simply react to extreme geophysical events when they occur. Knowing that a hazard can happen is an important beginning to a proper response. Knowing how to respond is the crucial next step. How can we protect our lives and property from natural hazards? Can we prevent disasters from happening?

Sims and Baumann (1974, 26) list some of the factors to consider when studying hazards: "The first ... [is] hazard experience; that is, for example, does it make a difference as to what one does or will do about a hurricane if one has been through a hurricane before. A second factor [is] the probability of a hazard's occurrence; that is, does it make a difference in what one will do about hurricanes if the likelihood of their occurring is high. A third factor [is] the extent of economic investment; that is, does how much one has to lose in property, say, crops or buildings, make a difference in how one perceives the threat of a hurricane."

Sims and Baumann go on to explain that while these three factors appear very reasonable, the assumption involved is that we all act rationally. Unfortunately, this is not the case: "Before the lava cools or the flood waters fully recede, people are back on the volcano's slope and the river's edge, rebuilding" (p. 26). It is very difficult to respond in a rational manner when you are personally involved. Moreover, individuals often have few choices about where they live; as the flood water recedes, so do the memories of the nasty experience.

Beyond the individual are the neighbourhood, community, regional government, provincial government, and federal government. If it is valid to ask how the individual perceives hazards, it is equally valid to ask how these institutions perceive them. Each level must be involved in the decision-making process to reduce and eliminate risks.

Five types of geophysical event are primarily responsible for loss of life and property damage by natural hazard in British Columbia: floods, wildfires, avalanches, debris flows and torrents, and earthquakes. Each will be discussed in terms of its physical characteristics, the destruction it causes, and the responses by individuals and governments. It should be recognized however, that many other extreme geophysical events have occurred in this province, including landslides (e.g., the Hope slide in 1965), wind storms (e.g., hurricane Frieda in 1962 and the 157-kilometre-per-hour winds that hit Vancouver in

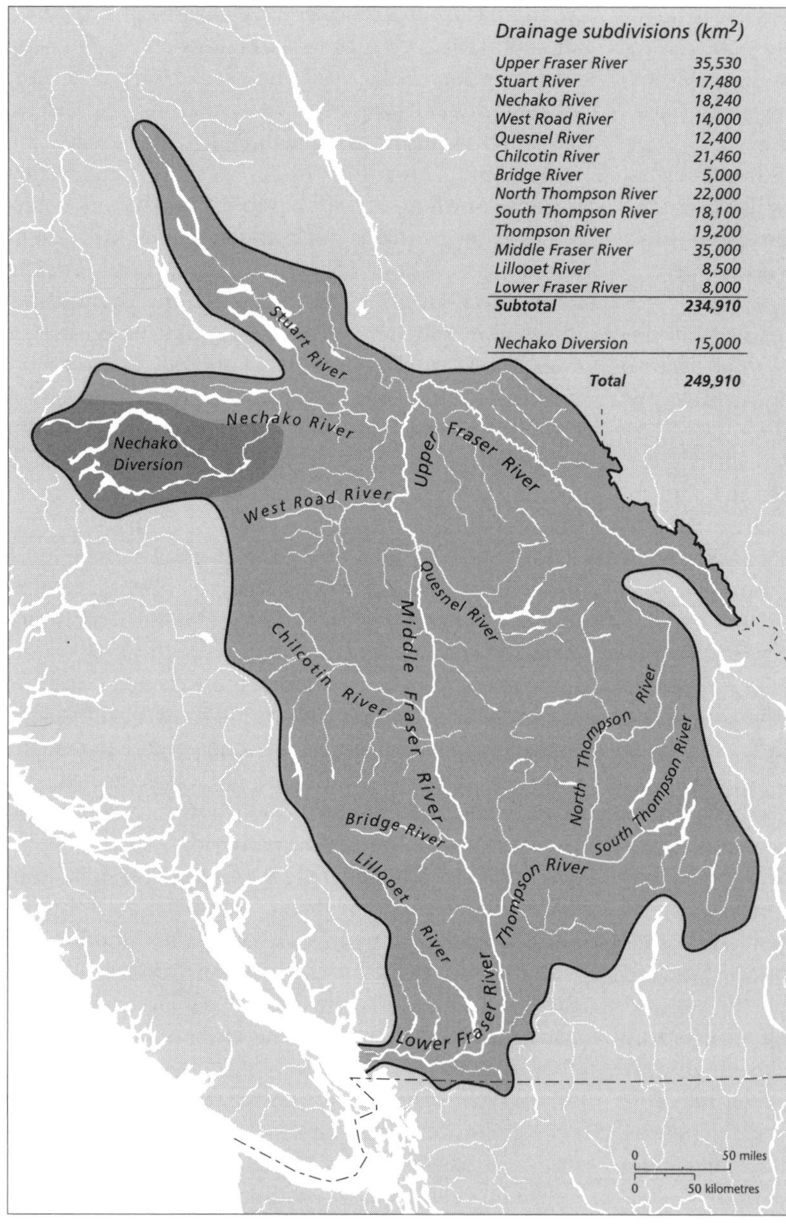

Figure 3.2 Fraser River drainage basin
Source: Modified from Fraser Basin Management Program (1994, 5).

2006), hail storms (e.g., in the Okanagan in 1946 and 1994), tsunamis (e.g., at Port Alberni in 1964), and volcanic eruptions (e.g., the Tseax cone eruption in 1775, which killed an estimated 2,000 Nisga'a) (Natural Resources Canada 2008c). All of these hazards require corrective and preventative measures.

Floods

The historical need for and use of water systems in British Columbia has spawned housing, industry, and transportation systems in close proximity. It is therefore unsurprising that floods have caused the greatest amount of property damage in the province (Foster 1987, 48).

Three types of flooding are associated with the river systems of British Columbia: **snow-melt flooding**, **flash flooding**, and **ice jam flooding**. Snow-melt flooding, often referred to as spring run-off flooding, occurs on river systems such as the Fraser, Columbia, Peace, and Skeena, which drain the interior of the province. The volume of water, or discharge, through these systems depends on the size of the drainage basin, the amount of snowpack or accumulation, and spring weather conditions. The Fraser and all its tributaries encompass nearly one-quarter of the land base of British Columbia (Figure 3.2). The system has a large drainage basin and represents a potentially huge volume of water. How much water runs off depends on the amount of snow that has accumulated over the winter and how quickly it melts.

The smaller river systems of Vancouver Island, Haida Gwaii, and the coastal region can produce conditions that result in flash flooding. For these river systems, the rapid rise of water is directly related to intense amounts of rainfall, which can cause the rivers to overflow their banks.

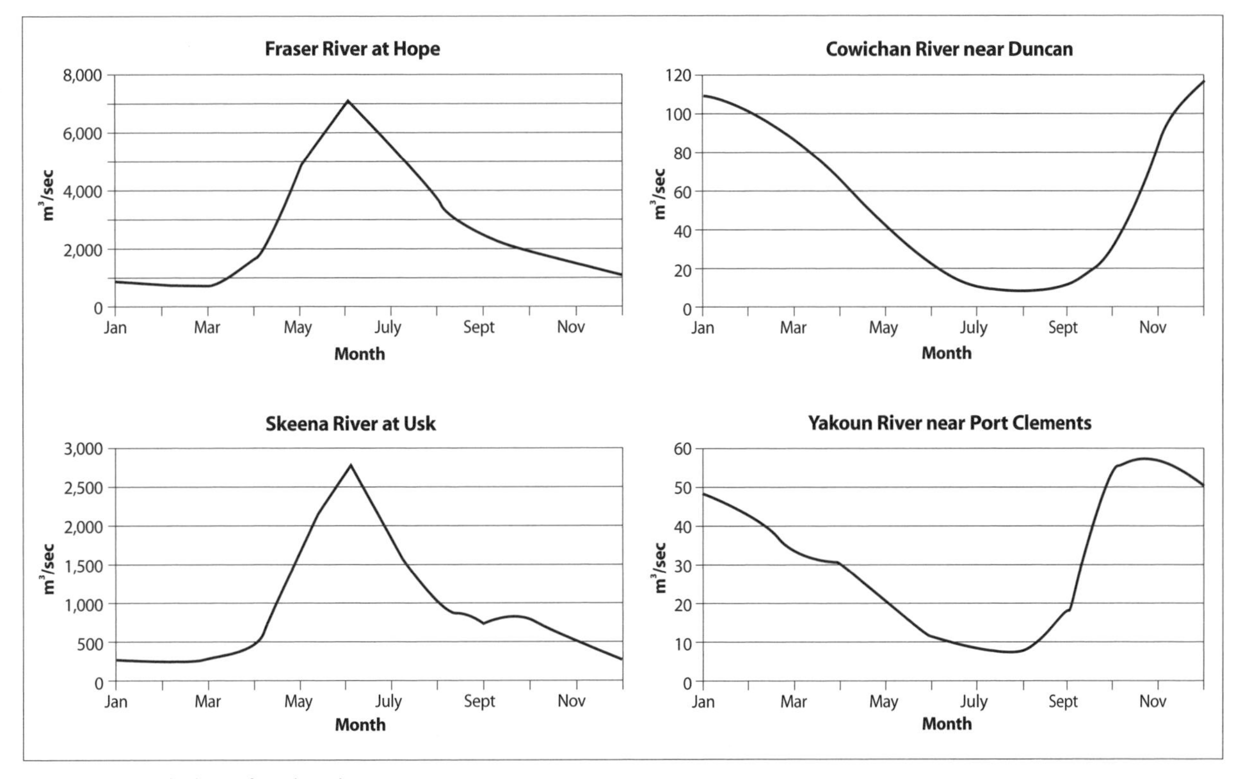

Figure 3.3 River discharge for selected rivers
Source: Data from Environment Canada (1991, 243, 350, 892, 1077).

Most northern rivers in British Columbia freeze over in winter, and this can result in ice jam floods. These may occur during freeze-up or unusual midwinter warming, but most frequently occur with the spring breakup of river ice. In all of these instances, ice fragments are set free only to jam into downstream ice that is still stuck firmly to the shore or piled against bridge piers. The jam then causes water in the channel to back up and flood the regions behind the dam of ice. This temporary dam may release suddenly, shifting the potential flooding and damage downstream.

Figure 3.3 uses several graphs to show the range of water discharge for selected rivers in British Columbia. There is a great difference between the volumes discharged by the large drainage basin of the Fraser River and the small drainage basins of river systems such as the Yakoun River and Cowichan River. Although volume of water is important, however, it is the pattern of discharge that determines flood potential at different times of year. Peak periods of discharge occur in the late spring for both the Fraser and Liard River systems, for example, producing snow-melt flood conditions in these interior drainage basins. Because large river systems drain so much of the interior of British Columbia, the potential for flooding during spring run-off affects most of the province. As the graphs for the Yakoun River in Haida Gwaii and the Cowichan River on Vancouver Island show, the greatest rates of discharge occur in the winter months, which is when flash flooding usually happens. For example, "About 300 households in Duncan and North Cowichan were issued evacuation orders last Friday after heavy rains, melting snow and a high tide led two local rivers to overflow their dikes and flood some low-lying neighbourhoods with up to one metre of water" (CBC 2009).

Documenting flood events is essential to forming a proper response. The volume of water discharged in any

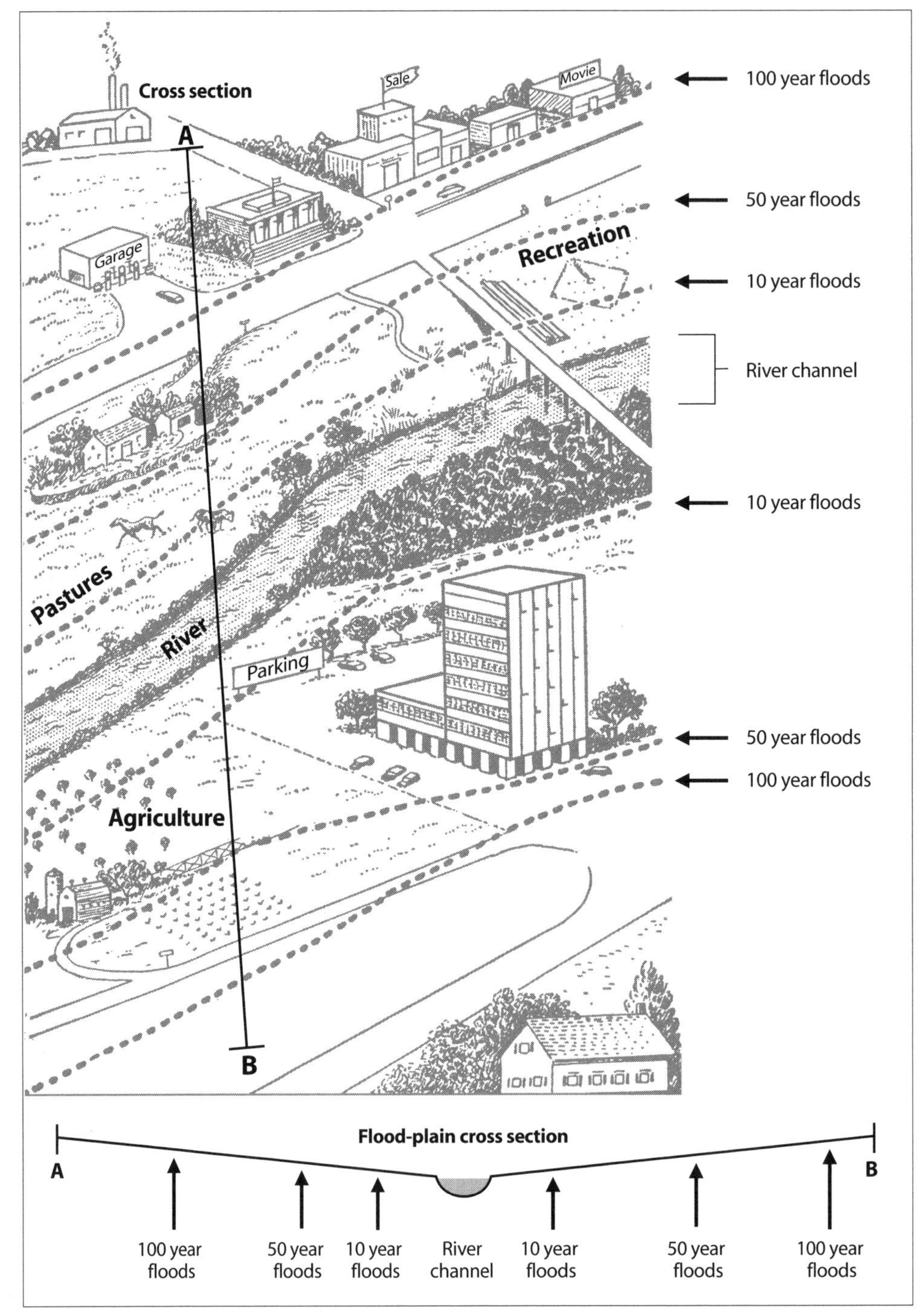

Figure 3.4 Flood plains and land uses
Source: Modified from Tufty (1969, 231), with permission.

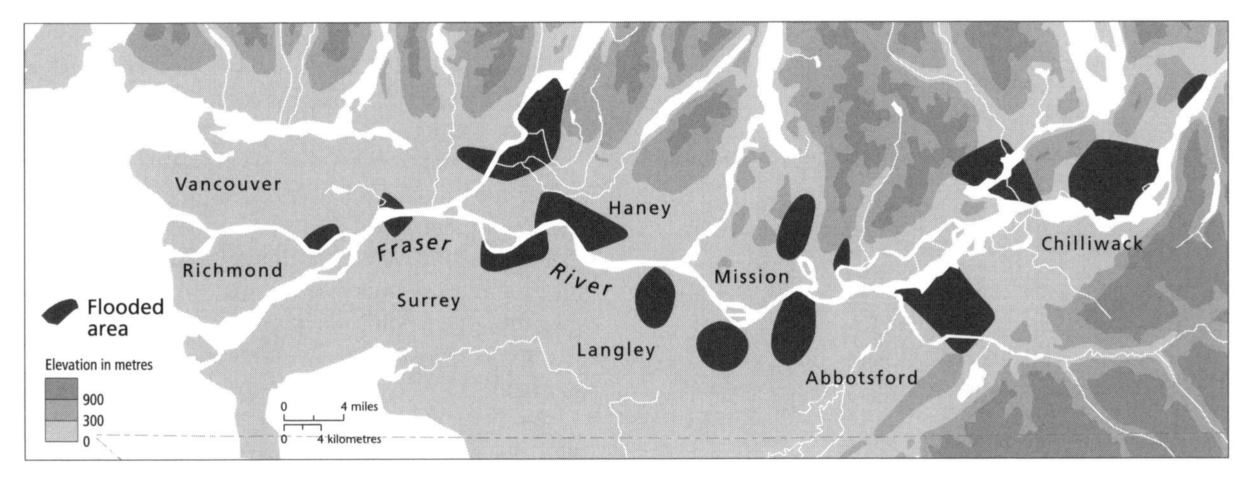

Figure 3.5 Lower Fraser Valley flooding, 1948

given year is measured in relation to benchmarks for any community. The flood plains affecting most flood-prone communities in British Columbia and the rest of Canada have been mapped. The extent of flooding depends on the level to which the river ultimately rises. The high water "contours" can be mapped to show the areas that will be inundated. Accumulating information about both magnitude and frequency of flooding allows for the production of a map of statistical probability, or what is referred to in Figure 3.4 as a ten-, fifty-, or hundred-year flood plain. The fifty-year flood plain implies the potential of the water rising to that level, or contour, once in fifty years. The flood plain mapping program financed by both the federal and provincial governments includes the 200-year flood plain. It is interesting to note that the "Great Flood" in Manitoba in 1997 reached the 500-year flood plain level. In terms of decision making, it is obvious from Figure 3.4 that the risk from flooding is reduced substantially by locating higher in elevation from the river's edge.

The 1948 flood in the Fraser Valley was catastrophic: "This flooding of 22,260 hectares (55,000 acres) left 200 families homeless, caused 10 fatalities and washed out 82 bridges. Together the federal and provincial governments provided approximately 20 million dollars to rehabilitate flood victims and repair and strengthen the diking system" (Foster 1987, 49). Figure 3.5 gives some idea of the extent of flooding in the Lower Fraser Valley alone. Flooding also occurred throughout the Fraser River drainage system in 1948.

The unusual condition most responsible for the 1948 flood was the prolonged spring season. Snow and cold weather continued well into April and early May. Then, as many farmers commented, "summer came." The hot weather melted the snow rapidly and the flooding began. Figure 3.6 shows the Fraser River discharge at Hope, measured in cubic metres per second, for 1948 as well as for several other high water years. The 1948 flood caused a great deal of damage and loss of life, but considerably higher flood waters were recorded in 1894. There is little record of the extent of the damage in the earlier flood, however, because few people lived in the Lower Fraser Valley then. This demonstrates that as the number of people on the flood plain increases, so does the risk. The population of the Lower Fraser Valley has exploded since 1948, and many more homes have been built in the areas most susceptible to flooding. The question becomes: Are we prepared for flood levels that would match or exceed those of 1894? A flood of the same magnitude has a statistical probability of recurring at Mission once every 140 years (Fraser Basin Management Program 1994, 27).

People sometimes get romantic notions about floods, in which rowing to work or school is regarded as an adventure. This myth needs to be dispelled. One of the greatest concerns related to flooding is the spread of diseases. Flood waters carry many pathogens, and each

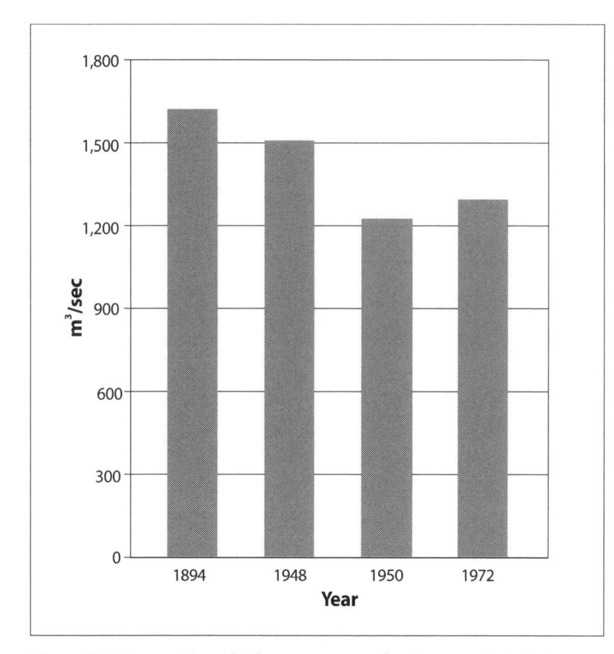

Figure 3.6 Fraser River discharge measured at Hope, 1894-1972
Sources: Data from Fraser Basin Management Program (1994, 27); Environment Canada (1991, 351).

house must be inspected by a health officer after the waters have receded. There is an enormous amount of damage to home furnishings, and the fine silt deposited by flood water usually renders any motor useless. Moreover, no price tag can be put on pets, photo albums, and many other items lost or destroyed in a flood. In the 1948 flood, the Fraser Valley was cut off from the rest of the province for most of June because the waters washed away the railways and roads. The army had to be called in to prevent looting, and the price of staple food products such as bread and milk had to be fixed because of shortages. Floods have a huge cost, and their risk needs to be reduced or avoided.

The most common way to guard against flooding is to build diking systems. In fact, too much reliance has often been placed on this form of technology. Dikes are subject to erosion, and to remain viable they must be continuously maintained. This is especially true in light of the fact that river drainage systems can change course; in some cases, dikes are part of the change. As communities in the drainage system grow in population, range land is overgrazed, forests are clear-cut, and more pavement is added. The cumulative impact of all these human activities is an increased rate of run-off in the spring, bringing greater pressure to bear on the diking system.

As more people move into the Lower Fraser Valley and interior communities, diking systems must be expanded. As more and more kilometres of dikes are built, more and more of the flood waters are contained, resulting in ever greater volumes of water and pressure on downstream dikes. Rivers carry enormous amounts of silt – the Fraser, in particular, has a "muddy" reputation – that are either deposited on the land in a flood or build up on the river bed and the delta as the river slows. When the diking system is increased, more sediment stays in stream and builds up the bottom of the river floor, thus raising the level of the water and again increasing pressure on the downstream dikes. The UBC Department of Geography (2004) has been monitoring the Lower Fraser River condition and is concerned "that the dikes might be too low in some locations between Hope and Mission as gravel moving from the mountains is deposited on the diked and confined channel." There is no such thing as a permanent dike. They are subject to erosion and must constantly be reinforced and maintained to reflect the changing dynamics of river systems.

Ice jam floods are also seen to have a spatial pattern when frequency and magnitude are recorded. Environment Canada (2009) recognizes that certain features "enhance the probability of ice jam formation: bridge piers, islands, bends, shallows, slope reductions, and constrictions." Possible corrective and preventive measures include ones applicable to all other flood situations, such as monitoring, warning, evacuation, diking systems, flood proofing, and so forth, and also measures specific to the event. Prior to breakup, cutting, drilling holes, or blasting the ice, covering the ice with dark material, and injecting ice with warm water are means to facilitate the flow and minimize the jamming of ice (Beltaos, Pomerleau, and Halliday 2000).

Figure 3.7 shows the variety of potential corrective and preventive measures to reduce or eliminate the risk from flooding. Corrective options are the many short- and long-term responses for people who are affected by flooding. Quite simply, how can people who live in a flood-prone area protect themselves and their property?

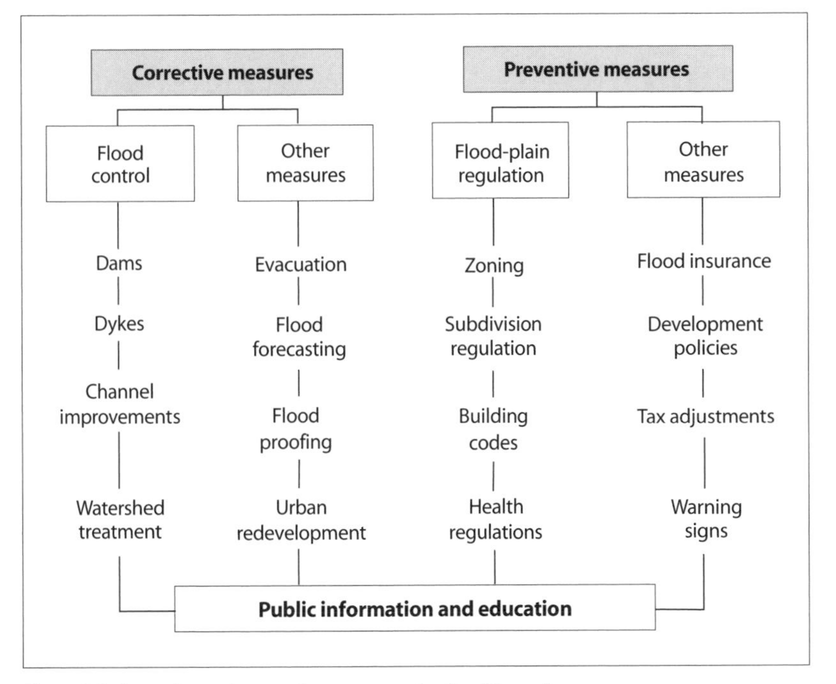

Figure 3.7 Corrective and preventive measures for flood hazards
Sources: Modified from Sewell (1965); Whittow (1980).

Immediate responses range from warning systems to sand bagging and evacuation. More long-range approaches involve various means of controlling water, including dams, dikes, and channel improvements such as dredging, and also the recognition that human activities in the watershed, such as clear-cutting, affect water run-off and flooding. Owners of individual homes in the flood plain can invest in flood proofing, thus keeping flood waters from entering basements or ground floors. Urban redevelopment may result in flood proofing of entire high risk sections of the community, or even building on pillars.

Preventive measures keep human activity out of flood plains. Many of these options require various government institutions to establish regulations and barriers. Zoning can be effective in keeping high risk areas as pasture lands, parks, or other uses that exclude housing. Regulation can achieve the same end by not allowing subdivision of flood plains. Building codes and health regulations may also stipulate rules that prevent construction.

There are other ways to prevent building in flood plains. A rather subtle form of prevention, for example, is the posting of warning signs. Flood insurance and tax adjustments can be scaled to risk as a preventive measure; the higher the risk, the higher the insurance premium, or the taxes, will be. Development policies may discriminate against construction within a flood plain or, if structures are to be built, require the developer to build up the land to an elevation above the 200-year flood level.

It is an interesting exercise to assess the theoretical range of options any community has with respect to reducing and preventing floods and match these options to decisions that have already been made. Public information and education, a component of both correction and prevention, is essential and arguably the most important element in all levels of decision making. Individual home owners, investors, politicians, and others must be aware of the options if they are to be implemented. As Rob de Loë (2000, 357) makes abundantly clear, "Floods are considered *hazards* only in cases where human beings occupy floodplains and shorelands. Therefore, the problem is one of human behaviour rather than the vagaries of the hydrological system."

Figure 3.7 was designed specifically to examine the range of options for flood conditions. As other extreme geophysical events are discussed, however, keep this framework in mind. Reducing or eliminating risk from avalanches, debris flows, and earthquakes involves many similar considerations.

Wildfire and Interface Fire

Forest fires are a common occurrence in British Columbia, especially in the dry interior, although no parts of the province are immune from this hazard. Figure 3.8 indicates the wide fluctuation in the number of fires from

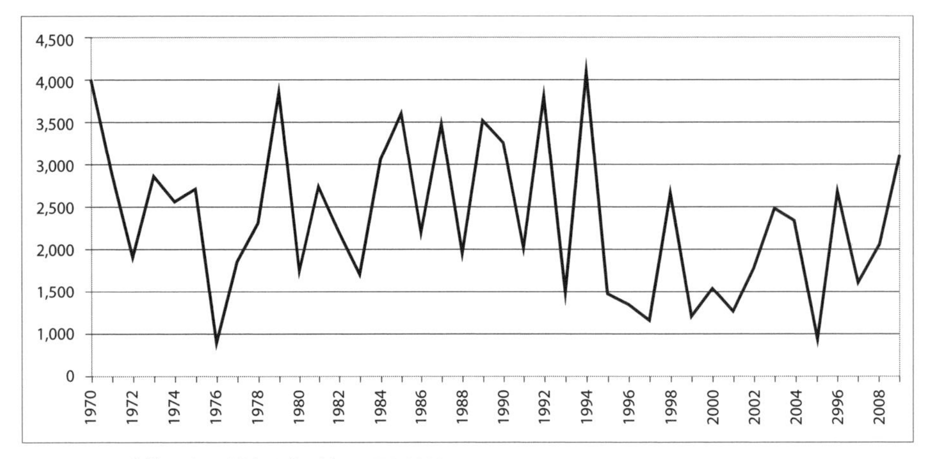

Figure 3.8 Wildfires in British Columbia, 1970-2009
Sources: BC Wildlife Management Branch (2009a; 2009b).

year to year and, while this is important, there is little relationship between the number of fires and the area destroyed by wildfire or the cost in terms of fighting fires and the damage inflicted by wildfires (Table 3.1). The year 2003 had a moderate number of fires, but records were set for the area burned and the cost of damage inflicted.

Table 3.1

Area destroyed and cost of wildfires in British Columbia, 1970-2009

Year	Area (ha)	Cost of fighting ($ millions)	Year	Area (ha)	Cost of fighting ($ millions)
1970	49,742	12.6	1990	52,575	60.2
1971	128,354	11.4	1991	11,249	32.2
1972	5,974	2.2	1992	17,212	69.9
1973	20,554	10.7	1993	1,376	25.2
1974	9,993	6.0	1994	20,737	90.9
1975	8,526	7.7	1995	26,888	38.5
1976	17,721	1.1	1996	2,670	43.0
1977	1,582	9.2	1997	286	19.0
1978	35,728	17.7	1998	76,574	153.9
1979	18,461	23.2	1999	11,581	21.1
1980	32,743	18.0	2000	17,673	52.7
1981	57,277	39.2	2001	9,677	53.8
1982	280,676	42.3	2002	8,581	37.5
1983	32,848	24.8	2003	255,466	700.0
1984	12,227	37.5	2004	230,000	156.0
1985	54,231	101.5	2005	34,588	47.2
1986	9,474	21.5	2006	139,265	159.0
1987	22,308	35.8	2007	29,440	98.8
1988	3,284	53.0	2008	13,233	82.1
1989	11,089	64.0	2009	241,066	320.0

Note: Cost does not include property losses.
Sources: Filmon (2003); BC Wildfire Management Branch (2009a; 2009b).

Firestorm 2003 was the worst on record in the province for interface fires – forest fires that affect personal property. "The interface fires of last summer destroyed over 334 homes and many businesses, and forced the evacuation of over 45,000 people. The total cost of the Firestorm is estimated at $700 million. The greatest cost of all was the loss of the lives of three pilots who died in the line of duty" (Filmon 2003). Many of the impacts of the fire were immediate and short term, such as dealing with evacuees, or those who lost homes and had to rebuild. Others are more long term. The loss of the sawmill, the main employment base in Louis Creek, was a major economic blow for the whole community. Another tragedy was the burning of many of the historic Kettle Valley Rail bridges at Myra Canyon near Kelowna, which devastated a growing tourist industry in the area built around this popular walking/cycling tour route. Through provincial and federal funding, the twelve trestle bridges were restored and back in service by 2008. Such intense wildfires also lead to short- and long-term ecosystem instability. Fire pollutes the air, destroys forests, and kills wildlife. Destabilization of soil can lead to mud slides and even flooding during rain storms, and it will take considerable time for forests and vegetation to regenerate. The year 2003 was catastrophic, particularly for the Kelowna area, where 238 homes burned, and for the communities of Louis Creek and Barrier north of Kamloops. Of course, it can happen again and did in the summer of 2009. There were more wildfires that summer than in 2003, and thousands of people were evacuated throughout the province. Fortunately, far fewer structures were destroyed.

Wildfires are a natural phenomenon in the province and several physical characteristics are responsible for their occurrence. Prolonged periods of dry weather and the presence of fuel are key factors in wildfire. A fire can be ignited by lightning or by human activity, which causes 48 percent of wildfires (BC Ministry of Forests, n.d., 3). Once a fire is in progress, wind is an unpredictable and sometimes deadly factor. As well, several relatively recent factors have increased the risk of wildfire overall and interface fires in particular. Mountain pine beetle has plagued British Columbia, mainly in the lodgepole pine forests of the southern and central interior. The combination of mild winters and mature forests has made many parts of the province vulnerable to this infestation, resulting in large areas of standing dead trees, which represent a significant increase in fuel.

Past forest practices included the burning of slash piles left over from logging activities; however, this activity has often been postponed or neglected on account of local opposition to smoke, increasing the available fuel. Moreover, British Columbia has been very effective in fire suppression: "Before suppression began 90 years ago, wildfires likely burned on average more than 500,000 hectares (1.2 million acres) of forest each year in B.C. Today, more than 2,500 wildfires burn less than 30,000 hectares (75,000 acres) on average each year" (BC Wildfire Management Branch 2010). While suppression has made more wood available for the forest industry to harvest, curtailing the natural activity of wildfire results in much longer burn cycles, and the longer the cycle the greater the amount of fuel buildup. Fire suppression and beetle-killed forests are major factors increasing the risk of large, extremely hot, wildfires.

One of the equations recognized in flood hazards was that as the population living on flood plains increased, so too did the risk. The same equation can be applied to the risk of interface fires. More and more people in British Columbia are opting to live in rural settings on larger lots with plenty of trees, or adjacent to forests. Whole subdivisions on the outskirts of cities project an image of being closer to nature, where cedar-sided homes are nestled into the trees. In an interface fire, these homes are extremely vulnerable, as the Kelowna experience testifies.

From a hazards perspective, the above information on wildfire and interface fire confirms a spatial pattern based on prolonged dry weather and fuel buildup as well as a sense of frequency and magnitude. The speed of onset can be exceedingly rapid, particularly under conditions of strong winds, which can cause fires to leapfrog a kilometre or more at a time. Once the fire has engulfed an area, its duration may not be long. Witnesses in Kelowna reported that from the time the fire reached their home twelve minutes sufficed to reduce it to ashes (CBC 2004). The severity of firestorm 2003 resulted in a commission, headed by Gary Filmon, past premier of Manitoba, to investigate the factors responsible for the firestorm and make recommendations for corrective and preventive measures (Filmon 2003).

Filmon advised reducing fuel through prescribed burning in the wet season of slash from logging activities and deadfall in parks. He also recommended that the standing fuel from beetle kill be reduced by encouraging forest companies to harvest these trees. Fire insurance companies are encouraged to use rates for risk reduction practices. The provincial and municipal/regional district levels of government need to consider land-use practices such as the installation of sprinklers on houses. Another recommendation included ongoing education for adults as well as schoolchildren. Filmon believed funding for these measures should be provided by both the provincial and federal governments because "this investment in prevention will undoubtedly result in a reduction in future damage costs under the Disaster Financial Assistance Arrangements" (Filmon 2003).

As a consequence of the Filmon Commission, "The Home Owners FireSmart Manual" (BC Ministry of Forests n.d.) and "FireSmart: Protecting Your Community from Wildfire" (Partners in Protection 2003) are available to the public. These manuals detail the many corrective and preventive measures that must be undertaken by individual homeowners as well as by those responsible for designing subdivisions. There is also a warning that, during an interface fire, fire fighters may not direct their efforts to properties that have not implemented FireSmart principles such as removing woodpiles and debris within ten metres and installing fire-resistant roofing and siding.

Wildfires, of which some will be interface fires, will occur in the future, and therefore it is important to have warning systems for early detection and the means of early suppression, especially if the fire is in proximity to homes. Water bombers, trained fire fighters, fire guards, back burning, spraying fire retardants on buildings, and evacuation procedures are some of the reactive measures taken once fire threatens a community. Banning campfires and even backcountry travel are critical measures to reduce human-caused wildfires.

Table 3.2

New snowfall and avalanche risk

Snowfall (cm)	Risk
15 to 20	Slight avalanche risk
30 to 60	Small avalanche risk, danger for skiers
60 to 90	Moderate avalanche risk, blocked roads and rails
90 to 120	High avalanche risk, structural damage possible
Over 120	Major disaster

AVALANCHES

Of all the hazards, avalanches have resulted in the greatest loss of life in the province, primarily because railways have been built through steep mountain passes that have a high incidence of avalanche. While many transportation systems and a few communities are still threatened, the greatest danger today is in the ski industry.

Colin Fraser's book *The Avalanche Enigma* (1966, 79) outlines four interrelated factors that produce avalanches:

1 When the *type of snow* varies from hard pellets to the typical "star" shape, pellets do not bind well and create unstable snow conditions.
2 The *rate of snowfall* is crucial, especially if it is accumulating at 2.5 centimetres per hour or more. Table 3.2 demonstrates how the accumulation rate of new snow serves as a guide to the stability of conditions.
3 *Terrain*, or slope, is an important factor in avalanche conditions. Slopes greater than 60° are usually too steep for large accumulations of snow but experience frequent, small avalanches instead. Slopes below 30° are usually not steep enough to pose any threat of an avalanche. Snow can build up on slopes between 30° and 60°, and their terrain is steep enough to be unstable.
4 *Change in temperature* of 6°C per hour or more, either warmer or colder, can create the unstable conditions that result in avalanche.

Avalanche risk should thus be considered within the context of the slope of the terrain, the rate of snowfall and type of snow that has previously accumulated, and rapid temperature change. Figure 3.9 classifies and illustrates the various forms of avalanche conditions. It should be noted that a dry, airborne-powder avalanche can reach speeds of up to 325 kilometres per hour and destroy practically everything in its path.

Mapping the spatial pattern of our highest risk areas is essential. The road network in British Columbia was mapped for risk after a tragic avalanche destroyed the North Route Café west of Terrace on 22 January 1974, claiming seven lives. The provincial government then commissioned a task force to assess all highways in British Columbia and recommend measures to reduce the risk from this extreme geophysical event (BC Ministry of Highways 1974).

Understanding the physical processes involved and accumulating statistics are the first steps toward corrective and preventive measures. Some of the measures used in British Columbia's road and rail systems are erecting roadside warning signs; making fairly extensive use of artillery fire and dynamite to prevent the buildup of large snowpacks; constructing tunnels, snow fences, or snow sheds over roads and rail lines; and building berms or catchment basins, which form a semicircle at the bottom of an avalanche path to prevent it from continuing on to a highway. Terrain management is another measure, and involves planting trees to stabilize snow conditions and/or building deflecting barriers. Where buildings are involved, regulatory mechanisms include building codes, restrictive zoning in high risk areas, tax incentives, and prohibitive insurance programs.

As mentioned earlier, the highest avalanche risk today is for skiers and those involved in outdoor snow activities generally. With the growing popularity of backcountry skiing, heli-skiing, and snowmobiling into untracked and often untested snow conditions, the risks are high. Even on managed ski slopes, adventurers who ski out of the prescribed boundaries run a serious risk. The managed,

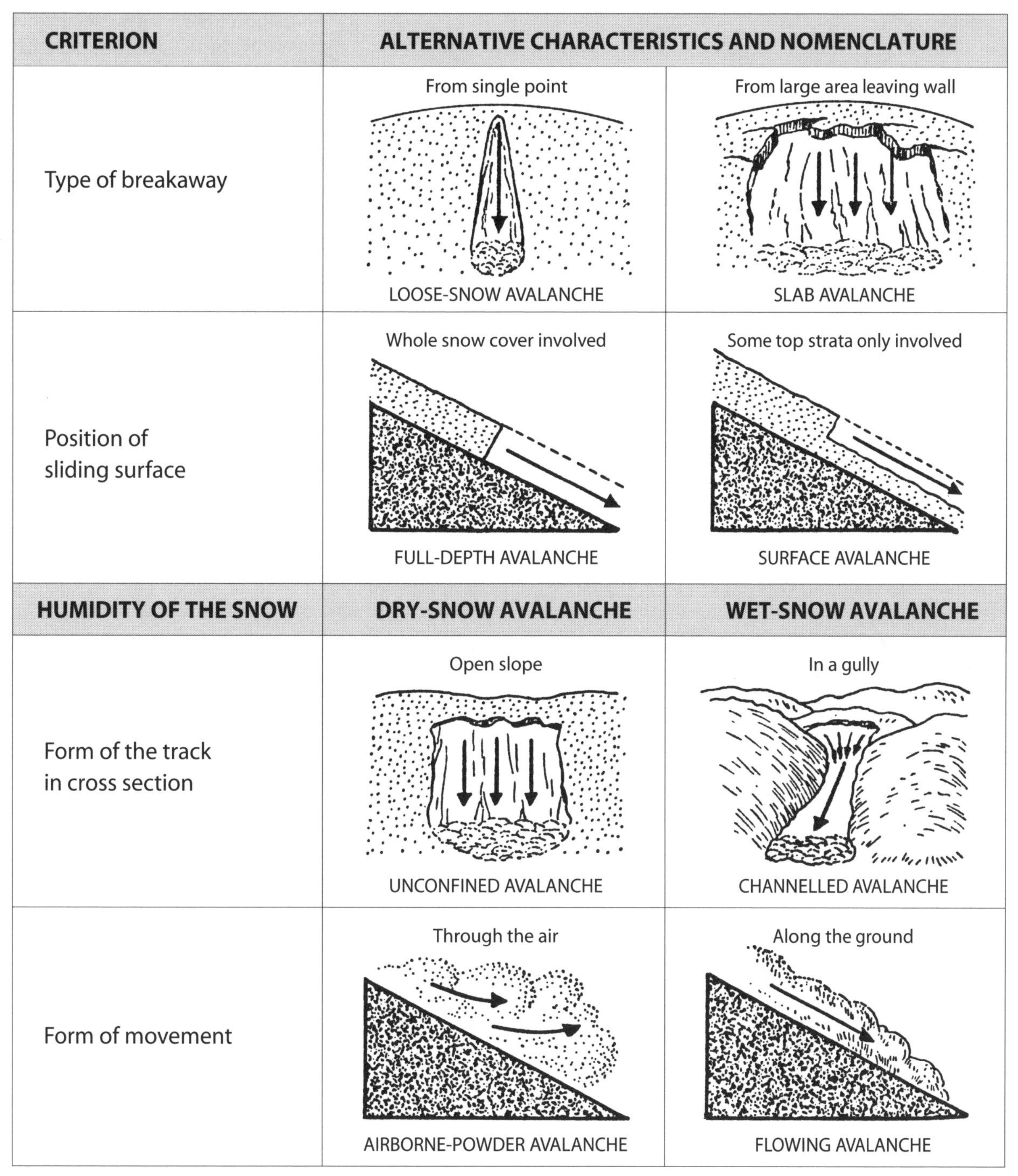

Figure 3.9 General types of avalanche
Source: Fraser (1966, 52).

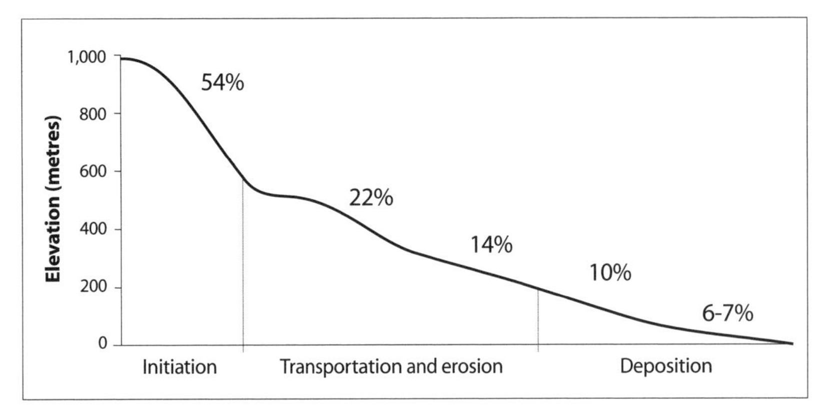

Figure 3.10 Typical debris torrent slope
Source: Modified from Thurber Engineering (1983, 8).

commercial ski slopes employ a host of mitigating measures within their boundaries: warning signs, trained avalanche rescue teams, artillery, closure of high risk runs, and so forth. For the heli-ski industry and other commercial and recreational organizations using the backcountry, the main survival techniques of avalanche preparedness are information, education, and electronic locators worn by each participant. Given the significant loss of life during these recreational activities, considerably more attention is required.

Debris Flows and Debris Torrents

Debris flows and torrents are often mistakenly referred to as landslides, mudslides, or lahars (mudslides from volcanic eruption). This is not surprising, as the terms *debris flows* and *debris torrents* are only recent designations for the types of gravitational event in which water, mud, rocks, trees, and all nature of debris flow either slowly or very quickly down a stream bed. The movement of this entire mass can have a very rapid speed of onset, providing little or no warning. The magnitude depends on the slope and topography of the creek bed and the amount of debris carried down the channel (see VanDine and Lister 1983).

Coastal locations are the most susceptible to this hazard because the many streams in these areas usually have their beginning on a steep mountain slope and exit into the ocean (Figure 3.10). These streams have extremely variable discharges that depend mainly on the amount of rainfall, which can be intense and last for weeks. Organic and inorganic matter accumulates in mountain stream beds through the normal erosion process and through human activities. The coastal areas have a long history of poor logging practices that have left behind many old roads and bridges collapsed into the waterways, allowing even more debris to build up in stream channels. Small dams can therefore form, with consequent ponding of water. With intense rainfall these dams fail, bringing a huge volume of water mixed with debris down the stream gullies, scouring out the stream bed in the process. The debris is carried downstream by gravity and is deposited on the more gentle slopes below – the slopes where people have built their communities and transportation systems.

Nine lives were lost in the Sea to Sky Highway (Highway 99) M Creek disaster in 1981. Here, a debris torrent destroyed the bridge crossing, and unaware travellers plunged into the chasm. Two years later in the same region, a debris torrent at Lions Bay claimed two lives as the debris jumped the stream bed and carried away a trailer in which the occupants were sleeping (Needham and Smith 1983). On Vancouver Island, the community of Port Alice has had numerous debris torrents, causing plenty of damage over the years. The risk is high for many communities and transportation systems, and requires both corrective and preventive measures. Table 3.3 outlines the natural hazards associated with the streams on the Sea to Sky Highway between 1906 and 1983.

Diking systems, deflection berms, catchment basins, and stream management have all been deployed as corrective measures to combat this hazard. Along the Sea to Sky Highway, much longer and higher spanning bridges have been installed to allow the debris to flow underneath. Zoning regulations, warning signs, and even insurance programs could act as preventive measures to bring about alternative and safer location decisions.

Table 3.3

Hazards related to streams on the Sea to Sky Highway

Recorded natural events, 1906-83

Streams	10 September 1906	28 October 1921	16 December 1931	29 December 1933	Fall 1960	Early 1960s	12 October 1962	22 December 1963	18 September 1969	1972	2 November 1972	7 November 1972	15 December 1972	23 May 1973	1976-7	December 1979	28 October 1981	31 October 1981	4 December 1981	6 October 1982	3 December 1982	11 February 1983
Disbrow			U													F						
Unnamed #1																						
Sclufield																						
Montizambert																						
Strid																						
Charles									D		D	D							D			
Turpin																						D
Newman									D										D	F		D
Lone Tree																						
Rundle																						
Harvey									U				F	F					F			
Alberta																					D	D
Magnesia					F		D										D					
M Creek																	D					
Loggers																						
Deeks										F												
Brunswick Pt.																						
Bertram																						
Kallanni						F											F					
Unnamed #7																						
Unnamed #8																						
Furry															F			F				
Unnamed #9																						
Daisy						F																
Thistle																						
Britannia	F	F		F				F														

Note: D = Debris flow; F = Flood; U = Uncertain origin.
Source: Thurber Engineering (1983, 4).

Earthquakes

The coastal area of British Columbia is an extremely complex and active earthquake zone. From northern California to just north of Vancouver Island, the Juan de Fuca Plate is subducting under the North American Plate. To the north of the Juan de Fuca Plate and in close proximity to Haida Gwaii, the Pacific Plate slips past the North American Plate, moving in a northerly direction until it reaches Alaska, where it subducts under the North American Plate. (See Chapter 2 for an explanation of plate tectonics.) These movements produce earthquakes and in some cases of high magnitude. The whole coast of British Columbia is one of the highest risk earthquake zones in Canada because of these tectonic processes.

The magnitude of earthquakes is usually measured on the Richter scale, on which each number represents a tenfold increase in the intensity of shaking. Table 3.4 matches the Richter scale with potential structural damage and gives some examples. The structural damage suggested here should be accepted with some caution, as the effects of an earthquake depend on many variables, including slope of the land, depth of soil, type of soil, nature of bedrock, building materials, height and design of building, and so on. Brick and stone buildings are less flexible than frame structures, for example, and high-rise complexes may not be able to withstand the horizontal motion of an earthquake. Structures located on landfill may be subject to soil liquefaction and lose their support. This information can be mapped for any community, clearly showing areas of higher and lower risk. (For the Vancouver area, see Macdonald and O'Keefe 1996.)

Table 3.4

Richter scale showing potential structural damage

Potential structural damage	Scale	Examples
Very little if any	1.0	Too numerous to list
	2.0	
	3.0	
	4.0	
	5.0	
Some structures	6.0	
	6.5	Seattle, 1975
Many structures	7.0	Vancouver Island, 1918
	7.3	Powell River/Comox, 1946
Most structures	8.0	Queen Charlotte Islands, 1949
	8.3	San Francisco, 1906
	8.5	Alaska, 1964
All structures	9.0	

Earthquakes occur frequently off the west coast of Vancouver Island and Haida Gwaii. Fortunately, most of these are of small magnitude and pose little risk. Although some are of considerable magnitude – for example, the 6.8 quake on 28 June 2004 north of Haida Gwaii and, more recently, a 6.6 quake on 17 November 2009 off the southern end of Haida Gwaii – even these caused little damage. Nevertheless, the west coast has a record of earthquakes of catastrophic magnitude and the potential for a megathrust earthquake of the Juan de Fuca Plate is of great concern (Heaton and Hartzell 1987; Charlwood and Atkinson 1983; Koppel 1989; Mayse 1992). The oceanic plate builds up tremendous pressure to subduct under the continental plate, and a megathrust is the rapid release of this pressure, resulting in an earthquake that could exceed 9 on the Richter scale. Evidence through First Nations' oral history, Japanese records of tsunamis, and coastal subsidence establishes a megathrust earthquake of approximately 9.0 on 26 January 1700 (Natural Resources Canada 2008c). The video recording titled *Quake Hunters* (1998) is a well-researched documentary that traces the evidence of this megathrust earthquake. A follow-up documentary titled *Shock Wave* (2010) focuses on the Cascadia subduction zone and assesses the impact of a tsunami on the west coast. The Natural Resources Canada Earthquakes Canada webpage (2008b) suggests "that huge subduction earthquakes have struck this coast every 300-800 years."

Other significant earthquakes include one of magnitude 8.1 that occurred on Haida Gwaii in 1949 and another calculated at 7.3 that was recorded in the Powell River/Comox area in 1946, causing property damage and one fatality. In March 2001, a 6.8 magnitude earthquake occurred just south of Seattle, Washington, and caused severe shaking at the southern end of Vancouver Island as well as southern British Columbia.

Seismic activity can also produce landslides, slumping, and destruction of telephone lines, hydro lines, roads, rails, and pipelines. These related hazards bring their own damage and threat to life. The great earthquake of 1964 in Alaska caused enormous destruction to the port of Valdez and the city of Anchorage, and the ensuing tsunami damaged port communities all the way to California. One of the hardest hit by the tsunami was Port Alberni on Vancouver Island, where millions of dollars in damage occurred.

The frequency, magnitude, and spatial pattern of earthquakes in the coastal region of British Columbia are grounds for justifiable concern for the people living there. The 1980s saw considerable effort go into the public information and education side of preparedness, particularly in the public school system, where earthquake drills are common practice. Earthquake preparedness is essential for when the "big one" occurs. The Provincial Emergency Program (PEP 1992) accomplishes this through simulation exercises and the publication of a booklet titled *Earthquake and Tsunami Smart Manual*, which is available to the public. Another important measure to reduce the risk is the regulatory mechanism of building codes, which are constantly upgrading building materials and structural design to counteract seismic activity (see Natural Resources Canada 2008a). This often requires retrofitting older buildings such as schools and other institutional, commercial, and residential sites, as well as structures such as dams. As Kenneth Hewitt (2000, 333) states: "who lives or dies in earthquakes depends more directly upon land use and, especially, the siting and design of buildings."

SUMMARY

The study of natural hazards is an interesting and important area of research that combines the skills of both the physical and the human geographer. Far more information is required to understand the physical processes that produce extreme geophysical events and to produce higher degrees of accuracy in predicting where, when, and at what magnitude events will occur. One of the most important elements in risk reduction is public information and education to ensure that informed location decisions are made, from the individual to the governmental level. As the population of British Columbia expands so too does the risk. Hazards research offers considerable hope that many risks can be avoided.

REFERENCES

BC Ministry of Highways. 1974. *Avalanche Task Force*. Victoria: Ministry of Highways.

Burton, I., R.W. Kates, and G.F. White. 1978. *The Environment as Hazard*. New York: Oxford.

Charlwood, R.G., and G.M. Atkinson. 1983. "Earthquake Hazards in British Columbia." *BC Professional Engineer* (December): 13-16.

de Loë, R. 2000. "Floodplain Management in Canada: Overview and Prospects." *Canadian Geographer* 44 (4): 355-68.

Environment Canada. 1991. *Historical Streamflow Summary British Columbia*. Ottawa: Inland Water Directorate, Water Resources Branch, Water Survey of Canada.

Foster, H.D. 1987. "Landforms and Natural Hazards." In *British Columbia: Its Resources and People*, ed. C.N. Forward, 43-63. Western Geographical Series vol. 22. Victoria: University of Victoria.

Fraser, C. 1966. *The Avalanche Enigma*. London: Murray.

Fraser Basin Management Program. 1994. *Review of the Fraser River Flood Control Program*. Vancouver: Fraser Basin Management Board.

Heaton, T.H., and S.H. Hartzell. 1987. "Earthquake Hazards on the Cascadia Subduction Zone." *Science* 236: 162-68.

Hewitt, K. 2000. "Safe Place or 'Catastrophic Society'? Perspectives on Hazards and Disasters in Canada." *Canadian Geographer* 44 (4): 325-41.

Koppel, T. 1989. "Earthquake: A Major Quake Is Overdue on the West Coast." *Canadian Geographic* 109 (4): 46-55.

Macdonald, I., and B. O'Keefe. 1996. *Earthquake: Your Chances, Your Options, Your Future*. North Vancouver: Cavendish.

Mayse, S. 1992. *Earthquake: Surviving the Big One*. Edmonton: Lone Pine.

Needham, P., and D. Smith. 1983. "Mudslide Engulfs Trailer: Two Dead at Lions Bay." *Vancouver Sun*, 11 February, A1.

Partners in Protection. 2003. "FireSmart: Protecting Your Community from Wildfire." Edmonton: Partners in Protection.

Provincial Emergency Program (PEP). 1992. *British Columbia Earthquake Response Plan*. Victoria: PEP.

Riis, N. 1973. "The Walhachin Myth: A Study of Settlement Abandonment." *BC Studies* 17 (Spring): 3-25.

Sewell, W.R.D. 1965. *Water Management and Floods in the Fraser River Basin*. Research paper no. 100. Chicago: University of Chicago.

Sims, J.H., and D.D. Baumann. 1974. "Human Response to the Hurricane." In *Natural Hazards: Local, National, Global*, ed. G.F. White, 25-30. New York: Oxford.

Thurber Engineering. 1983. *Debris Torrents and Flooding Hazards, Highway 99, Howe Sound*. Vancouver: Thurber Engineering.

Tufty, B., 1969. *1001 Questions Answered about Natural Land Disasters*. New York: Dodd, Mead.

VanDine, D.F., and D.R. Lister. 1983. "Debris Torrents: A New Natural Hazard?" *BC Professional Engineer* (December): 9-11.

Whittow, J. 1980. *Disasters: The Anatomy of Environmental Hazards*. Markham: Penguin.

FILMS

Omni Films for the Canadian Broadcasting Corporation. 2010. *Shock Wave*. Directed by Jerry Thompson. Produced by Omni Films in association with CBC-TV. 88.34 mins. Raincoast Storylines for the Canadian Broadcasting Corporation.

–. 1998. *Quake Hunters: Tracking a Monster in the Subduction Zone*. Written, directed, and narrated by Jerry Thompson. Produced by Terence McKeown. 45 mins.

INTERNET

BC Ministry of Forests, Protection Branch. n.d. "The Home Owners FireSmart Manual." www.pep.bc.ca/hazard_preparedness/FireSmart-BC4.pdf.

BC Wildfire Management Branch. 2009a. "Current Statistics." bcwildfire.ca/hprScripts/WildfireNews/Statistics.asp.

–. 2009b. "Fire Averages." bcwildfire.ca/History/average.htm.

–. 2010. Homepage. bcwildfire.ca.

Beltaos, S., R. Pomerleau, and R.A. Halliday. 2000. "Ice-Jam Effects on Red River Flooding and Possible Mitigating Methods." International Red River Basin Task Force, International Joint Commission. www.ijc.org/rel/pdf/icereport.pdf.

CBC. 2004. *Almanac*. 14 July. www.radio.cbc.ca.

–. 2009. "Vancouver Island Residents Question Flood Plan." 24 November. www.cbc.ca/canada/british-columbia/story/2009/11/24/bc-duncan-cowichan-flood-damage.html.

Department of Geography, University of British Columbia. 2004. "Fraser River Gravel Reach Studies – Overview." www.geog.ubc.ca/fraserriver/overview.html.

Environment Canada. 2009. "Causes of Flooding – Ice Jams." www.ec.gc.ca/eau-water/default.asp?lang=En&n=E7EF8E56-1#icejams.

Filmon, G. 2003. *Firestorm 2003 Provincial Review*. www.2003firestorm.gov.bc.ca.

Natural Resources Canada. 2008a. "Seismic Hazard Information in the National Building Code." earthquakescanada.nrcan.gc.ca/hazard-alea/zoning/haz-eng.php.

Natural Resources Canada. 2008b. "Seismic Zones in Western Canada." earthquakescanada.nrcan.gc.ca/zones/westcan-eng.php

Natural Resources Canada. 2008c. "Some Significant Earthquakes in Canada 1600-1900." earthquakescanada.nrcan.gc.ca/histor/15-19th-eme/index-eng.php.

Modifying the Landscape 4

The Arrival of Europeans

Aboriginal peoples were the first to discover and settle throughout North America. It is nevertheless useful to understand early European values and settlement patterns prior to discussing First Nations, primarily because the discussion in Chapter 5 focuses on the contrasting value systems of First Nations and European peoples and Native reactions to changes imposed by the British. This chapter therefore focuses on the eighteenth- and nineteenth-century struggles of colonization by the British (first against the Spanish and later against the Americans), and the historical processes that have modified the landscape of British Columbia.

When the British claimed the territory now known as British Columbia, they were initially motivated by the desire to control the valuable fur resources of the northwest coast. The fur resources of the interior of the province, an extension of the fur trade initiated from eastern Canada, also played a role in colonial control. Until the 1850s, Europeans had little impact on the land: fur trade forts were erected, few resources other than furs were exploited, basic transportation systems were developed, and British sovereignty was established.

The discovery of gold by the mid-1800s escalated the settlement process by non-Natives and was responsible for the establishment of the present northern and eastern provincial boundaries. The attitude of those arriving was fundamental to rapid changes to the landscape. The search for gold caused much destruction to streams, salmon habitat, and forests, and trespassed on First Nations territory. Permanent and temporary communities were built, transportation systems were developed, and predominantly British institutions evolved. By 1871 the gold rush was over, and British Columbia was facing tough economic times. It was then, however, that British Columbia joined Confederation, and with the union came the promise of a national railway.

EARLY EUROPEAN CLAIMS AND INFLUENCE

Europeans "discovered" America, as the textbooks say, and the reference is usually to Columbus landing in the Bahamas and then in Cuba. This was not the first discovery of North America by Europeans, although it may have been the most important in terms of its impact.

Through the technologies of navigation, the Europeans discovered new lands, and through superior military firepower, they expanded their empires. "Discovery" implied an attitude of ownership that translated into the control and colonization of the new lands, with the expectation of wealth from resources. This quest for wealth expanded mercantilism and eventually became global in scale. Territories were defined and redefined as European nations battled each other in claiming colonies. Hand in hand with this economic, political, and military approach to exploration came other views. Christian missionaries, who were not particularly tolerant of other religions, or of animism, arrived to convert the "heathens." European views of settling in a new land included rules regarding civilization, superiority, ownership of land, and use of the land.

The voyages of Columbus, and others who followed, were given credit for the European discovery of North and South America, but their main quest was to find a way to the riches of the Orient. Sailing from Europe to Asia could be accomplished in only two directions – either south around Africa and past India, or across the Atlantic and around South America. Relative to either route, the Pacific Northwest was very remote and isolated, which partly explains the rather late interest by Europeans.

European interest in the northwest coast of North America arose essentially out of two complementary motivations in the late 1700s. The first, and more complex situation, involved exploiting the resources of colonial territories in exchange for the valued silks, teas, and porcelain of China. The second quest was for the Northwest Passage and the prize money for its discovery.

By the 1700s, the Spanish had claimed most of South America, Central America, and a good portion of the present-day southwestern United States, and they had established forts in what is now known as California. The Spanish forts became the bases from which the resources of the colonies, primarily silver, were shipped to China. The Chinese had little use for European goods but were most interested in silver as an exchange item. The continual challenge for Europeans, who were reluctant to part with silver, was to seek other resources that the Chinese would find desirable. The discovery by the late 1700s that sea otter pelts from the Pacific northwest coast were a highly favoured commodity in China enveloped the northwest coast of North America in a fierce struggle over expansion of colonial territory.

The concept of **territory** combines geographic space with political control, or sovereignty, over that space. Laying claim to foreign territories was accomplished by establishing forts and by documenting, through a ship's log book, the voyages of the early explorers. The Spanish, with their silver trade out of South and Central America, assumed that they had control over the whole west coast of North America and became concerned about the Russians, who were building forts (by the mid-1700s) and laying claim to Alaska. The northernmost fort for the Spanish, however, was San Francisco. Spanish vessels had sailed up to Haida Gwaii and stopped at Nootka Sound on the west side of Vancouver Island, where some initial trading with Aboriginal peoples occurred in 1774. Captain Cook arrived at Nootka Sound in 1778 in his search for the much sought-after Northwest Passage, hoping to gain the £20,000 reward on offer, and laid claim to this "undiscovered" territory and its valuable fur resources in the name of Britain.

While the Spanish were very secretive about their voyages, Cook's publication of his journals implied that the British had been there first and thus claimed the territory. It was not long before other British vessels came to this region in search of the valuable sea otter pelts, and the conflict over sovereignty ignited.

The Nootka Sound Incident of 1789, in which a British officer was taken at sword point and three British ships were confiscated by the Spanish (and taken to San Blas in Mexico), brought to a head the rival claims. The skirmish may have been local, but its ramifications were global. The resolution of this territorial conflict required six years of protracted negotiations in Europe. Between 1789 and 1795, Captain Vancouver for the British and Captains Malaspina, Galiano, and Valdés for the Spanish made a number of voyages up and down the coast, surveying the landscape and recording their impressions of the Aboriginal peoples. The many Spanish place names in the area – Texada Island, Galiano Island, and Juan de Fuca Strait are examples from over a hundred names – remain as a legacy of this period (Archer 1981; Gibson 1981).

A map from the journal of Captain Vancouver shows a fairly detailed outline of the northwest coast of North America (Figure 4.1). He was impressed with the rugged topography and highly indented coastline. Interestingly, only the mouth and a portion of the Columbia River, and none of the other large rivers in British Columbia, are shown on the map. British and Spanish place names are evident as well.

As a result of the negotiations, Spain ceded the northwest Pacific region to Britain. Transfer to British territorial control, however, did not diminish the intense competition for valuable sea otter pelts. The Russians had the advantage by being first into the region and establishing coastal fur trade forts across Alaska and down the Panhandle. The Americans, or Yankee traders, also aggressively pursued this trade and in doing so threatened British control of the territory. As Table 4.1 documents, the British dominated the northwest coast trade until the mid-1790s, by which time the Americans were in the majority. The near extinction of the sea otter by the early 1800s reduced the intensity of territorial conflict.

The overland fur trade, mainly for beaver pelts, began to make its way into British Columbia with Sir Alexander Mackenzie's trek through the Rockies to the coast at Bella Coola in 1793. Like the sea otter trade, the overland fur trade provoked competition, but this time the main rivals were competing commercial empires rather than competing countries. The Hudson's Bay Company and the North West Company struggled for supremacy over the fur-bearing animal trade. The North West Company, headquartered in Montreal, was by far the more aggressive trader, as it established a greater number of forts throughout the Prairies and British Columbia.

As employees of the North West Company, the early explorers of British Columbia – Mackenzie, Simon Fraser, and David Thompson – surveyed the potential for fur trade forts and made contact with indigenous peoples. Company forts were built and the mainland territory of present-day British Columbia was divided into two regions: the Columbia District and New Caledonia (Figure 4.2). Much of the Columbia River could be navigated, whereas the Fraser was far too turbulent, and an overland trail was constructed along its route. By 1821, the bitter rivalry ended with the buyout of the North West Company and merger of the two companies; consequently, the Hudson's Bay Company gained the forts and a monopoly on the fur trade (Harris 1992).

The fur trade across Canada had been established for over 200 years before it reached British Columbia, and with the merger the Hudson's Bay Company was

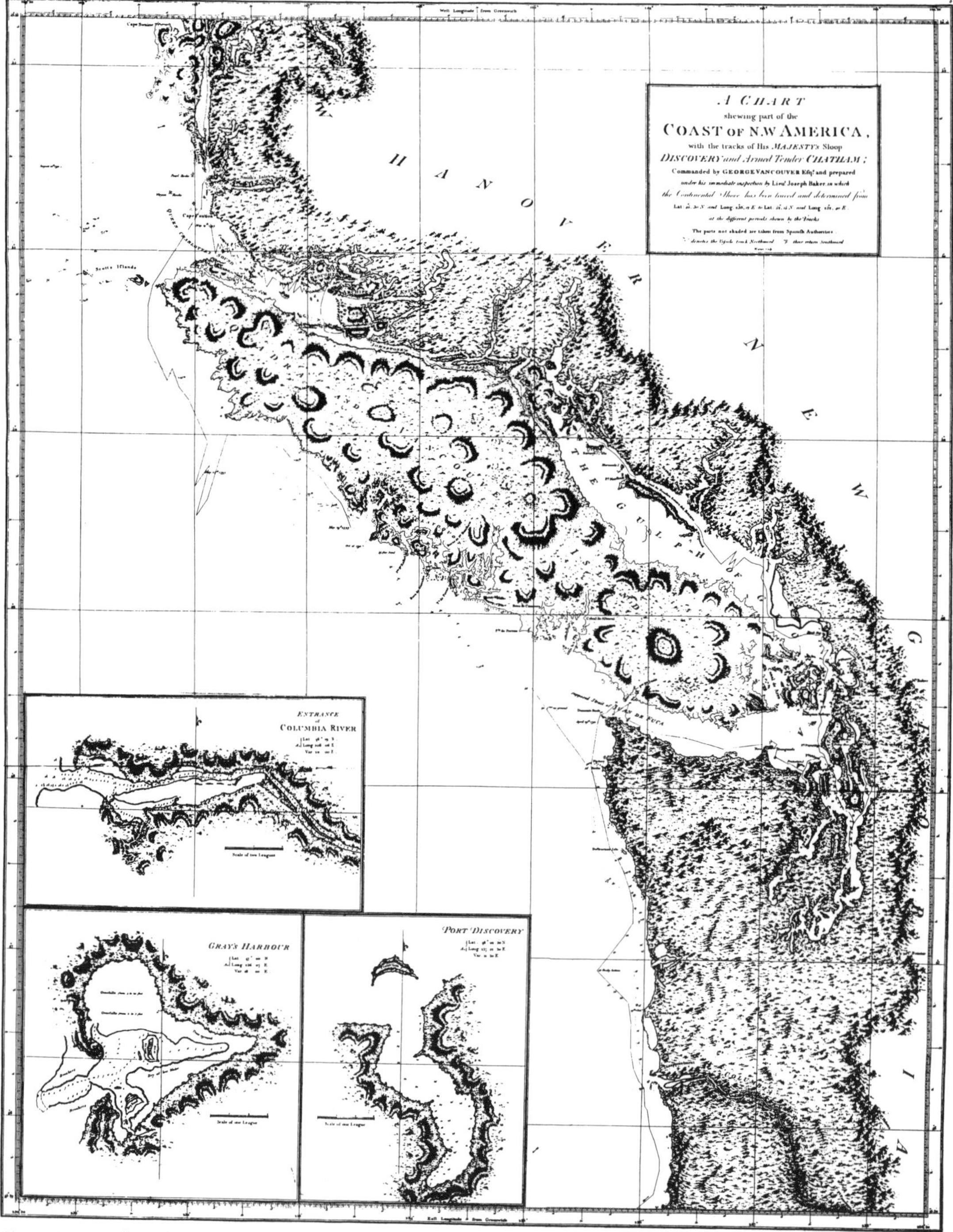

Figure 4.1 Map from the journal of Capt. Vancouver, 1798
Source: Royal BC Museum (2005).

Table 4.1

British and American trading vessels in the Pacific Northwest, 1785-1814

Years	British	American
1785-94	35	15
1795-1804	9	50
1805-14	3	40

Sources: Howay (1975).

in control of a continent-wide industry, albeit an industry in decline. Furs had been overharvested throughout the country, and fur traders ran into additional problems in British Columbia.

Birchbark canoes and extensive river systems were the main means of transporting furs to the Hudson's Bay forts and to Montreal. Many of the rivers in British Columbia were too rough to navigate, however, resulting in many portages. Moreover, in much of the province, there were no birch trees for making canoes. Overland trails and horses became the new, more dependable means of transportation. Some forts, particularly in the southern interior, became horse ranches to answer the need, but this system added extra costs to the transportation of furs.

Fur traders also encountered difficulties with respect to the sheer number of First Nations. Common practice for fur trade companies on the Prairies and elsewhere in Canada was to negotiate with tribal chiefs who, in turn, organized fur trapping. For the most part the territorial boundaries of these Native groups were fairly large, with few chiefs. In British Columbia, however, this system led to frustration for the trading companies. The region included approximately thirty or forty major ethnic groups, and these could be divided further into hundreds of clans, or houses, each with its own territorial boundaries and hereditary chief (Muckle 1998, 7). As a result, fur brigades made up of thirty or more heavily armed men, usually French Canadians, replaced Native trappers. The brigades would enter an area, do all the trapping, and transport the furs east. This system procured furs, but at a higher cost, and it also trespassed on Aboriginal lands.

As the fur trade struggled to remain economically viable, its other role, of asserting British sovereignty over

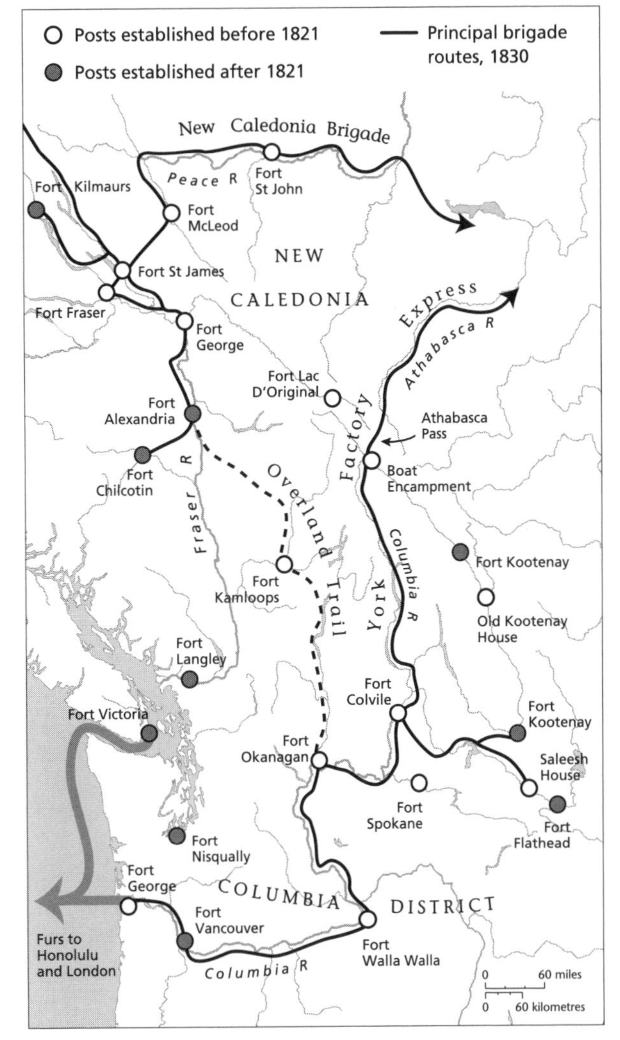

Figure 4.2 Fur trade forts and routes, 1805-46
Source: Modified from Harris (1997, 37).

the land, became important. The head of each fort, the fort factor, was responsible for the British institutions of marriage, law, and order, and was obligated to establish British rule over the land. The question of who had effective control over the land was still problematic, for the territory now referred to as British Columbia "was a bilateral free-trade zone, not a colony" (Harris 1997, 37).

An agreement between Britain and the United States in 1818 made the forty-ninth parallel the border between

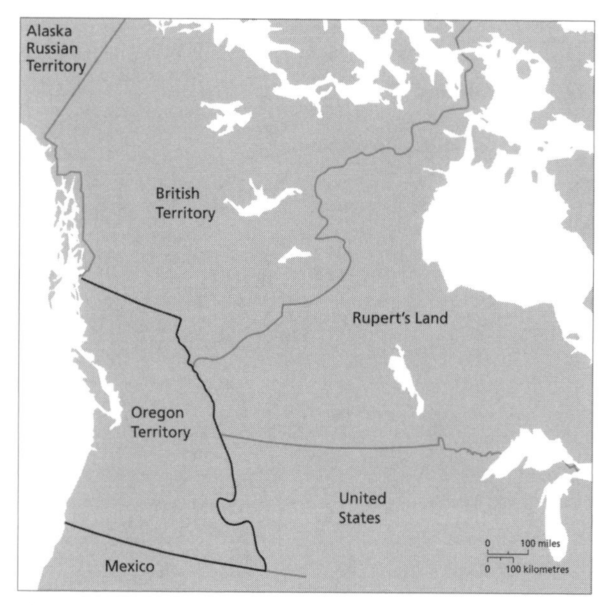

Figure 4.3 American claims to western North America, 1825
Source: Modified from Gentilcore (1993, Plate 21).

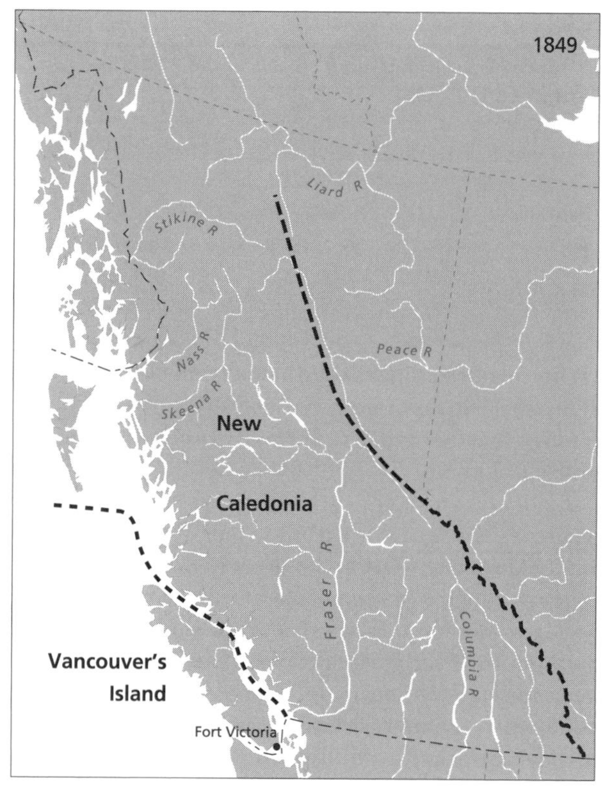

Figure 4.4 Colony of Vancouver Island, created 1849
Source: Modified from Kerr (1975, 32).

Canada and the United States, but it included the territory only as far west as the Rocky Mountains. Left open was Oregon Territory – the land from the end of the Alaska Panhandle (at 54° 40'), east to the divide in the Rockies, and south to the Hudson's Bay Company fur trade forts on the Columbia River (Figure 4.2). American westward and northward expansion, promoted in part by a philosophy of **manifest destiny**, led to a claim for the whole Oregon Territory (Figure 4.3). This conflict was brought to a climax with the US presidential election slogan "54-40 or Fight."

The **Oregon Treaty** of 1846 resolved the conflict by extending the border along the forty-ninth parallel from the Rockies to the Pacific and, for a time, kept the Americans from taking over **New Caledonia** (the southern half of British Columbia. Vancouver Island, which extends below the forty-ninth, remained British although a near war (the Pig War) over the ownership of the San Juan Islands took place in 1859. These islands were eventually awarded to the United States through arbitration in 1872. The Oregon Treaty cut off Hudson's Bay Company access to much of the Columbia River, the key river system used for navigation by the fur trade. The transportation of furs in the remaining British territory then had to rely much more on overland routes.

In 1849 the Hudson's Bay Company was given jurisdiction over the territory and a mandate to establish a colony that included Vancouver Island and Haida Gwaii (Figure 4.4). This region became more important with the discovery of gold on Haida Gwaii in 1850. The gold did not amount to much and attracted few miners, but it raised the prospect of finding more gold in the territory. A great deal more was found throughout the province, and gold ultimately shaped the northern and eastern boundaries of British Columbia and changed the geography of this region permanently.

THE GOLD RUSH: OPENING UP BRITISH COLUMBIA

Prior to the gold rush of 1858, most non-Natives in the territory were associated with the fur trade. Many were French Canadian or of British extraction, but there were also Kanakas (Hawaiians) and Americans. The discovery, and development by the Hudson's Bay Company, of coal on Vancouver Island had brought coal miners to the region in the late 1840s and early 1850s, but there were only a few hundred non-Natives in the colony of Vancouver Island and on the mainland.

The Fraser River and Cariboo gold rush, which began in 1858, should be viewed in the context of the 1849 gold rush in the San Francisco area. The San Francisco rush saw over 100,000 miners migrate into the Bay area and resulted in urbanization, roads, and a fair amount of chaos in the competition to find the elusive metal. As gold discoveries waned in northern California, the possibility of becoming rich from new discoveries on the Fraser attracted a great deal of attention. Some 25,000 to 30,000 miners pushed into the lower reaches of the Fraser panning for gold. The arrival of thousands of miners spurred the "opening up" of the region, with significant impact on the landscape.

Neither New Caledonia nor the separate colony of Vancouver Island was prepared for this onslaught of miners, many of whom were American. The arrival of so many Americans refuelled concerns about manifest destiny and prospects of border violation. In response, the colony of British Columbia, which included only the mainland, was created in 1858 (Figure 4.5). Chief Factor James Douglas of the Hudson's Bay Company had been appointed governor of the colony of Vancouver Island in 1851 by the British government. He was concerned for British sovereignty, as well as over the real possibility of war with Aboriginal peoples, whose lands were being trespassed upon and violated. He created a tax on Americans to remind them that they were on foreign land. For the First Nations, he negotiated a number of small treaties on Vancouver Island and established large reserves in the interior. The reserves were not created through treaties, however, and therefore were not accompanied by any formal rights.

As the gold seekers pushed up through the canyons of the Fraser and on into the Cariboo, where even more gold was discovered, new communities such as Barkerville sprang up, and a demand arose for transportation routes between Victoria and the interior (Figure 4.6).

Further discoveries of gold, on the Stikine River in 1862, resulted in the creation of the Stickeen (now spelled

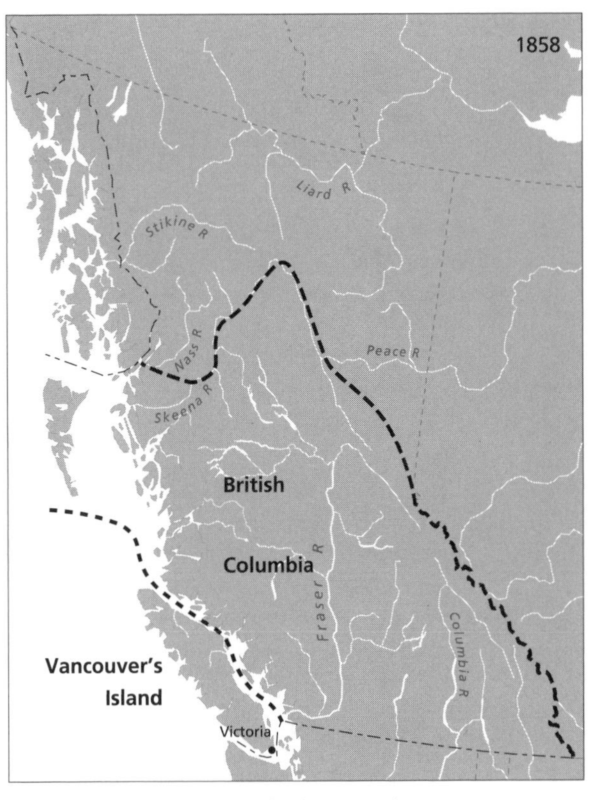

Figure 4.5 Colony of British Columbia, created 1858
Source: Modified from Kerr (1975, 32).

Stikine) Territory, with a northern boundary at the sixty-second parallel. In 1863 the British amalgamated the Stickeen Territory with British Columbia and, with the discovery of more gold on the Peace River and prospects in the Kootenays, pushed the boundaries east to the present-day configuration (Figure 4.7). Nevertheless, establishing boundaries and maintaining them are two separate processes, and in 1866 the two separate colonies of Vancouver Island and British Columbia amalgamated. The United States' purchase of Alaska in 1867 for $7.2 million left British Columbia facing American territory on two sides (Thomas 2001, 100). It was also of concern that many of British Columbia's inhabitants were American, and they were agitating for annexation.

When British Columbia joined Confederation in 1871, the issue of territoriality was finally put to rest: British Columbia was part of Canada. The new Dominion of

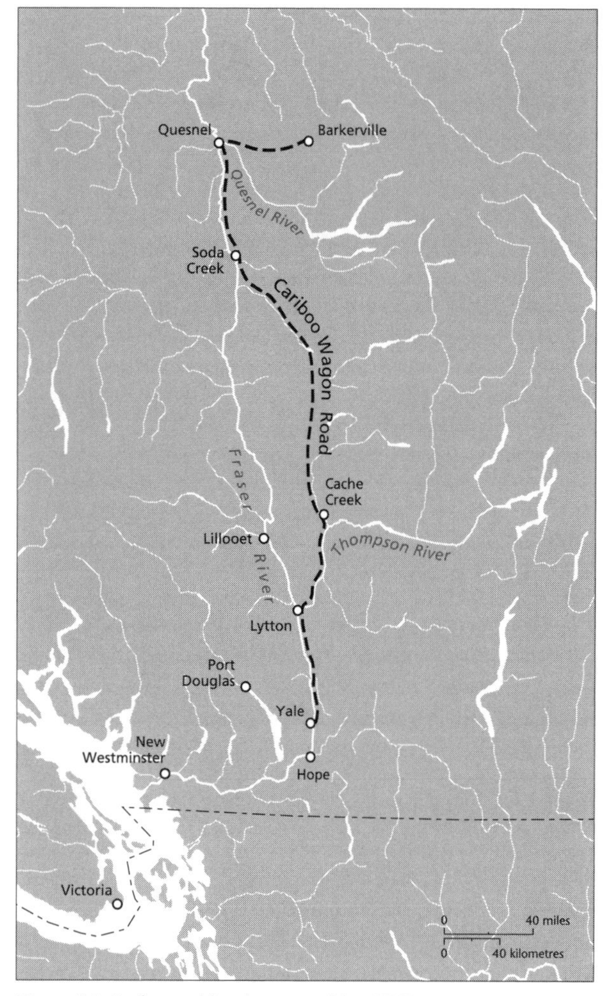

Figure 4.6 Cariboo gold rush communities, 1863

Canada dissolved the colony's debt, promised a trans-Canada railway and guaranteed federal transfer payments; but the federal government insisted that the colony of British Columbia transfer an area of land equivalent to twenty miles on each side of the railway. For this, the federal government selected 3.5 million acres of prime farmland in the Peace River region (Calverley Collection n.d.). It became known as the Peace River block and was not surveyed until 1907. By 1912, however, it was opened up for homesteads, and in 1930 it was transferred to British Columbia.

The gold rush had a profound impact on the landscape and on the indigenous peoples of British Columbia. The term **frontier mentality** describes the attitude of people who came to the frontier in search of wealth but had no intention of staying. Actions prompted by this attitude were particularly destructive to the environment. Most miners were single men from other parts of British North America, the United States, Europe, and China, lured by the possibility of striking it rich with gold. **Placer mining**, mining gold in its pure form, was very attractive for these individuals because it required little investment other than a shovel, a gold pan, and hard work. Gold, because of its weight, settles to the bottom of streams or on stream banks and sand bars, where it can be recovered quite easily. Over time, the technology of acquiring gold changed, and it was not long before gold pans gave way to small sluices, larger sluices, dredges, and hydraulics, all of which had a destructive impact on the rivers and streams that were the habitat of salmon, the staple food for many First Nations.

Gold-mining communities, thrown up in a haphazard fashion with frame structures and at extreme risk for fire, reflected this attitude of impermanence. Similarly, the destruction of the forests for fuel, building supplies, and mining left communities vulnerable to avalanches, mud flows, and flooding. The miners of the Cariboo provided a market for beef, and cattle drives, often originating in the western United States, competed to be the first into communities such as Barkerville in the spring. In the process, the natural grasslands of the south central interior valleys, where the highly nutritious bunch grass was the dominant form of vegetation, were seriously overgrazed. The vegetation that succeeded bunch grass, and is dominant through these dry interior valleys today, is sagebrush, which has deep, large roots and cannot be eaten by cattle.

Not all who were present in British Columbia during this early period had a frontier mentality. Many intended to remain in the region and demonstrated a desire for permanence. Initially, the economic and colonial interests of the Hudson's Bay Company had provided the impetus for the territory to remain British. Individuals such as Governor Douglas, who taxed American miners and established British law and order along with many other British institutions, represented the attitude of permanence. The

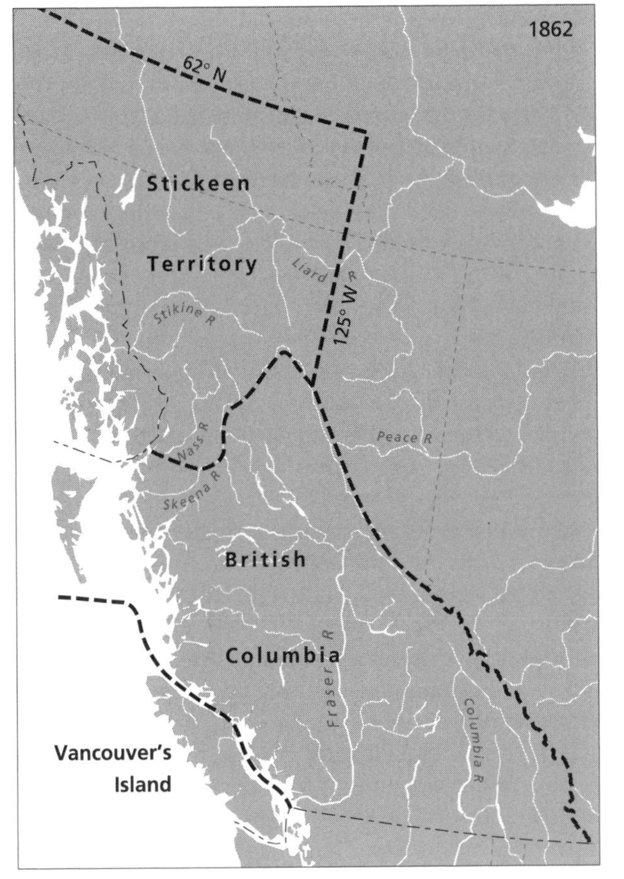

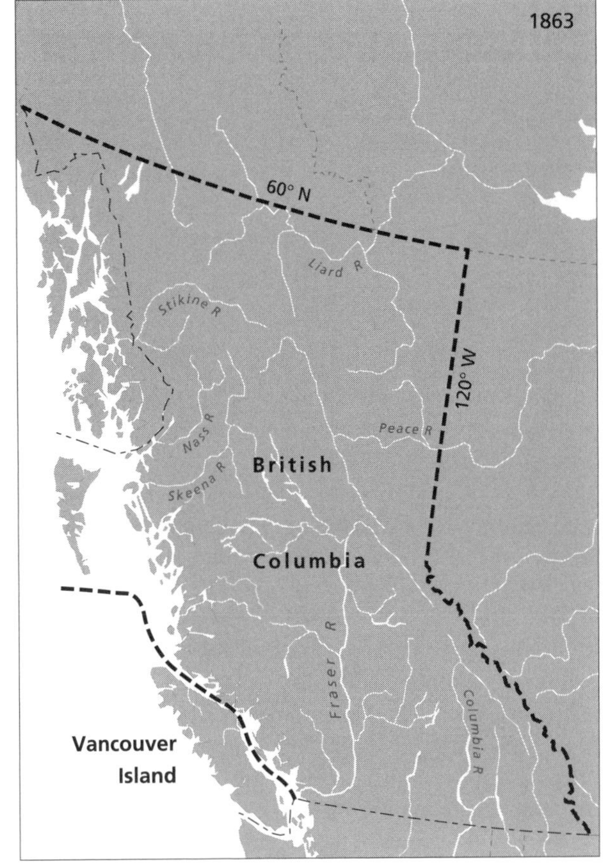

Figure 4.7 Boundary changes to British Columbia, 1862 and 1863
Source: Modified from Nicholson (1954, 32).

Royal Engineers, who provided a military presence, were responsible for planning, surveying, and building structures such as the Cariboo Road, and laying out the town site of New Westminster. The famous "hanging judge," Sir Matthew Baillie Begbie, left the reputation of a lawful British frontier as opposed to an unlawful American one, although historical geographer Cole Harris (1997) questions the truth of this interpretation.

Before long, families began to migrate to British Columbia, intending to call the region home. These families were often British and were imbued with the values of their homeland. They began the formidable task of remaking this wild landscape into a more familiar, tamer one.

Differences in attitudes toward, and uses of, the land did not end with the gold rush. Rather, philosophically opposing views over resource development and management were expressed through political parties and much rivalry.

When British Columbia joined Canada in 1871 (at once becoming the largest province in Canada), the act brought a sense of permanence to its boundaries and an expectation that the wilderness was to be "civilized" with British people and British values. Politically, the move brought two levels of government: federal and provincial. Economically, the province was strengthened as a frontier of expanding industrial capitalism, and the promise of a national railway connected to an international port

Table 4.2

1881 census population

Ethnic group	Population
Indians	26,849
Chinese	4,195
Whites	19,069
Africans	274
Total	50,387

Source: Uncorrected nominal returns. Modified from Galois and Harris (1994, 39).

fuelled land speculation and growth generally. Political scandals prevented the railway from being built in British Columbia until the 1880s, which was the next "boom" period (Chodos 1973). Provincial dependence on resource and infrastructure development and on an external demand for resources was the beginning of an enduring cycle of boom and bust.

The census of 1881 divided the population of British Columbia by ethnicity and location. Ethnically, the population was classified as "White," "Chinese," "Indian," and "African" (Table 4.2). The population of Africans, at 274, was considered insignificant as most were scattered throughout the mining areas of the Cariboo and Cassiar regions. The white population was concentrated in the southwestern corner of the province, and especially Victoria, although some lived in the mining camps of the Cariboo and Cassiar areas and a few were involved in fishing, forestry, and farming. The Chinese "tended to occupy locations that had been created by whites" (Galois and Harris 1994, 41) and therefore worked the placer mines, in the canneries, in the mills, or on the farms. Those in urban locations tended to be in service occupations.

Aboriginal people were the most numerous and were spatially separated from the other three groups. They experienced the greatest change with the arrival of non-Natives. Previously unknown diseases, a newly acquired problem of alcohol addiction, confinement on reserves, and missionary influences removed them from their traditional lands. By 1881 the once Native-dominated landscape was altered, reshaped, and dominated by a white society.

SUMMARY

The arrival of Europeans to the northwest coast of North America came late in the European phase of colonization. The existing system of global trade, colonization, and transportation by sailing vessels, however, made this geographic region one of the most isolated in the world. Britain was able to secure the coastal territory through negotiations with Spain, while the interior also came under British domination because of the overland fur trade companies operating out of Montreal and later Hudson Bay.

Until the Fraser River and Cariboo gold rushes, European settlement at fur trade posts was mainly temporary, and few resources other than gold were developed. The discovery of gold changed the landscape dramatically. Politically, British Columbia's boundaries had solidified, but the greatest changes came throughout the southern interior with the building of communities, farming, ranching, road building, and mining. The frontier mentality led to a great deal of destruction. Conversely, an attitude of permanence recognized the forces of institution building and landscape modification, often in the image of British value systems.

The search for gold had produced a frenzy of activity – permanent settlement, transportation systems, the immigration of many more people, especially Americans and, to an extent, Chinese – that challenged the existing British colonial system governing the territory. By the end of the 1860s, considerably less gold was being discovered, and British Columbia experienced the downward cycle of a bust economy.

Essentially, this historical account ends in 1871, when British Columbia joined Confederation. With this act, British Columbia was no longer a British colony but a separate province with elected members operating within the realm of powers outlined in the British North America Act. Confederation also put to rest the fear of annexation by the United States. Finally, Confederation came with the promise of another economic boom – the building of a national railway connected to an international port. For political reasons, the railway was delayed for ten years.

REFERENCES

Archer, C.I. 1981. "The Transient Presence: A Re-Appraisal of Spanish Attitudes toward the Northwest Coast in the Eighteenth Century." In *British Columbia: Historical Readings*, ed. W.P. Ward and R.A.J. McDonald, 37-65. Vancouver: Douglas and McIntyre.

Chodos, R. 1973. *The CPR: A Century of Corporate Welfare.* Toronto: James Lorimer.

Galois, R., and C. Harris. 1994. "Recalibrating Society: The Population Geography of British Columbia in 1881." *Canadian Geographer* 38 (1): 37-53.

Gentilcore, R.L., ed. 1993. Historical Atlas of Canada. Vol. 2: *The Land Transformed 1800-1891.* Toronto: University of Toronto Press.

Gibson, J.R. 1981. "Bostonians and Muscovites on the Northwest Coast, 1788-1841." In *British Columbia: Historical Readings*, ed. W.P. Ward and R.A.J. McDonald, 66-95. Vancouver: Douglas and McIntyre.

Harris, C. 1992. "The Lower Mainland, 1820-81." In *Vancouver and Its Region*, ed. G. Wynn and T. Oke, 38-68. Vancouver: UBC Press.

–. 1997. *The Resettlement of British Columbia: Essays on Colonialism and Geographical Change.* Vancouver: UBC Press.

Howay, F.W. 1975. Ed. by R.A. Pierce. *A List of Trading Vessels in the Maritime Fur Trade, 1785-1825.* Kingston: Limestone Press.

Kerr, D.G.G. 1975. *Historical Atlas of Canada*, 3rd ed. Toronto: Thomas Nelson.

Muckle, R.J. 1998. *The First Nations of British Columbia.* Vancouver: UBC Press.

Nicholson, N.L. 1954. *Boundaries of Canada, Its Provinces and Territories.* Ottawa: Queen's Printer, Department of Mines and Technical Surveys.

Thomas, P.F. 2001. "Geopolitical Development: An Overview." In *British Columbia, The Pacific Province: Geographical Essays*, ed. C.J.B. Wood, 94-131. Canadian Western Geographical Series vol. 36. Victoria: Western Geographical Press.

INTERNET

Calverley Collection. n.d. "8-27: The Peace River Block." www.calverley.ca/Part08-Agriculture/8-27.html.

Royal BC Museum. 2005. BC Archives. www.bcarchives.gov.bc.ca/index.htm.

5

First Nations and Their Territories

Reclaiming the Land

In British Columbia, "there is general agreement that the term First Nations refers to a group of people who can trace their ancestry to the populations that occupied the land prior to the arrival of Europeans and Americans in the late eighteenth century" (Muckle 1998, 2). The term recognizes that Native peoples have their own languages, customs, and territories and have occupied British Columbia for approximately 10,000 years. Various cultures and their territorial boundaries have changed over those many years, and the initial discussion here examines some of the cultural components of how these peoples used the land prior to the arrival of non-Natives.

Europeans made their contact from a number of directions in the 1700s and early 1800s, bringing with them new trade items, technologies, and value systems that included the organization and use of land, a philosophy of mercantilism, and a definition of civilization. The Europeans were few at first, but their numbers increased dramatically with the gold rush of 1858. The impact was catastrophic for First Nations. The seemingly endless stories of suppression, assimilation, and loss of land and life began to improve slowly only after the 1950s. The interrelated roles of the federal government, provincial government, court systems, public opinion, and Aboriginal organizations are examined here in terms of why land is the key issue in Aboriginal affairs and central to the modern treaty system.

PRE-CONTACT ABORIGINAL SETTLEMENTS

The generally accepted theory of North American settlement is that the first humans came from Siberia via the ice bridge of the Bering Sea some 11,000 to 12,000 years ago, when glaciers from the last ice age were receding. The last ice age was at its peak about 20,000 years ago, when vast sheets of ice covered polar regions of the world, including most of Canada and northern portions of the United States (see Chapter 2). These ice sheets gained their moisture from oceans; as a result, the "sea level was about 120 m lower than it is now, so that a land bridge existed between Siberia and Alaska" (Aguado and Burt 2001, 441). These physical conditions were responsible for a relatively flat plain that extended from Siberia to the Mackenzie delta, known as Beringia. The present-day Bering Strait became the land bridge, and Beringia, because of its dry climate, was free of ice. Archaeologists suggest that early Indian groups (Paleo-Indians) followed the herds of animals from northern Asia across to North America and travelled in one of two directions: a coastal migration and an inland migration via an ice-free corridor. The evidence of a coastal migration is much more difficult to obtain as the historical coastline is now covered by 120 metres of water. Figure 5.1 indicates the two routes into British Columbia and the location of several archaeological sites.

Climate change has not been consistent. The rise in temperatures 12,000 or so years ago was interrupted by perhaps 2,000 to 3,000 years of cooling (6,000 to 3,000 years ago), before temperatures rose again (Kump, Kasting, and Crane 1999, 235). The effects of all of these climatic changes – glacial meltwater that covered extensive amounts of land, a rise in ocean levels, isostatic rebound, and subsequent establishment of flora and fauna zones – resulted in a major transition of the land and of the people who lived off the land.

In most of Canada, the glaciers receded to their present location in the alpine Cordilleran region and the high Arctic some 7,000 to 6,000 years ago. Consequently, the

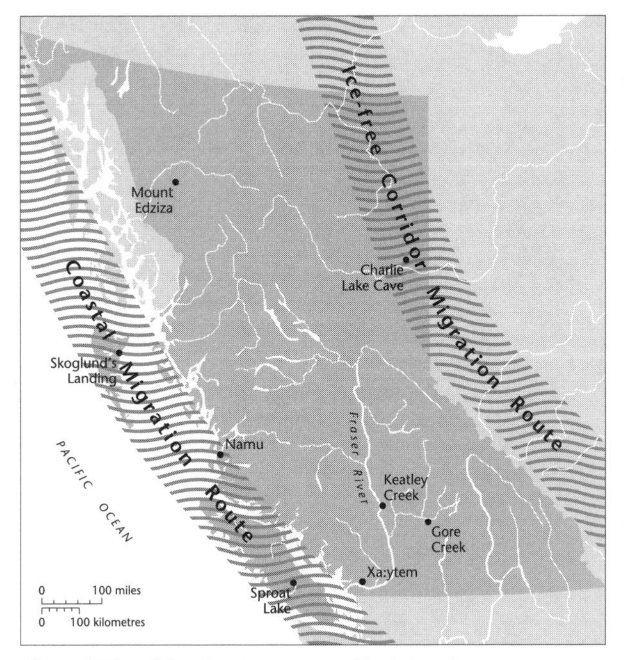

Figure 5.1 Possible migration routes to North America
Source: Muckle (1998, 20).

boreal forest migrated even farther north than its present location, salmon swam in the rivers flowing to the Pacific, and other plants and animals established habitats throughout the province. For the most part, the original Paleo-Indians were nomadic hunting-gathering societies that had to adapt to the many changes. They were, however, in sufficient numbers and possessed hunting technologies that resulted in the extinction of a number of species, including mastodon, several birds, and giant beaver (Wynn 2007, 22-23).

As the ice melted and receded, the land stabilized and these groups became seminomadic. In other words, though they moved from season to season depending on the availability of resources, they remained within regions and occupied the same sites for hundreds and perhaps thousands of years. This allowed the various culture groups to develop an intimate knowledge of their areas and to refine their tools and technologies, which led to an increase in population.

In the region of British Columbia, the transition of the first ethnic groups from a nomadic to a seminomadic existence occurred approximately 5,000 years ago, and coincides with technologies such as "barbed harpoons for taking sea mammals, fish hooks, weights for fish nets, ground slate knives and weapon points, and wood working tools" (McGhee 1985, 1467). Later, fishing became perfected with dip and gill nets, traps, spears, and the hook and line. Some groups, such as the Haida on Haida Gwaii, used large ocean-going cedar dugout canoes for the hunting of mammals, including whales. The Haida also cultivated tobacco (Glaven 2000, 91). A more sedentary way of life led to divisions of labour wherein complex organizational structures developed, reflected in elaborate art, shamanism, the building of longhouses, and political and social institutions such as the potlatch.

The population and density of Aboriginal peoples varied with geographic location. The highest and most densely populated regions were along the coast and along salmon-bearing streams in the interior. Density was considerably lower in the rest of the interior. Table 5.1 shows the historically different language groups and dialects, and Figure 5.2 matches some of these with their traditional territory. It should be noted that there is no consensus among either First Nations or anthropologists on the divisions of major ethnic groups or on the nations that belong to them. For example, the Sinixt Nation (a.k.a. Lakes or Arrow Lakes band) are not officially a band, but they trace their ancestry in the West Kootenays back some five thousand years, and "their culture is still alive, still connected with this land, and still determined to leave a legacy to the next generation" (Sinixt Nation n.d.). Different writers use different categorizations, territorial boundaries, and spellings. The map showing the linguistic

Table 5.1

First Nations language divisions

Ethnic division	Language	Dialect
Haida	Haida	Masset
		Skidegate
Tsimshian	Tsimshian	Tsimshian
		Gitxsan
		Nisga'a
Kwakwaka'wakw	Kwakwaka'wakw	Haisla
		Heiltsuk
		Southern Kwakwaka'wakw
Nuu'chah'nulth	Nuu'chah'nulth	Northern Nuu'chah'nulth
		Southern Nuu'chah'nulth
Nuxalk	Nuxalk	
Straits Salish	Comox	
	Pentlactch (extinct)	
	Sechelt	
	Squamish	
	Halq'emeylem/Hul'q'umi'num/Hən'q'əmin'əm'	
	Straits Salish	
Interior Salish	Nlaka'pamux	
	Stl'atl'imx	
	Secwepemc	
	Okanagan	Okanagan Lakes (extinct)
Kootenai	Kootenai	
Athapaskan	Tsilhqot'in	
	Dakelh	
	Sekani	
	Tahltan	
	Kaska	
	Dene-thah	
	Dunne-za	
	Tsetsaut (extinct)	
	Nicola (extinct)	
Inland Tlingit	Tlingit	

Sources: Duff (1965, 15); Muckle (1998, 117-22).

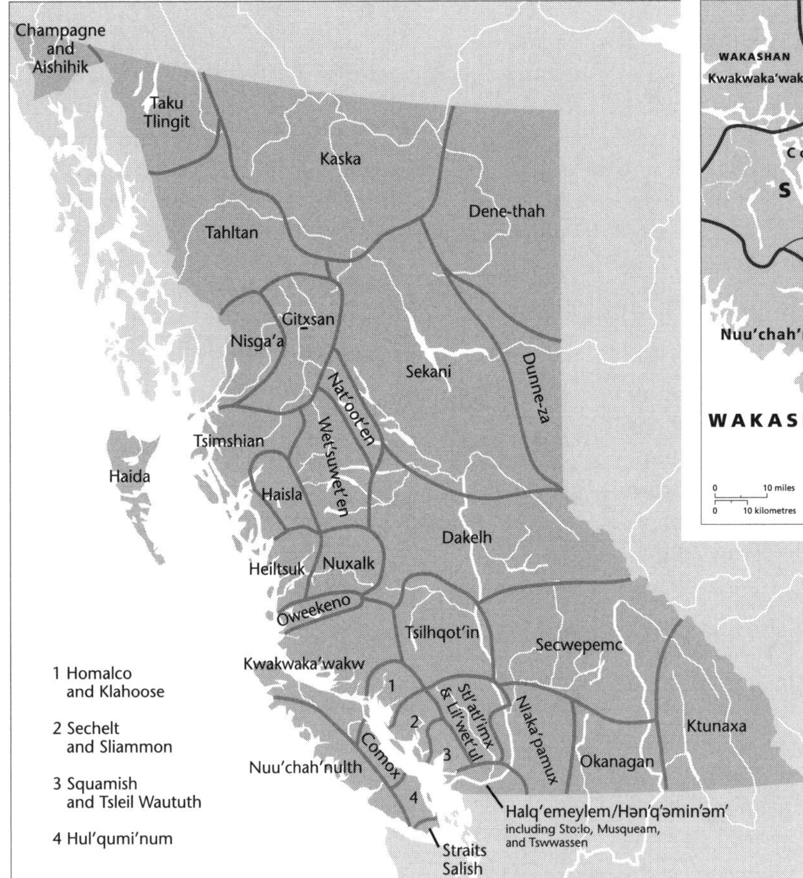

Figure 5.2 Historical territorialization of First Nations in British Columbia
Source: Muckle (1998, 7).

breakdown of Aboriginal peoples in the southwestern corner of British Columbia outlines the complex spatial organization, or territorial boundaries, of one area (Figure 5.3).

The Aboriginal means of production was based on very strong ties to the land and its resources, going beyond the basics of food, clothing, and shelter. Native spiritual beliefs, specifically animism, were intimately tied to the land, as was the social and spatial organization of the various peoples. Territorial boundaries based on language group were further defined in terms of ownership by individual houses or clans. McKee (1996, 4) elaborates on the Native concept of private property: "Private property was vital, and just as sophisticated as that of European nations at the time. Each household possessed land for village sites and for hunting and food gathering. Specific items and rights were held in common by the members of each household, including canoes, totem poles, ceremonial objects, and rights to hunt and fish in specific waters and harvest particular food species."

Figure 5.3 First Nations linguistic divisions in southwestern British Columbia
Source: Modified from Harris (1992, 40).

Today, **Aboriginal rights** are recognized by section 35 of the Canadian Constitution, but it is not clear what those rights include. Ownership or control of territory for each group is recognized as **Aboriginal title**. Aboriginal rights also imply the freedom to pursue a particular mode of production (through use of land, labour, resources, technology, and capital) in combination with forms of political, social, and spiritual organization that ensure survival. And Aboriginal rights are to take priority over the rights of others subject only to the needs of conservation (BC Ministry of Aboriginal Relations and Reconciliation 2010).

The Fraser River delta sustained many land and water resources – especially salmon – which were available at

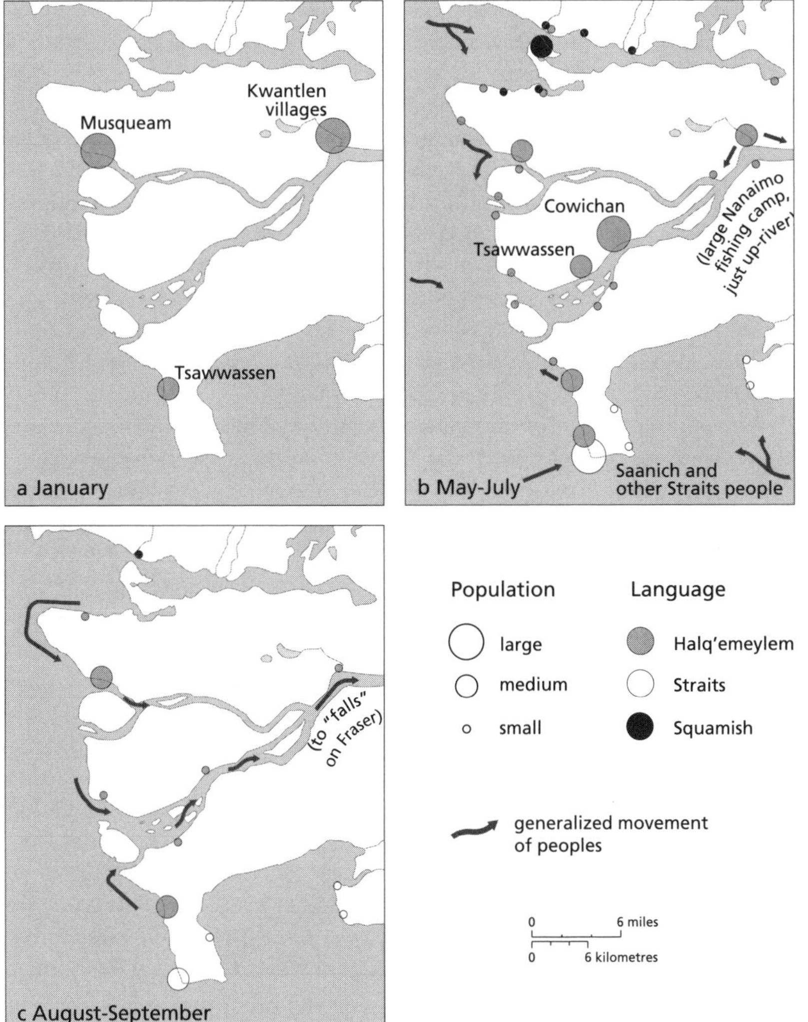

Figure 5.4 Semi-nomadic migrations in the Lower Mainland area
Source: Modified from Harris (1997, 39).

different times of the year, thus influencing the semi-nomadic movement of various Native groups throughout the year. Figure 5.4 outlines the typical migrations in the delta for portions of the year, illustrating how many groups travelled beyond their traditional boundaries to procure food there. As a consequence, there was considerable overlap of territorial boundaries. This has created some difficulty in the modern treaty process, which begins by attempting to recognize title to specific lands.

Aboriginal societies were based on oral tradition; history, legends, stories, and wisdom were held by the elders and passed on to the next generation. Stories, dances, and songs were important possessions in defining the resources of the land and the territorial limits of the land base. These communally based societies functioned through an elaborate division of labour in which each individual had a role in sustaining the group.

First Nations of the coast had complex social, spiritual, and political organizations. Work was organized based on gender and age. Sophisticated carvings (on totems, dugout canoes, and so forth), shamanism, and the potlatch were symbols of well-organized cultural groups within well-defined boundaries. The potlatch was an extremely important political institution practised almost exclusively by coastal First Nations. It was a form of government that organized and legitimized the decisions each nation made: "Potlatch government business is conducted at feasts. The house invites members of other clans to its feasts to witness the decisions taken, and to receive gifts that reinforce cooperation between the various houses. Potlatch feasts are held to commemorate deaths, to pay debts, to resolve disputes, and to confirm positions within the house leadership. They also provide a forum for discussing and resolving community problems. Potlatch feasts are organized according to strict procedures as to invitations, seating, speaking, gift giving, payments, dress and conduct" (Brown 1992, 4).

British Columbia, and particularly coastal British Columbia, had far more linguistic groups and clans, living in much higher densities, than anywhere else in Canada.

They developed very sophisticated cultures and complex social organizations based on territorial boundaries. Their value systems and their historical use of the land defined a geography that was seriously challenged and changed with the arrival of Europeans.

EUROPEAN CONTACT AND IMPACT

The history of early explorers romanticizes individuals who battled the unknown elements with rather primitive technologies, claiming land for European monarchs. These initial discoveries usually led to further exploration and exploitation of the resources of the land. All development was carried out in the belief that, because they were the first Europeans in a "new" land, they had the absolute right of ownership of the land. This presumption is implicit in the European place names given to the new territory.

This was a **Eurocentric** view, in which exploitation of other people, their land, and their resources was justified even when military force was used. It was also a racist view, in which superiority was based on skin colour. Christian missionary zeal went hand in hand with colonialism as it attempted to convert Native people from their supposedly inferior and heathen beliefs. Eurocentrism encompassed a view of history that defined civilization and proposed a set of civilizing rules. Within British Columbia this meant that one should speak English, adopt Christian values, live within the British common law system, accept the principle of private land ownership, and develop the land for agricultural use. These are largely the values of a capitalist system, and they produced a serious clash with First Nations values. According to Harris (1997) this resulted in a **deterritorialization** of Aboriginal land and way of life and a **reterritorialization** by non-Natives with a capitalist value system. One of the persistent struggles to this day for First Nations is assimilation, or the loss of their cultural values and a whole way of life related to the land. Aboriginal peoples often say that the land is the culture.

Initial European contact was made on the coast by sailors in the late 1700s and was not marked by a great deal of conflict, although conflict did occur. Rather, Native curiosity about the Europeans in ocean-going vessels and interest in trade goods was more common. A mutual dependence evolved: Europeans wanted Aboriginal peoples to acquire sea otter pelts, which were extremely valuable in China, and Aboriginal peoples wanted European goods.

The overland fur trade entered British Columbia from eastern Canada in the early 1800s and resulted in a new relationship between Europeans and Aboriginals. It was marked by European frustration over trade agreements and difficulties in transporting furs. Consequently, First Nations were often eclipsed by European fur brigades. The abusive attitudes and actions of Eurocentrism surfaced as fur trading posts were established and non-Natives began to trespass on Native land. (See Chapter 4 for more historical detail.)

Continuous contact between Natives and non-Natives in British Columbia was late in European history, occurring at the end of the eighteenth century and beginning of the nineteenth. And even this contact was minimal until the Fraser River gold rush of 1858. In eastern and central Canada, there was a much longer history of contact and conflict, and a very significant event eventually affected the whole of Canada – the Royal Proclamation of 1763.

Aboriginal Rights, Title, and Treaties

Through the proclamation the British recognized that Aboriginal peoples had occupied the land first and that they had a form of self-government over the territory. Most significantly, the proclamation established a formal process by which Aboriginal peoples received certain rights under British rule. The proclamation stated that First Nations should not be abused, that they should have the right to continue governing themselves, and that they should be compensated for the use of their land (Tennant 1990; Frideres 1988). This last point in particular defined a new geographic relationship as treaties were the means of compensation.

Aboriginal title refers to the right a tribe or band has to a given block of land it used and defined as its territory historically. Because these territories were not static, their boundaries overlapped. The Royal Proclamation recognized Aboriginal title, and the treaty process, implemented throughout much of Canada, extinguished, or ceded, Aboriginal title in exchange for compensation. Compensation usually took the form of small cash settlements, material goods, and reserve lands. The Royal

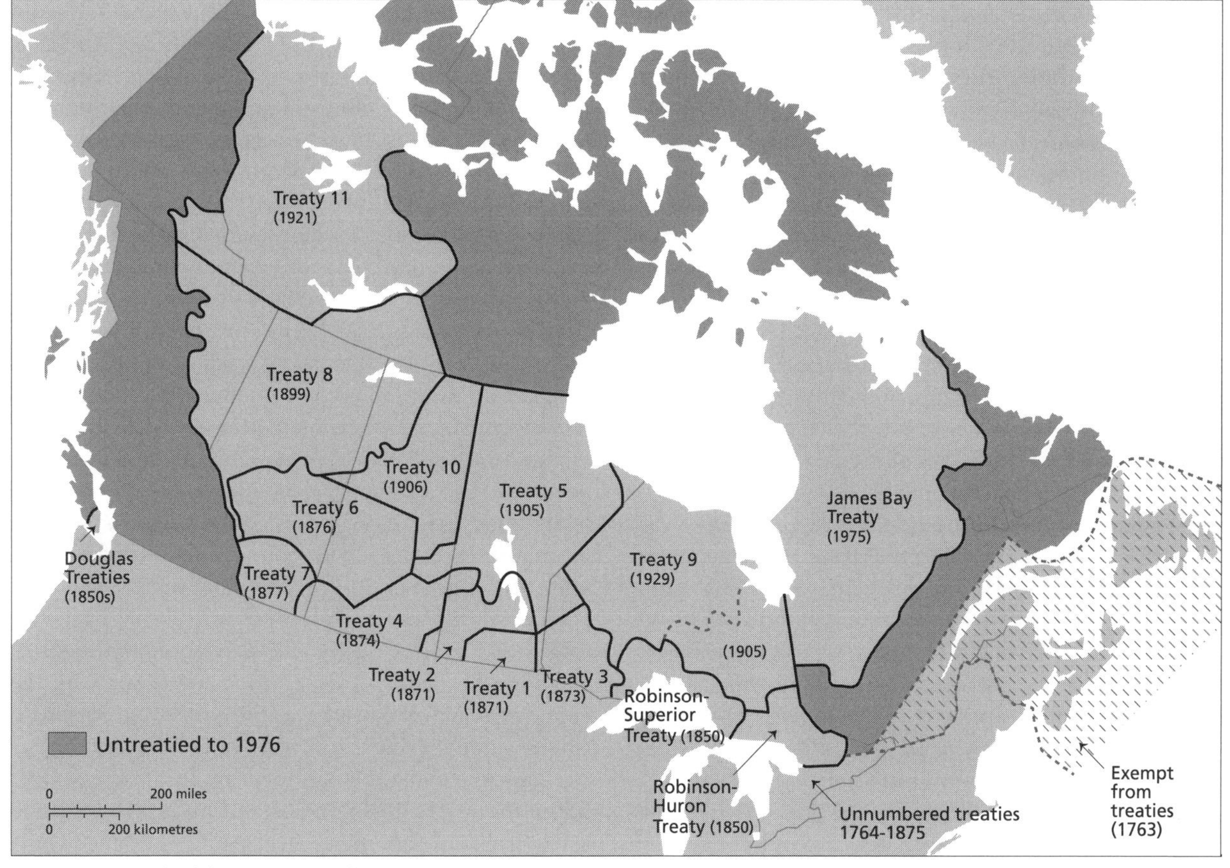

Figure 5.5 Treaty land divisions in Canada

Proclamation also recognized that Aboriginal peoples could continue to hunt, fish, and trap as they always had.

Figure 5.5 shows the various historical treaties throughout Canada; only two apply to British Columbia. In the 1850s, Europeans acquired land on southern Vancouver Island through the Douglas Treaties. Then in 1899, the signing of Treaty 8, which included the Peace River region of the province, was precipitated by the Yukon gold rush. Throughout the rest of British Columbia, the land was simply taken and occupied by non-Natives; no treaties were signed and no compensation was given.

James Douglas, the chief factor for the Hudson's Bay Company, became governor of the colony of Vancouver Island in 1851. He was concerned about the potential for conflict between the few Europeans who had settled in the region and the Native population, which greatly outnumbered them. By 1854, he had arranged fourteen small "purchases" of land from the local Aboriginal peoples on southern Vancouver Island, which became known as the Douglas Treaties. The gold rush of 1858 saw some 25,000 to 30,000 non-Natives enter British territory, mainly through Fort Victoria, and then push into the lower reaches of the Fraser River searching for gold. Douglas was then appointed governor of the mainland colony of British Columbia, as well as Vancouver Island, to deal with this potentially explosive situation. Historical documents show that through his term as governor, which lasted to 1864, he upheld the spirit of the Royal Proclamation and had every intention of accomplishing more treaties, but Britain was unwilling to supply funding (Cumming and

Mickenberg 1981). Douglas also created reserves without treaties that were large enough "to provide support for native people during the early stages of their adjustment to civilization and, later, as they became civilized, to allow them to enter the colonial economy and society as full, participating members" (Harris 2002, 35-36); he also extended rights of settlement to Aboriginal peoples similar to those extended to Europeans.

From the beginning of non-Native land settlement in British Columbia, the British had both experience with Native peoples (especially in the South Pacific and eastern Canada) and a clear view of how to deal with them. The British intention was to give them a Western education, convert them to Christianity, and integrate them into the economy. This policy of assimilation denied historical rights to the land. Douglas retired in 1864 and a new regime, with radically different views on First Nations, came into power. Joseph Trutch, the new chief commissioner of lands and works, remarked that "'the Indian was an obstruction to settlement and progress.' Indeed, in a report on Indian claims to certain lands that had been laid out as reserves for them on Douglas' instructions, Trutch commented: 'The Indians have really no right to the lands they claim, nor are they of any actual value or utility to them, and I cannot see why they should either retain these lands to the prejudice of the general interests of the Colony, or to be allowed to make a market of them either to the Government or to individuals'" (Cumming and Mickenberg 1981, 193).

With Trutch came an attitude that reversed the initial direction of Douglas in upholding the requirements of the Royal Proclamation. Aboriginal title was denied, there was no intention of compensation, and after the death of Douglas, soon after his retirement in 1864, large reserves were made considerably smaller (Harris 2002, 45). This condescending view and policy toward First Nations in British Columbia pre-dated the province's entry into Confederation in 1871 and persisted in subsequent provincial governments for the next 120 years.

Negotiations over Confederation focused more on providing a railway and eliminating British Columbia's debt than on understanding and addressing First Nations issues in the province, though the federal government became responsible for Aboriginal affairs, taking on what is known as a **fiduciary trust** (or obligation). In other words, the federal government had the legal obligation to look after the best interests of First Nations, which has been interpreted to include establishing reserves. This responsibility spoke more of European interests than of Aboriginal interests. To make matters worse, when British Columbia joined Confederation the provincial government gained control of Crown land (under the British North America Act). This land made up Aboriginal title and was required to create reserves. The federal government, confronted with demands for more land by First Nations, often acquiesced to existing provincial policies that instead denied them any rights to the land. Only after a near war with Aboriginal peoples in the late 1870s was the provincial government prepared, reluctantly, to increase the number of reserves. The size of reserves then became a crucial issue in negotiations. By 1880 the provincial policy had only confirmed the previous practice of small reserves, and "had completely side-stepped and ignored the issue of title" (Harris 2002, 68).

Wherever the federal government was involved in treaties across Canada, reserve size was frequently based on 160 acres per family. This size was consistent with the Homestead Act, which allowed European immigrants to claim a quarter section (160 acres or 65 hectares). Creating a reserve under this formula simply involved calculating the number of families and multiplying by 160 acres. In British Columbia, because the provincial government was opposed to increasing reserve land (which would reduce provincial Crown land) and because there were numerous bands to be considered, the federal government recommended 80 acres per family. The provincial government proposed 10 acres per family. The two governments then agreed on 10 to 20 acres (4 to 8 hectares) per family, which is why so many "postage stamp" sized reserves exist in the province today.

The anomaly is Treaty 8, initially agreed to in 1899, which includes the northeastern section of British Columbia. Here, reserves were created on the basis of "one square mile (640 acres or 259 hectares) for each family of five. Initial cash payments were given to the chiefs and councillors of the various tribes, and provision was made for education, farm stock, implements, and ammunition" (McKee 1996, 22). The story of Treaty 8 is well documented by Arthur Ray (1999), who describes the conditions for northern First Nations in the late 1800s. Because of the region's potential for gold, petroleum, and sulphur, government officials were interested in the proposal of a treaty. But the promotion of this region as a significant route to the Klondike transformed this interest into a necessity. As non-Native miners made their way from Edmonton north to the Yukon, they had little regard for local First Nations. Native horses were stolen, dogs were shot, caches of food taken, and traps destroyed; miners also depended on game for food. First Nations found all of these actions intolerable, but the most troubling was the depletion of an already scarce food supply and the subsequent winter conditions of starvation. A war nearly occurred in 1898 when some 500 First Nations gathered at Fort St. John; a treaty appeared to be the solution.

The federal government needed permission from the provincial government to engage in negotiating a treaty that included British Columbia land. When the federal government requested permission, however, the province declined to respond. The BC government did not want the war, but neither did it want to make a "precedent breaking decision" (Ray 1999, 38). The federal government therefore proceeded unilaterally with Treaty 8. This was the best deal for any First Nation in the province, although it should be noted that few Aboriginal peoples could read, write, or speak English at this time, and most believed these arrangements to be peace treaties, not treaties relinquishing their right to land.

Assimilation and Decimation

The impact of non-Native settlement on First Nations went far beyond the racism and discrimination described above. The term "four plagues" describes some of the key factors in deterritorialization: a new geography of dispossession in which Aboriginal peoples were no longer able to use the land as they had historically. The four plagues were diseases, alcohol, the reserve system, and the role of missionaries.

Diseases took a huge toll on Native peoples. Smallpox was a deadly disease that had devastating consequences for indigenous peoples throughout the world, and the people of the Pacific Northwest did not escape its consequences. In fact, the disease arrived before most Europeans. It spread overland from Mexico in 1782, killing one-third of affected communities in British Columbia, mostly in Salish territories (Union of BC Indian Chiefs 2005, 14). And successive waves of smallpox – for instance, in 1836-38, 1853-54, and 1862-63 – were equally ravaging. Smallpox, however, was not the only disease introduced in early times. In 1999, the body of a First Nations person was discovered in a glacier in the Tatshenshini region (in far northwestern British Columbia). After considerable DNA testing, it was determined that this young male (approximately eighteen or nineteen years old) had died over 200 years ago. He was carrying an iron knife and was infected with tuberculosis when he died (Pringle 2008, 74). The TB might well have come from Russian sailors who were in the area in the 1740s.

Smallpox epidemics claimed the greatest number of lives, although measles, influenza, tuberculosis, malaria, and venereal diseases accounted for many deaths. Alcoholism, although a disease, is noted separately because alcohol was deliberately offered to First Nations as a commodity of trade. The consequence to societies that had no experience of this drug was debilitating and responsible for further deaths.

The reserve system included only some of the historical village sites and a fraction of the traditional territory, or Aboriginal title. The system of deliberately confining reserves to small, fixed plots of land seriously restricted the seminomadic hunting and gathering activities that were the foundation of the Aboriginal economy. For societies in which land was integral to the way of life, reserves imposed a major alteration and deterritorialization.

Missionaries encouraged the reserve system because it rendered Aboriginal societies less mobile and thus easier to reach. In partnership with government, missionaries also became responsible for education. From today's perspective, the residential school is seen as one of the most devastating institutions imposed on First Nations. Children were separated from their parents and forced to attend schools where their Native languages were forbidden and the educational values were questionable. Recently reported stories of sexual abuse within residential schools are horrific. This social upheaval led to the breakdown of the family unit in Native society.

One simple way of defining a culture is through its language. The language represents a medium of expression for that culture, or the values that the culture holds. Envision a circle as representing all the words in the English language and a separate circle representing all the words

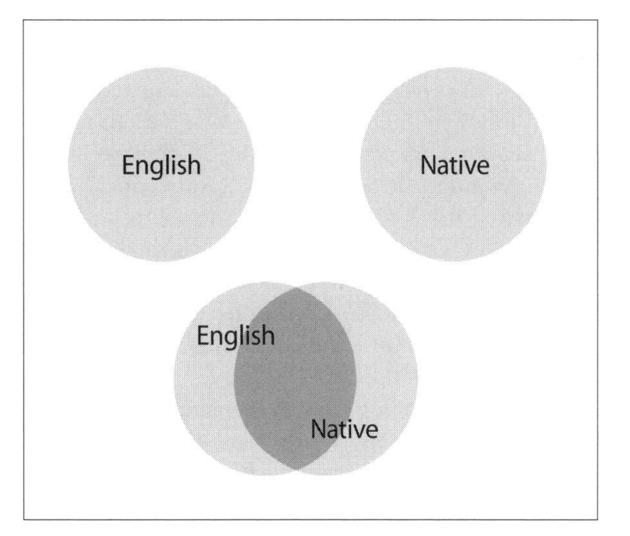

Figure 5.6 Language as a representation of culture

in another language (Figure 5.6). Even though the two circles may be the same size, they do not correspond perfectly when one tries to compare the meanings of the words and phrases of each language. Anyone who speaks two or more languages knows this well, since any language represents ideas and values that may not exist in another language. To take a language away, as the residential school system did, is a form of cultural genocide. Unfortunately, a number of First Nations languages are now extinct and others are in jeopardy.

All of these changes were devastating to First Nations. Frank Duerden (2001, 136) suggests a Native population of approximately 300,000 on the eve of contact in the late 1700s, but only 50,000 by 1881. Table 5.2 illustrates the huge impact of disease and other causes of death on the various populations. It should therefore be recognized that the 1835 population figures in Table 5.2 are an estimate, and further, that disease had already eliminated some of the population. Harris suggests that perhaps 90 to 95 percent of the BC Aboriginal population perished (Harris 1997, 30).

The young and the old were the most vulnerable to disease. Because Native societies were oral, the death of elders meant that their wisdom was disappearing with them. Epidemics shattered the elaborate division of labour necessary for communal societies. With so many dying, being pushed onto reserves, having their children taken away to residential schools, and being forced to exist under the rules of assimilation, Native societies were in chaos.

Many policies besides the missionary school system attempted to assimilate First Nations. The Indian Act of 1876 became the main instrument of assimilation and alienation from the land. It imposed a system of bands with an elected chief and council, thus abolishing the hereditary system. With the appointment of a non-Native Indian agent to relay the needs of each band to Ottawa, the new system was also paternalistic. When potlatches were outlawed in 1884, First Nations lost their chief political institution.

Enfranchisement was another divisive tactic, because essentially it offered the choice of becoming non-Indian. Aboriginal peoples were often persuaded to leave the reserves in exchange for certain rights. They could sign a document that gave them the right to vote, to enter liquor stores, to receive a small sum of money, and, in the early days, to receive a parcel of land. This process created the designations of non-status and status Indian, with profound implications for future generations. A non-status Indian had no rights to the reserve and its governance. If a status man were to marry a non-status woman, their children would be able to retain status, but until 1985 the children of a status woman married to a non-status man were not entitled to status.

Many other restrictions were imposed: dances and ceremonies were prohibited, a status card had to be carried at all times off the reserve, and in 1927 non-Native lawyers were not allowed to represent First Nations bands on any land claim issues (Tennant 1990). In many ways British Columbia, along with the federal government, had instituted an apartheid system.

The McKenna-McBride Commission (1912-16) reveals the frustration for the First Nations in dealing with both levels of government in British Columbia over reserve lands. John McKenna represented the federal government and Richard McBride, the premier of the province, represented BC interests. The intention of the commission was to sit down at the table and finalize the size of reserve lands once and for all. Its official mandate included both the expansion and the reduction of reserves, provided that the bands consented. This consent clause was ignored,

Table 5.2

Aboriginal populations for selected years

Ethnic division	1835	1885	Low	(year)	1963
Haida	6,000	800	588	(1915)	1,224
Tsimshian	8,500	4,500	3,550	(1895)	6,475
Kwakwaka'wakw	10,700	3,000	1,854	(1929)	4,304
Nuu'chah'nulth	7,500	3,500	1,605	(1939)	2,899
Nuxalk	2,000	450	249	(1929)	536
Straits Salish	12,000	5,525	4,120	(1915)	8,495
Interior Salish	13,500	5,800	5,348	(1890)	9,512
Kootenai	1,000	625	381	(1939)	443
Athapaskan and Inland Tlingit	8,800	3,750	3,716	(1895)	6,912
Provincial total	70,000	28,000	22,605	(1929)	40,800

Source: Duff (1965, 32).

and "laws were later passed stating that Indian consent was not really necessary after all" (Gunn 1976, 5). The following documents the extent of the cut-off lands: "Cut-offs were recommended by the McKenna-McBride Commission on 35 reserves amounting to a total of 36,000 acres. All but one of these reserves are located outside the urbanized Vancouver area. The one reserve within the Vancouver area is the Capilano lands of the Squamish Indians. A 130-acre piece of land was removed from the reserve on the north shore of Burrard Inlet near the mouth of the Capilano River (some reports claim 132 acres)" (p. 9).

Eighty thousand acres were added to other reserves, but these lands were of marginal value. The Vancouver area example above is the current location of the West Vancouver Park Royal Shopping Centre, where the land is valuable indeed. The McKenna-McBride Commission was only one process that cut off land from already small reserves. Reserve lands were lost to any number of transportation developments: highways, railways, hydroelectric transmission lines, and so forth. Lands adjacent to or within municipal boundaries were also targets for annexation. All these losses from existing reserves are referred to as cut-off lands in British Columbia.

The long litany of discrimination, assimilation, and basic injustice has been resisted by Aboriginal peoples. There was a near war in 1876 when the BC government took the position that it would not set more land aside for reserves, and more reserves were created. The story of the Nisga'a, who escorted non-Native surveyors off their land at gunpoint in 1887, illustrates not only resistance but also a clear view of Aboriginal title. Similarly, in 1898 the threat of waging war on the miners headed to the Yukon resulted in Treaty 8. Paul Tennant (1990) documents Native organizations that came together after the Indian Act (1876) to fight common injustices. One of the strengths of those organizations, whose leadership had been brought up under the missionary school system, was a common language – English.

DEVELOPMENTS AFTER THE SECOND WORLD WAR: NEW RIGHTS AND NEW THREATS

After the Second World War, new attitudes toward discrimination against all visible minorities resulted in changes to old policies. Status Indians were allowed to vote provincially in 1949, and by 1951 the bans on the potlatch, dances, and ceremonies were lifted. Residential schools were phased out in the 1960s. By 1988, bands were able to levy taxes on the leased portions of their lands and, in 1991, British Columbia finally elected a provincial government that recognized Aboriginal title.

Bill C-31, An Act to Amend the Indian Act, was passed in 1985 in an attempt to correct the problems that status and non-status designations caused. The Indian Act now legislates four categories of First Nations:

1 Status with band membership – Indians who have the right to both registered status and band membership;
2 Status only – Indians who have the right to be registered without the automatic right to band membership;
3 Non-status band members – Indians eligible to be registered under a band list in accordance with the Band Citizenship Code but who do not have the right to registered status;
4 Non-status Indians – Indians who are still not entitled to be registered (Joseph 1991, 69).

Some bands benefited by increased membership while others saw the act as another top-down set of rules imposed by Ottawa. A more serious concern is expressed by Susan Joseph (1991, 69): "The first generation cut-off clause dictates that only the first generation descendants of an individual are entitled to be registered. Second and succeeding generations will never be allowed status, nor will they be allowed to pass a right to status on to their children."

The transition to fewer discriminating policies and racist attitudes was not smooth. The 1960s, in particular, brought other, more serious threats to First Nations. This was the decade of megaprojects, during which the provincial government encouraged large, and often foreign, capital investments in resource development. The expansion of the forest and mining industries, combined with major infrastructure developments of hydroelectric dams, railways, and roads, put enormous pressures on the land. Many First Nations bands felt threatened by the assault on traditional lands that had never been surrendered. Court action and, in some cases, roadblocks and threats of violence ensued.

With the rapid economic growth of the 1950s and 1960s, a major struggle for First Nations was to get federal and provincial governments to recognize Aboriginal title and Aboriginal rights. Both levels of government were reluctant because of the economic implications. Consequently Aboriginal peoples turned to the court system and public opinion to persuade government to change its mind. The various actors involved – courts, governments, and public – are shown in Figure 5.7. The lines between public opinion and the two levels of government represent the connection between citizens and the federal and provincial governments they elect, indicating the importance of public opinion in government decisions. Individuals or groups have the right to challenge existing laws and conditions through the courts, and the judgment of the court becomes law until an appeal or new challenge occurs. The diagram recognizes the power of the Supreme Court of Canada to interpret laws in ways that affect both the provincial and federal governments. The 1969 White Paper forced First Nations peoples to become more familiar with these interconnections.

In the 1960s, the federal government under Prime Minister Pierre Trudeau was struggling to come to terms with the Indian Act and growing resistance by Aboriginal peoples. The resulting 1969 federal White Paper proposed

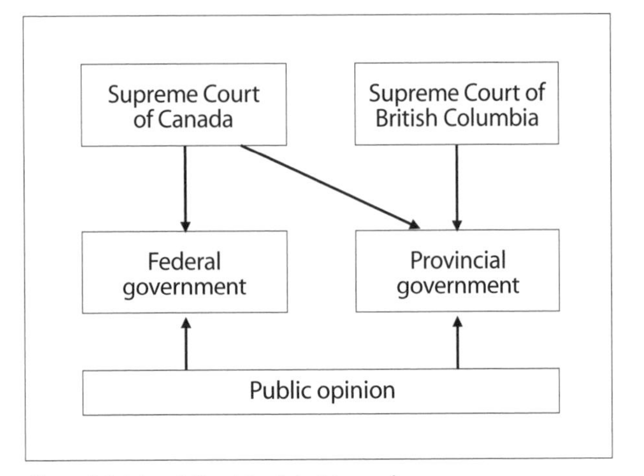

Figure 5.7 Interrelationship of decision makers

radical changes: the Indian Act would be abolished, reserve lands would be divided, and special status would end. In other words, there would be no recognition of Aboriginal title, no compensation for past injustices or expropriation of land, and no special status to recognize the inherent right to Aboriginal self-government. In short, the White Paper intended to turn all First Nations into "ordinary Canadians" – the ultimate in assimilation. Harold Cardinal, as president of the Indian Association of Alberta, commented: "For the Indian to survive, says the government in effect, he must become a good little brown white man. The Americans to the south of us used to have a saying: 'the only good Indian is a dead Indian.' The MacDonald-Chrétien doctrine would amend this but slightly to, 'The only good Indian is a non-Indian'" (1969, 1).

Aboriginal peoples across Canada reacted to form their largest national organization ever to reject the proposed policy. The federal government did withdraw the White Paper, but change in its policy was achieved through a decision by the Supreme Court of Canada. The *Calder* case in 1973 is famous because the Nisga'a Tribal Council took the federal government to court, asserting that their Aboriginal title had never been surrendered, or ceded. The Supreme Court ruled "that aboriginal title is rooted in the long-time occupation, possession and use of traditional territories. As such, title existed at the time of original contact with Europeans, regardless of whether or not Europeans recognized it" (BC Ministry of Aboriginal Affairs 1994). The court split evenly between six judges over the question of whether Aboriginal title still existed, and a seventh judge dismissed the case on a technicality. This became a landmark decision because its recognition of Aboriginal title produced an abrupt change in federal policy. The federal government began a process of negotiating **comprehensive claims**.

Comprehensive claims were, and are, conducted on unsurrendered lands within Canada. These are lands not previously subject to treaty. Comprehensive claims give rise to modern treaties in which negotiations can include land, resource management rights, cash compensation, education opportunities, economic development, and other rights and compensation, but initially not self-government. It took constitutional debates, military standoffs, and much confrontation before self-government was included as a possibility in 1995. Comprehensive claims are distinct from **specific claims**, which are for individual injustices done to specific bands. Cut-off lands, for example, are matters for specific claims.

The about-face of the federal government opened the floodgates for treaty negotiations. Nevertheless, claims did not proceed rapidly. By 1997, all except one had been settled in the Northwest Territories and Yukon, where provincial governments were not involved. In British Columbia, the provincial government stood fast until the 1990s on its position that Aboriginal title did not exist, making negotiations extremely frustrating.

Another important development occurred, however, on 11 June 2008, when Prime Minister Stephen Harper rose in the House of Commons to make an official apology to First Nations for the damage to Aboriginal culture caused by residential schools. This apology was followed up by a compensation package that "included an initial payout for each person who attended a residential school of $10,000, plus $3,000 per year" (CBC 2008). There is considerably more compensation for those who suffered physical or sexual abuse in the schools.

In this province, the conflict around resource exploitation on traditional territories continued, resulting in roadblocks and court cases. The Meares Island decision in 1985 by the British Columbia Court of Appeal upheld an injunction to prohibit MacMillan Bloedel from logging the area until a land claim by the Nuu'chah'nulth people was settled. Logging interests were blocked again at the Stein Valley near Boston Bar, and the logging of south Moresby Island in Haida Gwaii was so contentious that most of the disputed land became a national park. Industrial capitalism at the frontier was being blocked, and corporate profits, along with resource revenues to the provincial government, were being curtailed. Resource allocations for the future were uncertain.

The *Sparrow* decision (1990) put into question historical Aboriginal entitlement to fish versus commercial entitlement to salmon through the regulatory system of licences and run openings. Essentially, the Supreme Court of Canada overturned a lower court decision to convict an elder of the Musqueam Nation of illegal fishing, ruling that the Constitution Act provided "a strong measure of protection" for Aboriginal rights to fisheries. The decision stated that Aboriginal people "must be given priority to fish for food over other user groups." The federal government, which manages the fishing industry, interpreted the court decision to mean that a Native-only commercial fishery had priority over the non-Native commercial fishery. In an industry already suffering competing interests and declining stocks, the United Fishermen and Allied Workers' Union launched an appeal. The 1996 appeal decision reiterated that Aboriginal people had an unextinguished right to fish for food and that this right must be broadly interpreted. The federal government continued to have Native-only openings (with plenty of protest by non-Native fishers) until a further court ruling in 2003 (the *Kapp* decision) that the practice was in violation of the Charter of Rights and Freedoms. In 2004, however, this ruling was also overturned and Native-only openings continue (McRae and Pearse 2004, 9). In addition, in 2009 the BC Supreme Court ruled that the Nuu-Chah-Nulth have an Aboriginal right to harvest and sell all species of fish found within their territories (Hui 2009).

Many of the court cases of the 1980s and early 1990s appeared to be upholding the terms, conditions, and concept of Aboriginal title, but the *Delgamuukw* case set aside this pattern. Thirty-five Gitxsan (or Gitksan) and thirteen Wet'suwet'en hereditary chiefs went to the Supreme Court of British Columbia in 1984 to claim compensation for the loss of land and resources on their traditional lands. They also claimed the inherent right of Aboriginal self-government over these lands. Chief Justice

Allan McEachern did not accept traditional oral descriptions as evidence of historical land use and consequently ruled in 1991 that Aboriginal title had been extinguished at the time of Confederation. The news release by the Gitxsan and Wet'suwet'en in reaction to the decision shows their frustration: "This judge is attempting to push back justice for native people in Canada at least 20 years. This has been done by either ignoring or rejecting legal gains made in the last few years by native people all over the country. This judge goes as far as saying on page 300 of the document that native people, especially the Gitksan and Wet'suwet'en, are economically marginalized because they live on reserves and that the only hope of change can come if they leave and assimilate into the mainstream. This is the 1969 White Paper, a policy long rejected by successive federal regimes" (Ryan 1991, 1).

Needless to say, they appealed this decision, and in 1993 the BC Court of Appeal reversed much of the former judgment and recognized Aboriginal title, though it also ruled that those rights were non-exclusive. In December 1997, the Supreme Court of Canada ordered a new trial and ruled on several aspects of the *Delgamuukw* case, making it absolutely clear that Aboriginal title means ownership of the land. There was no recommendation on compensation though. For the Gitxsan and Wet'suwet'en, hope for justice lies either in negotiations with the treaty commission or in returning to the courts. Like the *Calder* case, the *Delgamuukw* decision has national implications. Court cases since *Delgamuukw* have added some clarity to the important issues of Aboriginal title and Aboriginal rights. For example, in 2004 Justice Lambert of the Supreme Court of Canada ruled in *Haida Nation v. BC and Weyerhaeuser* "that government has a duty to consult and possibly accommodate aboriginal interests even where title has not been proven"; and further, that treaties were not to extinguish "aboriginal rights and finality in agreements. Instead, the goal of treaty making is to reconcile aboriginal rights with other rights and interests" (BC Treaty Commission 2009b).

In the 1980s, court cases, blockades, and public opinion began turning against the provincial government's entrenched position. The public was becoming aware of and concerned about the environmental damage brought on by resource exploitation and was beginning to make the link between treatment of the land and treatment of the people who were attempting to stop the exploitation. The public demanded fair treatment of First Nations so that these issues would not continue for the next generations.

Premier Bill Vander Zalm created a Ministry of Native Affairs in 1990 in response to the pressure. Its mandate was to investigate Aboriginal title, while, ironically, making clear that the provincial government did not recognize Aboriginal title. Interestingly though, it stated that if Aboriginal title did exist, then the federal government would be 100 percent responsible for compensation. This ministry, and its mandate, did little to satisfy British Columbians.

Finally, in 1991 a change in government led to recognition of Aboriginal title. The British Columbia Claims Task Force was appointed and its recommendation to enter into a treaty process was accepted (BC Ministry of Aboriginal Affairs 1991). Bill 22 established the BC Treaty Commission in 1993. Modern treaties had finally come to British Columbia. It is worth noting that, had the New Democratic Party not won the 1996 election, the treaty process would have been dissolved by the other political parties as stated in their party platforms. Further, the Liberal Party of British Columbia (as the Official Opposition) challenged the legality of the Nisga'a Treaty in terms of Aboriginal self-government in 1999 – the ruling in July 2000 by Justice Paul Williamson was "that under the Constitution First Nations possess an inherent right to self-government" (Berger 2002, A11).

When the Liberal Party won the 2001 election (with all but two seats), it launched a province-wide referendum on treaty negotiations, prompting commentators to suggest the government was "trying to impose 19th-century ideas on a 21st-century problem (Hume 2002). Angus Reid (2002), who conducts many public opinion polls, stated, "The British Columbia aboriginal referendum is one of the most amateurish, one-sided attempts to gauge the public will that I have seen in my professional career." Clearly, provincial politics plays a major role in the treaty process; however, the courts play an even bigger role, and the Liberals have recognized the precedent of past court decisions and continued the treaty process.

MODERN TREATIES AND THE FUTURE

The BC treaty negotiations, or **modern treaty system**, are designed to overcome the lack of treaties and compensation for all bands and tribal councils in British Columbia. This system established a compensation formula between the federal and provincial governments in which the federal government is responsible for: 60 percent of negotiating costs; 75-90 percent of cash compensation; and 50 percent of the purchase price of third-party interests. The province is responsible for the remaining costs as well as for land and resource allocations (McKee 1996; BC Studies 1998-99; BC Treaty Commission 2009c). The negotiations themselves have a six-stage process:

Stage 1: Statement of intent by the First Nation to negotiate a treaty
Stage 2: Treaty negotiations preparation
Stage 3: Negotiation of a framework agreement
Stage 4: Negotiation of an agreement-in-principle
Stage 5: Negotiations to finalize a treaty
Stage 6: Treaty implementation

Throughout the province, various bands and tribal councils are at different stages of the process: six are at Stage 2; three at Stage 3; forty-three at Stage 4; seven at Stage 5, and one at Stage 6 as of 2009. These sixty First Nations represent approximately two-thirds of all Aboriginal people in the province. Conversely, one third have not engaged in the process at all. A current list of the bands and tribal councils that have signed on and which stage each group has reached can be found on the BC Treaty Commission website (under Negotiation Update). The rights of third parties are a major concern from a non-Native perspective. Third parties include all the users of land within any claim area: resource companies, farmers, other private landholders, local governments, and organizations that have a vested interest. Simply put, third parties are concerned that their interests will be jeopardized by a treaty. The mandate of the treaty process is to protect the right of continued use by third parties; bands and tribal councils receive cash and access to resources in compensation.

From the outset of the BC Treaty Commission negotiations, reports have been commissioned to assess the projected benefits of settling treaties in the province. The most recent report, *Financial and Economic Impacts of Treaty Settlements in British Columbia* (Thornton 2009), takes into account the three final agreements and earlier financial assessment reports. Its authors conclude that "if all 60 First Nations currently in the BC treaty process completed treaties by 2025, they (BC's economy) could receive a net financial benefit of $10.28 billion." Settling the land claims issue and bringing finality to land use and ownership is obviously in the provincial government's best interest, though the benefits may take many years to materialize.

Although it followed a similar process, the 2000 Nisga'a Treaty was negotiated outside the BC Treaty Commission guidelines and rules because it had been in progress with the federal government since 1976. It is a unique document, and while it claims not to set a precedent, the various issues it addresses have established the basis of future treaty negotiations for many bands and tribal groups throughout British Columbia. This was the first completed treaty in the province.

Figure 5.8 shows the traditional lands, or Aboriginal title, of the Nisga'a, who are based on the Nass River system. The agreement allocates 8 percent of the traditional area as a land base for the Nisga'a people, and it includes both surface and subsurface rights. This land base is centred on the existing communities. The Nisga'a land is no longer reserve land, the Indian Act no longer applies, and the tax exemption status on reserve lands will be phased out over time (e.g., the sales tax exemption ended in 2008, and other taxes, including the income tax exemption, will end in 2012). There was a cash compensation for the other lands of $190 million and a further $11.5 million for the Nisga'a to become involved in the commercial salmon fishery.

The treaty includes "an annual treaty-entitlement of salmon, which will, on average, comprise approximately 18 percent of the Canadian Nass River total allowable catch" (Indian and Northern Affairs Canada 2010). Tangent to this treaty is a separate Own Source Revenue Agreement that includes a twenty-five-year (renewable after fifteen years) Harvest Agreement for a commercial allocation amounting to 13 percent of sockeye and 15 percent of pink salmon on the Nass River (McRae and

Pearse 2004). This is precedent setting, for it transforms salmon from a common property resource into a private property resource. The document gives the Nisga'a responsibility within their territory for wildlife management as well as for establishing environmental standards. There is also provision for self-government and the establishment of justiciability, which means the right to establish policing and court systems.

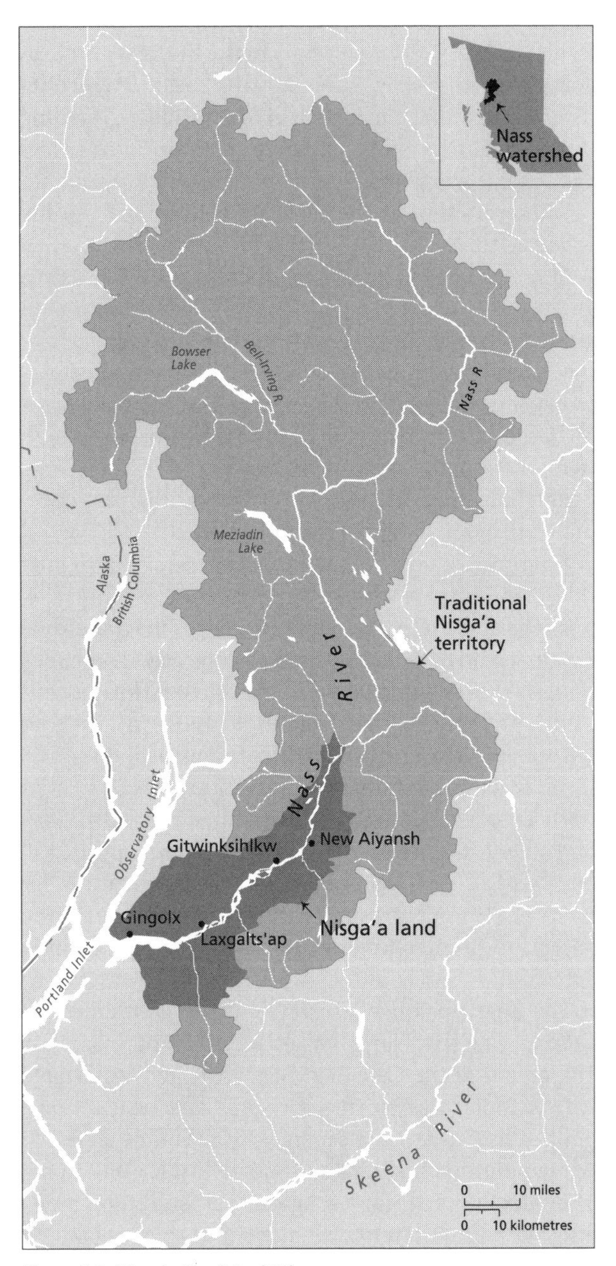

Figure 5.8 Nisga'a Aboriginal title
Source: GeoBC (2010).

Ratification of the Nisga'a treaty has put an end to the long history of injustice and illegal use of Nisga'a lands. It also ends the road blockades and confrontations over land use, and brings the Nisga'a much closer to equality with the rest of Canadian society. Not having followed the six stage process, however, may result in future legal confrontations. One of the conditions in moving through the stages is that the various First Nations claiming territory must settle overlaps in Aboriginal title boundaries between themselves. This did not occur in the Nisga'a Treaty process, and compensation was awarded based on Aboriginal title boundaries that included the whole Nass River watershed (Sterritt et al. 1999). The Gitanyow, Gitxsan, and Tahltan all have documentation as to their historical claim to parts of the Nass watershed, but the Gitanyow and the Gitxsan are still at Stage 4 and the Tahltan have not yet engaged in the treaty process.

As of 2010, one treaty has been completed via the six-stage process. The Tsawwassen First Nation treaty was signed on 3 April 2009, seventeen years after the establishment of the BC Treaty Commission. It is also the first urban treaty, a fact that complicated the issue, and it was opposed by various organizations. The original Tsawwassen reserve is a small (290 hectare) parcel of land south of Vancouver that is bisected by a highway that carries millions of vehicles each year to and from Vancouver Island (as well as a number of Gulf Islands) (see Figure 5.9). The industrial coal port and container terminal of Roberts Bank lies just to the north of the ferry terminal, bringing considerable rail traffic to the region. Agricultural land surrounds the reserve, which is in the agricultural land reserve (ALR) (see Chapter 12 for more detail).

Treaty negotiations added 334 hectares to the original reserve, for a total of 624 hectares. Like the Nisga'a, the Tsawwassen will own subsurface rights, their land will no longer be considered reserve land, and they will no longer fall under the Indian Act. Band members will own their own homes, and tax exemptions will be phased out over time. Tsawwassen will have self-government and the ability to develop their land and establish an economic path

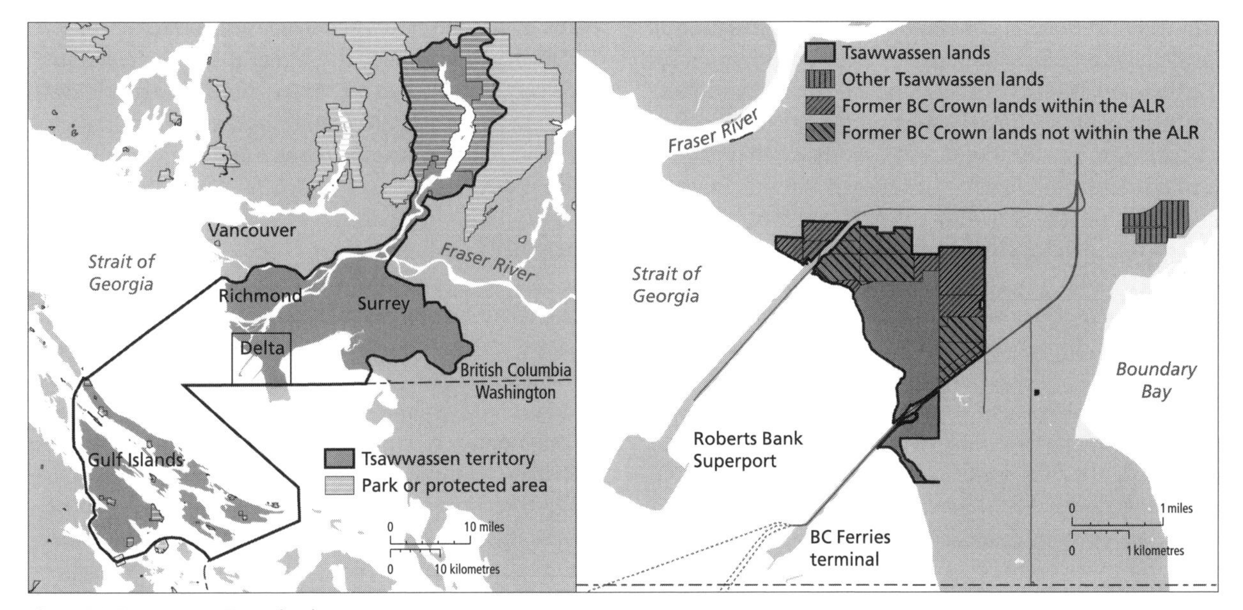

Figure 5.9 Tsawwassen Treaty lands
Note: "Tsawwassen Territory" refers to the historic region (the Aboriginal title) that the Tsawwassen claim as their territory. "Tsawwassen Lands" refers to the land that the Tsawwassen now have governance over, as designated through the treaty negotiations.
Source: BC Ministry of Aboriginal Relations and Reconciliation (2009).

for their 347 (as of 2008) band members. In compensation for ceding Aboriginal title, they will be paid cash compensation of $20.7 million. In addition, "Tsawwassen First Nation can exercise its harvesting rights with respect to migratory birds, wildlife, and plant gathering. Slightly different geographic areas, the Tsawwassen Fishing Area and the Tsawwassen Intertidal Bivalve Fishing Area, are where Tsawwassen First Nation can exercise its harvesting rights with respect to fish" (Indian and Northern Affairs Canada 2009b). Moreover, as in the Nisga'a negotiations, the government and the Tsawwassen came to agreement outside the treaty for a commercial allocation of a percentage of sockeye, chum, and pink salmon on the Fraser River.

Not all were pleased with the treaty. The surrounding municipality of Delta, along with farm organizations, were in opposition because the treaty removed 207 hectares from the ALR, and the Tsawwassen will likely convert the land for industrial and commercial uses. Other First Nations in the region also voiced their opposition to the treaty because they believe the Tsawwassen were compensated for Aboriginal title that other bands claim. Commercial fishers are concerned about the allocation of salmon and fear that other bands on the Fraser will gain similar concessions when they complete their treaties. Opposition also came from within the Tsawwassen First Nation when some members voiced concern about the loss of tax exemptions (Gordon 2008, 59).

As stated above, not all Aboriginal groups have signed on to the treaty process, and as of 2010 some fifty-six bands or tribal councils are not part of the process for a "variety of reasons: they disagree with treaty process as set out; they are not yet ready to negotiate; they are waiting to see what will be accomplished before joining; they are not prepared to borrow money to negotiate; and so on" (Kenzie Andrews, BC Treaty Commission, e-mail to author, 2004). Incidents such as a blockade at Gustafsen Lake in 1995 that led to an armed confrontation for much of that summer, or the violent protests over the expansion of Sun Peaks Resort (near Kamloops and on Secwepemc territory) in 2001, and many other protests since this time, reflect separate and hostile views. Such confrontations stem from an overwhelming frustration at the government's inability to deal with issues facing Native peoples, such as Aboriginal title, inadequate housing, lack of employment opportunities, the inordinately high suicide rate among Native youth, and continued resource exploitation where treaty negotiations have not been finalized (Willems-Braun 1997). Millions of dollars can be raised to subsidize resource industries operating on contentious, untreatied lands or to mobilize an army to put down a Gustafsen Lake uprising, but the funding needed to improve a band's conditions receives little attention.

The delay in gaining each of the steps in the treaty process is being viewed by such groups as the Sechelts, Gitxsan, and others as negotiation in bad faith, and consequently they have threatened to go back to the courts to have settlements imposed. Another strategy with considerable merit and success has been the negotiation of interim measures agreements or, more recently, incremental treaty agreements. These agreements resolve contentious issues such as resource harvesting, the setting aside of sacred land, and land-use planning, and they become "important building blocks for treaties" (BC Treaty Commission 2003, 6).

From the perspective of First Nations involved in negotiations, the treaty process is long overdue in righting the wrongs of the past. The historical reserves, "which were often too small, too remote and of too poor quality" (Ripmeester 2003, 77), restricted First Nations in making a living and, as Cole Harris (2003, 86) states, "It also seems clear that a relentless politics of assimilation has failed, and that the only alternative to it is a politics of difference." Frank Duerden (2001, 145) makes the point that land claims are not about First Nations claiming the land; their land was never surrendered. The process is about governments claiming, and negotiating, for land. Use of the land and its resources, management rights, cash compensation, and self-government are all ways to establish a solid foundation for preserving cultural values and an economic base for the future.

Examples of First Nations strategies for economic development exist across Canada (Indian and Northern Affairs Canada 2009a). The specific location of the Aboriginal group, its access to resources, and how its development plans fit into the global economic arena must all be considered. Although even these might not be "solutions," they certainly lead in a positive direction (Harris 2002, 323), and successful treaties, self-government, and First Nations economic ventures will benefit all in British Columbia.

SUMMARY

The story of First Nations dates back more than 10,000 years, when the transformation occurred from nomadic hunting and gathering to seminomadic occupation of specific territories, as the glaciers receded and the land stabilized. It is a story of peoples who used the land and its resources for long periods before Europeans arrived in British Columbia. This traditional geography was drastically interrupted by the arrival of Europeans, and a new geography of predominantly British and capitalist values was imposed on the land. The actions of the newcomers, both intentional and unintentional, had a catastrophic impact on First Nations. Their numbers were reduced to perhaps 5 percent of former populations, and their way of life was altered forever with the creation of reserves, residential schools, and a never-ending set of policies governing who they were and how they lived.

Despite these injustices, Aboriginal peoples did not give up the struggle for their culture and their land. It was not until the early 1990s that the process of land claims was seriously considered in British Columbia, but two bands have achieved a treaty – the Nisga'a (albeit not via the six-stage modern treaty process) and the Tsawwassen. As of 2010, there have also been two final agreements (i.e., the completion of stage five of the treaty process by reaching an agreement on the treaty, but the treaty still needs to be ratified by the First Nations' members). The Maa-nulth First Nations Final Agreement, which includes five First Nations on Vancouver Island, was signed on 9 April 2009; however, this treaty will not be implemented until 2011. On 17 June 2010, the Sliammon (Tla'amin) First Nation near Powell River signed a final agreement, but a date has not yet been established for implementation.

For other bands and tribal councils, modern treaties are close to being finalized, and interim measures agreements and incremental treaty agreements are resolving many contentious land-use conflicts. Although land claim settlement will not cure all ills, with accompanying self-government it will establish individual dignity and an economic base for the future.

REFERENCES

Aguado, E., and J.E. Burt. 2001. *Understanding Weather and Climate*. 2nd ed. Upper Saddle River, NJ: Prentice Hall.

BC Ministry of Aboriginal Affairs. 1991. *The Report of the British Columbia Claims Task Force*. 28 June. Victoria.

–. 1994. "Landmark Court Cases." Victoria: Communications Branch.

BC Studies. 1998-99. Special issue, *The Nisga'a Treaty*. No. 120.

BC Treaty Commission. 2003. *Annual Report 2003: Where Are We?* Vancouver.

Berger, T. 2002. "Why I Won't Be Voting." *Vancouver Sun*, 15 April, A11.

Brown, D. 1992. "Aboriginal Rights." Educational report for the Carrier-Sekani Tribal Council, Prince George.

Cardinal, H. 1969. *The Unjust Society*. Edmonton: Hurtig.

Cumming, P., and N. Mickenberg. 1981. "Native Rights in Canada: British Columbia." In *British Columbia: Historical Readings*, ed. W.P. Ward and R.A.J. McDonald, 184-211. Vancouver: Douglas and McIntyre.

Duerden, F. 2001. "Geography and Treaty Negotiations." In *British Columbia, The Pacific Province: Geographical Essays*, ed. C.J.B. Wood, 130-45. Canadian Western Geographical Series vol. 36. Victoria: Western Geographical Press.

Duff, W. 1965. The *Indian History of British Columbia*. Victoria: Royal British Columbia Museum.

Frideres, J.S. 1988. *Native Peoples in Canada: Contemporary Conflicts*. 3rd ed. Scarborough, ON: Prentice-Hall.

Glaven, T. 2000. *The Last Great Sea*. Vancouver: Greystone.

Gordon, K. 2008. "No Reservations." *Canadian Geographic* 128 (2): 48-62.

Gunn, A. 1976. "The Lost Lands." *The Province* (Vancouver), 20 January, 5.

Harris, C. 1992. "The Lower Mainland, 1820-81." In *Vancouver and Its Region*, ed. G. Wynn and T. Oke, 38-68. Vancouver: UBC Press.

–. 1997. *The Resettlement of British Columbia: Essays on Colonialism and Geographical Change*. Vancouver: UBC Press.

–. 2002. *Making Native Space*. Vancouver: UBC Press.

–. 2003. "Focus: Making Native Space: A Review Symposium." *Canadian Geographer* 47 (1): 73-87.

Joseph, S. 1991. "Assimilation Tools: Then and Now." *BC Studies* 89 (Spring): 64-81.

Hume, S. 2002. "Campbell's Referendum Folly Hasn't a Legal Leg to Stand On." *Vancouver Sun*, 14 March, A15.

Kump, L.R., J.F. Kasting, and R.G. Crane. 1999. *The Earth System*. Upper Saddle River, NJ: Prentice Hall.

McGhee, R. 1985. "Prehistory." In *The Canadian Encyclopedia*, vol. 1, 1466-9.

McKee, C. 1996. *Treaty Talks in British Columbia: Negotiating a Mutually Beneficial Future*. Vancouver: UBC Press.

McRae, R.M., and P.H. Pearse. 2004. *Treaties and Transition: Towards a Sustainable Fishery on Canada's Pacific Coast*. Report for Ministry of Fisheries and Oceans and Government of British Columbia. Ottawa.

Muckle, R.J. 1998. *The First Nations of British Columbia*. Vancouver: UBC Press.

Pringle, H. 2008. "The Messenger." *Canadian Geographer* 128 (6): 70-78.

Ray, A. 1999. "Treaty 8: A British Columbia Anomaly." *BC Studies* 123 (Autumn): 5-58.

Reid, A. 2002. "Treaty Referendum Is No Laughing Matter." *Vancouver Sun*, 5 April, A15.

Ripmeester, M. 2003. "Focus: Making Native Space: A Review Symposium." *Canadian Geographer* 47 (1): 73-87.

Ryan, D. 1991. "A Travesty of Justice." 8 March. News release. Wet'suwet'en Territory, Smithers. Sterritt, N. 1998-99. "The Nisga'a Treaty: Competing Claims Ignored!" *BC Studies* 120 (Winter): 73-97.

Sterritt, N.J., S. Marsden, R. Galois, P.R. Grant, R. Overstall. *Tribal Boundaries in the Nass Watershed*. Vancouver: UBC Press.

Tennant, P. 1990. *Aboriginal Peoples and Politics: The Indian Land Question in British Columbia, 1849-1989*. Vancouver: UBC Press.

Union of BC Indian Chiefs. 2005. *Stolen Lands, Broken Promises: Researching the Indian Land Question in British Columbia*. 2nd ed. Vancouver: UBCIC.

Willems-Braun, B. 1997. "Colonial Vestiges: Representing Forest Landscapes on Canada's West Coast." In *Troubles in the Rainforest: British Columbia's Forest Economy in Transition*, ed. T.J. Barnes and R. Hayter, 99-127. Canadian Western Geographical Series vol. 33. Victoria: Western Geographical Press.

Wynn, G. 2007. *Canada and Arctic North America: An Environmental History*. Santa Barbara: ABC-CLIO.

INTERNET

BC Ministry of Aboriginal Relations and Reconciliation. 2009. "Tsawwassen First Nation Final Agreement Appendices English Final – March 10, 2009." www.gov.bc.ca/arr/firstnation/tsawwassen/.

–. 2010. Treaties and Other Negotiations. www.gov.bc.ca/arr/treaty/default.html.

BC Treaty Commission. 2009a. "Negotiation Update." www.bctreaty.net/files/updates.php.

–. 2009b. "Aboriginal Rights." www.bctreaty.net/files/issues_rights.php.

–. 2009c. "Financial Issues." www.bctreaty.net/files/issues_financial.php.

CBC. 2008. "PM Cites 'Sad Chapter' in Apology for Residential Schools." 11 June. www.cbc.ca/canada/story/2008/06/11/aboriginal-apology.html.

GeoBC. 2010. "BC Geographical Names: Nisga'a Names, Nisga'a Lands." archive.ilmb.gov.bc.ca/bcnames/g2_nl.htm.

Hui, S. 2009. "Nuu-chah-nulth Have Aboriginal Right to Harvest and Sell Fish, BC Supreme Court Rules." Straight.com, 3 November. www.straight.com/article-268466/nuuchahnulth-have-aboriginal-right-harvest-and-sell-fish-bc-supreme-court-rules.

Indian and Northern Affairs Canada. 2009a. "Economic Development." www.ainc-inac.gc.ca/ecd/index-eng.asp.

–. 2009b. "Frequently Asked Questions – Tsawwassen Final Agreement. www.ainc-inac.gc.ca/ai/mr/nr/j-a2009/faq000000272-eng.asp.

–. 2010. "Fact Sheet: The Nisga'a Treaty." www.ainc-inac.gc.ca/ai/mr/is/nit-eng.asp.

Sinixt Nation. n.d. "The People of the Arrow Lakes." sinixt.kics.bc.ca/.

Thornton, G. 2009. *Financial and Economic Impacts of Treaty Settlements in British Columbia.* Ministry of Aboriginal Relations and Reconciliation. www.gov.bc.ca/arr/reports/thornton.html.

6 The Geography of Racism

The Spatial Diffusion of Asians

The Chinese were the first Asians to arrive in British Columbia and may even have arrived before Europeans. There has been some speculation that Buddhist monks reached the shores of western North America in approximately AD 458 and later in AD 594 (Caley 1983; Lai 1987). Certainly Chinese were involved in the initial building of the British Nootka Sound Fort in 1788. It was the gold rush of 1858, however, that resulted in significant immigration of the Chinese to British Columbia. The gold rush was also the beginning of attitudes and policies of racism and discrimination that increased in intolerance to the point of hostility and were applied to all Asians in the province.

Spatial diffusion, introduced in Chapter 1, is the movement over space and through time of people, goods, and ideas. The forces or influences that assist in this movement are the *carriers*, and the forces that oppose or block the diffusion process are referred to as *barriers*. The spatial diffusion of people, the focus of this chapter, is known as relocation diffusion. "Push" and "pull" factors assist in assessing the movement of people. Push factors are the various forces – political, economic, religious, and so forth – that sway, or even force, people from their present location. Pull factors are the many influences that attract people to a specific location. Through this concept we can assess the relocation of Chinese, Japanese, and East Indians to British Columbia at a number of geographic scales. Both push and pull factors were involved in the move from their home countries, and once they arrived and "settled," other factors pushed and pulled these people throughout the province and Canada. Each group had its own unique push and pull factors, although racism was a common force. The discussion in this chapter is broken down as follows:

- Chinese immigration and diffusion from 1858 to 1907
- Japanese immigration and diffusion to 1907
- East Indian immigration and diffusion to 1907
- Chinese immigration and diffusion from 1908 to 1947
- Japanese immigration and diffusion from 1908 to 1941
- East Indian immigration and diffusion from 1908 to 1947
- Japanese relocation and diffusion from 1942 to 1949
- Asian immigration and diffusion to 2010

CHINESE IMMIGRATION AND DIFFUSION FROM 1858 TO 1907

The Fraser River gold rush was the second major gold rush in North America, the first being in the Bay area of California in 1849. As Chapter 4 described, many of the early miners came to British Columbia from the San Francisco area, and some of those were Chinese. Others came directly from China. Statistics are not particularly accurate for those early days, but David Lai (1987, 337 and 339) suggests that there were about 1,000 Chinese here by 1862, and 1,759 by 1868. There may have been more, considering that the north half of Barkerville was known as "Chinatown," and Barkerville was a community of around 4,000 at its peak in 1865. Indeed, there may have been as many as 6,000 or 7,000 Chinese in the province in the early 1860s (Con et al. 1982, 14).

The lure of gold was clearly a pull factor, and the possibility of becoming rich and returning to China was an obvious motivation for relocation. The actual decision to leave family and friends to go to a very remote, foreign, and not entirely friendly land, however, does deserve some analysis and consideration of the conditions in China at this time.

China generally, and southern China specifically, can be described as in a state of chaos throughout the second half of the nineteenth century. The fixed amount of agricultural land simply could not support the number of people. In many cases, the subdivision of family lands to sons was no longer possible, as the plots were too small to support even one family. As a result, second, third, and following sons became landless and left home to gain income by selling their labour. At the overall political level, the xenophobic Ch'ing dynasty was being seriously challenged by Europeans, who used their military superiority to introduce opium to southern China and to demand trade. To make matters considerably worse, the Taiping Rebellion in southern China from 1850 to 1864 claimed an estimated 30 million lives (Hucker 1975). Needless to say, all this provided some incentive to leave the country. The decision to go was facilitated by a family, or lineage, system that often pooled the meagre family resources to send landless sons to the new country with the expectation that they would remit any income. The push factors were powerful indeed.

The Chinese who came to British Columbia in those early days were mainly Cantonese-speaking single men from Guangdong province in southeastern China. They intended to earn whatever they could in this strange and foreign land and send surplus income to their families in China, always with the intention of returning themselves with sufficient money to buy land and support a family.

Within British Columbia another spatial diffusion process subsequently occurred, influenced initially by economic opportunity. Many Chinese became prospectors and followed the trails up the Fraser and into the Cariboo. Others became involved in service industries in the communities of Victoria, Nanaimo, New Westminster, Lillooet, Quesnel, and Barkerville. As the Cariboo gold finds waned, new gold interests opened up on the Stikine, Peace, and Columbia Rivers. The Chinese located in these regions also.

European attitudes of superiority were inherent in colonization, treating all other visible ethnic groups as inferior. British Columbia was no exception. Added to this base of racism and discrimination was the economic nature of the new British colony. It depended on external demand for its natural resources, a condition that led to "boom" and "bust" times. During the bust periods, racism and discrimination tended to increase because of intense competition for scarce employment. After the boom period of the gold rush, the Chinese were desperate for employment and often without the option of returning to China. They were willing to work for very low pay. White labour versus Chinese labour became a significant part of the growing bitter racial conflict by the late 1860s.

Institutional racism was, in many ways, the most serious form of racism because it resulted in legislation that discriminated on the basis of race. These laws expressed the attitudes of the dominant white population through their elected politicians. In British Columbia, anti-Chinese sentiment came early. In 1872 the new provincial government proposed charging a **head tax** of fifty dollars on all incoming Chinese to limit immigration. Immigration came under federal jurisdiction, however, and the federal government rejected this proposal. In 1878 the provincial government excluded the Chinese from any provincial works, and thus began the institutional barriers to employment. Anderson (1989) documents well

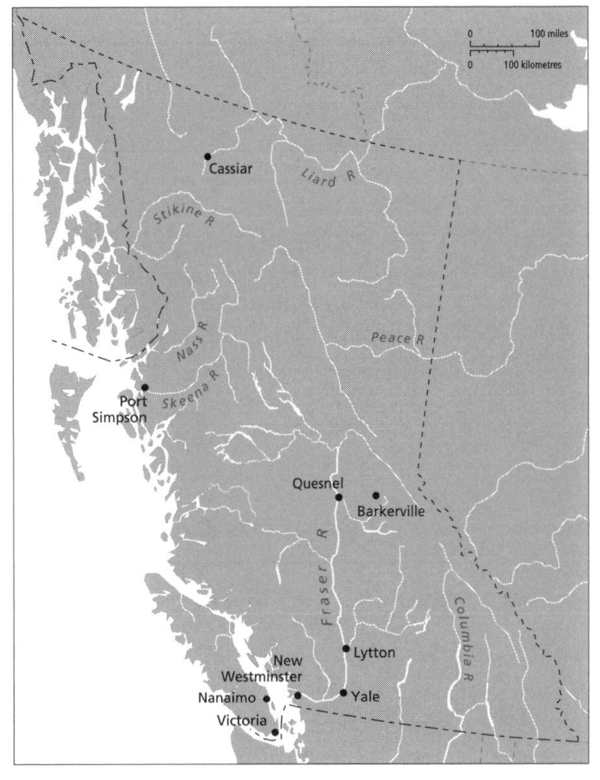

Figure 6.1 Main locations of Chinese in British Columbia, 1881
Source: Modified from Galois and Harris (1994, 42).

the historical roots of anti-Asiatic sentiments and the many forms of institutional racism practised in Canada, in British Columbia, and in Vancouver.

The building of the Canadian Pacific Railway (CPR) was another boom time that promoted a major influx of Chinese immigration – and a racial backlash. The census of 1881 probably underestimates the number of Chinese in British Columbia. Table 6.1 shows the range of their employment, with CPR construction workers heading the list. Figure 6.1 shows the spatial distribution of the Chinese throughout the province. The main influence on this distribution was a combination of old employment opportunities (mining and service industries) and new employment opportunities (railway construction).

The main construction years in British Columbia were between 1881 and 1885. The railway line was divided into

Table 6.1

Chinese occupations in British Columbia, 1884

Occupation	Number of persons	Population percentage
Labourers		
Railroad	2,900	27.6
Store employees	302	2.9
Ditch diggers	156	1.2
Others	296	2.8
Farming, logging, and mining		
Miners (mostly gold)	1,709	16.3
Coal miners	727	6.9
Farm labourers	686	6.5
Wood cutters	230	2.2
Fuel cutters	147	1.4
Vegetable gardeners	114	1.1
Production processes		
Fishers	700	6.7
Sawmill workers	267	2.5
Boat builders	130	1.2
Brick workers	85	0.8
Service and recreation		
Cooks and servants	279	2.7
Launderers	156	1.5
Barbers	71	0.7
Prostitutes	70	0.7
Managerial		
Merchants	120	1.1
Restaurant keepers	11	0.1
Sales		
Peddlers	67	0.6
Occupations not stated		
Married women	55	0.5
Girls	33	0.3
Boys under 17	529	5.0
New arrivals	602	5.7
Professional and technical		
Doctors	42	0.4
School teachers	8	0.1
Total	10,492	100.0

Note: Total does not equal 100 percent, due to rounding.
Source: Lai (1975, 16).

many individual contracts, and each contract meant an increase in demand for labour in a province that had a limited population of white workers. This provided the CPR with the rationale for importing Chinese labour. Andrew Onderdonk, the contractor for the CPR, brought in 15,701 Chinese from the United States and China over these years, though many either went into other professions or left the province after their short contracts were up. They were paid half the wages of white labourers, and many were promised return passage to China (Chodos 1973; Gunn 1975; Li 1988; Lai 1987). This promise was never fulfilled.

The end of the contracts created a serious condition of surplus labour and unemployment. The Americans anticipated the crisis and compounded the problem by trapping the Chinese in British Columbia: the US Chinese Exclusion Act, passed in 1882, did not allow the Chinese in British Columbia access to the United States. These events resulted in further diffusion of the Chinese throughout the province and, in some cases, back to China (Table 6.2).

With the significant increase in Chinese immigrants, many of whom were unemployed and viewed as a threat, more racial legislation was enacted. A provincial law passed in 1885 denied Chinese the vote, and the provincial government was able to convince the federal government to impose a fifty-dollar head tax on all Chinese immigrating to British Columbia. An examination of Table 6.2 indicates that the combination of high unemployment and a head tax had the desired effect of limiting immigration, but only in the very short run.

By the 1890s, the number of Chinese paying the head tax had increased considerably. Consequently, it was raised to $100 in 1901 and $500 in 1904, a great amount of money in those days. Again the tax reduced immigration – only 8 arrived in 1905, and 22 in 1906. By 1908, the number was up to 1,482, and another 7,078 came in 1913 (Lai 1987, 345). The increasing numbers reflected the continuing conditions of chaos and lack of economic opportunity in China. British Columbia, for all its racism and discrimination, provided the Chinese with a substantially better economic environment than did China.

Racism often led to Chinese people being placed in hazardous work situations as well as being subject to physical violence and property damage. The many deaths

Table 6.2

Chinese immigration and emigration, 1886-1900

	Entries	Exits	Net immigration	Estimated population
1886	212	829	-617	11,400
1887	124	734	-610	10,800
1888	290	868	-578	10,100
1889	892	1,322	-430	9,600
1890	1,166	1,671	-505	9,100
1891	2,125	1,617	508	9,129
1892	3,282	2,168	1,114	9,110
1893	2,258	1,277	891	9,800
1894	2,109	666	1,443	10,400
1895	1,462	473	989	11,000
1896	1,786	696	1,089	11,500
1897	2,471	768	1,703	12,200
1898	2,192	802	1,390	12,600
1899	4,402	859	3,543	13,500
1900	4,257	1,102	3,155	15,000

Source: Con et al. (1982, 296).

from unsafe working conditions in the building of the CPR are well documented (Berton 1970). Anti-Chinese riots occurred in a land-clearing contract for the Brighouse Estate in Richmond in 1887, and coal miners went on strike in Nanaimo over the hiring of Chinese. By far the most publicized violence was the Labour Day riot of 1907. Here, anti-Asiatic feeling was whipped up by fire-and-brimstone speakers calling for a white Canada. Most of the establishments of Vancouver's Pender Street Chinatown were damaged as a result, as well as many of the buildings in "Little Tokyo," the Japanese district on Powell Street (Adachi 1976).

While many British Columbians were unaffected by the presence of Asians in the province, newspapers, politicians, and special interest groups were quick to stereotype and blame Asians for many economic problems. Nevertheless, they were appreciated by those who hired them and profited by their hard work.

JAPANESE IMMIGRATION AND DIFFUSION TO 1907

To know the story of the Japanese in British Columbia requires some understanding of Japanese history and international relations. Japanese arrived in the province much later than the Chinese, had different push and pull factors, and experienced different forms of institutional racism.

In the Tokugawa Period (1600-1867), Japan was closed to Western trade and influence. The shogun, or leader of Japan, not only rejected outside relations but also prohibited Japanese people from leaving the country. Indeed, the size of fishing boats was limited by law specifically to prevent people from building ocean-going vessels. First the United States and then Europe used military superiority to impose unequal trade treaties on Japan. Japan responded with a revolution that overthrew the shogun system and installed an emperor, beginning the period referred to as the Meiji Restoration (1868-1912).

Unlike China, Japan adopted Western technology, science, a constitution, and industrial development almost overnight, which resulted in major restructuring of the society and economy. Agricultural workers were displaced, the population grew rapidly, the rigid class structure diminished, and taxes were very high, setting the stage for revolts and emigration. By 1885, emigration was encouraged.

Industrialization required raw materials such as iron ore, coal, and other mineral resources, of which Japan had very little. In a quest for these materials, Japan embarked on its own imperialism by exerting control over Taiwan and Korea, with unconcealed designs on mineral-rich Manchuria, by the 1890s. Russia also had designs on Manchuria and sent troops in to secure the rail line that ran from Russian territory through Manchuria to Port Arthur. Russian and Japanese troops soon clashed, creating tension between the two countries.

Russia was attempting to become a world military force in the 1890s and early 1900s by building up a modern naval fleet and army on the European side of the continent. This posed a threat to British naval interests. Japan's movement toward Manchuria was viewed by Russia as a threat to its port of Vladivostok. In an attempt to control Russia, Britain and Japan signed an alliance in 1902. Ultimately, war broke out between Russia and Japan (1904-5). Japan's decisive victory on both land and sea elevated its global status to a modern military state, ended the unequal trade treaties imposed earlier, and cemented relations between Japan and Britain.

The Anglo-Japanese Alliance meant that the British colony of Canada could not impose restrictive head taxes on the Japanese as it had on the Chinese. Nevertheless, the Japanese were denied the vote in 1895, and disenfranchisement meant restrictions on employment opportunities for both immigrant Japanese and future generations born in Canada.

The 1901 census records show only 4,738 Japanese in British Columbia. Most were poor, single men lured by the opportunities of fishing or employment in the coal mines and lumber mills. Few at this time could afford land for farming. Besides, most intended to work for a few years only and then return to Japan (Adachi 1976). Few Japanese immigrated to British Columbia, or anywhere, between 1901 and 1904 because of the impending war with Russia. Once the war was over, however, immigration to British Columbia was renewed at approximately 2,000 per year.

Events in 1907 increased Japanese immigration to British Columbia, and racism and discrimination increased as a consequence. Hawaii was the most popular destination for emigrating Japanese. Between 1885 and 1907, approximately 180,000 Japanese made their way to Hawaii. Many returned to Japan after accumulating some income, but many others went from Hawaii to the United States and some to British Columbia. In 1907 the United States passed an act that prohibited the Japanese from leaving Hawaii to go the United States. As a result, British Columbia saw a large increase in Japanese immigrants – over 7,000 that year (Adachi 1976). This coincided with a substantial increase in Chinese and East Indian immigration. White British Columbians demanded that immigration be halted.

Rumours of the new Grand Trunk Pacific Railway demanding even more Asian labour, along with a plan by the CPR to bring in 1,000 Japanese for an irrigation project in southwestern Alberta, added fuel to anti-Asiatic feelings and provoked the Labour Day race riot of 1907. The Chinese boarded up their establishments on Pender Street in response to the mob of rioters. The Japanese organized themselves into a resistance force, and the physical confrontation that occurred ended the riot (Adachi 1976, 74). Following the riot, the Chinese went on strike, and eventually the City of Vancouver paid $26,990 in compensation for the damages (Li 1988, 32). The event made headlines around the world and became an embarrassment to Canada because of the Anglo-Japanese Alliance.

EAST INDIAN IMMIGRATION AND DIFFUSION TO 1907

The emigration of people from India to British Columbia was almost exclusively Sikh prior to 1960. The Sikhs had their own unique struggles within India that produced the push factors for emigration. Particularly distinctive here was that the Sikhs, unlike the Chinese or Japanese, came from another colony of Britain. They, like British Columbians, were British subjects. This held little merit for the majority of British Columbians, however, who did not want to see any Asians in Canada.

Sikhism had evolved out of Hinduism, but not without the confrontation and violence of religious and political separation. India's colonial period was marked by resistance to British imperialism on the part of the two main religious groups, Hindus and Muslims. The Sikhs sided with the British and as a consequence were rewarded with land in the Punjab region of northern India for their "homeland." By the 1900s, chaos there had reached extreme levels; the Punjab region was being heavily taxed, religious and political conflicts with India continued, famines struck, and so did major outbreaks of diseases (Sandhu 1972).

Sikh troops visiting the province in 1897 brought back the word that British Columbia was opening up and that there was a demand for labour in the lumber industry, a field that Sikh workers were familiar with in the Punjab. There were few Sikhs in British Columbia by 1904 – perhaps 100 or fewer – but 387 came in 1905, 2,124 in 1906, and 2,623 in 1907 (Minhas 1994). The attraction of employment in the lumber industry of British Columbia had been discovered.

Repeating the pattern of the Chinese and Japanese immigration, most of the Sikhs were poor, single men and most became employed in the sawmills of the Lower Mainland. Spatial diffusion occurred throughout British Columbia as employment opportunities on Vancouver Island and in the interior became available in land clearing, road building, and other labour-intensive activities.

By 1907 the number of Sikhs in British Columbia had gained the attention of, and alarmed, the dominant anti-Asiatic white society. Yet because Sikhs were British subjects, the range of racially discriminating laws was limited. They were denied the vote in 1907, and disenfranchisement put a ban on work in the professions. The riot of 1907 did not directly involve Sikhs, but it spelled out very clearly the sentiment of exclusion that dictated that Asian immigration should be curbed. The only question was how the immigration of British subjects could be limited.

CHINESE IMMIGRATION AND DIFFUSION FROM 1908 TO 1947

The "solution" to Chinese immigration, the $500 head tax, was not working. In fact, it was becoming a windfall in revenues for the Canadian government (Table 6.3). Chinese immigration lessened during the First World War because considerably fewer ships were travelling between China and British Columbia, but the numbers increased significantly when shipping resumed as the war ended (Figure 6.2). In 1919 alone, 4,066 Chinese arrived. As Chinese immigration to British Columbia climbed, so too did the demand for Chinese exclusion.

Table 6.3

Revenue from head taxes on Chinese entering British Columbia

	Number paying head tax	Number exempt	Revenue ($)
1886-94	12,197	264	624,679
1895-1904	32,457	430	2,374,400
1905-14	27,578	4,458	13,845,977
1915-24	10,147	2,807	5,678,865
Total	82,379	7,959	22,523,921

Source: Li (1988, 38).

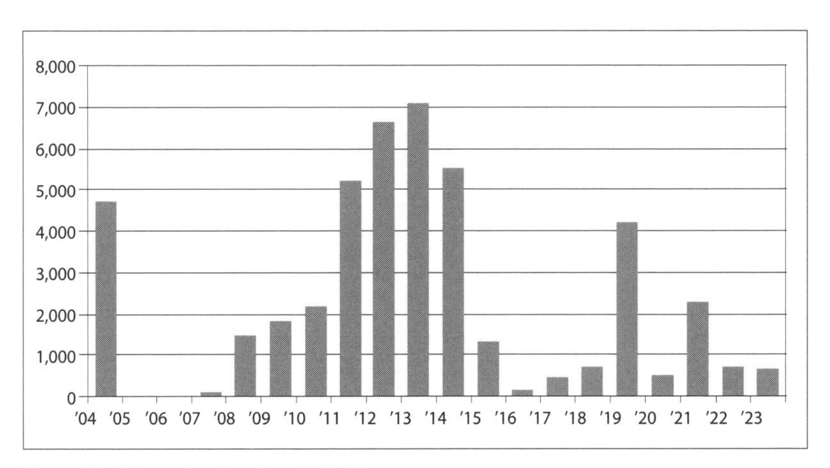

Figure 6.2 Chinese immigration, 1904-23
Sources: Modified from Lai (1987); Urquhart (1965).

The Chinese were employed in the service industries and resided in the Chinatowns of Vancouver, Victoria, New Westminster, Cumberland, Nanaimo, and many of the old mining and railway communities of the interior. Some were still employed in mining, some in the lumber mills, and a great many in canneries, and agriculture became an important way of life for others. The Chinese were beginning to see British Columbia as home, and this was of grave concern to those with anti-Asiatic sentiments. In the federal election of 1921, all the candidates from British Columbia, regardless of political party, ran on platforms to end Chinese immigration (Lai 1987, 347). There were approximately 24,000 Chinese in British Columbia by this time, forming a significant minority group. The Chinese Immigration Act, passed in 1923, had a profound impact on immigration and on Chinese people already in the province. Approximately 12 Chinese immigrated between 1924 and 1946.

The act prohibited Chinese from re-entering Canada if they left for a period of more than two years, and it denied entry to Canada to wives and children left in China by men who had already emigrated (Lai 1987, 348). The new rules created serious hardship. A number of Chinese returned to China permanently; others began to spread across Canada. Many settled in Toronto, but the largest concentration still formed Vancouver's Chinatown.

JAPANESE IMMIGRATION AND DIFFUSION FROM 1908 TO 1941

The embarrassment of the 1907 race riot resulted in Canada's labour minister being sent to Japan to make amends and to work out immigration limits. The Gentleman's Agreement of 1908 created four categories of Japanese immigrants: A, prior Canadian residents; B, domestic servants for Japanese residents; C, contract labourers; and D, agricultural labourers. A limit of 400 per year was set for classes B and D but no restrictions on classes A and C.

A fifth "class" of Japanese immigrant was not restricted by this agreement – wives. Marriages were arranged by parents in Japan. The single men in British Columbia recognized that they were here to stay and requested their parents in Japan to find suitable wives for them. This became known as the "picture bride" era. It became very common for relatives in Japan to send a picture of a prospective bride to British Columbia, and if she met with approval, to send the woman over on the next vessel. The wedding ceremony would then be conducted, often in a Christian church.

This process caused considerable racist sentiment in British Columbia on two counts. First, the concept of an arranged marriage appeared immoral to British, Christian morals. Second, the natural consequence of many of these marriages was children, increasing an Asian population that was perceived as unassimilable and undesirable.

Powell Street became the commercial centre for the Japanese, and the enclave of Steveston in south Richmond preserved their language, religion, and customs. The main employment in Steveston was fishing, which the Japanese began to dominate fairly early on. By 1901, for example, the Japanese held 1,958 fishing licences out of a total of 4,722 issued. With two people to a boat, or licence, over 4,000 Japanese were engaged in the fishing industry (Adachi 1976, 47). Fishing interests also led to the diffusion of the Japanese to the central and north coast fishing grounds of Rivers Inlet and up to the Skeena River region. Forestry, mining, and labouring jobs attracted others throughout British Columbia. The majority of Japanese, however, remained in the Lower Mainland region.

The demand for exclusion of the Japanese continued, even though a number of Japanese living in this country fought for Canada in the First World War, and some died. Those who survived were promised the vote, but opposition was so strong that they did not receive it until 1931 (Adachi 1976, 155). By 1922 the alliance forged between Britain and Japan had expired and so, apparently, had the need for treating the Japanese in British Columbia with diplomacy. Japanese domination of the fishing industry, a source of much concern, became the target of the Duff Commission in 1922, which reduced Japanese fishing licences by 40 percent. When over 1,200 fishers were driven from employment, the Japanese focused on farming. Many farmed the marginal lands of the Fraser Valley, while others gravitated to the Comox, Okanagan, and Cariboo valleys. In 1923, class B and D immigrants were restricted to 150 per year, and by 1928 all classes, including "picture brides," were restricted to that number. Similar to the Chinese, the door to Canada for the Japanese was nearly closed.

The impact of not being able to vote became more and more restrictive, particularly for the nisei, or second generation Japanese, who were born here. Educated in British Columbia's schools and, in some cases, graduating from the University of British Columbia, the professions were out of reach for them because they were not on the voters' list. In 1936 the Japanese in British Columbia organized and lobbied Ottawa for the federal vote. Unfortunately, most of the 1930s was influenced by the Depression. As well, on the international scene Japan was providing a military challenge to China and, indirectly, to Britain and the Western world. The backlash affected all Japanese, even those born in Canada. No franchise was granted. In fact, when the Second World War began in 1939, Japanese Canadians who volunteered for military duty were rejected and all Japanese over the age of fifteen were forced to be fingerprinted and registered.

The bombing of Pearl Harbor on 7 December 1941 by the Japanese and the capture of Hong Kong on Christmas Day, resulting in Canadian casualties and hundreds of prisoners of war, solidified the Pacific coast's worst fears – imminent invasion by Japan. All Japanese in western North America become suspect and were classified as enemy aliens. This fear had a devastating impact on the Japanese in British Columbia, who were forcibly moved from the coast and lost goods and property.

EAST INDIAN IMMIGRATION AND DIFFUSION FROM 1908 TO 1947

The Canadian authorities were well aware that Sikhs were British subjects and that they belonged to a minority group seeking independence in India, a conflict they did not want to spill over into Canada. They were also amply aware of the anti-Asiatic feelings of those British Columbians who wanted all Asians banned. The new rules established for Sikh immigration in 1908 were as effective as the Chinese Immigration Act. Two key rules were put in place. First, immigrants from India had to have $200 in their possession when entering Canada, a great sum of money for the mainly poor peasants immigrating to find employment. Second, immigrants had to arrive by continuous passage, even though with shipping technology at this time no vessels travelled directly between India and British Columbia. Because of these restrictions, Sikhs were no longer able to enter Canada legally. Many left the province, some going to the United States and others returning to India.

These discriminatory rules were not without challenges, the most publicized being the *Komagata Maru* incident in 1914. This Japanese ship carrying approximately 400 Sikhs arrived in Vancouver on 23 May but did not comply with the continuous passage rule or, in most cases, the $200 rule. Refused docking privileges, food, and water, the *Komagata Maru* remained in Vancouver's harbour for eight weeks while protracted legal negotiations resulted in riots and bitterness. The federal government finally provided food and water, but few of the passengers were allowed to stay in British Columbia and the *Komagata Maru* was escorted out of Canadian waters by a navy gunboat. By the end of the First World War, only a thousand or so Sikhs remained in Canada. The door was effectively closed on this Asian group.

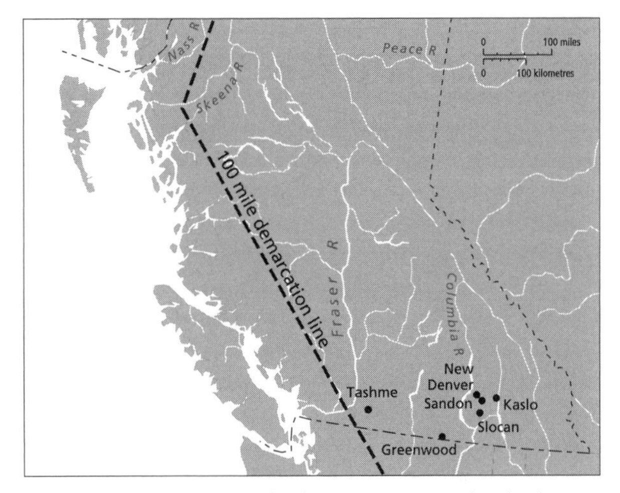

Figure 6.3 Relocation camps for the Japanese in British Columbia

JAPANESE RELOCATION AND DIFFUSION FROM 1942 TO 1949

Most Japanese in British Columbia lost everything in the Second World War. They were suspected of spying and consorting with the enemy; their fishing boats were confiscated along with automobiles and radios. All were auctioned off at fire sale prices. Japanese-language newspapers and schools were outlawed, and many employers laid off their Japanese employees. The racist politicians of British Columbia used the threat of a Japanese invasion as an excuse to incarcerate all Japanese. Mass evacuation from the coast of British Columbia was seen as the "solution" to the Japanese problem.

A hundred-mile demarcation line from the coast of British Columbia was established and all Japanese were to be relocated east of it (Figure 6.3). Initially, the most suspect were rounded up and sent to a prisoner-of-war camp in northern Ontario. Over 2,000 men were sent to road construction camps in the interior, while families were rounded up in the cattle exhibition area of the Pacific National Exhibition grounds in Vancouver. Most relocation was to nearly bankrupt ghost towns in the Kootenays – Sandon, Greenwood, New Denver, Slocan Valley, and Kaslo – as well as to Tashme, a name derived from the first two letters of Taylor, Shirras, and Meade, members of a security commission created under the War Measures Act. Tashme was only twenty-three kilometres from Hope and became the site of the Hope-Princeton road construction project, which 2,300 people spent three years building in supervised confinement (Evenden and Anderson 1973, 42). The first winter was exceptionally cold and uncomfortable, with two families per uninsulated cabin. Insulation and weather proofing were added for the next winters.

The war effort produced a shortage of labour everywhere, and politicians saw the Japanese as pools of labour not only for road building but also for the harvest of sugar beets on the Prairies and in Ontario. Several thousand Japanese relocated to these destinations, although Alberta expressed concern over inheriting British Columbia's so-called Asian problems (Table 6.4; Adachi 1976, 415).

As the war progressed, the Japanese settled into the detention camps in a more permanent way. The Canadian government recognized that the Japanese still had equity in the buildings and property from which they had been

Table 6.4

Japanese evacuees, 31 October 1942

	Number of evacuees
Detention camps	
Greenwood	1,177
Slocan Valley	4,814
Sandon	933
Kaslo	964
Tashme	2,636
New Denver	1,505
Subtotal	12,029
Road construction camps	
Blue River/Yellowhead	258
Revelstoke/Sicamous	346
Hope/Princeton	296
Schreiber	32
Black Spur	13
Subtotal	945[a]
Sugar beet projects	
Alberta	2,588
Manitoba	1,053
Ontario (males only)	350
Subtotal	3,991
Other permits and situations	2,520
Total	19,485

a Between March and June 1942, 2,161 Japanese were placed in road construction camps. Many were allowed to have their families join them by October 1942.
Source: Adachi (1976, 415).

Table 6.5

National status of Japanese Canadians in 1941

	Number	Percentage
Japanese nationals	5,564	25.2
Naturalized Canadians	3,223	14.6
Born Canadians	13,309	60.2
Total	22,096	100.0

Source: Adachi (1976, 414).

evicted on the coast. These possessions became the next round for the auctioneers, and in the process the Japanese lost most of what they had acquired before 1941. Homes, farms, and commercial enterprises were all sold off.

As the war drew to an end, provincial politicians began to pressure the federal government to prevent the camps from becoming permanent locations. By the end of 1944, the decision was made to close the detention camps, and the Japanese were given a choice. They could relocate east of the Rockies, since the coast was still out of bounds, or be repatriated back to Japan. Rejected by Canada, many Japanese initially opted to be deported to Japan; when it came time to leave, however, only about 4,000 actually went back to war-torn Japan. Those who remained underwent another relocation process; the greatest number took up residence in Ontario, mostly in Toronto.

Finally, in 1949, four years after the end of the war, the Japanese were allowed to return to the coast of British Columbia to live, and they were given the vote. One of the tragedies of this period, as can be seen in Table 6.5, is that the majority of incarcerated Japanese were born in Canada.

ASIAN IMMIGRATION AND DIFFUSION TO 2010

The end of the Second World War ushered in a period of affluence and a change of thinking about minority groups in Canada. A new acceptance led to renewed immigration and the franchise. The Chinese Immigration Act was repealed in 1947, and limited immigration resumed. For East Indians, the restrictive rules on continuous passage also ended in 1947, at which time they received the vote. The Japanese were allowed back to the Pacific coast and received the vote in 1949. These were important changes in breaking down institutional racism, but a quota system for immigrants by race was in place, making the immigration process restrictive and racist still.

International pressures ended racial discrimination as a policy of Canadian immigration by 1962, but the non-racist point system that currently operates was not put in place until 1967. The mix of immigrants changed considerably in both numbers and background with the modifications to the immigration rules. Many more Chinese and East Indians, especially, began to immigrate to Canada. The class of new immigrants also changed substantially. These newcomers were no longer dominated by poor farmers, fishers, and labourers as in the old days, nor did they have to confine themselves to the Lower Mainland of British Columbia.

In terms of spatial diffusion throughout Canada, it is interesting to see that, while each of these groups initially established themselves in British Columbia, they spread out across Canada, particularly after the Second World War, and relocated mainly to Ontario. The Chinese, Japanese, and East Indians have very different histories in British Columbia but share a common history of suffering from racism, distrust, and discrimination. Ontario stands out as being far more receptive to racial differences, and consequently, more Chinese, Japanese, and East Indians live in Ontario than in British Columbia.

For the Chinese, the residential segregation that created the many Chinatowns throughout British Columbia has largely disappeared from the landscape. The exceptions are Vancouver and Victoria, where Chinatowns exist mainly as tourist centres and as the location of some Chinese institutions, goods, and services but provide little in the way of housing. On 22 June 2006, Prime Minister Stephen Harper offered an official apology for the head tax and a compensation package of approximately $20,000. Not all were happy with the compensation conditions, however, as they applied only to those who had paid the tax, not sons or daughters. In 2006 there were only about twenty Chinese Canadians still alive who had paid the tax (Chinatown History 2008)

The East Indian population in Canada is nearly half a million (Statistics Canada 2006 Census) and the Sikh population is now greater than 300,000. East Indians have likewise spread across the country, with over 100,000 in the Toronto area and about the same number in the Lower Mainland of British Columbia. Prime Minister Harper also offered an apology for the 1914 *Komagata Maru* incident, but not in the House of Commons. The apology is therefore not recognized as official, and it was not accompanied by compensation.

For the Japanese in Canada, 1989 was an important year because Prime Minister Brian Mulroney issued an official apology for their treatment during and after the Second World War. Moreover, a compensation package of approximately $21,000 was offered to every Japanese Canadian born or residing in Canada prior to 1948. A reflective message, however, is offered by Sunahara (1981, 169): "In a country that prides itself on its democratic tradition, it is sobering to note that everything done to the Japanese Canadians was, and still is, legal under Canadian law."

REFERENCES

Adachi, K. 1976. *The Enemy That Never Was*. Toronto: McClelland and Stewart.

Anderson, K.J. 1989. "Cultural Hegemony and the Race-Definition Process in Chinatown, Vancouver: 1880-1980." *Environment* 6 (2): 127-49.

Berton, P. 1970. *The National Dream: The Great Railway 1871-1881*. Toronto: McClelland and Stewart.

Caley, P. 1983. "Canada's Chinese Columbus." *The Beaver* (Spring): 4-11.

Chodos, R. 1973. *The CPR: A Century of Corporate Welfare*. Toronto: James Lorimer.

Con, H., R. Con, G. Johnson, E. Wickbers, and W.E. Willmot. 1982. *From China to Canada*. Toronto: McClelland and Stewart.

Evenden, L.J., and I.D. Anderson. 1973. "The Presence of a Past Community: Tashme, British Columbia." In *Peoples of the Living Land: Geography of Cultural Diversity in British Columbia*, ed. J.V. Minghi, 41-66. BC Geographical Series no. 15. Vancouver: Tantalus.

Galois, R., and C. Harris. 1994. "Recalibrating Society: The Population Geography of British Columbia in 1881." *Canadian Geographer* 38 (1): 37-53.

Gunn, A. 1975. "Our History of Ethnic Prejudice." *The Province* (Vancouver), 18 November, 5.

Hucker, C.O. 1975. *China to 1850: A Short History*. Stanford: Stanford University Press.

Lai, D.C. 1975. "Home Country and Clan Origins of Overseas Chinese in Canada in the Early 1880s." *BC Studies* 27 (Spring): 3-29.

–. 1987. "Chinese Communities." In *British Columbia: Its Resources and People*, ed. C.N. Forward, 335-57. Western Geographical Series vol. 22. Victoria: University of Victoria.

Li, P.S. 1988. *The Chinese in Canada*. Toronto: Oxford.

Minhas, M.S. 1994. *The Sikh Canadians*. Edmonton: Reidmore Books.

Sandhu, K.S. 1972. "Indian Immigration and Racial Prejudice in British Columbia: Some Preliminary Observations." In *Peoples of the Living Land: Geography of Cultural Diversity in British Columbia*, ed. J.V. Minghi, 29-40. BC Geographical Series no. 15. Vancouver: Tantalus.

Sunahara, A.G. 1981. *The Politics of Racism*. Toronto: James Lorimer.

Urquhart, M.C., ed. 1965. *Historical Statistics of Canada*. Toronto: Macmillan.

INTERNET

Chinatown History. 2008. "Chinatown History about Head Tax Redress." chinatownhistory.com/HeadTaxRedress.html.

Statistics Canada. 2006 Census. "Cumulative Profile, 2006 - Canada, Provinces and Territories" (table), www.statcan.gc.ca/.

7

Resource Management in a Changing Global Economy

Resources have been, and continue to be, the strength of British Columbia. The value of these resources has fluctuated over time and new resources have been discovered, bringing changes, and sometimes conflict, to regional economies. This chapter begins by defining resources and describing several categories of use, or models, for examining them. Resources are managed at the three levels of government: federal, provincial, and municipal. The responsibilities of the three may overlap and cause conflict.

Not only is there a political dimension to resource management but there are a host of economic considerations in terms of the way in which resources are used, or valued. These have changed over time and are influenced by the interrelated and shifting roles of labour, technology, capital, land, and other factors affecting the production of goods and services. Here again, all levels of government play significant roles. British Columbia's reliance on resources and on the changing factors of production and markets has left its mark of boom and bust economies. Today's global economy produces spatial differences in British Columbia in terms of what resources are used, how resources are used, and how people of British Columbia make a living.

Resource harvesting and processing have been largely responsible for the population increase in the province, although population dynamics are also related to natural increase and net migration. Assessing these many factors produces population projections for the province and its regions, thereby forecasting economic, political, and social changes.

It should also be recognized that resource management and researching demographic change are important career areas for students of geography.

DEFINING RESOURCES AND CATEGORIZING THEIR USES

A simple definition of a **resource** is any naturally occurring substance that is of value to society. The key to this definition is not in naming an enormous number of substances but in understanding their function in society. Implied is that resources are culturally defined. Herring roe, for example, is a significant resource in this province, bringing in several millions of dollars to BC fishers even though very little roe is consumed in British Columbia. It is highly valued in Japan, however, where it is considered a delicacy. Different cultural groups value materials, or resources, in very different ways.

Also implied in the definition is that the value of a resource can change. One needs only to consider the value of coal from the beginning of the twentieth century to now. Coal used to be the most important energy source for cooking and heating homes, running locomotives, and firing steam-driven engines in industry. By the 1950s and 1960s, petroleum, in the form of diesel fuel and gasoline, was replacing the transportation uses of coal, while electricity and natural gas were replacing its residential and industrial uses. These were not good times for coal miners in the Fernie area of southeastern British Columbia. Then, in the 1970s, new technologies and demands by new markets increased the value of coal. Japan initially became the main purchaser of BC coal, using coking coal in the iron and steel industry and thermal coal in creating electricity. The new communities of Elkford and Sparwood in the Fernie area and Tumbler Ridge in the Peace River area symbolize new interests and employment related to the metamorphosis of this resource.

The same resources may have more than one value. Fresh water, for example, has numerous values. Anglers, swimmers, boaters, environmentalists, and others value fresh water in its natural setting from very different perspectives, although all these views may be complementary. The use of fresh water for hydroelectric energy production or as a system to discharge effluent from industry or domestic sewage is more contentious. Competing values are the essence of the need for resource management.

Resources can be categorized in a number of ways that help in understanding the dynamics of their use and management. One way is to separate renewable from nonrenewable resources. Renewable resources are living, or biotic, resources, which can be further divided into plants and animals. The various species of trees, agricultural crops, all other plant life, domestic and wild species of animals, and humans are all classified as renewable resources. A cautionary note is required with this classification: these resources are renewable *only if they are properly managed*. The extinction of plants and animals is a concern throughout the world. Noted Canadian geneticist David Suzuki has enlightened many Canadians about the threat to biodiversity due to decimation of species and reduction in resource options.

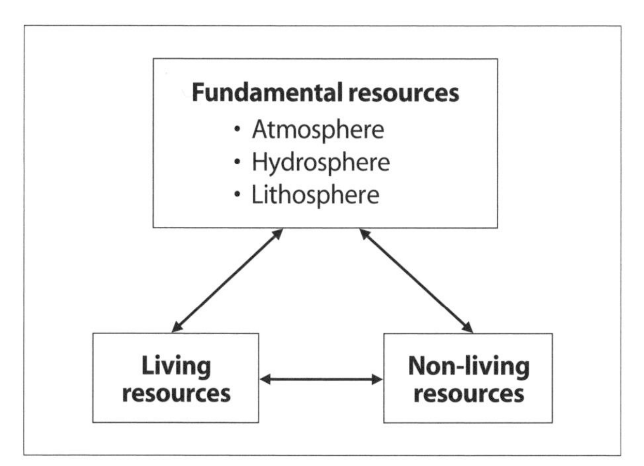

Figure 7.1 Ecosystem model of resource use

Nonrenewable resources are nonliving, or abiotic, such as minerals and fossil fuels. Nonrenewable implies that there is a fixed or finite amount. In practical terms, it is not a simple matter to determine the fixed amount of a nonrenewable resource such as oil or natural gas, much less gold or copper. More frequently we view these resources within a more specific, regional context and make projections about how long a particular mine or oil well will produce until it is exhausted. Even then, these estimates are often good only for a particular time and a particular technology. The concept of a finite amount of any resource is important to the management process.

Figure 7.1 is an ecosystem model of resource use, illustrating that our use of resources affects the environment and the use of other resources. The term *fundamental resources* refers to air (atmosphere), water (hydrosphere), and land (lithosphere). The mix of these three essential spheres, along with energy from the sun, produces photosynthesis. They are thus fundamental to life and are therefore known as life-giving resources.

The dynamics of the ecosystem model are revealed by looking at the example of how people rely on and use more and more fossil fuels for energy and materials. Using these nonliving resources results in ever greater amounts of waste (pollution) such as sulphur dioxide, which is given off to the atmosphere, one of the fundamental resources. In the atmosphere, sulphur dioxide (SO_2) combines with water vapour (H_2O) and falls as acid precipitation (H_2SO_4), which affects the land (lithosphere) and the resources of the land, including living resources. Increased acidity from acid rain can have a serious impact on drinking water, forests, agriculture, and fish living in the streams and lakes (hydrosphere). Similarly, the increased production of carbon dioxide (CO_2) globally is responsible for climate change, and it will potentially have catastrophic effects on this province and the rest of the world (see Chapter 2 for more detail). The ecosystem model demonstrates the cyclical nature of resource use, thereby recognizing environmental impact as a "cost" in the production of any given commodity. This assessment incorporates the real cost of commodities and raises the concept of sustainability of resources.

In the field of resource management it is essential to view the whole process of how resources are developed into products. A struggle in the past that continues to this day is recognizing and assessing the impact and costs associated with resource development: the packaging that increases the costs for solid wastes, the effluents from industries discharged into our water and land, and the potential environmental effects, including human health, of using products such as herbicides and pesticides. What level of effluent or discharge is "safe," and who pays the costs of cleanup? Considering that many of the consequences to human health and the environment take a great deal of time to assess, it is not surprising that these negative factors have been labelled as externalities to the production process. Corporations have therefore not had to include environmental repercussions among the costs of production.

The term **sustainability**, or **sustainable development**, is a relatively new concept, dating since the 1980s. Fundamentally, sustainable development means that resources should not be exploited to a level where they will not be available for future generations. This concept rests on a knowledge base and recognition that the destruction or extinction of species brings ruin to an economy and a way of life and that, in some cases, this destruction occurs through government policy. Sustainability recognizes ecological relationships; thus, it is not an easy concept to put into practice, particularly when increasing populations create a corresponding increase in demand for resources and products. Phil McManus (2000, 813) suggests that a debate continues as to where the emphasis on sustainability is placed: on development (i.e., continuing a standard of living), or on ecological processes. The message of this geography of British Columbia is the need to gauge resource use and its impacts for each region of the province. Sustainability, however, must also be viewed globally.

Subsequent chapters examine individual resources that have been and continue to be important to British Columbia. A common theme is to see how resource use has developed and changed. As conditions of society change, so do its need for and use of resources. Technologies have changed, ways of making a living have changed, understanding of and attitudes toward the environment have changed, and British Columbia has become much more urban. Change provokes new conflicts over resource use. Traditional forestry practices, for example, run headlong into other, non-timber values such as old-growth preservation, wildlife, water quality, and tourism. More and more, all resources require management.

MANAGING OUR RESOURCES: THE ROLE OF GOVERNMENTS

The federal, provincial, and municipal levels of government have the responsibility for managing resources. The distribution of powers between the federal and provincial governments was laid out at Confederation in the British North America Act in 1867 (renamed the Constitution Act of 1867) and now in the Constitution Act (1982). The conditions and responsibilities of municipalities and regional districts in British Columbia are established by the Municipal Act under the provincial government. It should be kept in mind that the responsibilities of each entity are constantly being challenged and changed.

The federal government is responsible for relationships with other countries and for things external to Canada. This includes all foreign trade. Controversial issues such as the sale of CANDU reactors to foreign countries or fresh water to the United States or softwood lumber agreements are in the hands of Ottawa managers. The federal government also has jurisdiction over oceans and the resources of the oceans. For British Columbia this includes not only any potential offshore oil and gas reserves but also, and most important, the fishing industry. Chapter 9 describes how the management of fish stocks, salmon especially, has been fraught with difficulty. Aboriginal affairs are another federal government responsibility, and Chapter 5 describes the modern treaty process, which in many ways is about the management of resources. Most of the main infrastructure in British Columbia – ports, railways, the Trans-Canada Highway, pipelines, and airports – has been established by the federal government. As well as being crucial to the movement of resources, this infrastructure represents a substantial use of resources in its construction and operation. Many other areas of federal responsibility influence resource use, ranging from research and development funding to health, education, and social welfare.

The provincial government's responsibilities and involvement in managing resources are very broad. The following is a partial listing of the relevant ministries in 2010:

Ministry of Aboriginal Relations and Reconciliation
Ministry of Advanced Education and Labour Market Development
Ministry of Agriculture and Lands
Ministry of Education
Ministry of Energy, Mines and Petroleum Resources
Ministry of Environment
Ministry of Forests and Range
Ministry of Health Services
Ministry of Healthy Living and Sport
Ministry of Housing and Social Development
Ministry of Labour
Ministry of Small Business, Technology and Economic Development
Ministry of Tourism, Culture and the Arts
Ministry of Transportation and Infrastructure

Clearly, there is some overlap between provincial and federal government jurisdiction. One of the reasons salmon stocks are so difficult to manage, for example, is that salmon spend critical periods in freshwater streams before migrating to the ocean. Although the freshwater streams are regulated under the federal Fisheries Act, the provincial government has allowed other resource developments that damage the quality of salmon-bearing streams, such as dams, logging, mining, urban development, and the discharge of effluent from any number of sources. Similarly, modern treaty negotiations require agreement between provincial and federal levels of government as well as the First Nations.

Local government in British Columbia is made up of incorporated municipalities and, since 1965, regional districts. Their jurisdiction over land and other resources is subordinate to provincial authority. Local government plays a significant role in resource management through the process of land-use zoning. Zoning influences where people work, live, and engage in recreation. Communities also build infrastructure that is essential to tourism, such as parks and recreation facilities. Again, overlap between local and senior levels of government can lead to conflicts.

In the mid- to late 1980s, fish farming took on the appearance of a gold rush as many companies wanted to raise salmon in net pens. Most of this activity occurred on the Sunshine Coast, mainly in the Sechelt Inlet. A classic multijurisdictional struggle to manage this new resource developed. Because the net pens were located in salt water, the federal Department of Fisheries and Oceans was involved, but as it was necessary to anchor the net pens to the foreshore, the provincial Ministry of Lands became involved with foreshore leases. At the local government level, the Sunshine Coast Regional District expressed concerns over many resource use conflicts, from pollution to the privatization of the foreshore. These issues were in its backyard, but it lacked the jurisdiction or political power to resolve the conflicts. Eventually all three levels of government were able to cooperate to make decisions on resource uses in Sechelt Inlet.

Political managers are responsible for setting the "rules" for resource development. Governments establish permits for resource development, including rates of harvest and allowable levels of discharge of wastes into the environment. The various ministries listed above do much more than set the rules, however; they promote and encourage resource development through infrastructure projects, laws, and, in some cases, direct subsidization. This often powerful role of government is usually justified in terms of creating employment and gaining taxation revenues. Another way of viewing resources and resource management is from the perspective of economic viability. The way resources have been used to develop the economy of British Columbia is best seen from a historical perspective that situates the province with reference to the rest of Canada and the world.

RESOURCES, DEVELOPMENT, AND THE BC ECONOMY

Staples Theory

The Canadian economic historian Harold Innis used **staples theory** to describe the development of Canada (Watkins 1963; Barnes and Hayter 1997). His theory is based on the exploitation of resources, or staples, in Canada. Its main assumption is the existence of an external demand for these resources. Innis identified five resources in the development of Canada: fish, furs, timber, wheat, and minerals. These five represent both a historical and an east-to-west pattern of development. Minerals, the last resource on the list, are the exception to the directional trend. The first resource was cod, taken off the coast of Newfoundland beginning in the late 1400s. Next, the continental resources were developed from east to west with the fur trade, followed by timber and agriculture. Minerals tended to be discovered sporadically from region to region. Through these resources and the economic activities tied to them, the Canadian economy developed.

The terms *backward, forward,* and *final demand linkages* are used in staples theory to describe the types of economic activity, including implications for employment, associated with each staple. Backward linkage refers to all the conditions necessary to export a resource. The most important backward linkage for any resource is the collection of transportation systems because it can influence so many other economic activities. When the country was first developed, this meant building and running port facilities with warehouses, boat repairs, and all the employment related to loading resources onto ships for export. Over time, ports were connected to canals, railways, and road systems. Backward linkages also include the employment created from building these facilities as well as the construction and manufacturing of rails, boats, trains, trucks, or any of the inputs to export a resource.

Forward linkage is the process of adding value to any resource through further processing or manufacturing prior to export. In British Columbia, there has always been concern about the export of raw logs versus the forward linkage of milling the logs into dimension lumber.

Obviously, if the sawmilling activity occurs here, then more jobs are created in British Columbia. Much higher value added to wood can be gained by manufacturing furniture, doors, or even musical instruments, with consequent greater benefits to employment and to government revenues. Once a resource is tagged for export, a community is usually involved, whether at the port or at the location of resource extraction. Over time, and with the accumulation of backward and forward linkages, some of these communities have become major centres.

Final demand linkages are defined as the demand for production of goods and services for the local or domestic market. In smaller communities, consumer goods may have to be imported. Nevertheless, as the population increases, the ability to reach thresholds for local production also increases.

The accumulation of all these linkages is referred to as the multiplier effect. A pulp mill or mine, for example, locates in a community, adding 500 workers (forward linkage). These workers in turn need housing, food, and many other necessities and luxuries, and so the local economy grows to fulfill this increased demand (final demand linkages). One more pulp mill or mine in the province may be all it takes to stimulate the manufacture of pulp or mine machine components (backward linkages). All this economic activity can mean considerably more than 500 people working. To be kept in mind as well, though, is the main assumption of this theory, that there is an external demand for the resource. Periods of recession and depression show that external demand has not always been sustained. When the mill or mine shuts down, the multiplier effect works in reverse and many more people, apart from those in the mill or mine, face unemployment. When a mill or mine is the only industry, its closure may jeopardize an entire community.

British Columbia's history of dependence on external demand for its resources continues today. The data in Figure 7.2 show that overall imports of goods and services exceed exports for the years shown. British Columbia continues to rely on the export of resource-based goods (energy, metals, and forest products in particular) and the import of higher value-added products (Table 7.1). Staples theory is therefore a useful framework to assess the effects of various forward, backward, and final demand linkages in terms of employment and social benefits. These effects are not necessarily easy to determine, however, because the technologies of resource processing change along with accessibility, competition, markets, and so forth.

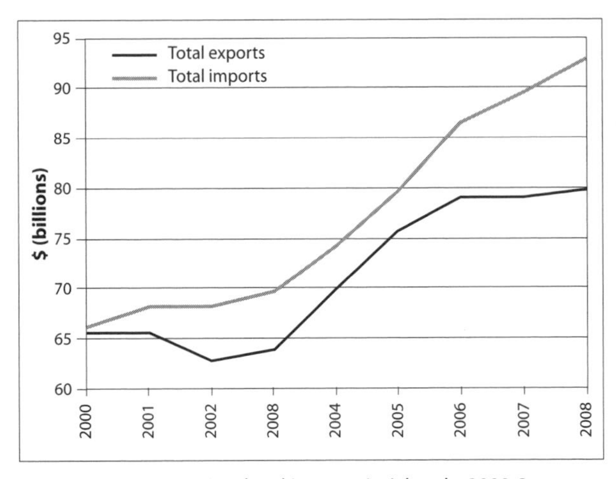

Figure 7.2 BC international and interprovincial trade, 2000-8
Source: BC Stats (2009b).

Table 7.1

British Columbia international merchandise trade, 2009

Merchandise	Exports ($ millions)	Imports ($ millions)
Agricultural products including animal products, prepared foods, beverages, and tobacco	2,335.4	3,998.7
Mineral products including energy, articles of stone, glass, jewellery, and base metals and base metal products	8,579.2	5,754.3
Wood products including pulp and paper	11,692.2	2,412.7
Chemical and plastic products	966.9	4,312.0
Textiles and footwear	320.2	1,867.4
Machinery, electronics, photographic products, medical products, musical instruments, etc.	3,046.0	8,328.3
Vehicles and transportation	695.8	6,452.1
Other (mainly miscellaneous manufactured products)	977.9	3,628.3
Total	28,613.6	36,753.8

Source: Statistics Canada (2009, table 386-002).

Fordism, the Industrial Economy, and the Transition to the Service Economy

One of the most fundamental changes to resource processing occurred early in the twentieth century when Henry Ford originated the assembly line process. Since then the term **Fordism** has been used to define mass production of standardized goods, usually at a centralized assembly plant. Assembly line technology transforms resources into consumer goods. In British Columbia, Fordist methods were employed by the resource industries in the canning of salmon, the concentration of minerals, the manufacture of two-by-fours and other dimension lumber, and the production of pulp and later paper, plywood, and oriented strand board (particle board or OSB).

After the Second World War, a period of prosperity created the greatest ever demand for consumer goods. The principles of Fordism were refined, and an increased demand for the forest and mineral resources of British Columbia resulted in a major expansion of these industries in particular. Investments in manufacturing plants, however, tended to be made in central Canada and other global metropolitan locations. Many single resource communities were created and became dependent on one particular resource.

The 1960s and 1970s saw the rise of oligopoly conditions, in which large corporations with little competition began to control the resource sectors of British Columbia and elsewhere. They employed new scientific management principles that included capital-intensive technologies to produce ever greater amounts of goods from resources, but these new machines and techniques began to replace labour. The old, labour-intensive ways of processing resources, from sawmilling to mineral extraction, were fast disappearing.

The struggle for provincial governments, and to a lesser degree for the federal government, was to encourage – usually by providing the infrastructure – corporate investment in the resource sector to produce more employment. As long as new plants were being opened, the potential multiplier effect of new employment resulted, and greater revenues from resource taxation accrued to government. Once the plant and equipment were in place, however, new capital-intensive technologies eventually reduced the number of workers required. More and more it became the government's role to become involved in establishing a social safety net, along with educational retraining for displaced workers. It should be remembered, though, that up until the early 1970s almost the whole country experienced good economic times with plenty of job opportunities, and governments had money to spend.

Unions were caught in an even greater dilemma. The new technologies often necessitated increased productivity to remain competitive in an industry, making it difficult for unions to oppose the equipment that reduced their rank-and-file members. The trade-off for unions was to ensure better working conditions and wages – for those workers who remained.

Figure 7.3 illustrates the interlocking relationship of various groups within British Columbia; change in one sphere will influence change in the others. For example, the introduction of an automated greenchain (an assembly line of freshly cut dimension lumber) at a sawmill immediately increases the productivity of the mill but leaves many workers seeking (un)Employment Insurance or retraining and fewer rank-and-file members in the Industrial, Wood and Allied Workers of Canada.

British Columbia and the Global Economy

Figure 7.4 illustrates the condition of British Columbia's resource-based and resource-dependent economy, in which our resources achieve only low value-added processing (e.g., pulp, lumber, mineral concentrates). The corporations that own the resource industries in British Columbia reside in heartland areas beyond its borders

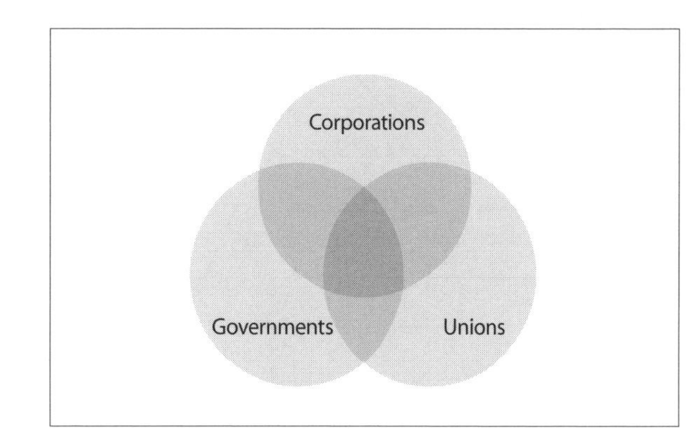

Figure 7.3 Interrelationship of corporations, governments, and unions

such as southern Ontario, United States, Britain, or Japan. A corporation decides where to locate its head office and research facilities and where to develop new products and implement new technologies. The related high paying and often professional jobs are not located in the hinterland. Similarly, the production of high value-added products – mainly consumer goods produced from semiprocessed resources – and the suppliers of transportation, goods, and services are also located outside British Columbia. Moreover, because resource industries within the province are extraprovincially controlled, the profits generated are exported, and the province must import goods, services, and high value-added consumer goods.

Thomas Gunton (1982, 10) uses this model to express concern over encouraging large foreign corporations to invest in the resource sector of the province: "Overall ... a staple region such as B.C. which is dominated by externally controlled firms interested in obtaining a secure supply of resources for externally located operations will not develop strong regional linkages. Instead the economy will develop only lower order processing and service activities tied to the regional market. It will lose out on much of the 'footloose' employment such as head office management, research and development and higher order processing which will be located outside the region simply because the firm is externally controlled." Gunton suggests that British Columbia should develop far more linkages with its resource base.

Barnes et al. (1992, 178-80) confirm these trends and suggest that British Columbia has developed in four unique ways in comparison to the rest of Canada. First, industries employing Fordism were decentralized into the resource frontier. Second, these Fordist industries gained for British Columbia the fewest forward and backward linkages of any province in Canada. Third, the provincial government played the most facilitative role of any government in Canada in providing infrastructure and attracting corporations. Fourth, British Columbia has experienced the largest boom and bust cycles of any region in Canada.

In good economic times – in British Columbia, from the end of the Second World War to the early 1970s – the cyclical changes and their impact on workers were handled by relatively wealthy federal and provincial government programs. Accomplishments during this period included the establishment of community colleges with training programs and a major expansion of the social safety net. The economic engine, not just for British Columbia but also for the western world, was based on the manufacturing industry; and this **industrial economy** led to increased urbanization.

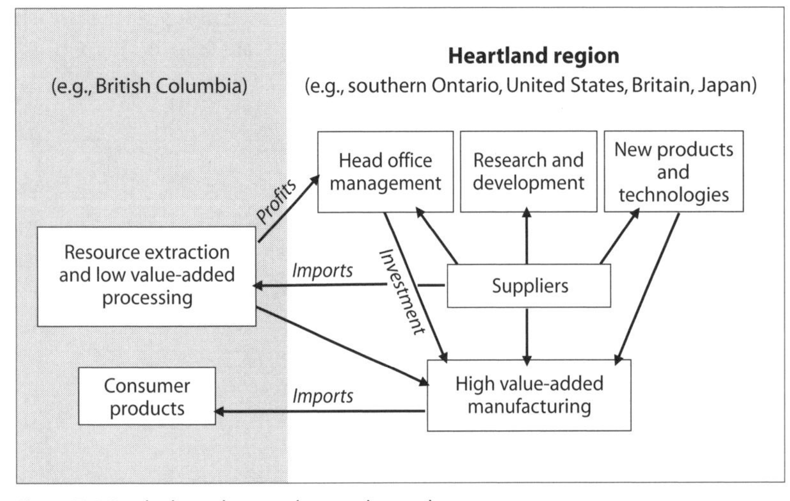

Figure 7.4 Staple dependency and external control
Source: Gunton (1982, 14). Reproduced with permission.

Global conditions changed, and these changes had serious implications for economies that relied on resource development and manufacturing. By the 1970s, the world was moving into **post-Fordism**, wherein the industrial economy was surpassed by the **service economy** in terms of providing employment. Early in the decade, currencies that had been pegged to fixed gold rates became destabilized, and the energy crisis created global unease. Increased pressure for global trade – along with major economic recessions from 1981 to 1986, in the early 1990s, and again in the late 1990s, early 2000s, and 2008-9 – indicated that the traditional economy was not working.

Two trends transformed the production process and caused a major restructuring of resource-based economies such as British Columbia's. Time-space convergence (discussed in Chapter 1), or time-space collapse, refers to the ability of transportation technologies to shrink the world in terms of patterns of movement. The advent of supertankers, jet air-cargo carriers, coaxial cables, satellites, telephones, fax machines, Internet, e-mail, and other technologies has conquered distance, permitting a global scale of organization in the movement of goods and services.

The second trend is the fragmentation of the production process, whereby the phases of a given process are divided and assigned to various locations in the most cost-efficient manner. Two questions apply here: What are the components of any good or service, and where are the cheapest places in the world to produce these components?

Time-space convergence and process fragmentation have brought about the multinational or transnational corporation: "IBM, for example, one of the truly giant multinationals, actually has no single plant outside the United States in which a complete product is manufactured. Each phase of the production process – ranging from management control and raw material production, through simple and more complex component manufacture, to research, design, and final assembly – is usually located in a different place, with the market perhaps in still another country" (Galois and Mabin 1987, 10). Another result of these trends is the international division of labour. Labour is the least mobile global factor, and "cheap" labour regions such as Mexico and China have emerged. Globalization of production has shaken the world order of nations in terms of the production of goods and services.

The global marketplace is in a state of uncertainty, particularly with respect to traditional production processes that produce standardized, one-size-fits-all materials. Post-Fordism brings with it the concept of flexible specialization, by which goods can be produced to fit individual consumer demands. This is often accomplished by contracting out to small firms throughout the world. The term *restructuring* is often heard these days and refers to the ways in which resources and technologies are combined to produce goods and services in the global marketplace. Restructuring has a somewhat negative connotation: when corporations, or whole economies, restructure, it implies major changes, including employment loss and uncertainty for many.

A serious recession from 1981 to 1986 caused many traditional Fordist operations to restructure in British Columbia. MacMillan Bloedel's sawmill in Chemainus on Vancouver Island is a typical example of the impact and conditions of restructuring. This fairly large mill produced dimension lumber mainly for the American housing market. That market was in a seriously depressed state in the early 1980s. The mill closed in 1982, laying off 642 workers, a major blow to the town's economy. A new mill opened in 1985, organized on the principles of flexible specialization. It employed specialized equipment to cut lumber to fit the market demands of Japan and Europe as well as the United States, and required only 145 employees. Meanwhile, the company contracted out a number of the functions that it had previously run itself under the old Fordist system, such as a dry kiln operation (Barnes et al. 1992; Rees and Hayter 1996). For the community of Chemainus, these were tough times marked by high unemployment and outmigration in search of jobs elsewhere. As a response, Chemainus turned to the tourism industry, where it was promoted as the "Little Town That Did." The history of Chemainus was depicted in large murals on the walls of the downtown stores, successfully attracting tourists. New, service-oriented jobs, often at minimum wage, along with contract employment in fields paying much less than before, were part of the restructuring.

Restructuring of the BC economy was not restricted to the primary resource industries. The service sector, whose products are often knowledge and business transactions, has been subject to both new technologies and flexible specialization. One needs only to consider banking and the automated banking machine, or the number of firms using contracted message services. Both large and small firms have experienced layoffs, or what the media have commonly referred to as "downsizing" and "reengineering," as a result of restructuring. Consequently, all communities, whether small resource-based ones or large metropolitan centres such as Vancouver, are affected by these global economic systems (Norcliffe 1994). Table 7.2 compares unemployment rates in British Columbia and Canada. The collapse of Asian economies in the late 1990s, followed by recessions in the early 2000s, clearly influenced BC's economy, which did not recover until metal and energy markets took off after 2005. The more recent recession, which began in 2008, was buffered (especially for the Lower Mainland) by construction and preparation for the 2010 Winter Olympics.

Not all these changes are negative. The twenty-first century has been referred to as the "Pacific century," and Asian countries are becoming the new leaders in industrial growth, development, and investment. The Pacific Rim nations include over 50 percent of the world's population, and British Columbia is well positioned to engage in these markets (Figure 7.5).

BC experience in resource production and in infrastructure and urban development have become potential knowledge-based exports. British Columbia is politically

Table 7.2

Unemployment in British Columbia and Canada, 1999-2009

	British Columbia	Canada
1999	8.3	7.6
2000	7.1	6.8
2001	7.7	7.2
2002	8.5	7.7
2003	8.0	7.6
2004	7.2	7.2
2005	5.9	6.8
2006	4.8	6.3
2007	4.2	6.0
2008	4.1	5.8
2009	6.9	8.3

Source: BC Stats (2010b).

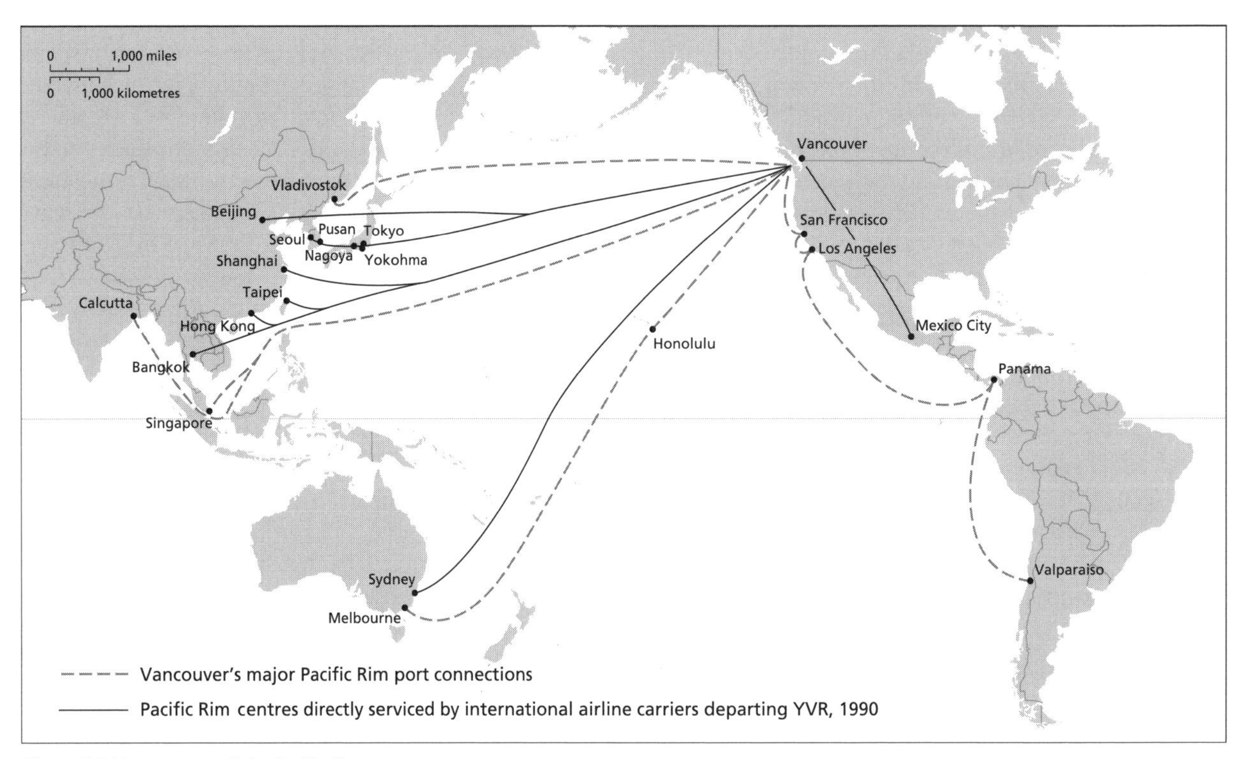

Figure 7.5 Vancouver and the Pacific Rim
Source: Modified from Barnes et al. (1992, 191).

stable and comes complete with modern banking and financial institutions, excellent educational facilities, and a highly educated population: attributes that attract immigration and investment, especially in the service industries. As well, this province still has many resource-based commodities required throughout the world.

The economic sector of British Columbia is experiencing a spatial differentiation, or two geographies. The service economy mentioned above, with its new knowledge- and information-based employment, is located mainly in a rather small, triangular geographic region encompassing the area from Victoria to Nanaimo and, most important, the Greater Vancouver area. This region contains over 70 percent of the provincial population and is rapidly growing. Approximately 2.6 million people lived in the Lower Mainland alone in 2010, and the figure is expected to surpass 4 million by 2024 (BC Stats 2010f). Ley, Hiebert, and Pratt (1992), Hutton (1997), Binkley (1997), Edgington (2004), and others suggest that the growth is linked to Vancouver's transformation into a cosmopolitan city that is becoming unhinged from, or less dependent on, staples from its provincial hinterland.

The growth of the Lower Mainland is also symbolic of increasing economic opportunity and development of manufacturing industries, tourism, the movie industry, the high-tech industry, and many other services. Table 7.3 is a profile of employment within British Columbia as well as the Vancouver Census Metropolitan Area (Greater Vancouver Regional District, or GVRD). Vancouver has nearly 55 percent of all jobs within the province, but within the goods-producing sector, not surprisingly, Vancouver falls below the BC average, particularly with respect to the primary industry categories. It is in the

Table 7.3

Employment in British Columbia and Greater Vancouver Regional District, 2009

	British Columbia		GVRD		
	Labour force ('000s)	% in sector	Labour force ('000s)	% in sector	% of province
All industries	2,259.4	100.0	1,234.2	100.0	54.6
Goods-producing sector	446.5	19.8	209.9	17.0	9.3
Agriculture	34.3	1.5	8.1	0.7	0.4
Forestry, fishing, mining, oil, and gas	40.3	1.8	5.8	0.5	0.3
Utilities	12.7	0.6	7.0	0.6	0.3
Construction	195.3	8.6	97.2	7.9	4.3
Manufacturing	163.8	7.2	91.8	7.4	4.1
Service-producing sector	1,812.9	80.2	1,024.2	83.0	45.3
Trade	369.1	16.3	199.8	16.2	8.8
Transportation and warehousing	115.2	5.1	66.8	5.4	3.0
Finance, insurance, real estate, and leasing	143.1	6.3	91.9	7.4	4.1
Professional, scientific, and technical services	169.3	7.5	117.4	9.5	5.2
Business, building, and other support	93.9	4.2	56.6	4.6	2.5
Educational services	162.3	7.2	96.9	7.9	4.3
Health care and social assistance	260.3	11.5	115.1	9.3	5.1
Information, culture, and recreation	119.4	5.3	74.4	6.0	3.3
Accommodation and food services	172.1	7.6	86.4	7.0	3.8
Other services	102.0	4.5	53.3	4.3	2.4
Public administration	106.3	4.7	48.6	3.9	2.2

Source: BC Stats (2010c).

service-producing sector that Vancouver makes most of its gains. These statistics do not indicate employment categories such as the film, television, and high-tech industries, most of which are located in the Vancouver area.

The high-tech industry is made up of both manufacturing (e.g., computer components and wireless technologies) and services (e.g., engineering services, digital media, and medical services). Table 7.4 indicates changes in revenues and employment for the years 2000 to 2007. Although employment in this sector has faced some setbacks, overall revenues have increased, mainly on the service side. The industry is still fairly small within Canada, and British Columbia ranks fourth behind Ontario, Quebec, and Alberta. In 2007 there were over 8,500 establishments in British Columbia and, perhaps not surprisingly, two-thirds were located in the Lower Mainland.

The film and television production industry is both a national and an international industry. Vancouver has emerged as a major production destination within Canada because of its film studios, expertise, and favourable location (particularly being in the same time zone as and a two-hour flight from Hollywood). British Columbia production fluctuates in terms of the number of productions and their value to the province (Table 7.5). In this price-sensitive industry, as the Canadian dollar rises in relation to the American dollar, there is less incentive for foreign productions to come to Canada. As well, because there is intense competition between provinces to attract this industry, tax incentives are offered. As of 2007-9, British Columbia was in second place, with a 31 percent share of Canadian film and television productions. Ontario led the way with a 37 percent share (Canadian Film and Television Production Association 2009).

Much of the rest of the province forms the second geography, where there is a greater dependency on resource extraction and processing. For example, forestry-related employment fluctuates in relation to a number of factors: the price of lumber, pulp, and paper; the value of the Canadian dollar; softwood lumber agreements; the cost of labour and other inputs of production and competition from foreign producers; and the efficiency of plant and equipment. As Table 7.6 shows, the most recent recession has had a devastating impact on all aspects of this industry and, as a consequence the communities and regions that rely on forestry. Table 7.7 gives the average unemployment rates for 2009 for the economic regions of the province as well as selected communities. Forestry dependent regions such as the Cariboo have been hit hard by mill shutdowns and closures.

Table 7.5

Film and television production, 2004-9

	2004	2005	2006	2007	2008	2009
Total	194	211	230	202	260	239
Domestic	116	118	144	138	174	155
Foreign	78	93	86	64	86	84
$ spent in BC (millions)	801.20	1,233.67	1,227.87	943.34	1,206.77	1,313.45

Source: BC Film Commission (2009).

Table 7.4

High-tech manufacturing and services industries, revenues and employment, 2000-7

	2000	2001	2002	2003	2004	2005	2006	2007
Total revenues (millions)	11,782	11,471	12,245	13,323	14,001	15,404	16,706	18,103
Manufacturing	2,819	2,524	2,224	2,464	2,548	2,404	2,741	2,622
Services	8,963	8,947	10,021	10,859	11,453	13,000	13,965	15,480
Total employment	67,070	71,580	67,110	65,950	68,010	71,140	77,440	81,140
Manufacturing	15,040	15,180	13,830	12,590	12,160	13,280	14,480	14,610
Services	52,030	56,410	53,280	53,360	55,850	57,860	62,960	66,530

Source: BC Ministry of Technology, Trade and Economic Development and the Ministry of Advanced Education and Labour Market Development (2008).

Table 7.6

Total employment in forestry and forestry-related manufacturing industries, 1999-2009

	Forestry ('000s)	Manufacturing ('000s)	
		Wood	Pulp and paper
1999	30.0	42.4	42.4
2000	35.5	45.7	45.7
2001	24.7	48.9	48.9
2002	25.3	43.8	43.8
2003	27.7	49.0	49.0
2004	21.5	46.9	46.9
2005	21.6	45.8	45.8
2006	21.7	44.8	44.8
2007	24.3	34.3	34.3
2008	17.4	29.3	29.3
2009	13.9	27.2	27.2

Source: BC Stats (2010b).

Table 7.7

Unemployment rates by economic region and selected communities, 2009

Region and communities	Unemployment (%)
North Coast/Nechako	10.4
Northeast	6.8
Dawson Creek	9.6
Cariboo	12.0
Prince George	11.5
Thompson/Okanagan	8.7
Kamloops	8.3
Vernon	6.8
Kelowna	8.7
Kootenay	8.7
Vancouver Island/Coast	7.2
Nanaimo	7.6
Victoria	6.4
Mainland/Southwest	7.2
Abbotsford	7.8
Vancouver	7.0

Source: BC Stats (2010a).

The BC economy thus relies on a combination of the older Fordist production systems and the new post-Fordist ways of doing business. The economic structure is thus much more complex than it used to be, and there are many more variables responsible for economic growth. The development of human resources and knowledge, British Columbia's favourable location for Asia-Pacific trade, a greater involvement in the tourism industry, and the encouragement of new investment locally and from abroad are among the key components. Globalization has also changed the role of government as manager of resources: "Governments, facing pressures to privatize and deregulate and liberalize, increasingly find themselves with diminished control over the processes and strategies of production, except insofar as they can represent the vested interests of their corporations in the development of international rules of corporate behavior and/or privilege" (Wilkinson 1997, 131). Unfortunately, economic and population growth is not evenly distributed throughout British Columbia.

POPULATION DYNAMICS AND SOME ECONOMIC IMPLICATIONS

Population change in British Columbia (discussed in Chapter 1) is a function of natural increase (births minus deaths) and net migration (immigration minus emigration). There are many influences on each of these factors.

Birth rates (fertility) can vary because of religious, cultural, and economic conditions. For example, the "baby boom" after the Second World War was a reversal in the downward trend for birth rates that lasted for approximately twenty years. Today, the fertility rate is exceptionally low for a variety of reasons: both parents working and establishing careers, increased education, the high costs of raising children, availability of birth control methods and abortion clinics, and the fear of bringing up children in this turbulent world. On the other hand, death rates have fallen for a host of reasons largely due to improved health conditions (new and better facilities and medicines) and knowledge about health. These health improvements have conquered major diseases, which has resulted in far fewer young people dying, decreased infant mortality, and an increased life expectancy (Figure 7.6).

Immigration is a function of federal government policy, and the policy and practice in Canada was to encourage people from Europe and the United States to immigrate. Asians and other nonwhites were discouraged and

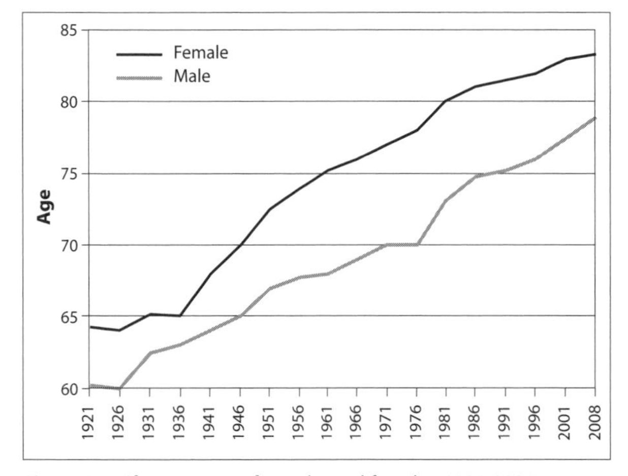

Figure 7.6 Life expectancy for males and females, 1921-2008
Source: BC Stats (2005c; 2009c).

Table 7.8

Immigration by source countries, 2008

World region	Immigrants	Source country	Immigrants
Asia	31,890	China	9,914
Europe	5,736	India	5,484
North and Central America	3,341	Philippines	4,996
Africa	1,101	South Korea	2,902
South America	728	United States	2,657
Australasia	588	England	2,316
Oceania and Islands	311	Taiwan	2,085
Caribbean	213	Iran	1,281
Not stated	84	Japan	601
Total	43,992	Singapore	592

Source: BC Stats (2009a).

Table 7.9

Interprovincial net migration, 1980-2009

Year	Net migration	Year	Net migration	Year	Net migration	Year	Net migration
1980	41,067	1988	21,614	1996	22,025	2004	7,786
1981	39,008	1989	29,421	1997	9,880	2005	7,212
1982	7,046	1990	39,984	1998	-10,029	2006	12,799
1983	-611	1991	34,108	1999	-14,484	2007	16,776
1984	6,424	1992	38,004	2000	-14,610	2008	8,379
1985	1,967	1993	40,099	2001	-8,286	2009	7,499
1986	-3,727	1994	37,871	2002	-8,556		
1987	9,493	1995	29,291	2003	-4,591		

Sources: Statistics Canada (2004), table 051-0018; BC Stats (2009d, 2010d).

discriminated against with barriers such as the Chinese Immigration Act in 1923 (see Chapter 6). A nonracial, points-based system was introduced in 1967, which led to a much more multicultural Canadian society, and in the "late 1980s the Canadian government adopted a new immigration policy that led to the admission of much higher numbers of immigrants" (Hiebert 2000, 26). Today immigration to British Columbia is heavily weighted toward Asian migration, as can be observed in Table 7.8. The table also lists the top ten countries from which people migrated to British Columbia in 2008.

Interprovincial migration is largely a function of economic conditions, whereby either the "push" of job loss and generally poor economic prospects or, conversely, the "pull" of good economic times lures people into or out of the province. Table 7.9 is a general indicator of financial activity in British Columbia. A recession in the mid-1980s and a similar occurrence at the end of the 1990s and early 2000s resulted in out-migration to provinces such as Alberta, where the oil and gas industry continues to boom, attracting considerable investment and providing employment opportunities. The rising price for commodities such as metals, coal, oil, and natural gas in the mid-2000s, along with construction for the 2010 Olympics, have resulted in positive net migration.

From past trends in natural increase plus net migration, the future population of British Columbia can be projected. Figures 7.7 to 7.9 are age-sex pyramids giving pictures of the population at 1971, 2003, and 2021. These pyramids can be a powerful predictive tool, giving some sense of future population characteristics and people's demands. For example, in British Columbia there were many more young people than old people in 1971, but by 2003 the majority of people (the baby boom generation) were considerably older. This trend toward an aging population is going to continue into 2021. An aging population has many implications for the province, from the health care system through to housing and tourism demands. This trend also implies that many people will be retiring, putting increased demands on pension

Figure 7.7 BC age-sex pyramid, 1971
Source: Data from BC Stats (2005d).

Figure 7.8 BC age-sex pyramid, 2003
Source: Data from BC Stats (2005d).

Figure 7.9 BC age-sex pyramid, 2021
Source: Data from BC Stats (2005a).

funds, but opening up many positions for the younger generation.

The regional population can also be projected. Table 7.10 begins with data from the 2006 census and forecasts the increase in population for each of the eight economic development regions (Figure 7.10) in the province to 2031. Overall, a population increase of approximately one and a half million is expected; however, this population is not evenly distributed, with regions such as the Lower Mainland increasing more rapidly. Other regions – the Kootenays and northern regions – are projected to have more modest growth. This difference reflects the dual economies in the province: the resource-dependent hinterland is expected to grow more slowly than the core, or heartland, which is much more influenced by the service economy.

SUMMARY

The early settlement and development of British Columbia was very much tied to resources such as gold and other minerals, forests, fish, energy, and some agriculture; all produced commodities mainly for an external market. The values and uses of these resources have changed over time and so have production processes, corporations, governments, and markets. Categorizing resources as renewable or nonrenewable, or using an ecosystem model of resource interaction, recognizes potential limitations and conflicts and the need for resource management. The three levels of government, often with overlapping responsibilities, do not have an easy task as they weigh the many political, economic, and environmental issues associated with resource exploitation.

Staples theory helps to unravel the many economic links between resources and related development. The multiplier effect of increased employment opportunities results from gaining more and more linkages. This occurs provided that the main assumption, the existence of an external demand, holds true. If this fails, the multiplier effect may work in reverse.

The global economic system is not static, nor are the technologies used in transforming resources. Fordism was adopted by industrial firms in British Columbia, where employment and development expanded well into the 1970s. Changing economic conditions from the 1970s on saw the creation of multinational corporations that were able to organize the production of goods and services on a global scale, with a global division of labour. Post-Fordism is marked by uncertainty in resource production, restructuring for many firms, and a shift to a service-based economy. New opportunities have arisen also, particularly with respect to the economies of the Pacific Rim.

British Columbia still has resources, and they are especially important to the employment and economy of the interior and the north end of Vancouver Island. The core area of the province, Victoria/Nanaimo/Vancouver, along with the major service centres in the interior (e.g., Kelowna, Kamloops, and Prince George), are capturing new employment opportunities based on knowledge, information, and amenity migration. British Columbia

Table 7.10

Projected population by economic development region, 2006-31

Year	Vancouver Island/ coast	Mainland/ southwest	Thompson-Okanagan	Kootenay	Cariboo	North coast	Nechako	Northeast	British Columbia total
2006	744,686	2,530,432	503,018	143,406	157,632	58,783	39,996	65,627	4,243,580
2011	790,467	2,756,471	547,321	150,838	159,947	57,953	40,216	69,404	4,572,617
2016	834,209	2,987,796	579,855	153,562	161,485	57,808	41,124	73,646	4,889,485
2021	879,669	3,221,659	612,238	157,108	164,302	57,955	41,739	77,711	5,212,381
2031	960,206	3,642,682	669,020	163,517	170,828	58,314	42,572	85,341	5,792,480

Source: BC Stats (2010e).

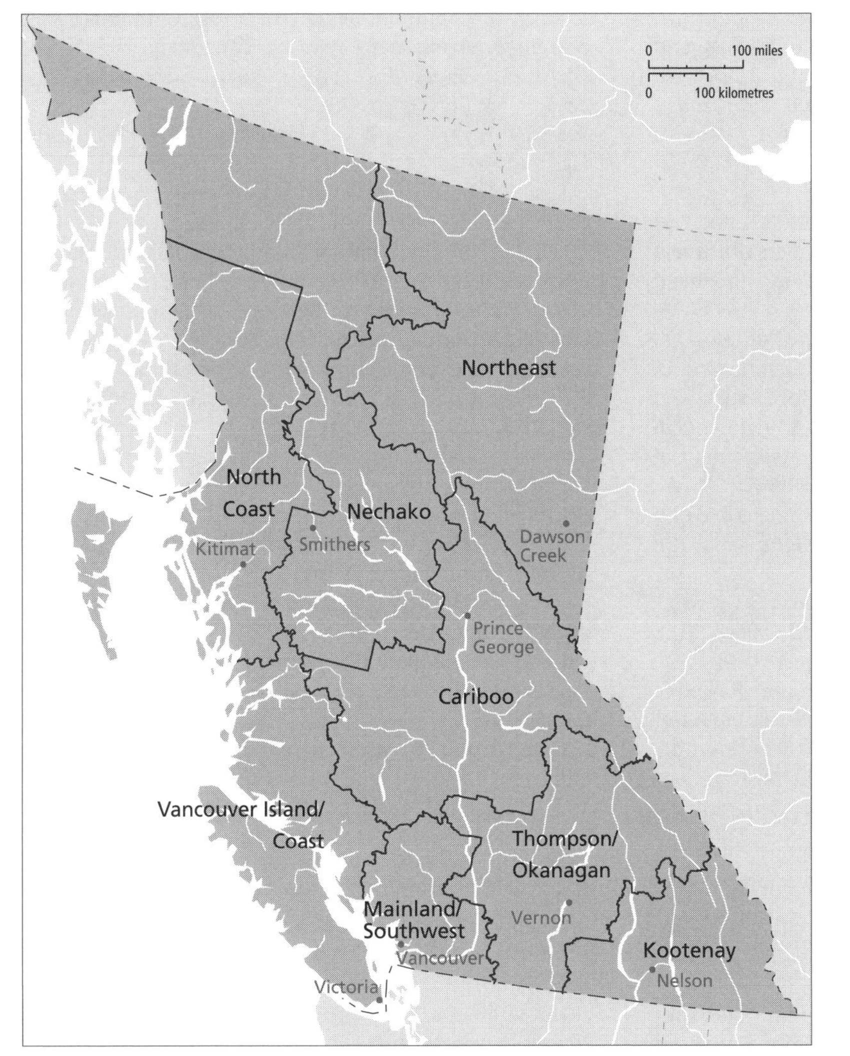

Figure 7.10 Economic Development Regions of British Columbia
Source: Data from BC Stats (2005b)

thus has essentially two economies, producing even greater challenges for the managers of our resources.

The past economic, political, and social trends can be employed to project population growth. Birth rates, death rates, immigration, and emigration (at the international and interprovincial scales) can be taken into account to calculate future population profiles by age and sex. British Columbia's population is aging, and this trend has many economic and political implications. As well, population projections can be applied to the various regions of the province. The uneven distribution of population presents challenges, particularly for regions dependent on the export of resources.

REFERENCES

Barnes, T.J., and R. Hayter, eds. 1997. *Troubles in the Rainforest: British Columbia's Forest Economy in Transition.* Canadian Western Geographical Series vol. 33. Victoria: Western Geographical Press.

Barnes, T.J., D.W. Edgington, K.G. Denike, and T.G. McGee. 1992. "Vancouver, the Province, and the Pacific Rim." In *Vancouver and Its Region*, ed. G. Wynn and T. Oke, 171-99. Vancouver: UBC Press.

Binkley, C.S. 1997. "A Crossroad in the Forest: The Path to a Sustainable Forest Sector in British Columbia." In *Troubles in the Rainforest: British Columbia's Forest Economy in Transition*, ed. T.J. Barnes and R. Hayter, 15-35. Canadian Western Geographical Series vol. 33. Victoria: Western Geographical Press.

Edgington, D.W. 2004. "British Columbia and Its Regional Economies: An Overview of Research Issues Presentation." *Canadian Journal of Regional Science* 27 (3): 303-16.

Galois, R.M., and A. Mabin. 1987. "Canada, the United States, and the World-System: The Metropolis-Hinterland Paradox." In *Heartland and Hinterland: A Regional Geography of Canada*, 2nd ed., ed. L.D. McCann, 39-67. Scarborough, ON: Prentice-Hall.

Gunton, T. 1982. *Resources, Regional Development and Provincial Policy: A Case Study of British Columbia.* No. 7. Ottawa: Canadian Centre for Policy Alternatives.

Heibert, D. 2000. "Immigration and the Changing Canadian City." *Canadian Geographer* 44, 1: 25-43.

Hutton, T.A. 1997. "Vancouver as a Control Centre for British Columbia's Resource Hinterland: Aspects of Linkage and Divergence in a Provincial Staple Economy." In *Troubles in the Rainforest: British Columbia's Forest Economy in Transition*, ed. T.J. Barnes and R. Hayter, 233-61. Canadian Western Geographical Series vol. 33. Victoria: Western Geographical Press.

Ley, D., D. Hiebert, and G. Pratt. 1992. "Time to Grow Up? From Urban Village to World City, 1966-91." In *Vancouver and Its Region*, ed. G. Wynn and T. Oke, 234-66. Vancouver: UBC Press.

McManus, P. 2000. "Sustainable Development." In *The Dictionary of Human Geography*, 4th ed., ed. R.J. Johnston, D. Gregory, G. Pratt, and M. Watts, 812-16. Oxford: Blackwell.

Norcliffe, G. 1994. "Regional Labour Market Adjustments in a Period of Structural Transformation: An Assessment of the Canadian Case." *Canadian Geographer* 38 (1): 2-17.

Rees, K., and R. Hayter. 1996. "Enterprise Strategies in Wood Manufacturing, Vancouver." *Canadian Geographer* 40 (3): 203-19.

Watkins, M.H. 1963. "A Staple Theory of Economic Growth." *Canadian Journal of Economics and Political Science* 29 (2): 141-58.

Wilkinson, B.W. 1997. "Globalization of Canada's Resource Sector: An Innisian Perspective." In *Troubles in the Rainforest: British Columbia's Forest Economy in Transition*, ed. T.J. Barnes and R. Hayter, 131-47. Canadian Western Geographical Series vol. 33. Victoria: Western Geographical Press.

INTERNET

BC Film Commission. 2009. "Production Statistics." www.bcfilmcommission.com/database/rte/files/2009%20Stats%20Package.pdf.

BC Ministry of Technology, Trade and Economic Development and the Ministry of Advanced Education and Labour Market Development. 2008. *Profile of the British Columbia High Technology Sector – 2008 Edition.* www.bcstats.gov.bc.ca/data/bus_stat/busind/hi_tech/HTP2008.pdf.

BC Stats. 2005a. "British Columbia Forecast – 04/12." Table 7, Male population by age; Table 8, Female population by age. www.bcstats.gov.bc.ca/.

–. 2005b. "Development Regions of British Columbia." Development region maps. www.bcstats.gov.bc.ca/.

–. 2005c. "Life Expectancy at Age 0 – Male and Female, 1921-2004." www.bcstats.gov.bc.ca/.

–. 2005d. "Provincial and Territorial Populations: 1971-2004 (age and Sex, July 1)." www.bcstats.gov.bc.ca/.

–. 2009a. "BC Immigrant Landings by Source." www.bcstats.gov.bc.ca/data/pop/mig/imm08t4a.pdf.

–. 2009b. "BC International and Interprovincial Trade Flows." www.bcstats.gov.bc.ca/data/bus_stat/bcea/tradec.asp.

–. 2009c. "Life Expectancy." www.bcstats.gov.bc.ca/data/pop/pop/dynamic/lifeexpectancy.asp.

–. 2009d. "Quarterly Components of Population Change for British Columbia." www.bcstats.gov.bc.ca/pubs/mig/mg91093.pdf.

–. 2010a. "British Columbia Unemployment Rates, 2009." www.bcstats.gov.bc.ca/data/lss/lfs/ur09.pdf.

–. 2010b. "Earning and Employment Trends: December 2009." www.bcstats.gov.bc.ca/pubs/eet/eetdata.pdf.

–. 2010c. "Employment by Industry for BC, Development Regions, and Metro Areas: Annual Averages." www.bcstats.gov.bc.ca/data/dd/handout/EMPREGN.pdf.

–. 2010d. "Net Population Movement For British Columbia: Jan. 2009 to Dec. 2009." www.bcstats.gov.bc.ca/pubs/mig/mf2009.pdf.

–. 2010e. "Population projections." www.bcstats.gov.bc.ca/data/pop/pop/dynamic/PopulationStatistics/.

–. 2010f. "Regional Populations Estimates and Projections." www.bcstats.gov.bc.ca/data/pop/pop/dynamic/PopulationStatistics/index.asp.

Canadian Film and Television Production Association. 2009. *An Economic Report on the Canadian Film and Television Production Industry.* www.cftpa.ca/newsroom/pdf/profile/profile2009-en.pdf.

Statistics Canada. 2004. "Interprovincial in-, out-, and net-Migrants, by Province or Territory, annual (persons)." CANSIM Table 051-0018. www.statcan.gc.ca/.

–. 2009. Interprovincial and international trade flows at producer prices, annual (dollars).CANSIM Table 386-0002. www.statcan.gc.ca/.

Forestry

8

A Dominant Export Industry in Difficult Times

Commercial forestry has a long history in British Columbia, although there are major differences between the coastal and interior forest industries. Dominant tree species, their physical characteristics, and accessibility to the resource and to the market all differ. The coastal forest industry predominated, in terms of the volume of wood cut for export, until changes in corporate organization and technology in the 1960s allowed the interior forests to fulfill growing demand. Technological changes have always had a major impact on all aspects of forestry, influencing how the resource is harvested, at what rate, which forests are cut, where mills are located, and how wood is processed and transported, as well as employment, taxation, and the economic well-being of the various regions and the province as a whole.

The provincial government is the manager of this resource and as such has implemented a number of forms of **tenure,** or legal contract, to allow private corporations to harvest forest lands. Types of tenure, tree species, location, transportation, and end use all become part of the complex equation that determines the revenues and costs of timber harvesting. With the provincial government in control of 95 percent of the forest land in the province, these revenues have been significant to the tax base of British Columbia.

Harvesting and managing the forests has also brought a series of concerns and issues. Vicky Husband (1995) of the Sierra Legal Defence Fund notes that "it took 85 years to cut the first half of what we have cut in British Columbia, and only 15 years for the last half." The rate of harvesting raises the question of whether there will be forests for the future, while harvesting techniques raise issues about the environment and other values of our forests. Simply put, should trees be viewed as fibre only? The various types of tenure provoke concerns about the role and domination of large corporations, about how to attain higher value-added for timber resources, and about privatization. Bitter trade wars with the United States, British Columbia's major importer of softwood lumber, also put pressure on the traditional forms of tenure and raise the question of how "free" free trade agreements are. Another challenge has been forest fires and pests such as the mountain pine beetle, which is demolishing the forests of the interior and potentially adding greenhouse gases to the environment. Meanwhile, new treaty negotiations may result in First Nations having a larger share of timber and more of a voice in management for some regions of British Columbia. However, regardless of whether large corporations, First Nations, or communities have access to forests, there needs to be an adequate external demand and world market price for the industry to operate.

COAST VERSUS INTERIOR: A BRIEF HISTORY

The distinction between the coast and the interior is based largely on physiographic and climatic differences, which in turn result in different tree species domination. The coastal region encompasses the whole strip from the Coast Mountains west to include Vancouver Island and Haida Gwaii (Figure 8.1). As Chapter 2 discussed, this region is characterized by wet, mild winters and more moderate summer weather than in the interior. The dominant tree species is western hemlock, although large stands of western red cedar and Douglas fir (below 51° latitude) are

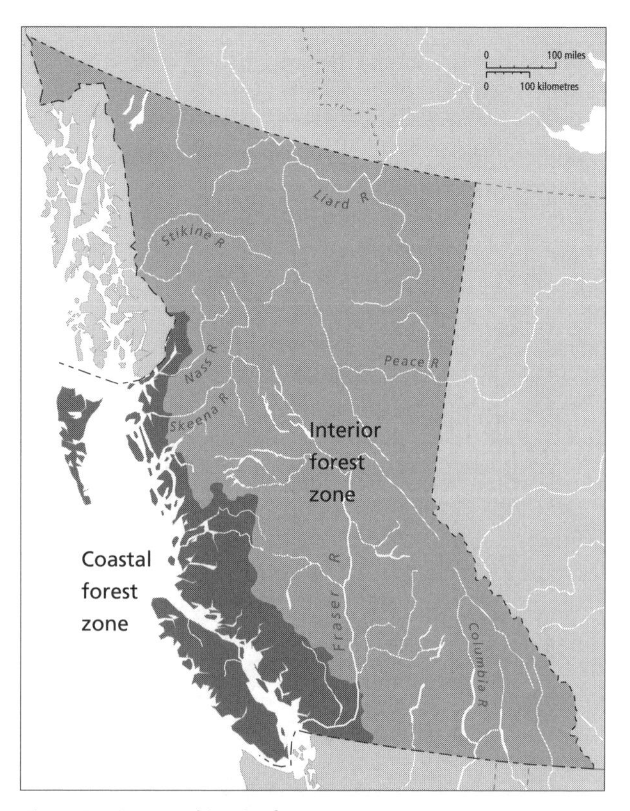

Figure 8.1 Coast and interior forest zones

also typical. The coastal trees grow to an enormous size, yielding a huge volume of wood per hectare.

In the interior, the climate is more extreme in both temperature and precipitation. Hemlock, cedar, and fir grow where the conditions are appropriate, but the dominant tree species are lodgepole pine and spruce. These trees are not nearly as large as those on the coast and they grow farther apart, so the volume of wood per hectare is considerably less. That this region is seven times larger than the coastal forest region is a compensating factor.

Logging for commercial purposes dates back to the 1820s and '30s, mainly for spars for sailing ships. With the California gold rush of 1849, the industry expanded into lumber production and trade began with San Francisco, Hawaii, South America, and China (Hayter and Galois 1991; Hayter 2000). Major lumber production for export accelerated in the 1860s as a number of sawmills were established on southern Vancouver Island and around Burrard Inlet on the mainland. The great advantage of the coastal region, besides very large trees, was its proximity to the ocean for transportation. Logs were dragged to the tideline and floated to sawmills, and the lumber was loaded onto ships and exported to foreign markets. With this advantage, and the growing demand for forest products, the coastal forest industry became dominant.

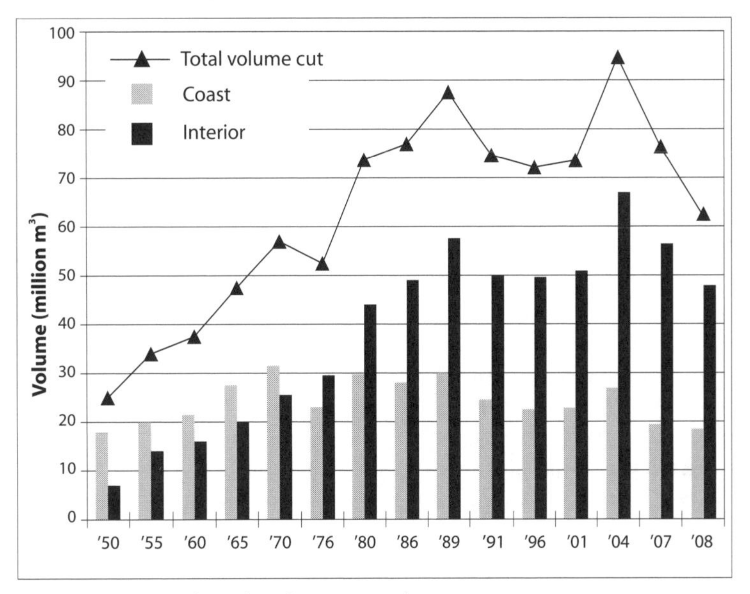

Figure 8.2 Volume of wood cut from coast and interior, 1950-2008
Sources: Barker (1977); Council of Forest Industries (2001); National Forestry Database Program (2004a, table 5.1; 2010); BC Ministry of Finance (2009).

The early commercial forest industry in the interior served a largely local demand. As the gold rush led prospectors into the Cariboo region in the 1860s, for example, there was a demand for wood for heating and cooking fuel as well as for housing and commercial construction, including pit props, sluices, and flumes. Once the gold claims were exhausted, many communities were abandoned and the demand for forest products ceased. Over time, with the discovery of silver, coal, and other minerals, new mines sprang up in various locations throughout the interior of British Columbia, sparking renewed commercial forest activity.

The completion of the Canadian Pacific Railway in 1886 provided a more stable, expanded market in the Prairies and central Canada for the interior mills located on the railway line in places such as Lytton, Kamloops, and Revelstoke. Later railway developments saw the Kettle Valley Line, Grand Trunk Pacific (now CNR), and Pacific Great Eastern (later BCR and now CN) make inroads into different regions of British Columbia and slowly open up forestry activities. Figure 8.2 shows the difference between the coast and interior volumes of wood cut. Transportation developments were important to the interior forest industry, but it took a host of other changes before the interior began to outproduce the coast.

CHANGING TECHNOLOGIES IN A CHANGING INDUSTRY

On the coast, transporting huge logs to the sawmills was the major problem initially. The first solution was to have teams of oxen drag the logs over skid roads to the tideline. There were sixteen to twenty oxen per team, and the skid roads were made of small logs, often greased with fish oil to reduce friction, and placed on a cleared stretch of ground leading to the ocean. The whole logging and transporting process was labour intensive and slow moving. This produced a linear pattern of logging, because it was more efficient to move farther along the coastline than to move inland. Mills such as Sue Moody's, located on the shores of present-day North Vancouver, utilized the logs of Howe Sound and the Sunshine Coast by the 1890s. It was much easier to tow logs by water to sawmills fifteen to twenty kilometres away than to drag them over skid roads for a kilometre or more.

Different species of trees have different end uses and markets, although there is certainly some overlap. Western red cedar has a very straight grain; it is light and water resistant, making it ideal for roofing and siding. The demand for cedar shakes and shingles increased in the 1890s and led to the introduction of two other means of transportation: horses and flumes. The enormous cedar trees could be cut up and split into four-foot (1.2 metre) shingle and shake bolts in the bush. These much smaller sizes could then be stacked on sleds and, because cedar is relatively light, horses could pull the loads. Horses are considerably faster than oxen, and more manageable, which reduced the time of a round trip and pushed the linear pattern farther inland.

Flumes produced a new pattern of logging. These wooden troughs supported on trestles sloped gradually from the mountain side to the ocean, giving access to cedar stands farther inland. Water from creeks was diverted down the wooden flumes, and in this way the shake bolts were floated down to the ocean, where they were gathered for the shake-and-shingle mills.

By the early twentieth century, mechanical power in the form of the steam engine began to replace animals and flumes, gaining new efficiencies in harvesting and producing another linear pattern of logging. Logging railways were constructed throughout Vancouver Island and the southern portion of the mainland, often using river and stream valleys. Railways were significant in penetrating considerable distances inland from the coast.

Another important innovation was the steam donkey, or donkey engine. This was a powerful, steam-driven winch that used cables attached to pulleys at the top of a spar tree. Spar trees were created by cutting off all the branches and the top of a tree located next to the rail line. Pulleys were then attached to this tall pole and cables run from the donkey engine to the fallen timbers. This system could haul logs from a wide perimeter to a central location, where it could then lift them on to railcars.

Through all these changes in transportation technology, the common element was to get the logs to water. On the coast, they were usually bundled into log booms (sometimes called rafts) in the ocean and floated to the mill. Also, particularly in the interior, rivers and lakes were used to transport logs and became an important factor in the location of mills.

The linear pattern of logging was largely broken with the introduction of the combustion engine. Bulldozers could use switchback roads to gain access to many areas that the railways could not, and few forests were so isolated that they could not be reached by road. Trucking became the main means of transporting the forest resource by the 1940s, and truck logging is still with us today.

The combustion engine was used in other ways to gain efficiencies in logging. The double-bladed axe and Swede saw gave way to the chainsaw. The donkey engine was replaced by the mechanical spar pole, which could be driven to the logging site. Hydraulics hoisted the telescoping steel pole into place, and a combustion engine was used to winch the logs. In the interior, where logs are smaller in diameter, the feller-buncher became the tool of choice by the 1980s. This machine, with metal tracks, cuts the tree, limbs it, and stacks it onto a logging truck – all with one operator.

As the most accessible forests of the coast were depleted, it became necessary to go farther and farther afield and higher up the mountain sides to supply the bulk of mills located on the southern coast of British Columbia. New methods of booming logs were required. The simple rafts of the old days were not sufficient to withstand the rough weather and ocean conditions that were common in regions such as Haida Gwaii. The Davis raft used a fairly successful bundling technique to prevent raft break-up, but it was replaced by a much hardier bundle boom and self-loading and self-dumping barges, which were faster and posed even less possibility of log loss (Hardwick 1960, 4).

In areas too remote for road-based logging or where it is desirable to avoid building roads, helicopter logging can transport the forest resource to a central location. To employ helicopters in such work processes is expensive, so the value of the species being harvested has to be high.

Technology has changed throughout the entire industry. The mills themselves have been transformed from labour-intensive to capital-intensive production plants, employing fewer and fewer workers while increasing production. Initially, lumber was cut by hand, or if a stream were nearby a water wheel would be used. The steam engine transformed many of these processes, and by the beginning of the First World War, electric motors ran the sawmill blades. All aspects of milling – the booming grounds, debarking logs, sawmilling, grading, and the greenchain (the assembly line of freshly cut lumber) – underwent automation. Hydraulic barkers, new bandsaws with laser directional beams, and the automated greenchain had restructured sawmilling by the 1980s.

Lumber has always been the main end-use product for the forest industry of British Columbia, but shakes and shingles are important products as well. In the early 1900s, the technology to turn wood fibre into pulp and paper economically resulted in the location of pulp mills in coastal British Columbia to fulfill a global demand. These early locations expanded production, and by 1931 Powell River and Ocean Falls were the major producers of pulp (Figure 8.3). The existing mills continued to expand and by 1951 new coastal pulp mills began to appear in locations such as Prince Rupert, Duncan Bay (Campbell River), Port Alberni, Harmac (Nanaimo), and Victoria. The locations were often chosen for their proximity to sawmills; their useful wood by-products could be used

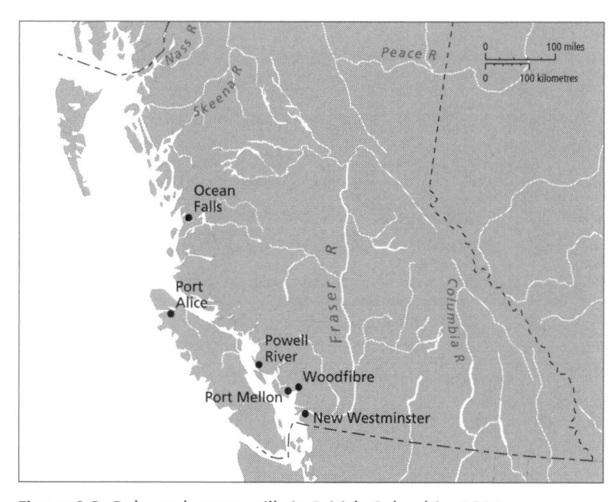

Figure 8.3 Pulp-and-paper mills in British Columbia, 1931
Source: Modified from Farley (1979), 61.

in the production of pulp. As global demand for virtually all wood products escalated by the 1960s, new and more efficient technologies of pulp-and paper production turned to the largely untouched forests of the interior.

Multinational corporations, often foreign, were encouraged to invest in the forest industry – and they did. These corporations believed in the consolidation and integration of all components of forestry. Weyerhaeuser, for example, an American company, established a new, state-of-the-art pulp mill in Kamloops in the 1960s, but its operations were much broader than the production of pulp. The company purchased a number of sawmills, thus consolidating many forestry functions under one corporate umbrella. Integration allows a corporation, which owns the rights to the raw material, much more control over its end uses. Owning sawmills in conjunction with pulp mills makes a great deal of economic sense, as the "wastes" from sawmilling can be chipped up for pulp or used as hog fuel for much needed electrical and steam production at the pulp mill. Figure 8.4 shows the pulp-and-paper mills in British Columbia in 2010.

The location and number of sawmills in the province escalated from the 1930s to the 1960s, in part because of the increased demand after the Second World War but also because of the expansion of road and rail systems that allowed the forests of the interior to be harvested for lumber (Figures 8.5 and 8.6). By the 1970s production had increased, but the consequence of consolidation and integration was fewer, more efficient, mills (Figure 8.7). For example, "wood production in the Prince George area rose from 380 million board feet (mid-1950s) to 1,250 million board feet, but the number of sawmills fell from approximately 1,200 to 400" by the mid-1960s (Lewis 2002, 189).

During the late 1960s and 1970s, new investments focused mainly on the interior, where some new communities such as Mackenzie were created. It was not long before the volume of wood cut surpassed that of the coast (Figure 8.2). The old mills on the coast were now under serious pressure to compete, and new investments were required to keep them competitive. In the case of Ocean Falls, the uncompetitive old mill was initially "rescued" by the provincial government in the early 1970s. A change in government in the mid-1970s resulted in closure of the pulpmill, and the community became a ghost town. New investments also led to the consolidation and integration of the forest industry on the coast; old mills were upgraded, a new mill was built in Kitimat, and another built on the west coast of Vancouver Island, around which the community of Gold River was created. As a result of the economic pressures that emerged in the 1990s and early 2000s, old coastal pulp mills, such as the ones at Powell River, Port Alice, Port Mellon, and Port Alberni, were restructured, and the mills at Gold River, Prince Rupert, Squamish, Campbell River, and Kitimat closed.

The technologies of making pulp have changed considerably from the exclusively chemical processes common before the Second World War to the Kraft and thermomechanical processes following the War. The Kraft process of creating pulp had replaced the sulphite process by the 1950s and '60s because a "wider fibre base may be used, and lignum, comprising 25 to 30 percent of wood volume, is recovered and used as fuel, reducing costs and stream pollution" (Hayter 2000, 53). The thermomechanical means of creating pulp has the advantage of being able to use a wider variety of wood types, as well as sawdust from sawmills, but it requires a great deal of steam and heat, which add energy costs. The chemical processes break down the wood fibres to produce an even stronger pulp. In either method, the last stage – bleaching – is often the most contentious because it requires chlorine. Chlorine releases dioxins and furans into the waste water, producing toxic levels of pollutants. Figure 8.8 is a simplified flow diagram of the conversion of wood to pulp.

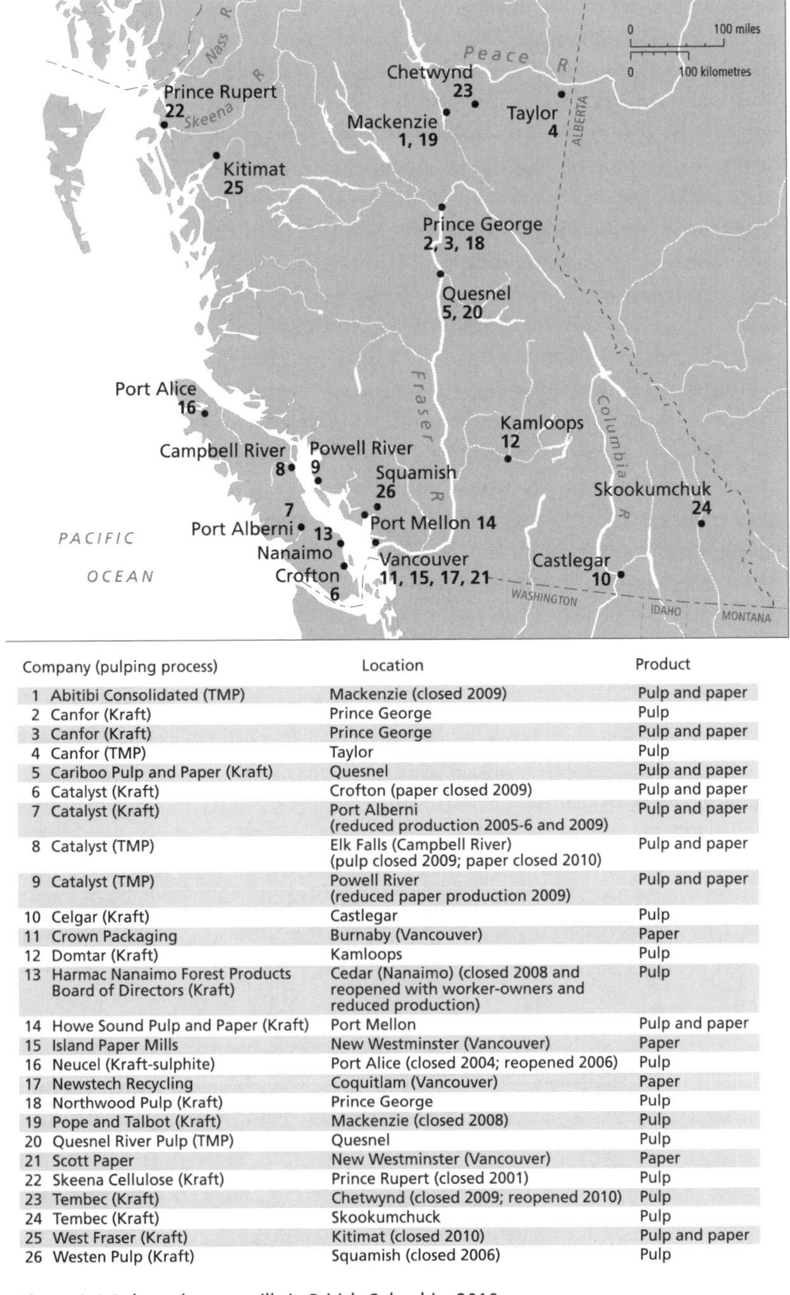

Company (pulping process)	Location	Product
1 Abitibi Consolidated (TMP)	Mackenzie (closed 2009)	Pulp and paper
2 Canfor (Kraft)	Prince George	Pulp
3 Canfor (Kraft)	Prince George	Pulp and paper
4 Canfor (TMP)	Taylor	Pulp
5 Cariboo Pulp and Paper (Kraft)	Quesnel	Pulp and paper
6 Catalyst (Kraft)	Crofton (paper closed 2009)	Pulp and paper
7 Catalyst (Kraft)	Port Alberni (reduced production 2005-6 and 2009)	Pulp and paper
8 Catalyst (TMP)	Elk Falls (Campbell River) (pulp closed 2009; paper closed 2010)	Pulp and paper
9 Catalyst (TMP)	Powell River (reduced paper production 2009)	Pulp and paper
10 Celgar (Kraft)	Castlegar	Pulp
11 Crown Packaging	Burnaby (Vancouver)	Paper
12 Domtar (Kraft)	Kamloops	Pulp
13 Harmac Nanaimo Forest Products Board of Directors (Kraft)	Cedar (Nanaimo) (closed 2008 and reopened with worker-owners and reduced production)	Pulp
14 Howe Sound Pulp and Paper (Kraft)	Port Mellon	Pulp and paper
15 Island Paper Mills	New Westminster (Vancouver)	Paper
16 Neucel (Kraft-sulphite)	Port Alice (closed 2004; reopened 2006)	Pulp
17 Newstech Recycling	Coquitlam (Vancouver)	Paper
18 Northwood Pulp (Kraft)	Prince George	Pulp
19 Pope and Talbot (Kraft)	Mackenzie (closed 2008)	Pulp
20 Quesnel River Pulp (TMP)	Quesnel	Pulp
21 Scott Paper	New Westminster (Vancouver)	Paper
22 Skeena Cellulose (Kraft)	Prince Rupert (closed 2001)	Pulp
23 Tembec (Kraft)	Chetwynd (closed 2009; reopened 2010)	Pulp
24 Tembec (Kraft)	Skookumchuck	Pulp
25 West Fraser (Kraft)	Kitimat (closed 2010)	Pulp and paper
26 Westen Pulp (Kraft)	Squamish (closed 2006)	Pulp

Figure 8.4 Pulp-and-paper mills in British Columbia, 2010

Note: "TMP" refers to thermomechanical pulping process. "Kraft" refers to pulp produced from chemical processes using sulphur compounds (hence the rotten egg smell).

Sources: Price Waterhouse Coopers (2007); The Vancouver Sun (2009); Pulp and Paper Canada (2010); CBC (2010).

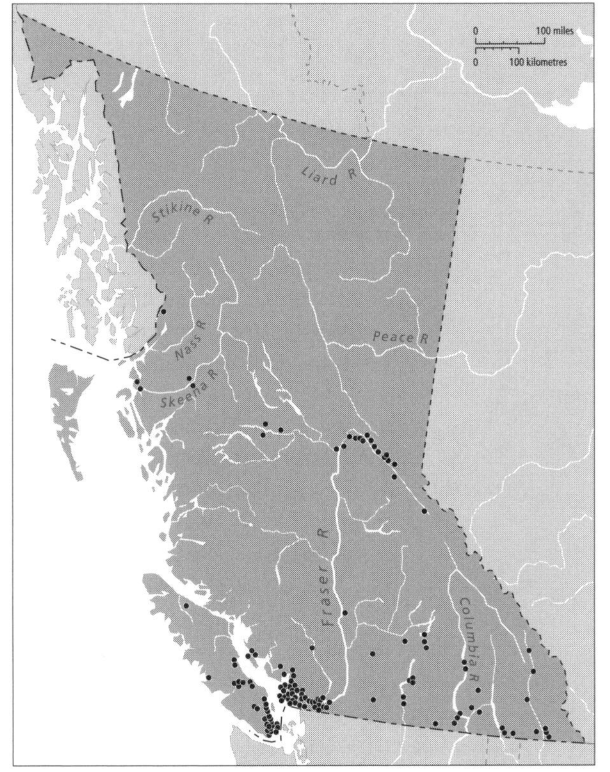

Figure 8.5 Sawmills in British Columbia, 1931
Source: Modified from Farley (1979, 65).

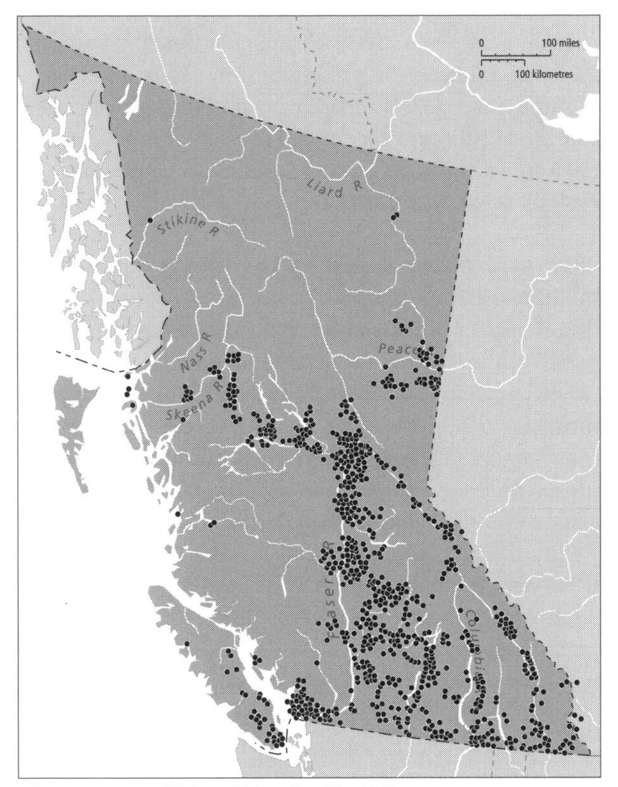

Figure 8.6 Sawmills in British Columbia, 1961
Source: Modified from Farley (1979, 65).

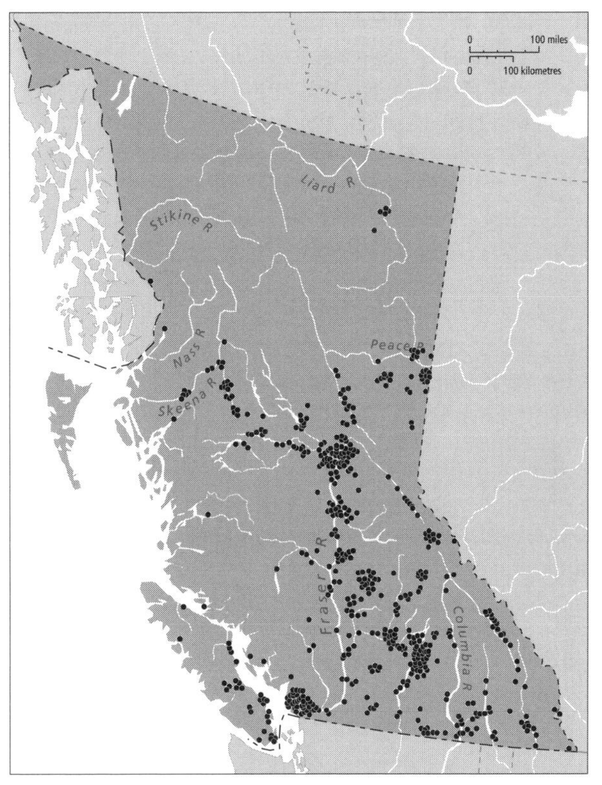

Figure 8.7 Sawmills in British Columbia, 1971
Source: Modified from Farley (1979, 67).

The types of pulp have also expanded to meet the global demand for a growing range of paper products. The Port Alice mill, at the north end of Vancouver Island, for example, produces sulphite pulp, which is used for a variety of specialty products, such as dry and moist food products. On the other hand, pulp products for the European market must be chlorine free. Considerably more pulp than paper is produced in British Columbia, and newsprint constitutes a significant component of paper production. However, more and more people are reading the news online.

Figure 8.9 charts the value of British Columbia's total forest exports and presents a bleak picture for this industry. Softwood lumber exports, particularly to the United States, have been a mainstay for forestry, but the drastic reduction in housing starts south of the border has resulted in many mill closures. The average annual value

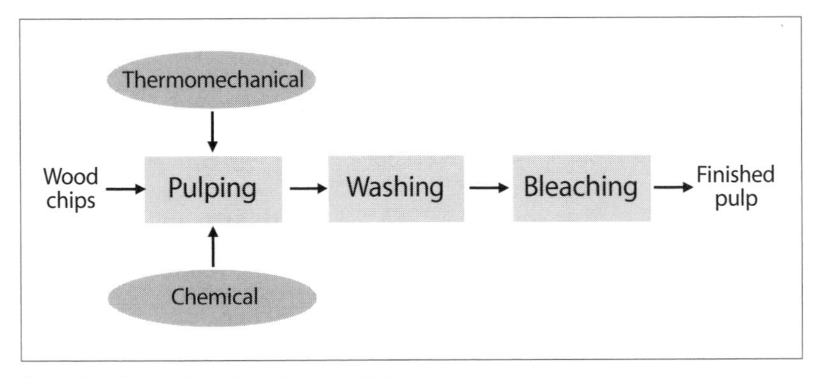

Figure 8.8 The creation of pulp from wood chips
Source: Modified from BC Ministry of Forests (1994, 3).

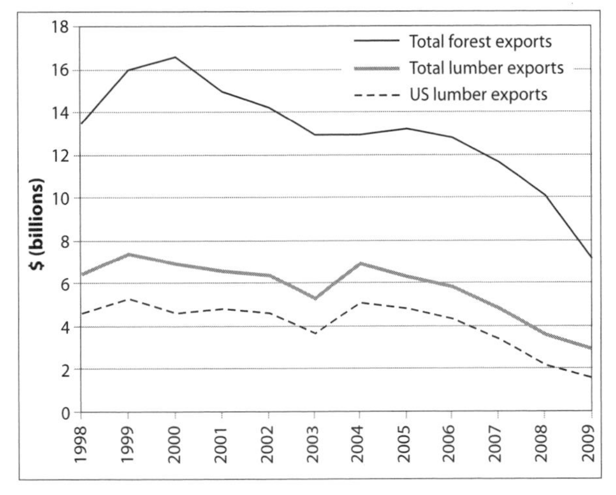

Figure 8.9 Forest product exports, 1998-2009
Note: The values for 2009 are estimates.
Source: BC Ministry of Finance (2009); Statistics Canada (2010, table 228-0034).

of key forest products listed in Table 8.1 indicates just how far forest commodity exports have fallen. Lumber prices have fallen off to a point where few sawmills can make a profit. Pulp and newsprint prices have fluctuated, but the high value of the Canadian dollar, along with the closure of sawmills (a main supply of wood chips), has led to higher fibre and transportation costs and intense international competition. Figure 8.4 documents the number of pulp-and-paper mills that have reduced their production or closed permanently.

With multinational corporations bringing in huge investments, the consolidation and integration of the forest industry raised other concerns. The new capital-intensive equipment created efficiencies to harvest trees and produce forest products, but once the major expansion of new mills was over, the number of workers required decreased, particularly for the traditional forest-related jobs – logging, producing lumber, and pulp and paper. It is no wonder so many forestry-related jobs have been lost if we also take into consideration the influence of a major economic recession, old plant and equipment that makes it difficult to compete with new producers in regions with much cheaper labour, and a rising Canadian dollar. Natural Resources Canada (2009) lists sixty-seven mill closures and a loss of 12,656 jobs between January 2003 and April 2009 (see Table 7.6 for sector job losses). Other issues still persist, such as forest clear-cutting, which is a cost-effective technique; however, it systematically eliminates old growth and produces many environmental conflicts. In some regions, it also challenges Aboriginal title and recent treaty negotiations. Tied to these concerns is the whole question of sustainability of the forests for the future.

PROVINCIAL MANAGEMENT OF THE FOREST RESOURCE: BACKGROUND

Historically, the forests were regarded as an inexhaustible resource. Even though it was decided in 1865 to retain Crown ownership of forest land, some of the best coastal forests were bartered away for tax revenues and railway development (Barker 1977, 88). Unfortunately, little was done in terms of reforestation or addressing the issues of sustainability until relatively recently.

Tenure is a legal right, granted to private companies, to harvest provincial forest lands, and it produces revenues for the province from the forests. From 1870 to 1905, forest companies paid lease fees and were granted blocks of forest land in perpetuity. A timber licensing system also operated from 1888 to 1909, which not only alienated Crown forest lands but also allowed the untaxed transfer of licences. Only once the timber was harvested was a licence holder subject to tax. The folly of this system

Table 8.1

Value of selected forest products, 1998-2009

Commodity	1998	1999	2000	2001	2002	2003	2004	2005	2006	2007	2008	2009
Lumber (US$/1000 board feet)	287	342	256	247	235	270	394	355	296	249	219	182
Pulp (US$/tonne)	515	523	681	543	463	523	616	611	674	793	853	716
Newsprint (US$/tonne)	595	513	564	588	468	501	549	608	667	597	687	569

Sources: BC Ministry of Finance (2009); RBC Commodity Price Monitor (2010).

was that it produced a great deal of speculation and consumption of forest land (Haig-Brown 1961). Leasing and licensing operated side by side during this period.

Crown grants were another means whereby provincial forests came under the control of private interests. Many of these grants took the form of cutting rights given to private railway companies as encouragement to construct new lines. One of the most famous, or infamous, railway grants was the Esquimalt and Nanaimo Railway charter, which placed one-quarter of Vancouver Island's forests in private hands. This was some of the best forest land in British Columbia because of its sixty-year growing cycle.

Whether land leases, timber licences, or Crown grants, these forms of tenure privatized forest land in British Columbia. And while the actual amount of land in these tenures amounted to only about 4 percent of the land base of the province, it accounted for most of the wood harvested until the 1940s and still accounts (it is down to 3 percent) for a significant amount today (Figure 8.10). These private forest lands are some of the best and most productive in the province.

Contemporary forms of tenure come from the recommendations of the Sloan Commission of 1945. The commission resulted in amendments to the Forest Act in 1947 and two forms of tenure were developed: tree farm licences (TFLs) and public sustained yield units (PSYUs). The TFL is an area-based tenure that allows a private forest company to obtain a relatively long-term renewable licence (initially twenty-one years and now twenty-five) to a block of forest land. The company is responsible for road building, planning, and reforestation within the licensed area. TFLs of the 1950s and '60s were mainly in the coastal forest region, where the large companies dominated, and with this form of tenure they retained a great deal of control over the land. As the industry expanded to the interior, so too did the TFL form of tenure. By 2009 there were thirty-three TFLs in the province.

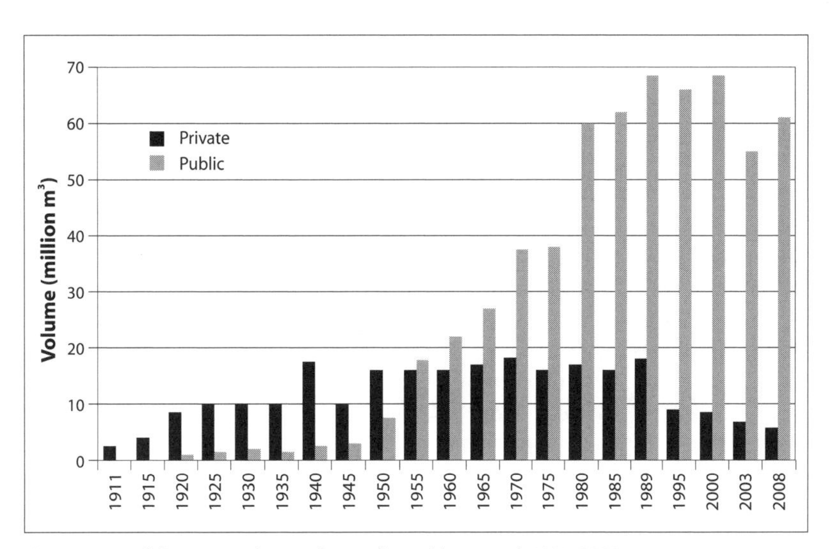

Figure 8.10 Public versus private volume of wood harvested, 1911-2008
Sources: Hammond (1991, 76); Canadian Council of Forest Ministers (1997, 8); National Forestry Database Program (2004a, 2010).

PSYUs are a volume-based tenure in which the Ministry of Forests and Range plays the lead role. The ministry establishes the volume of wood to be harvested by private companies under a series of licences and harvesting contracts, and is responsible for building the roads and for reforestation. In the 1950s and '60s, this form of tenure applied mainly to the interior, where there were few large forestry companies. Edgell (1987, 113) explains the rationale for these two new forms of tenure under the 1947 changes to the Forest Act. Their purpose was to:

1 ensure industrial access to guaranteed long-term timber supplies
2 stimulate capital investment in processing plants and therefore assure economic stability and development
3 bring forests under "sustained yield management."

Certainly the concept of sustained yield management was long overdue, and the commission was the turning point. It was finally realized that the forests would not last forever if they were not replanted and better managed than they had been in the past. The two forms of tenure eventually raised the question of who the better manager was, the company with the TFL or the ministry with the PSYU.

Further amendments to the Forest Act occurred in 1978. All lands held under the old timber licences (a term that by now included all historical tenures except Crown grants), once harvested, would revert to the Crown, and in this way the historically alienated forest land would be clawed back and in the control of the province. Under this amendment, the thirty-one TFLs (in 1978) were extended to twenty-five-year terms. The biggest change was to reorganize provincial forests into thirty-six (thirty-seven as of 2003) management regions, referred to as timber supply areas (TSAs) (Figure 8.11). Licences and harvesting contracts under the PSYU tenure were replaced by other licences. The forest licence (FL) was the most common tenure and gave the holder the right to harvest a stated volume of timber each year (BC Ministry of Forests 1988, 3). The FL initially retained the volume-based conditions of the PSYU with a fifteen-year renewable term. This has now been changed to a twenty-year renewable term, but some contracts are nonrenewable. Other forms of tenure that replaced the PSYU included the timber sale licence, the small business forest enterprise program (now timber sale license), the woodlot licence, and several pulpwood agreements (in the interior only). TFLs and FLs represented 90 percent of the annual allowable cut in the province in 1993 and thus dominated the forms of tenure (Hayter 2000, 72). The **allowable annual cut** (renamed "wood supply" in 2004) is the rate of harvest, or volume of wood allowed to be harvested, for each tenure in each timber supply area.

Production increased from the 1960s on, particularly in the interior of the province, and new questions about tenure surfaced as the provincial government tried to reduce its bureaucracy and permit privatization. The financial blow to the industry and to the BC economy brought on by the recession of 1981-86 triggered a move to reduce bureaucracy. This was achieved by significant reductions in ministry staffing, which then raised questions about adequate assessment and monitoring of the forest resource. A 1982 provincial Ministry of Forests policy proposed permitting forest licences and timber sale licences, both of which are for relatively short duration and under considerable government control, to be converted to tree farm licences, which, in the public's view, were much closer to privatization because of the twenty-five-year renewable conditions of this tenure. By 1988 the necessary legislation was in place to allow this conversion.

Reaction from the public, which by this time was both aware of and concerned about logging practices and the apparent lack of forest management, forced an end to this initiative. With tenure proposals to increase the number of tree farm licences off the agenda and a new provincial government in 1991, the public expected that the government would come to terms with many of the outstanding issues of this industry. A host of round table processes during the 1990s (see below) did little to change forest tenure, but a major increase in parks and protected areas throughout the province reduced available forest land. For example, "protected areas on the coast more than doubled in the 1990s" (Pearse 2001, 8).

Forestry produces provincial revenues primarily through taxes, some direct and others indirect. Stumpage is the most important and controversial source of direct revenues and the most complex. It is a direct tax placed on logs, but it varies depending on the species, size and quality, and end use. Wood for pulp, for example, can

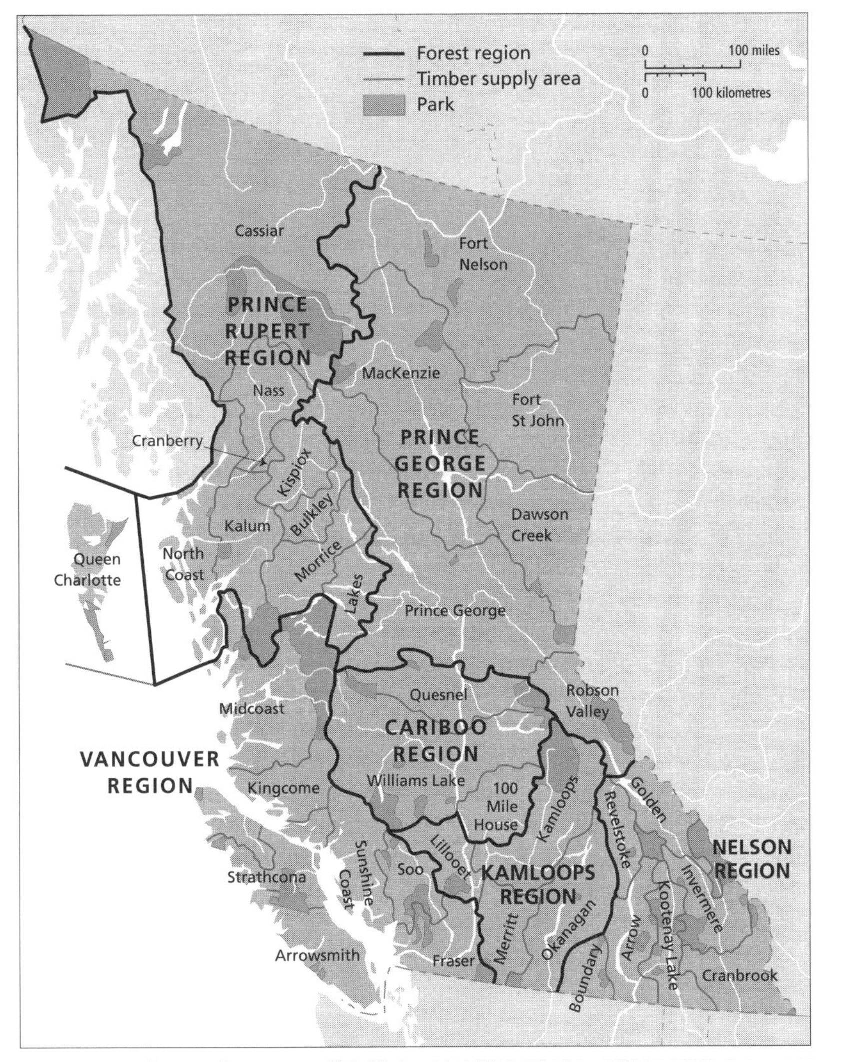

Figure 8.11 Timber supply areas, tree farm licences, and forest regions, 2001

come from a number of species of trees, and since quality is not an issue because the wood is for chipping, the stumpage rate is low. Large Douglas firs, used for lumber, bring in considerably more stumpage.

Besides the market value of the wood, stumpage takes into consideration physical conditions such as difficulty in gaining access to timber and volume of wood available per hectare. Old-growth stands of trees on the coast have a much higher volume of wood per hectare than those of the interior; consequently, more roads and more effort are required in interior forests for the same volume harvested on the coast. Stumpage rates are therefore lower to reflect the increased costs and extra labour required.

Finally, the stumpage rate varies with the type of tenure. The tree farm licence, by which a company is responsible for inventory and basically all aspects of managing an area, carries a lower stumpage rate than do volume-based tenures. The volume-based licence holders (for example, those with forest licences) have fewer responsibilities and costs.

Calculating stumpage involves a great many variables and, not surprisingly, several controversial issues. An ongoing battle over stumpage rates and types of tenure has been waged with the United States. A good deal of BC lumber, over 50 percent each year, is sold for housing and other wood products in the United States and is therefore in direct competition with Washington and Oregon lumber. American lumber companies complain that BC stumpage rates, inherent in long-term tenures, lead to unfair competition. A number of court cases have resulted, and in 1986 the United States imposed a 15 percent countervail tax on all BC lumber crossing the border. In 1990 British Columbia officially increased its stumpage rate by 15 percent and the countervail tax was dropped. Stumpage rates have been increased again since then, causing concern to BC companies but satisfying American producers. The 1996 Canada-United States Softwood Lumber Agreement established a quota of softwood lumber from each lumber-producing province in Canada but, as noted, when this agreement expired in 2001 a 19.3 percent countervail duty plus a 12.5 percent anti-dumping tariff were imposed by the United States. Finally, in 2006 a new softwood lumber agreement was signed by the federal government (see discussion below).

Stumpage accounts for about 90 percent of the direct forestry revenues to the province, clearly the most important government revenue source from this industry. The remainder is made up of royalties, scaling fees, direct grants by the federal government to the provincial Ministry of Forests and Range, mainly for silviculture, and several other charges.

Various indirect revenues go to all levels of government, mainly in the form of taxes on the industry and its employees. Because they are earning an income, employees pay income and property taxes and many other taxes for consumer goods. The industry uses a great deal of fuel and energy, and taxes are paid on these resources. Companies are also charged land taxes and business licence fees. The large consumption of water, by pulp-and-paper mills especially, results in a substantial water tax revenue to the province. All these indirect taxes are important sources of government revenue. It must be recognized, however, that in a recession this multiplier effect works in reverse.

The province also incurs costs with respect to the industry. The Ministry of Forests and Range bureaucracy is responsible for administering the industry. Engineering and building forestry roads, fighting forest fires, and battling pests and diseases that destroy forests all offset revenues. Reforestation, the goal of silviculture, is another expense but is essential to providing future forests.

It is necessary to distinguish, as the Forest Act does, between basic silviculture and intensive silviculture. Basic silviculture includes the research and development required to ensure the regeneration of healthy forests. Collecting seeds, growing seedlings in similar climate and soils to their eventual location, conducting controlled burns, or scarification, of clear-cut areas, and then planting the seedlings are all part of basic silviculture. Intensive silviculture involves caring for the seedlings as they grow. Table 8.2 shows the rather dramatic increase in planting of seedlings by the 1980s and '90s. This increase was necessary because of the backlog of clear-cut areas not previously restocked. More recent increases are directly related to the massive forestry losses caused by the mountain pine beetle infestation.

Table 8.2

Reforestation of seedlings, 1960-2008

	Number of seedlings (millions)		Number of seedlings (millions)
1960	6	1996	261
1975	65	2000	205
1989	210	2007	276
1993	215	2008	250

Sources: Council of Forest Industries (1994, 45; 2001, 38); National Forestry Data Base Program (2004b, Table 6.1); Council of Forest Industry (2001); Parfitt (2008b).

In the mid-1980s, the mayors of Vancouver Island used a "carrot patch" analogy to try to persuade the provincial and federal governments to spend money on intensive silviculture. They said that basic silviculture is analogous to preparing the garden and planting rows of carrots in the spring. If you do not tend the garden, however, you cannot expect a bountiful crop in the fall. Probably you will have a bunch of small, overcrowded carrots, all competing for limited nutrients. Replanting seedlings in the forest will result in the same disaster if they are not tended.

Table 8.3 illustrates very clearly the impact of residual spacing to weed out competition (the equivalent of thinning the small carrots), adding fertilizer, and then conducting commercial thinning (harvesting some of the trees earlier in the cycle, the equivalent of picking the carrot crop in stages). Limbing the lower branches is another intensive silviculture practice. As the table shows, it is feasible to double or triple the volume of wood per hectare. A subsidiary benefit is that the larger trees that result can be used for higher value-added wood products, bringing in even more direct revenues to the government. Unfortunately, most attention and money spent on reforestation has focused on basic silviculture. The time required for intensive silviculture, unlike for carrots, is over sixty years for the best forest lands and over eighty years for average forest lands. Governments, from whom this commitment is required, are elected for only four years.

Trevor Jones (1983) suggests that the costs of and revenues from logging should be assessed for each region. Are the direct revenues from logging a region such as the Stein Valley, for example, higher than the cost of building roads and bridges, administration, reforestation, and so on? Jones uses the benefit-cost approach to warn the

Table 8.3

Potential yield through intensive silviculture practices

Forest application	Trees/ha	Volume of wood harvested at 50 years (m^3/ha)	Diameter at 1.3 m above ground level (cm)
Natural untreated stands	±2,500	350	30
Planted stands with improved seeds	±1,500	560	38
Natural or planted stands with residual spacing (15 years)	±750	700	43
Fertilizer applied at 225 kg nitrogen at 25 years		840	46
Commercial thinning at 35 years	370	1,050	50

Sources: Modified from Travers (1993, 213); M'Gonigle and Parfitt (1994).

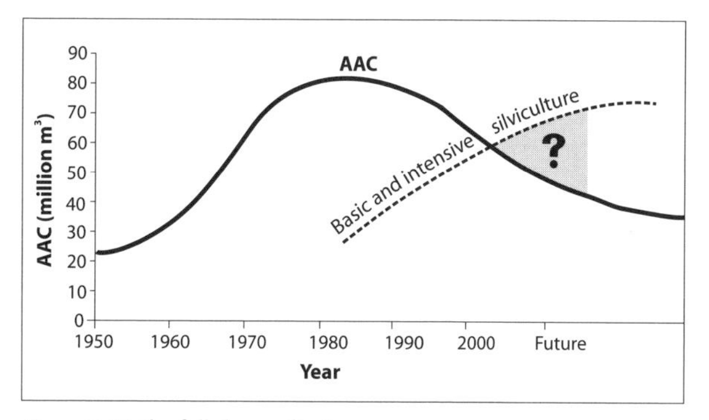

Figure 8.12 The fall-down effect

provincial government that "if the Forest Service's costs exceed its direct revenues for a logging operation on Crown land, then the operation is subsidized. This is not necessarily bad, because the forest company may make a profit, employment is created, and the Crown will obtain indirect revenues such as personal and corporate income taxes ... However, the subsidy has to be paid for through the indirect revenues, which are normally used to pay for society's general needs – highways, health facilities, education and other provincial services; external affairs, armed forces, coast guard and other federal services" (p. 6).

The film *The Fall-Down Effect* is old but still worth watching (1980). Economist Mike Halleran gives a good visual description of the history of logging, technological change, and the conflicts that are shrinking the forest base and making less wood available for the future. Figure 8.12 summarizes the fall-down effect. The dotted line beginning in the 1980s and leading into the future represents the increase in volume of wood per hectare obtained through basic and intensive silviculture practices. The question mark represents the government's level of investment in these programs. With large investments the fall-down effect is substantially reduced. The allowable annual cut (AAC) increased by approximately 300 percent between 1950 and 1980. As discussed earlier in this chapter, this increase is related to the consolidation and integration of the forest industry, which saw new corporate structures, harvesting techniques, investments, and a major expansion into the interior to meet an ever growing demand for forest products. Maintaining a high AAC is in the provincial government's interest because it means increased revenues. As Halleran asks, however, can we maintain this level of cutting? The graph shows an abrupt decline, or fall-down, in the AAC from the 1990s on.

Several conditions are responsible for the fall-down effect. In the coastal forest region, the base of old-growth trees has largely been cut, and most harvesting in this region today is of second-growth timber (Drushka 1999). These trees are considerably smaller and produce much less volume of wood per hectare. Another factor is the shrinking of the forest base. By the 1980s, many groups were demanding that forests be set aside for uses other than wood for fibre. Conflicts centred on old growth, wilderness, wildlife, water quality, recreation, grazing land, hydroelectric dams and the reservoirs they create, and Aboriginal title. Some of these issues have been addressed by creating parks in environmentally sensitive areas such as watersheds, but other issues have yet to be resolved. All these "other" uses of the forest require reduction of the AAC.

Areas where harvesting has been the most intense are the first to experience the fall-down effect. The north end of Vancouver Island, for example, had experienced a 38 percent reduction in its AAC by the early 1990s, and the Sunshine Coast a 24 percent reduction. More recently, the AAC has been increased to utilize the wood devastated by the mountain pine beetle before it decays. There will, however, be massive fall-down effect once the infected pine is finally harvested. The Pearse Commission addressed the crisis in the coastal forest industry in 2001 and states at the outset of its report, "It is sufficient to point to one well-supported fact; the Ministry's present Allowable Annual Cut of timber in the coastal region exceeds the supply of harvestable timber that will be available over the next few decades" (p. 7).

There are ways to reduce the fall-down effect. Basic and intensive silviculture could considerably increase the volume of wood for the future. Herb Hammond (1991) and others suggest that there must be a fundamental change in philosophy toward the way we grow and harvest trees. Hammond advocates holistic forest practices that avoid clear-cuts and use selective logging techniques; these could increase yields and reduce many environmental problems and conflicts. On the other hand, a report by the provincial government shows second growth in the interior at 35 percent to 55 percent higher than expected and suggests that the figures may be more impressive in the coastal zone, where growing conditions are better (Wilson 1998). These figures mean that the supply of wood can become available much more quickly than forecast, reducing the timeline of the fall-down effect. These trends were further confirmed in the government's second *Timber Supply Review:* "in addition to our new forests growing faster, the productivity of our forest lands was much higher than we anticipated" (Pedersen 2003). Its projection is that forest regeneration is between 20 and 30 percent higher on managed forest lands than on unmanaged forest lands.

MANAGING PRESENT AND FUTURE FORESTS IN A CLIMATE OF UNCERTAINTY

Many of the problems associated with forestry in this province have been economic. Since the 1970s, however, the environment has also become a major consideration. The relationship between forest practices and the environment has to be considered because managing forests to maximize economic revenues in a sustainable and environmentally sensitive manner depends on forward-looking forest policies. Economic, environmental, and political factors are interrelated, and each of these factors is assessed below.

Economic Issues in a Changing Economic Environment

The forest industry has been the backbone of British Columbia's economy for a long time. Until the 1970s it was widely proclaimed that fifty cents of every dollar spent in the province had been generated by this industry (Farley 1972, 87). Forestry is still crucial to both the provincial economy and individual communities and regions. The British Columbia Round Table found in 1993 that "94,000 British Columbians were directly employed in the forest sector. The livelihoods of as many as 140,000 more people depended largely on the forest sector" (p. 43). Unfortunately, a long history of exploitation (made easier by sophisticated technologies and rules that encourage overharvesting) and global competition have increased costs and caused recessions that have drastically reduced demand and world market prices. These developments have brought conflict and crisis to some regions and a climate of uncertainty to the whole industry (Marchak 1983; Clapp 1998; Fulton 1999; Hayter 2000; Pearse 2001; Parfitt 2010). By 2009, direct employment in this industry (46,787) had decreased nearly 50 percent from 1993 figures (Natural Resources Canada 2010a). Figure 8.13 outlines the continued geographic distribution of dependence on forestry-related employment throughout the province.

The forests of British Columbia are a component of a dynamic global marketplace in which the multinational corporations that have effective control over the resource make decisions based on world market prices for wood products – and both the prices and the demand for wood products fluctuate considerably. Moreover, British Columbia's competitive position in the global arena has been seriously challenged in some areas. The province used to be one of the cheapest places in the world to produce pulp, for example, but by the mid-1990s it had become one of the most expensive ("Coastal Mills" 1997). Countries such as Indonesia, where fibre is inexpensive and available and where the cost of labour is a fraction of that in British Columbia, put the older, less productive coastal mills in particular in jeopardy. Table 8.4 shows that British Columbia produced some of the most expensive lumber in the world in 2000. Hayter and Barnes (1997) and Hayter (2000) have documented the economic viability of producing wood products from sawmills, pulp mills, and plywood plants, along with mills that produce

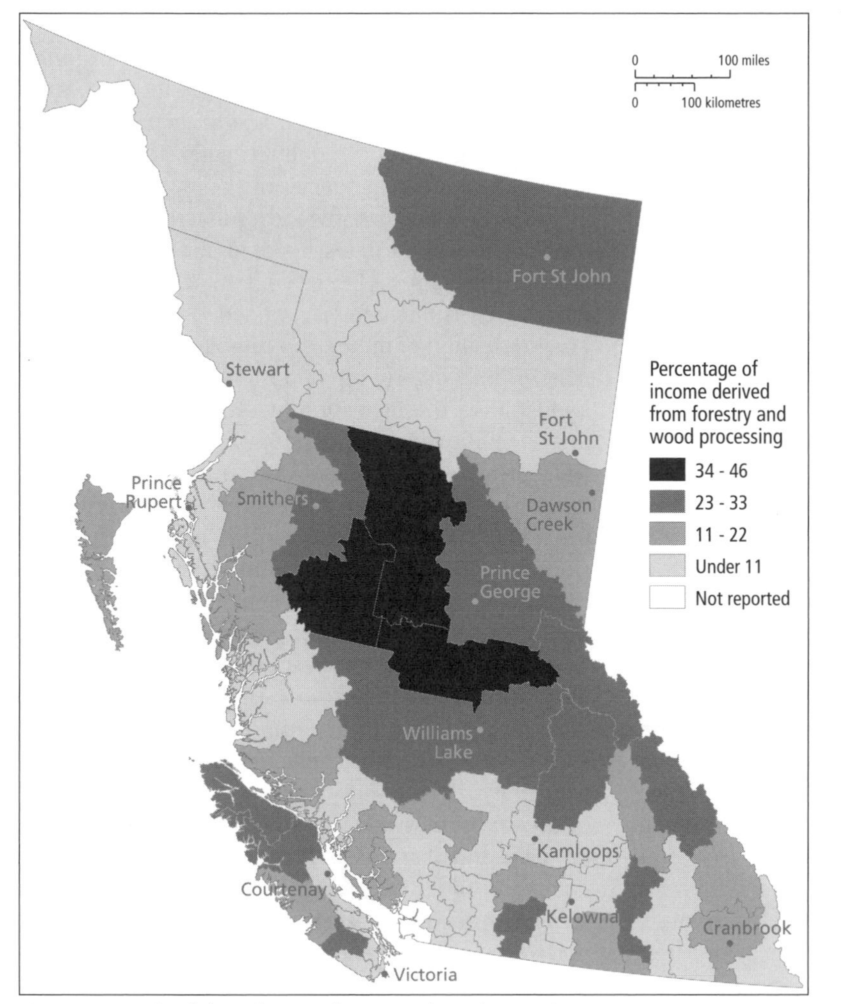

Figure 8.13 Regional dependence on forestry and wood processing, 2006
Source: BC Stats (2009).

fibreboard, particleboard, and shakes and shingles, and show that competition and uncertainty have led to major restructuring. Flexible specialization has meant fewer employees, more contracting out, and lower pay. In short, the old centralized Fordist methods of production have been challenged.

Binkley (1997, 22) points out the dilemma faced by this industry: "Productivity in the British Columbia forest sector is squeezed between a rising floor of raw material costs and a fixed ceiling for product prices set by international competitors in the forest product industry and by the cost of substitute products." Recession and the collapse of the US housing market (in 2008) has put the whole industry in a downward spiral. Diminished demand for lumber, lower lumber prices, and the closure of sawmills has reduced the supply of wood chips and hog fuel for pulp-and-paper mills, which, along with a higher Canadian dollar and international competition, has resulted in closures for pulp-and-paper mills. All of these developments have had a major negative impact on forest dependent communities. For example, Mackenzie, an instant town (see Chapter 15), was created in 1966 to service the industry. Its planer mill, pulp mill, paper mill, and four sawmills represented employment for over a thousand people. By 2008, however, all of these mills had been shut down, revealing the tenuous nature of communities that rely on one resource. (The pulp mill was sold to an Asian company – Sinar Mas – in April 2010. It will reopen in the fall, putting some 200 people back to work.) The list of closed pulp-and-paper mills in Figure 8.4, in addition to over seventy other plants (sawmills, OSB mills, planer mills, plywood mills, etc.), tells the equally devastating story of workers who have lost jobs in those communities (United Steelworkers 2010).

Making matters worse is the billions of dollars in subsidies given to US mills: "US companies are benefiting from a 50-cent-a-gallon subsidy on alternative fuels meant for highway vehicles. Pulp and paper companies in the US, however, discovered they were eligible if they added some diesel to their black liquor – a by-product of the pulping process that is used to generate heat and energy. The subsidy has reduced their costs by about $200 a tonne, turning unprofitable mills into the world's

Table 8.4

Cost of producing lumber by world supply regions, 2000

Region	Labour	Other direct costs	Total costs
	($/1,000 board feet)		
Chile	49	153	202
South Africa	92	139	231
Brazil	82	162	244
New Zealand	74	193	267
Canada Prairies	122	150	272
Sweden	64	235	299
Canada East	126	178	304
Australia	130	208	338
BC interior	*124*	*219*	*343*
US South	104	313	417
US Pacific northwest	109	328	437
US inland	124	326	450
BC coast	*228*	*331*	*559*

Source: Modified from Pearse (2001, 5).

lowest-cost producers overnight. It has resulted in extreme distortions in pulp and paper markets" (Hamilton 2009). The Canadian reaction was to do something similar by creating the Green Transformation Program, which has allocated hundreds of millions of dollars to BC's pulp mills (most mills burning black liquor and wood wastes to create electricity qualify as green projects).

Adding value to wood products is another component in the economic equation. The restructuring of the forest industry in the mid-1980s was, in part, a response to dependence on the mass production of dimension lumber (e.g., two-by-fours), destined mainly for the United States. The Japanese market, on the other hand, requires metric sizing of lumber products and different dimensions (and sometimes species of wood) than the United States. The market for softwood lumber remains concentrated in the United States, but pulp-and-paper products are more diversified (Table 8.5). Table 8.5 also illustrates the reduced export of forest products to Japan since 1997. The production of dimension lumber, regardless of the market, adds value to the wood, and this forward linkage adds jobs to the BC economy. Far more forward and final-demand linkages are possible, however, and these will increase employment and revenues. M'Gonigle and Parfitt (1994, 90) point out the unfortunate fact that approximately 50 percent of the lumber exported to the United States was used in remanufacturing plants in the early 1990s. By the mid-2000s, remanufacturing in British Columbia had not greatly improved (Sierra Club 2006). In other words, the lumber exported from British Columbia is turned into other wood products, from wooden doors and flooring to furniture and guitars, that gain even greater value. The challenge is to capture remanufacturing in British Columbia.

The export of raw logs is an ongoing issue and at the opposite end of remanufacturing, namely, it represents the export of all value-added employment. Until 2002 only logs from private land, not Crown land, could be exported. As Table 8.6 indicates, logs were cut and exported

Table 8.5

Softwood lumber and pulp and paper exports by destination, 1994-2009

	United States				Japan				European Union			
	Lumber		Pulp and paper		Lumber		Pulp and paper		Lumber		Pulp and paper	
Year	$ millions	%	$ millions	%	$ millions	%	$ millions	%	$ millions	%	$ millions	%
1994	4,737	62.0	2,192	40.9	2,277	29.8	841	15.7	404	5.3	1,123	21.0
1997	5,077	64.9	2,090	41.6	2,144	27.4	723	14.4	342	4.4	1,032	20.5
2000	4,620	67.2	2,572	37.1	1,684	24.5	872	12.6	305	4.4	1,585	22.8
2003	3,673	69.2	1,944	41.5	1,106	20.8	414	8.8	234	4.4	785	16.8
2008	2,194	60.8	2,120	45.1	719	19.9	258	5.5	437	7.1	351	7.5
2009	2,432	58.9	1,585	45.6	753	18.2	180	5.2	289	7.0	241	6.9

Sources: BC Stats (2004a, 5, 7, 9); BC Ministry of Finance (2009, 2010).

Table 8.6

Log exports, 1997-2009

Year	Log exports (tonnes)	Year	Log exports (tonnes)
1997	166	2004	3,429
1998	835	2005	4,740
1999	1,752	2006	4,271
2000	2,335	2007	3,293
2001	2,811	2008	2,580
2002	3,945	2009	2,418
2003	4,024		

Source: Natural Resources Canada (2010b).

at an ever increasing rate until the 2008 recession. According to the United Steelworkers Union (2010), raw-log exports has been the direct cause of mill closures in this province. Blair Redlin (2010) offers real numbers to clarify the issue: "These exports and large amounts of unprocessed wood waste are estimated to cost the BC economy some 5,800 direct jobs per year."

Managing Environmental Issues

Compounding the problem of uncertainty for forest workers and managers alike are environmental conflicts over practices such as clear-cutting, the emission of greenhouse gases, logging in watersheds, and the destruction and potential forest fires from mountain pine beetle infestation. On a global scale, organizations such as Greenpeace have successfully lobbied European nations and California to boycott BC lumber with old-growth content.

Increases to the allowable annual cut have reduced much of the old growth on the coast and produced some huge clear-cuts both on the coast and in the interior. Environmental issues have brought the greatest public awareness and criticism of provincial management of the forest industry. The New Democratic Party, elected in 1991, established a number of policies to try to extinguish environmental "brush fires." The Council on Resources and Environment (CORE) was created with a mandate to resolve environmental conflicts by increasing the number of parks throughout the province. In 1991 parks amounted to 6 percent of the BC land base; the objective of CORE was to increase parks, evenly distributed among regions, to at least 12 percent of the province. This objective was a significant challenge within geographical regions such as Vancouver Island, the Kootenays, and Cariboo. CORE first tackled Vancouver Island, where it recommended setting aside 13 percent of the land for parks. Logging interests were not happy nor, as it turned out, were environmentalists. Clayoquot Sound, where some of the last stands of old growth remain, was not included in the park allotment. Blockades in 1993, for which many protesters were jailed and fined, drew international attention to and condemnation of British Columbia's logging practices. CORE's process did not resolve the conflicts, and negotiations over Clayoquot continued until the end of the 1990s.

CORE was replaced by the more scaled-down Land Use Coordination Office in 1995, and the focus shifted to strategic environmental and land-use planning at varying geographic scales. The government initiated land and resource management plans (LRMPs) to resolve major land use issues. For issues related to small- or medium-sized regions, it initiated sustainable resource management plans (SRMPs). With the election of the Liberals in 2001, no new LRMPs were to be undertaken, but those already established were to be completed. According to a December 2006 report by the Ministry of Agriculture and Lands: "Over 85 percent of the provincial Crown land base is now covered by 26 regional land use plans and LRMPs – including 102 SRMPs completed and a remaining 93 plans underway, for a total of 195."

These plans have created more parks and protected areas to preserve the wilderness, they have provided benefits to First Nations, and they protect endangered species and water quality. Yet conflict continues in some parts of the province. Watershed issues could be resolved through the creation of a Quality of Water Act under the Ministry of Environment rather than the Ministry of Forests and Range. This legislation would send the message that water quality and quantity and the timing of flows form the most important values in any watershed and that all other activities, including cutting trees, must defer to them.

A Forestry Practices Code was enacted in 1994 to set limits on clear-cuts, increase fines and penalties for environmental violations, and pay more attention to stream setbacks and damage to biodiversity, wilderness, and other aspects of the environment (BC Ministry of Forests 1994). Unfortunately, the code was overly bureaucratic, and forest companies protested the amount of paperwork. Other groups, such as Eco-Justice (formerly the Sierra Legal Defence Fund), questioned the effectiveness of the legislation. Clark Binkley's (1997, 25) specific criticism of the code is that it promotes the rapid development of road systems: "Since the very worst environmental problems associated with timber production in British Columbia come from the failure of roads, it is ironic that new legislation purporting to protect environmental values actually demands more road building." In 1998 the provincial government promised a reduction in paperwork, which caused some to question whether environmental standards would be relaxed.

The Liberal government replaced the Forestry Practices Code with the Forest and Range Practices Act in 2004. This legislation is referred to as a "results-based code." However, there has been plenty of criticism of this act: "Under this new 'Results-Biased' Code, companies are now drafting new plans every five years instead of every year that are devoid of any specific information. They even have the option of devising their own environmental objectives if they think the existing ones are too restrictive" (Shuswap Environmental Action Society 2006). There is a basic conflict in having the Ministry of Forests and Range, which has a mandate to encourage the harvesting of trees, act as the gatekeeper of environmental standards.

Environmentalists are also concerned about the principles that underpin the calculation of the allowable annual cut. Hammond (1991); Drushka (1999); Drushka, Nixon, and Travers (1993); M'Gonigle and Parfitt (1994); Binkley (1997); Fulton (1999); Marchak (1999); and others see the AAC as a process that liquidates old growth and replaces it with uniform plantations, thereby creating serious reductions in biodiversity and sustainability. This gets right to the heart of whether the provincial government, as manager of the forests, is more concerned with a high AAC and revenues than with an ecosystem based on sustainability. The land base of British Columbia amounts to 91.6 million hectares, and 57.9 million hectares (over 63 percent of the province) are classed as forest land. However, the British Columbia Forest Practices Code states that only 26 million hectares is viable "commercial forest land," indicating a considerable discrepancy between the amount of productive forest land and the amount of land that the Ministry of Forests and Range wants to influence and control (Osberg and Murphy n.d.). A partial solution may be to assess forest land in the same manner as agricultural land (see Chapter 12) and to put the best forest land in reserves that minimize conflict among potential uses.

Harvesting techniques, or prescriptions, are equally important to the environmental condition of forests. Kimmins (1997) suggests that the technique of clear-cutting may be appropriate in fire-prone interior forests but completely inappropriate in many coastal zones. This observation not only raises the issue of clear-cutting versus selective logging, it also calls into question road-based logging and methods of hauling logs and other logging practices that affect the environment. The method of cutting must be sensitive to the individual ecosphere, and environmental conflicts must be reduced or eliminated.

The Clayoquot Forest Harvest Agreement, completed in the summer of 1999 and now called Tree Farm Licence 57, may point to yet another model for harvesting timber. In this environmentally sensitive region, a joint venture corporation has been created between the local First Nations (Nuu'chah'nulth), who own 51 percent, and MacMillan Bloedel (now Weyerhaeuser). The agreement includes a logging plan that has the consensus of environmental groups. Tim Aitkins (1999, 7) stresses the agreement's importance: "Greenpeace and other environmental groups have agreed to endorse and promote this project as a global model of ecologically sustainable forestry. The groups will also help develop markets for the eco-friendly, processed timber products as well as nontimber economic ventures." Certification, or a guarantee that wood has been harvested in a sustainable manner, is bringing pressure to bear on all corporations in the province to amend their clear-cutting practices. As of 2010, British Columbia had over 62 million hectares of its forest land classed as certified (Natural Resources Canada 2010b).

Unlike past bouts of insect infestation (e.g., spruce bud worm), the mountain pine beetle, according to a Natural Resources Canada report (2010c) "is causing widespread mortality of the lodgepole pine forests, the province's most abundant commercial tree species. At the current rate of spread, 50 per cent of the mature pine will be dead by 2008 and 80 per cent by 2013." This means "a billion or more pine trees are now dead in the interior of the province" (Parfitt 2010), and this standing fuel represents the potential for massive forest fires and the release of vast amounts of greenhouse gases into the atmosphere. Ben Parfitt's report recognizes the imminent threat of climate change and recommends new accounting ("Carbon Cut Calculation, or CCC, replacing the existing Allowable Annual Cut, or AAC") that will maximize the carbon stored in British Columbia's forests. Included among his ten proposed steps are the following: delaying cutting, preserving more forests, eliminating wood waste, and establishing "carbon plantations." Old-growth forests in particular are vital to storing carbon. Jim Pojar (2010) suggests that British Columbia's forests "average 311 tonnes per hectare." However, "old-growth forests steadily accumulate carbon for centuries and store vast quantities of it, up to 1100 tonnes per hectare in BC temperate rainforests – some of the highest storage capacities in the world. BC's forest ecosystems are estimated to store 18 billion tonnes of carbon." Pojar agrees with Parfitt that there is a greater need for protected areas and for a major reduction in clear-cutting.

Political Policies and Forest Management

A thornier issue, because it is highly political and often viewed as the root cause of so many of the problems, is obsolete forms of tenure and their control by large corporations. Forest licences and tree farm licences have been the main forms of tenure. Until recently, they have accounted for nearly 80 percent of BC's forests, and control of these long-term tenures remains in the hands of large, mainly foreign corporations. For example, "the top 10 lumber producers in BC today control about 75 per cent of the lumber manufacturing in the province" (Sierra Club 2006). These tenures, which give corporations a great deal of control over provincial forests, are the target of those concerned about issues such as declining employment and underproduction of high value-added products and the neglect of First Nations title to forest land. In addition, the United States claims that the tenures are a form of subsidy. M'Gonigle and Parfitt (1994), along with the Sierra Club (2006) and Parfitt (2010), argue that the large multinationals have sewn up the wood supply and left insufficient wood available for small, innovative remanufacturing mills that want to produce higher value-added products. Along the same lines, the Truck Loggers Association has expressed concern that the multinational corporations have considerable interest in turning saw logs into pulp chips. The stumpage for chips is low and so are the labour and value-added components.

The conditions of tenure have changed over time, and although tenures were intended to produce a sense of security for forest company investment (especially through the twenty- and twenty-five-year renewable clauses of FLs and TFLs), corporations recognize that they are still leases and thus very much subject to changes in government policy. The consequence, according to Drushka, Nixon, and Travers (1993, 14), is that "silviculture [has been] an obligation instead of an economic opportunity." Moreover, corporations' operating according to the bottom line often dictates technologies that promote clear-cutting and streamlining production such as high-volume dimension lumber, OSB, pulp, or other wood products, but each mill has only one focus.

Changes to tenure conditions also affect local communities: "With the demise of the old appurtenance clauses that linked tenure to the maintenance of specific mills, companies can dispose of wood as they see fit" (M'Gonigle and Parfitt 1994, 96). Even though a region may still have timber, it is no longer guaranteed that the local mill will receive it. Corporations are at liberty to designate the wood to locations outside the region, thus jeopardizing the survival of local mills and local employment.

Are there new and different forms of tenure that address these many issues? As noted earlier in this chapter, Hammond (1991), who has practised holistic forestry over a lifetime, suggests turning the forest into a smaller area-based tenure for individuals sensitive to principles of sustainability and the ecosystem. Others suggest the Swedish model, in which much of the forest is privately owned and managed. M'Gonigle and Parfitt (1994) and Parfitt (2010) argue for much more community control of the forests. Some of the greatest pressure to change, though, has come from British Columbia's biggest export market – the United States.

The tenure system, which has led to corporate concentration of the forest industry, and the taxation system on softwood lumber (stumpage) underlie the ongoing disagreements between the United States and Canada. The United States charges that the BC forestry system unfairly subsidizes softwood lumber production, which, in turn, allows the lumber to be dumped into the US market, driving US producers into bankruptcy. The last bitter dispute was sparked on 31 March 2001, when Canada's softwood lumber agreement with the United States expired and no new agreement could be negotiated. With no agreement in place, the United States unilaterally put an 18.79 percent levy on all Canadian softwood lumber coming across the border and an 8.43 percent anti-dumping duty. Canada claims that these duties, in excess of 27 percent, are unfair and that "both the WTO and NAFTA have found that the US Department of Commerce has made errors in calculating both the countervailing and anti-dumping duties. More important, both tribunals also found that imports of Canadian lumber were not harming the American industry" (BC Stats 2004b, 3).

Part of the difficulty in comparing the US and BC forestry management systems stems from the fact that most US timber comes from private forest lands and is sold on the open market. In Canada each province has its own set of management rules and regulations. However, since this is an international dispute, the entire country is involved. In British Columbia, most forest land is Crown or public forest land, and private companies are entitled to cut timber through the various tenures (e.g., tree farm licences) and are required to pay the complex stumpage fees for the timber cut. To complicate the issue further, there is little comparison between US lumber from rapid-growing pine and the slower-growing species of softwoods (e.g., Douglas fir) north of the border, which are easier to cut, nail, and handle and which are preferred by US builders. There are also differences in reforestation procedures and requirements, which further complicate the comparison.

The frustration for British Columbia and other lumber-exporting provinces in Canada is that appealing to the World Trade Organization (WTO) under the NAFTA rules is time consuming and costly, in terms of both legal fees and unilateral duties and tariffs. Even when Canada wins a decision, the United States appeals and the challenges continue. The United States relaxed duties somewhat in 2004 (22.69 percent) and again in 2005 (10.81 percent) before Canada signed another agreement in 2006. Meanwhile, the United States collected some $5 billion in duties.

The most recent agreement (signed in June 2006) has a seven-year span. It stipulates that only $4.3 billion of the $5 billion in duties should be returned. Other conditions include a ceiling on exports. "Canadian-sourced lumber would also be kept to no more than its current 34 per cent share of the US softwood market" (CBC 2006). British Columbia's historical share is approximately 50 percent of this quota. A world market price trigger for export duty charges allows Canada to "collect an export tax on softwood lumber exported to the United States if the price drops below $355 a thousand board feet" (CBC 2006). Unfortunately, that price has not been reached during this agreement (see Table 8.1).

The BC government has responded to these disputes in a number of ways. It introduced Bill 31 in 2003, which resulted in the Forestry Revitalization Plan, an attempt to address the issue of providing an open market for log sales. Policy states that "about 20 per cent of the long-term logging rights held by the largest tenure holders in BC will be taken back by government. This timber will then be reallocated to other parties to support market-based pricing of public timber, and to increase the role of First Nations and small tenures in forestry" (BC Ministry of Forests 2004). BC Timber Sales is the venue for much of this wood, which is sold to the highest bidder. In addition, there have been substantial allocations to community forest licenses, including First Nations communities. "As of April 2009, approximately 900,000 hectares were being managed as community forests with 33 active community forests in the province and another 18 communities in the application process" (BC Ministry of Forests and Range 2009b).

The softwood lumber agreement, which was signed by the federal rather than the provincial government, concerns other groups. For example, the $4.3 billion in rebates went to forest companies operating in Canada, half of which operate in British Columbia. Ben Parfitt (2008a) has shown that tax incentives in the agreement encouraged BC forest companies to take the rebates and reinvest the money in the United States – "Canfor Corporation, West Fraser Mills and Interfor spent more than US$620 million to purchase or upgrade US sawmills." Still others criticize the government and the agreement for providing "new incentives to export logs in unprocessed form, since the Agreement places export taxes on most softwood lumber products, but exempts softwood logs from such taxes" (Redlin 2010). And, as is demonstrated in Table 8.6, log sales have escalated since the end of the 1990s. Parfitt (2010) suggests "taxing all log exports to discourage shipment of unprocessed logs and encourage the maximum amount of local manufacturing." People's Voice (2008) advocates going even further and banning the export of logs.

Adding value to BC wood is a major objective and a major challenge for government policy makers. When the forest economy fluctuated in the 1990s, the government created Forest Renewal BC in 1995 and gave it a mandate to retrain and employ forest workers for a variety of new forest-related activities. Stumpage rates were increased to finance these projects, which included reforestation, value-added initiatives, the restoration of environmentally damaged areas, the retraining of forest workers, and assistance to forest-based communities. This policy change was influenced directly by flexible specialization, the need to employ displaced forest workers, and a desire to encourage the more labour-intensive value-added sector. "A significant portion of publicly owned timber was allocated to companies on the basis of social returns" (Sierra Club 2006). This program continued under the guidance of the new Liberal government, but the name was changed to the Forest Investment Account in 2001 and some of the conditions were changed. The allocation of wood for local value-added forest companies ended and, as a result of corporate concentration, small companies, the status of most value-added firms, were unable to bid on the huge volume of timber that make up most timber sales.

The economic crises that overtook the forestry industry from the 1980s on culminated in a serious economic blow with the 2008 recession. By 2009, the government had implemented the Wood First Act, which includes a Wood Enterprise Centre and a Value for Wood Secretariat. The goal of the legislation is to ensure that more wood is used for BC buildings. It includes "amendments to the BC Building Code that allow for six-storey, wood-frame construction." The Act also seeks to "facilitate technology transfer, promote training, and expand markets for further-manufactured products" (BC Ministry of Forests and Range 2009a).

Are these initiatives sufficient to secure value-added manufacturing? A report from the Canadian Centre for Policy Alternatives (Parfitt 2010) proposes other policy directions such as "targeted research and development funds and tax write-offs to companies making higher-value, finished wood products ... which are not subject to export taxes under the Canada–US Softwood Lumber Agreement." Perhaps more radical is a policy "requiring the holders of new or transferred long-term forest tenures to manufacture products, which would encourage higher-value forest product manufacturing."

Establishing tenures and AAC, reacting to US pressures to reform the stumpage and tenure system in British Columbia, and developing policies to encourage value-added manufacturing are only one part of the political equation. The conflict over Crown forest land and Aboriginal title, which includes most Crown forest land, has been just as, if not more, important. For First Nations, increases to the AAC in the 1970s, '80s, and '90s challenged their traditional territories. Some bands and tribal councils engaged in open confrontations and blockades; others sought redress in court action, and many were successful in gaining injunctions. Logging was stopped on Meares Island, Stein Valley, and southern Moresby Island by this method. Willems-Braun (1997) offers a good overview of how the provincial government and forest companies have denied Aboriginal title and shut out First Nations during this period. The new treaty process, discussed in Chapter 5, is designed to resolve many outstanding problems for First Nations. Although only the Nisga'a and Tsawwassen treaties have been signed, with an allotment of timber negotiated for the Nisga'a, interim measures agreements for other bands, along with community forest tenures, have addressed some of these issues.

The forest industry in much of Vancouver Island, Haida Gwaii, and the interior (regions where many communities depend on the industry) is in a state of crisis. The forest base is declining, employment is at an all time low, the mountain pine beetle has destroyed the pine forests of the interior, environmentalists and First Nations are often in conflict over forest practices, and government decisions to increase parks are often viewed as representative of southern, urban values. Hope lies in reducing log sales, diversifying export trade, and being less dependent on the US housing market. The community forest tenures and the greater involvement of First Nations in forest management also hold promise for the future of the industry, especially if First Nations and community forest tenures are encouraged to allocate their wood into locally owned, value-added products.

SUMMARY

Until recently forestry has been the dominant industry in British Columbia from the perspective of generating revenues and employment and sustaining communities. It favoured the coastal region initially because of the advantages of using the ocean as a means of transportation. Technological developments affected all aspects of the industry and allowed wood to be harvested from more remote locations and processed for many end uses, the most important being lumber, pulp, and paper. The most significant changes occurred from the 1960s on, when the industry expanded into the interior and opened up whole new areas, and communities, in the province. These new technologies also consumed a huge volume of wood each year, putting into question the sustainability of this renewable resource.

As the manager of this resource, the provincial government retains the ownership of 96 percent of the forests and has used various forms of tenure to allow private companies to harvest and process timber from public lands. In the early days, some of the best forest land on the coast was given away through Crown grants and other forms of tenure. These private forest lands may be small in comparison to the whole province, but they dominated the AAC until the 1940s and still generate a significant portion of it.

It was not until the Sloan Commission in 1945 that sustained yield and reforestation became important concepts. The two main forms of tenure today come from this commission – tree farm licences (area based) and public sustained yield units (volume based) – although PSYUs were reorganized into timber supply areas in 1979 with a number of licences to harvest timber.

The large volumes of wood harvested until the 1980s provoked concerns about economic stability, the fall-down effect, and a host of land use and environmental conflicts. New forums and policies – treaties, CORE, LRMPs, the Forest and Range Practices Act (formerly the Forestry Practices Code), and the Forest Investment Account (formerly Forest Renewal BC) – have addressed many issues and made adjustments to park land and harvesting methods, but many concerns persist. The geography of forestry shows the domination of large corporations, which control much of the timber. It also shows how many Vancouver Island and interior communities depend on the forest resource although they have little influence in how it is used and managed. The softwood lumber dispute with the United States has had an enormous impact on the industry. Another change has been a shift in the timber supply from major companies to First Nations and community forests. Still the issue of raw log export remains contentious, as does the greater reliance on corporations to monitor forestry practices. Furthermore, policies to encourage value-added manufacturing remain elusive.

Viewing the forests from the ecosystem perspective of sustainability challenges the basis of the present tenure system and the idea that forests have value solely as fibre. Forests are a renewable resource, but renewable only when they are properly managed.

REFERENCES

Aitkins, T. 1999. "Clayoquot: A Model for the Future." *Greenpeace* 7 (3): 7.

Barker, M.L. 1977. *Natural Resources of British Columbia and the Yukon*. Vancouver: Douglas, David, and Charles.

BC Ministry of Forests. 1988. "Background Information to the Government's Policy to Replace Forest Licenses with Tree Farm Licences." Victoria: The Ministry.

–. 1994. *British Columbia Forestry Practices Code*. Victoria: The Ministry.

Binkley, C.S. 1997. "A Crossroad in the Forest: The Path to a Sustainable Forest Sector in British Columbia." In *Troubles in the Rainforest: British Columbia's Forest Economy in Transition*, ed. T.J. Barnes and R. Hayter, 15-35. Canadian Western Geographical Series vol. 33. Victoria: Western Geographical Press.

British Columbia Round Table on the Environment and Economy. 1993. *An Economic Framework for Sustainability*. Victoria: The Round Table.

Canadian Council of Forest Ministers. 1997. *Compendium of Canadian Forestry Statistics*. Ottawa: Natural Resources Canada, Canadian Forest Service.

Clapp, R.A. 1998. "The Resource Cycle in Forestry and Fishing." *Canadian Geographer* 42 (2): 129-44.
"Coastal Mills Are Now World's Highest-Cost Pulp Producers." 1997. *Vancouver Sun*, 21 November, A10.
Council of Forest Industries (COFI). 1994. British Columbia Forest Industry Fact Book 1994. Vancouver: COFI.
Drushka, K. 1999. "British Columbia's Forests: A New Way to Grow." *Vancouver Sun*, 31 March, D1, 2, 19, 20.
Drushka, K., B. Nixon, and R. Travers, eds. 1993. *Touch Wood: B.C. Forests at the Crossroads*. Madeira Park, BC: Harbour.
Edgell, M.C.R. 1987. "Forestry." In *British Columbia: Its Resources and People*, ed. C.N. Forward, 109-37. Western Geographical Series vol. 22. Victoria: University of Victoria.
Farley, A.L. 1972. "The Forest Resource." In *Studies in Canadian Geography: British Columbia*, 87-118. Toronto: University of Toronto Press.
–. 1979. *Atlas of British Columbia: People, Environment, and Resource Use*. Vancouver: University of British Columbia Press.
Fulton, J. 1999. "British Columbia's Struggling Forest Industry." *Vancouver Sun*, 20 January, A12.
Haig-Brown, R. 1961. *The Living Land*. Toronto: Macmillan.
Hamilton, G. 2009. "Ottawa to Announce $1 billion in Forestry Aid; BC Expected to Get up to $440 Million of the Package Aimed at Making the Industry 'Greener.'" *Vancouver Sun*, 17 June, D1.
Hammond, H. 1991. *Seeing the Forest among the Trees*. Vancouver: Polestar.
Hardwick, W.G. 1960. "Changing Logging and Sawmill Sites in Coastal British Columbia." In *Occasional Papers in Geography* no. 2, 1-7. Vancouver: Tantalus.
Hayter, R. 2000. *Flexible Crossroads: The Restructuring of British Columbia's Forest Economy*. Vancouver: UBC Press.
Hayter, R., and T. Barnes. 1997. "The Restructuring of British Columbia's Coastal Forest Sector: Flexible Perspectives." *BC Studies* 113 (Spring): 6-34.
Hayter, R., and R. Galois. 1991. "The Wheel of Fortune: B.C. Lumber and Global Economics." In *British Columbia Geographical Essays: Geographical Essays in Honour of A. MacPherson*, ed. P. Koroscil, 169-201. Burnaby: Department of Geography, Simon Fraser University.
Husband, V. 1995. Interview by Peter Gzowski. Canadian Broadcasting Corporation, *Morningside*, 5 December.
Jones, T. 1983. *Wilderness or Logging? Case Studies of Two Conflicts in B.C.* Vancouver: Federation of Mountain Clubs of British Columbia.
Kimmins, H. 1997. *Balancing Act: Environmental Issues in Forestry*, 2nd ed. Vancouver: University of British Columbia Press.
Lewis, K.J. 2002. "Forestry and the Forest Industry in the Central Interior of British Columbia." *Western Geography* (Canadian Association of Geographers) 12: 185-215.
M'Gonigle, M., and B. Parfitt. 1994. *Forestopia: A Practical Guide to the New Forest Economy*. Madeira Park, BC: Harbour Publishing.
Marchak, P. 1983. *Green Gold: The Forest Industry in British Columbia*. Vancouver: UBC Press.
–. 1999. *Falldown: Forest Policy in British Columbia*. Vancouver: David Suzuki Foundation.
Pearse, P. 2001. *Ready for Change: Crisis and Opportunity in the Coast Forest Industry*. Victoria: Royal Commission, Ministry of Forests.
Willems-Braun, B. 1997. "Colonial Vestiges: Representing Forest Landscapes on Canada's West Coast." In *Troubles in the Rainforest: British Columbia's Forest Economy in Transition*, ed. T.J. Barnes and R. Hayter, 99-127. Canadian Western Geographical Series vol. 33. Victoria: Western Geographical Press.
Wilson, M. 1998. "Munro: Trees Ready Early." *The Province* (Vancouver), 31 May, A49.

FILMS

Collins, G., and G. Cuthbert, directors. 1980. *The Fall-Down Effect: A Discussion on the British Columbia Timber Supply Analysis*. Selkirk College, Castlegar, BC.

INTERNET

BC Ministry of Agriculture and Lands. 2006. "A New Direction for Strategic Land Use Planning in BC." www.llbc.leg.bc.ca/public/pubdocs/bcdocs/408673/strategic_landuse_planning.pdf.
BC Ministry of Finance. 2009. *2009 British Columbia Financial and Economic Reivew*. www.fin.gov.bc.ca/tbs/F&E review09.pdf.
BC Ministry of Finance. 2010. *2010 British Columbia Financial and Economic Reivew*. www.fin.gov.bc.ca/tbs/F&E review10.pdf.
BC Ministry of Forests. 2004. "Updated Contracting Rules in Timber Harvesting." Backgrounder, 21 June. www.for.gov.bc.ca/mof/plan/bill13.htm.
BC Ministry of Forests and Range. 2009a. "Act Introduced to Expand Wood Use in Public Buildings." www2.news.gov.bc.ca/news_releases_2009-2013/2009FOR0059-000379.htm.
–. 2009b. "Community Forests." www.for.gov.bc.ca/hth/community/index.htm.
BC Stats. 2004a. "Exports (BC Origin) 1994-2003." www.bcstats.gov.bc.ca/releases.

–. 2004b. "Lumber Dispute Big Issue for Small Business." *Small Business Quarterly* 3rd Quarter, 2004. www.bcstats.gov.bc.ca/pubs/sbq/sbq04q3.pdf.

–. 2009. "British Columbia Local Area Economic Dependencies – 2006." www.bcstats.gov.bc.ca/pubs/econ_dep/2006/2006_2.pdf.

CBC. 2006. "Softwood Lumber Dispute." 23 August. www.cbc.ca/news/background/softwood_lumber/.

–. 2010. "B.C. Town Braces for Mill Shutdown." 26 January. www.cbc.ca/canada/british-columbia/story/2010/01/26/bc-mill-shutdown-kitimat-mackenzie.html.

Council of Forest Industries (COFI). 2001. *Fact Book 2000*. www.cofi.org/reports/factbooks.htm.

National Forestry Database Program. 2004a. "Net Merchantable Volume of Roundwood Harvested by Category and Province/ Territory, 1940-2003." Table 5.1. nfdp.ccfm.org/compendium/ products/tables.

–. 2004b. "Silviculture Statistics by Province/Territory, 1975-2003, Number of Seedlings Planted." Table 6.1. nfdp.ccfm.org/compendium/silviculture/archive.

–. 2010. "Wood Supply." nfdp.ccfm.org/supply/quick_facts_e.php.

Natural Resources Canada. 2009. "Forest-Dependent Communities in Canada." canadaforests.nrcan.gc.ca/indicator/communities.

–. 2010a. "Canada's Forests: Statistical Data – BC Economy." canadaforests.nrcan.gc.ca/statsprofile/economicimpact/bc.

–. 2010b. "Canada's Forests: Statistical Data – BC Trade." canadaforests.nrcan.gc.ca/statsprofile.

–. 2010c. "Mountain Pine Beetle." mpb.cfs.nrcan.gc.ca/index_e.html.

Osberg, M. and Murphy, B. N.d. " British Columbia Forest Practices Code." www.fao.org/docrep/w3646e/w3646e0a.htm.

Parfitt, B. 2008a. "Softwood Lumber Agreement Rebates Not Invested in BC." Canadian Centre for Policy Alternatives. www.policyalternatives.ca/newsroom/news-releases/softwood-lumber-agreement-rebates-not-invested-bc.

Parfitt, B. 2008b. "Stumped: B.C. Reforestation." *BC Business*. August 1. www.bcbusinessonline.ca/bcb/top-stories/2008/08/01/stumped-b-c-reforestation?page=0%2C0.

–. 2010. "Managing BC's Forests for a Cooler Planet." Canadian Centre for Policy Alternatives. www.policyalternatives.ca/sites/default/files/uploads/publications/reports/docs/ccpa_bc_managingforests.pdf.

Pedersen, L. 2003. *Allowable Annual Cuts in British Columbia: The Agony and the Ecstasy.* UBC Faculty of Forestry Jubilee Lecture. Vancouver, March 20. www.for.gov.bc.ca/hts/pubs/jubilee_ubc.pdf.

People's Voice. 2008. "BC's Forest Jobs Crisis." www.peoplesvoice.ca/articleprint21/02)_BC'S_FOREST_ JOBS_CRISIS.html.

Pojar, J. 2010. "A New Climate for Conservation Nature, Carbon and Climate Change in British Columbia." West Coast Environmental Law. wcel.org/resources/publication/new-climate-conservation-nature-carbon-and-climate-change-british-columbia-ful.

Price Waterhouse Coopers. 2007. *Report on the Economic Impact of the BC Pulp and Paper Industry.* www.llbc.leg.bc.ca/public/pubdocs/bcdocs/431477/final_pwc_report_to_task_force_nov_07.pdf.

Pulp and Paper Canada. 2010. "Elk Falls Shutdown becomes Permanent." www.pulpandpapercanada.com/issues/story.aspx?aid=1000377726.

RBC Commodity Price Monitor. 2010. January. www.rbc.com/economics/market/pdf/cpm.pdf.

Redlin, B. 2010. *BC Commentary: Communities in Crisis – A Case Study of Campbell River.* Canadian Centre for Policy Alternatives. www.policyalternatives.ca/sites/default/files/uploads/publications/reports/docs/bccwinter10.pdf.

Shuswap Environmental Action Society. 2006. "Forestry Issues." www.seas.ca/content.asp?p=887.

Sierra Club. 2006. "British Columbia Value-Added Wood Policies: A Case Study." www.sierraclub.ca/national/programs/biodiversity/forests/scc-bc-value-added.pdf.

Statistics Canada. 2010. "Domestic Exports, Customs-Based, by Province of Origin, Monthly (Dollars)." CANSIM Table 228-0034. www.statcan.gc.ca/.

United Steelworkers. 2010. "More Raw Log Exports Are the Last Thing Needed Say Steelworkers." 22 January. Steelworkers: District 3 – Western Canada. www.usw.ca/program/content/6345.php.

Vancouver Sun. 2009. "Crofton B.C. Pulp Mill Closure Prompts Calls for Industry Aid." www.workingforest.com/content/crofton-bc-pulp-mill-closure-prompts-calls-industry-aid.

The Fishing Industry

Managing a Mobile Resource

Fish, like trees, are a renewable resource, but one very difficult to manage because it is mobile and not easy to see. A great many species of fish live off the coast of British Columbia, which "offers over 90 different species of wild and cultured fish, shellfish and plants harvested from fresh and marine waters" (BC Stats 2009, 9). This chapter concentrates mainly on the complex story of the salmon fishing industry, but the issues discussed here apply to the range of commercial species: over-harvesting, international disputes, intense competition, changing international markets, technological improvements, a deteriorating aquatic ecosystem, the introduction of aquaculture (fish farming), and poor management. Pacific salmon once dominated the commercial catch in British Columbia. As Table 9.1 indicates, the value of the salmon fishery, which was nearly 72 percent of the total value in 1951, was a mere 3 percent of the total value in 2008. By contrast, other commercial catches such as shellfish, halibut, cod, and other groundfish have increased in value. As well, the products of aquaculture, especially farmed Atlantic salmon, have increased in value immensely.

The five species of Pacific salmon have different physical characteristics and economic value, but they are all born in fresh water, migrate to the ocean, and return to fresh water, where they spawn and die. Understanding the fresh- and saltwater environments is central to our ability to manage the salmon and ensure survival.

Like other resource extraction industries in British Columbia, the fishery has a long history of technological change. The invention of canning solved the problem of preserving and exporting the salmon, and canneries were once located at the mouths of salmon-bearing rivers up and down the coast of the province. The canneries went through many changes, moving from a labour-intensive process to the capital-intensive Fordist process relatively early in the twentieth century. Consolidation of canneries by large corporations also occurred at this time. New markets and methods of preserving salmon developed over time, reducing the need for canneries.

The process of salmon fishing also underwent major technological change. Low technology gillnetters, which used oars and sails, were eventually supplanted by the most modern fishing fleet in the world, with a huge capacity to catch fish. This transformation had a major impact on cannery location. Historically, boats had little ability to travel any distance and were therefore tied to canneries. As fishing vessels developed the ability to travel farther and farther from the canneries, and technologies were developed to preserve fish on board, canneries in remote coastal locations were no longer needed. Abandonment of canneries and communities became the norm.

Table 9.1

Commercial and aquaculture catch by value, 1951-2008

		1951		1981		1991		2001		2008	
		Value ($'000)	%	Value ($'000)	%	Value ($'000)	%	Value ($'000)	%	Value ($'000)	%
Wild fish	Groundfish	4,611	11.7	27,873	11.8	102,741	20.7	118,875	19.0	102,162	15.5
	Pelagic	34,141	86.4	193,958	82.1	222,272	44.8	85,020	13.6	45,095	6.8
	Salmon	28,401	71.9	158,067	66.9	172,440	34.8	37,131	5.9	21,462	3.2
	Shellfish	712	1.8	11,145	4.7	46,279	9.3	123,001	19.7	88,753	12.3
	Total wild fish	39,495	100.0	236,181	100.0	380,181	76.7	339,706	54.3	236,009	35.7
Aquaculture	Salmon	–	–	–	–	110,913	22.4	269,400	43.1	406,200	61.5
	Shellfish	–	–	–	–	4,021	0.8	15,700	2.5	15,700	2.4
	Total farmed fish[a]	–	–	–	–	115,472	23.3	285,600	45.7	424,500	64.3
Overall total		–	–	–	–	495,653	100.0	625,306	100.0	660,509	100.0

a Includes some trout and other farmed species.
Source: Fisheries and Oceans Canada (2009d, 2010).

Salmon is a food resource, and its markets therefore fluctuate because of world events such as war and depression and also according to perceptions of how a particular food affects health. These considerations are reflected in changing world market prices and demands, putting the economic viability of fishing into question.

Most salmon are caught in the ocean, making management a federal responsibility. Management includes preservation, enhancement of stocks, and protection of habitat. That salmon migrate into international waters, often return through American waters, and finally end their journey in freshwater streams, where the provincial government allows conflicting resource extraction activities, makes the management task extremely difficult. The balancing act for the federal government is in allowing interested parties to harvest a certain number of salmon while simultaneously allowing a certain number of fish to escape upstream to spawn and start the cycle again. The industry has many interests intensely competing for a share of this resource.

A seemingly endless array of commissions and hearings reflects the difficulties in management. Unfortunately, the conflicts and issues of the past persist today, with even greater concerns about the sustainability of salmon. The loss and deterioration of freshwater habitat has prompted salmon enhancement programs that range from rebuilding spawning beds to establishing hatcheries. The use of hatcheries to raise and release salmon fry artificially is controversial in itself, but the development of the salmon farming industry (aquaculture) has provoked even greater friction. The price and demand for salmon has increased, creating pressure for larger harvests. Fish wars with the United States have resulted in overfishing. First Nations are demanding traditional rights to a food fishery and want a share of the commercial catch. The sport fishery, which is the basis of a growing, lucrative tourism industry, wants more fish. Trollers, gillnetters, and seiners, shown in Figure 9.1, make up the commercial fleet, and they have a large investment in boats and gear that needs to be paid off; they too want more salmon.

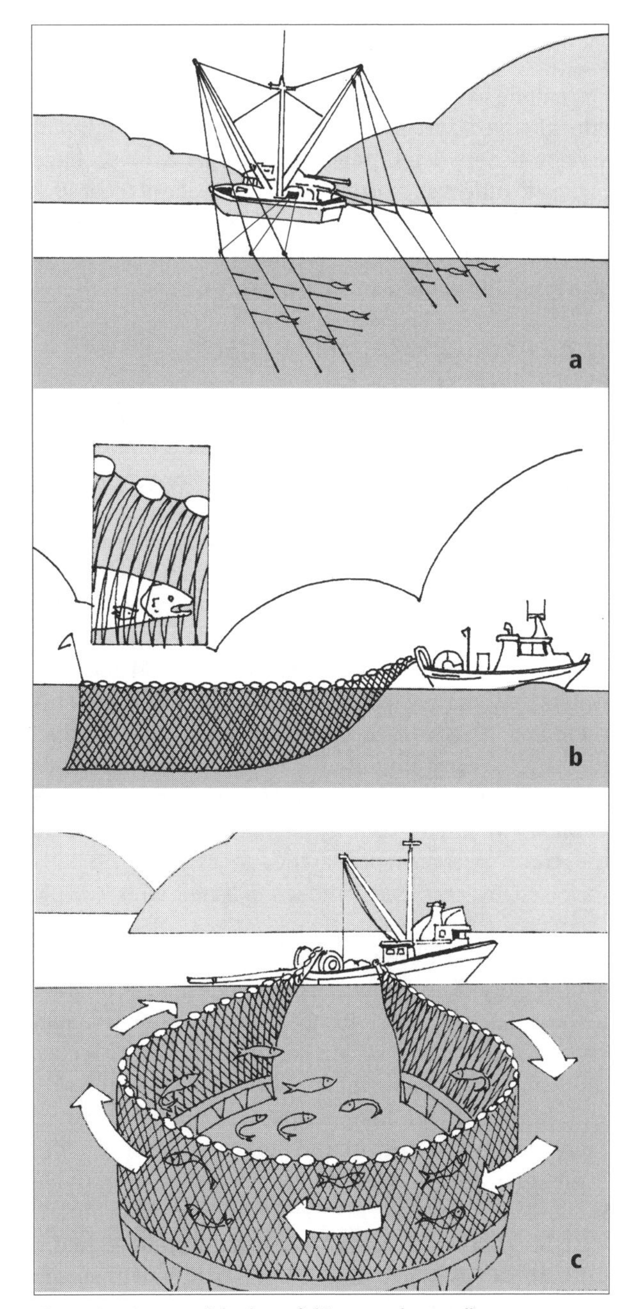

Figure 9.1 Commercial salmon fishing vessels: a, troller; b, gillnetter; c, seiner
Source: Fisheries and Oceans Canada (1999), 53-4. Used with permission.

Clearly, there are not enough salmon to satisfy the demands of all the interests. The provincial government is demanding a larger role in management, and so are local fishing communities and Aboriginal groups. A more fundamental question of management is whether this resource should be privatized. This question has precedence in the halibut fishery; there are considerable differences in the life cycles of halibut and salmon. As the wild salmon fishery deteriorates, commercial harvests have turned to other fish, particularly as the price of most species, especially shellfish and some groundfish (e.g., halibut) increases. It is necessary, therefore, to assess the sustainability of these other species.

SPECIES, CYCLES, RACES, AND HABITAT OF PACIFIC SALMON

There are five species of Pacific salmon, commonly known as sockeye, chinook (spring), coho, pink, and chum. Sockeye is the best known in terms of external demand and commands the highest price because of its rich red meat. Chinook, also referred to as spring or king salmon, is the largest of the five species, weighing up to thirty kilograms. Coho from northern and southern British Columbia differ in size. Large coho in southern waters weigh approximately four kilograms, whereas in northern waters they can reach over ten kilograms. Coho and chinook are referred to as the hook-and-line species because they are the most sought-after by anglers. Both species are essential to the sport fishery and tourism. Pinks and chum are referred to as the schooling varieties because they travel together at or near the surface in schools. This makes them accessible to the seiner fishers who encircle them with nets.

Salmon are anadromous: they are born and spend part of their early life in freshwater lakes and streams, migrate to the ocean, and then return to the fresh water to spawn and die. Although the cycle is common to all five species, there are variations within it. Sockeye spawn in rivers with lakes in them, for example, and spend a year or more there before migrating to the ocean. Sockeye in the Fraser follow a four-year cycle while the cycles of those in the Skeena and Nass River systems range from four to six years (Barker 1977, 115). Some of the chinook in the Columbia River system before it was dammed migrated 1,400 kilometres or more and were on a seven-year cycle. Coho often have a three year cycle, and pinks follow a two-year cycle. Chum and some pink salmon rarely migrate any distance upstream, preferring the gravel bed at the mouth of the river.

One way to view the relationship between the various salmon species and the rivers and streams where they are born is in terms of ownership. Because most salmon return and spawn in exactly the same stream in which they were born, that stream "owns" those salmon, and the salmon from it are referred to as a race, run, or stock of salmon bearing its name. Sockeye born in the Adams River, for example, are referred to as the Adams River race, which has been the largest run of sockeye salmon in the world. The Quesnel River also has a large race of sockeye, which by coincidence is on an identical four-year cycle to the Adams run. Both the Adams and Quesnel Rivers are tributaries of the Fraser. Consequently, every four years there is an enormous run of sockeye salmon on the Fraser. Of course, a river or stream may have several races of different salmon species each year.

Since salmon return to their birth streams, the physical condition of those streams is essential to the continuation of salmon stocks. Figure 9.2 shows the main salmon-bearing rivers in British Columbia.

The federal government manages the salmon resource and through the Fisheries Act is responsible for both saltwater and freshwater habitats of salmon. Unfortunately, the province has jurisdiction over land-based resources such as forestry, mining, agriculture, and hydroelectric production, and its encouragement of development has permitted many freshwater habitats to suffer serious damage, some of it irreversible. The gold rush on the Fraser and into the Cariboo turned streams upside down and inside out, destroying spawning beds. Later mining ventures and forestry increased siltation, added pollutants, and warmed the freshwater environment. Dams bring an end to salmon migration, which often means the extinction of races of salmon. Increased urban development not only adds wastes to our river systems but often results in streams being culverted and physically changed.

One specific catastrophe was the Hell's Gate slide in 1913. Hell's Gate is an extremely narrow restriction in the Fraser River just north of Yale that the salmon must navigate to reach the spawning beds of much of the Fraser River basin (Figure 9.2, insert). The construction of the Canadian Pacific Railway in the 1880s on one side of the Fraser, and then the Canadian National Railway in 1912-13 on the other side, added to the already difficult passage by dumping huge amounts of blasted rock into the Fraser. These problems were compounded when the rock loosened from blasting gave way in a rockslide in the late summer of 1913. Few salmon were able to overcome this barrier (Meggs 1992; Hume 1997). Tragically, 1913 was the fourth year of the sockeye cycle, and the rockslide occurred just as the salmon were about to return to their spawning grounds. Figure 9.3 shows the large numbers of sockeye salmon every four years prior to 1913 and the consequences of the slide. After the slide, it took over 30 years to install fish ladders, in 1946, when "eleven concrete fishways [were] designed in Seattle and built for $24 million cost shared equally by Canada and the United States" (Hume 1997). Even though fish ladders allow salmon to get past Hell's Gate, rebuilding these stocks to historical levels is a difficult process that is still in progress.

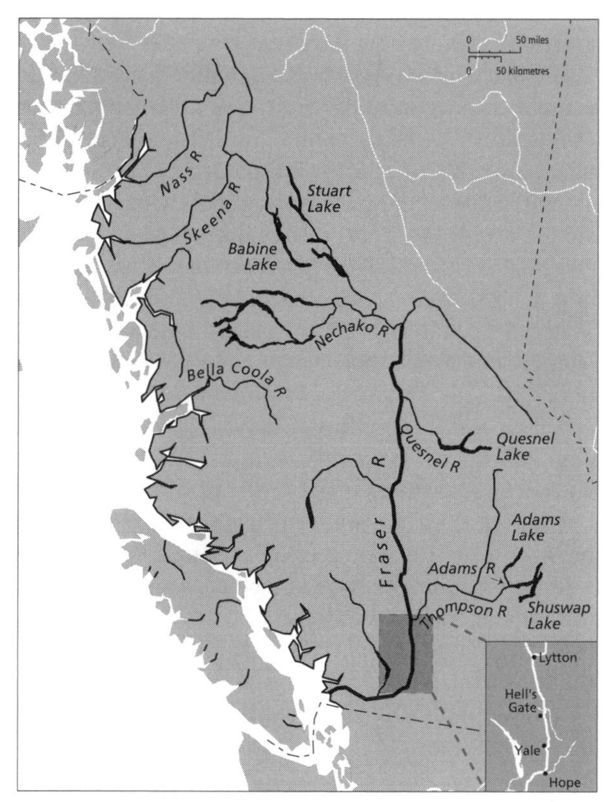

Figure 9.2 Major salmon-bearing rivers and lakes

The saltwater environment presents other factors to consider in managing the salmon. Seals and sea lions take their share of the salmon, and in the past war was waged on both of these mammals. Forester and Forester (1975, 217) give us some idea of how much of a threat these animals were perceived to be, according to the BC Department of Fisheries 1916 annual report: "Notwithstanding the fact that 749 seals and 2,875 sea lions were destroyed during the season of 1915, it was only a drop in the bucket, especially as in the Fraser River, where the depredations from seals appear to be greatest, only forty-eight of these mammals were destroyed, and it is prophesied by certain fishermen that unless a greater destruction takes place the spring-salmon fishery is doomed." They go on to explain how bounties, patrol officers with machine guns, and homemade bombs were employed to exterminate these salmon predators, but with limited success.

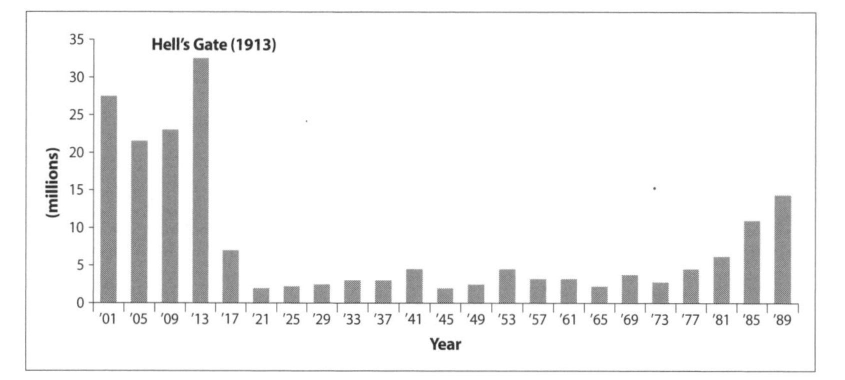

Figure 9.3 Commercial catches of Fraser sockeye salmon, 1901-89
Source: Roos (1991, 412, 413).

Human interference has brought on some serious imbalances in the saltwater ecosystem. Whales, the natural predators of seals and sea lions, were drastically reduced in numbers by a pervasive whaling industry. Natural causes and the herring fishery radically reduced herring, the main food supply for salmon, in the mid-1960s. Today, the seal population appears to be very large but environmental constraints prevent seal hunting.

Recent evidence suggests that the El Niño effect is responsible for bringing warmer water to the Pacific Northwest, and with it, mackerel, which normally do not migrate this far north. The mackerel become another consumer of young salmon. Of course, the greatest consumers of salmon in the saltwater environment are the many fishing interests, both foreign and local.

With all these circumstances taken together, one wonders how any salmon make it back to their spawning grounds. Mary Barker (1977, 115) points out the survival-of-the-fittest conditions for salmon: "3,000 eggs may be deposited by a sockeye, but only 100 will become fingerlings and travel to the sea. Of these, less than twenty will return as adults."

TECHNOLOGICAL CHANGES TO PRESERVING AND CATCHING SALMON

Salmon have always been an important resource to First Nations people, who preserved them by smoking and sun drying. When the Hudson's Bay Company forts, such as Fort Langley on the Fraser, became interested in salmon for export, they used the traditional European process of salting and barrelling. The commercial development of this industry, however, required a means of preserving salmon without altering its taste the way salting did. The invention of canning was the key, and canneries began to appear at the mouth of the Fraser River in the 1870s.

Duncan Stacey (1982) defines two periods: the manual canning era from 1871 to 1903, and the mechanical canning era from 1903 to 1913. Technology was responsible for this transition from labour-intensive to capital-intensive methods of processing fish. The manual canning era, as the name suggests, required much of the canning process to be carried out by labourers. As Stacey explains, "In the earliest canneries each can was cut out of sheet tinplate, formed, and soldered, by hand. By 1890 a number of machines had been introduced to punch out body pieces, tops, and bottoms and to apply solder, but these were still aids to the hand process" (p. 4).

In 1876, 9,847 cases of salmon were exported, with twenty-four one-pound tins to a case (Stacey 1982, 4). In 1905 the figure was 837,489. These statistics may suggest that a major change in technology was responsible for the changes in this twenty-nine-year period. In fact, part of the cause lies in a market trend after 1888. Prior to 1888, only sockeye salmon were harvested. There was no market for the other, lighter-coloured salmon species with less oil content. Commercial fishing was therefore only a five-to-six-week occupation, and in 1876 there were just three canneries on the Fraser. From 1889 onward, however, the market was expanded to include all five species of salmon. The fishing season lasted five to six months, and by 1901 there were forty-nine canneries.

The mechanized era brought a number of changes to canning, the two most significant being efficiency on the production line and corporate concentration. The introduction of electricity for lights and motors and Fordist assembly line techniques replaced the high demand for labour. Stacey (1982, 19) tells us that "hand-butchering gangs gave way to butchering machines, manually soldered cans began to be replaced by the mechanized, solderless, or sanitary, can, and all other sections of the canning line experienced varying forms of mechanization." The Smith Butchering Machine, developed in 1907, was the most important mechanical invention of this era. With the assistance of three people it could clean and cut up sixty to seventy-five fish per minute, replacing butchering gangs of about thirty people (p. 21). Those replaced were frequently Chinese, Japanese, and First Nations labourers. Consolidation occurred very early in this industry, when the British Columbia Packers Association acquired twenty-nine of the forty or so canneries on the Fraser in 1902 (p. 19). Consequently, this large corporation was able to invest in the capital equipment necessary for adopting Fordist technologies.

The assembly line process of canning fish still exists, but few canneries remain. With the fresh and fresh frozen markets and products such as smoked salmon becoming increasingly popular since the 1970s, canneries have given way to fish-processing plants. Canneries were abandoned in isolated coastal locations such as Rivers Inlet, Namu, and Klemtu as changes to fishing boat technology and refrigeration allowed vessels to travel the whole length of the BC coastline. By 1996, fish processing plants were concentrated in the Vancouver, Vancouver Island, and Prince Rupert areas.

In the nineteenth century, gillnetting was the most popular method of fishing in the salmon industry. Gillnets were set up in the mouths of the rivers to intercept the salmon as they went upstream to spawn. Gillnets would be set from Columbia River boats. Similar to the east coast dory, this boat is approximately six metres long, pointed at both ends, and operated by two people with oars and a sail. Typically, the gillnetter would row or sail upstream, set a net, drift downstream, pull the gillnet in by hand, and row upstream to repeat the process. When the boat was full it was time to travel to the cannery to unload the catch. This process changed little prior to 1903. The "big" invention was the placement of a roller at the stern of the vessel to facilitate hauling in the net.

Because they were propelled by oar or sail, these low technology gillnetters were able to range only short distances from the canneries where they had to unload their catch. With the proliferation of canneries and fishers, waters became crowded and the competition intense. Some competition was relieved with the steam-powered tender boat, used by some of the larger canneries to tow the gillnetter farther out with the expectation of intercepting the salmon.

For the individual fisher, the invention of the gasoline engine in 1907 heralded a major change: "The mechanized gillnetter could make more sets since it could move more quickly upriver to start a new drift. The gasoline engine also enabled the fishery to increase the fishing area by working farther offshore. It increased fishing time since it took less time to travel to and from the grounds and vessels could fish in rougher weather" (Stacey 1982, 26). The gasoline motor was also responsible for a vast improvement in the purse seine fishery, whereby a motorboat was used to set out a very long net that could rapidly encircle schools of salmon (Figure 9.1).

It should be noted that a rather unusual set of rules over the use of motors divided the fishery into two regions. In the northern fishery, mainly in the Skeena River and Prince Rupert region, the canneries were much more in control of the whole fishing industry. They owned the vessels and licences and did not wish to invest in motors. Nor did they want competition from independent fishers, so they had motors legally banned in 1911 (Stacey 1982, 26).

The ability of motorized fishing boats to travel greater distances from the cannery posed another problem: perishability. Once salmon are caught and hauled in, the deterioration process begins. This problem was resolved by 1915 with a diesel-powered packer that used ice as a means of refrigeration. The motorized gillnetters could now unload their catch at sea onto these relatively large vessels. The diesel-powered packer was designed to stack the salmon in a hold without crushing them.

Fishing is a highly competitive industry in which new technologies are constantly sought to gain advantage. Many inventions arising from the First and Second World Wars to identify submarines, to communicate, and to lift war materials diffused down into the fishing industry. Vessels installed echo sounders, asdic, sonar, radar, radio telephones, hydraulics, and other devices. Along with these sophisticated technologies came synthetic nets, methods of freezing salt water, and a host of other inventions. Individual vessels got larger, their ability to travel great distances to catch salmon improved, and so did their capacity to harvest salmon. Yet increased capacity was not matched by increases in the resource. In fact, there have been great concerns that the salmon fishing industry may meet the same end as the cod fishing industry on the east coast.

MARKET FACTORS AFFECTING A FOOD RESOURCE

Salmon, like most resources, is subject to many factors that influence supply and demand, and these are reflected in the world market price. Differences in demand and value apply to the five species, with sockeye being the most lucrative. Global recessions and depressions have reduced the demand and price for salmon generally. On the supply side, the natural cycle of salmon causes fluctuations, making for a lack of continuity of markets and income for fishers. Because it is a food resource, other unique influences affect the demand for salmon.

The Yukon gold rush of 1898 is an example of unpredictable changes in the demand for canned salmon. This gold rush was met with fear by the authorities, who thought that the stampede of miners entering the Yukon with little knowledge of this northern environment would starve to death. One of the requirements put in place for miners entering the territory was that they must be in possession of 1,100 pounds (500 kilograms) of provisions. This meant a boom for canned BC salmon since would-be-miners frequently purchased their supplies in Victoria and Vancouver and tinned salmon was relatively easy to pack.

Wars have also had a positive influence on the demand for this resource, which is a high source of protein and easily stored. Both world wars saw an increase in salmon processed to feed the troops. Statistics for 1930 and 1939 show

Table 9.2

BC salmon exports, 1930 and 1939

	1930 (cases)	1939 (cases)
To the British Empire		
United Kingdom	189,227	406,943
Australia	227,423	278,047
New Zealand	62,404	64,470
British South Africa	35,202	71,997
Fiji	11,015	15,020
British India	6,113	8,225
Other British countries	22,217	0
Total	553,601	844,702
To other countries		
France	219,854	108,283
Belgium	58,721	9,306
Italy	152,044	0
Colombia	5,011	9,252
Other South American countries	57,319	0
Portuguese East Africa	8,723	6,806
All other countries	73,526	43,363
Total	575,198	177,010

Source: Lyons (1969, 433).

the impact of the Depression of the 1930s and the market adjustments due to the start of the Second World War (Table 9.2). The end of the war had an unexpected impact also. As a result of the great expenses Britain had incurred in the war, BC salmon became a "luxury" item. By 1948, the British market for canned salmon had collapsed.

Food resources are subject to rumours and misinformation that can have a negative impact on their market. The "canned beef scandal" of 1906 is a typical example. Contaminated beef, tinned in Chicago, found its way to the British market. The backlash to this disastrous affair was that consumers in Britain rejected all imported canned food, including salmon from North America. As well, salmon producers have had to combat rumours that sockeye salmon is dyed to achieve its distinctive red colour and, worse yet, that it is linked to salmonella. This deadly virus, associated with poultry, was discovered in the late 1930s by Dr. Daniel Salmon. Although the virus had nothing to do with salmon, the word association was strong enough that many consumers were not going to take the risk. Several cases of a potentially lethal disease known as botulism, resulting from improper canning, did occur in the 1970s. When food is involved in a perceived risk to health, the demand for that food will plummet.

Today, the fresh and fresh-frozen salmon market is marked by another controversy: wild versus farmed salmon. Commercial trollers and salmon farmers often compete for the same market, and those catching wild salmon claim that farmed salmon is an inferior product that should be labelled for consumer information, as it has been in adjacent Alaska since 2004. Bumper stickers stating "Real salmon don't eat pellets!" reflect, at least, the war of words in this battle.

MANAGING SALMON: THE POLITICS OF REGULATION

British Columbia entered Confederation in 1871 with the condition that the federal government had jurisdiction over oceans and their resources. Coincidentally, the first canneries were being established in British Columbia at this time.

From a commercial perspective, there was little concern over the resource until 1888. Only sockeye were being caught prior to that time, but by 1889 the market had expanded to all species of salmon, increasing the season to five or six months. Regulations were put in place for the 1889 season that included the following:

1. Introduction of licenses with a restriction of 500. 350 for canneries and 150 to "outside" fishers.
2. Aboriginals prohibited from selling salmon.
3. Nets restricted to tidal waters and banned from fresh water.
4. Nets not permitted to obstruct more than one-third of a river.
5. Minimum size of mesh for nets – 6 inches.
6. Ensnarement, or fish traps, illegal.
7. Anchoring nets on shore, illegal.
8. No form of seining in the Fraser River permitted.
9. Closure: Saturday 6 p.m. to Monday 6 a.m. (quoted in Lyons 1969, 192).

Salmon have (until recently) been viewed as a **common property resource**, a resource for all to use, a view that brings with it, unfortunately, the possibility for abuse.

The introduction of regulations kept salmon as a common property resource but defined who was allowed to compete for it (licence holders) and the conditions under which they could compete. In this form of resource management, the federal government acts as the gatekeeper and operates under the assumption that it understands migrations of salmon and that the regulations are sufficient to allow salmon to escape upriver to spawn. The move from a completely unregulated industry to a fully regulated one, however, caused many in the industry to question both the assumption and the regulations. Whenever the public demonstrates sufficient concern over any rights or rules, the usual government response is to strike a commission to investigate and recommend changes. The first of many commissions on salmon occurred in 1890 as a reaction to the imposition of regulations in 1889.

The Wilmot Commission of 1890 was rather unsympathetic to the concerns of the fishing industry on the west coast and recommended two small changes to the regulations: a reduction in minimum mesh size from six inches to five and three-quarter inches, to enable a minor increase in catch; and a reduction in the closure to twenty-four hours, from Saturday 6 p.m. to Sunday 6 p.m. These changes were minimal, and the fishers demanded another commission to have the regulations relaxed. A new commission was struck two years later. The most significant new recommendation in 1892 was the removal of licence restrictions. Anyone could now purchase a licence to fish, with two long-term results: "The traditional fishing grounds became overcrowded and a change took place in the type of license ownership. The traditional ownership of the fishing license by the cannery gave way to non-cannery, or private, ownership of licenses" (Stacey 1982, 13). For the managers of this resource, more fishers meant more competition. Competition encouraged new technologies, which increased the capacity for each vessel. So many fishers entered the industry that for the first time there was some anxiety about sustainability.

The federal government made some efforts to increase the number of salmon through hatcheries. The first one was located at Port Mann on the lower Fraser River in 1883, and others followed. Meanwhile, streams were being damaged through mining and logging activities, but the federal government had little say about these provincial activities.

The questions and recommendations of the Duff Commission in 1922 showed other considerations in managing salmon. The three questions addressed by the commission were:

1 Should motor boats be allowed in the northern waters?
2 Do the Japanese have too many licences?
3 What should be done about the drastically reduced Fraser sockeye?

Recommendations brought down in 1924 included permission to use motor boats in northern waters and a 40 percent reduction in licences to the Japanese. The commission stated that it could do nothing about the Fraser River sockeye as it was an international problem (Lyons 1969, 352). The significance of the 1913 Hell's Gate disaster was not mentioned.

The need to manage salmon from an international perspective can be seen from Figure 9.4, which indicates the migration boundaries of the various species of salmon. Many of these runs of salmon migrate through international waters and the jurisdictional waters of Alaska and Washington State before returning to rivers in British Columbia to spawn. It should be noted, however, that national water jurisdiction has not remained static. Before 1964, jurisdiction extended to only three miles (4.8 kilometres) from the shoreline. The limit changed to nine miles (14.5 kilometres) in 1964, to twelve miles (19.3 kilometres) in 1971, and finally to 200 miles (322 kilometres) in 1977. It should be noted, however, that even with these changes most salmon continued to migrate through US water before entering Canadian waters, and Canada contends that the salmon born in the fresh waters of British Columbia belong to Canada. This argument has led to many disagreements with the United States.

When the border between British Columbia and Washington State was established in 1846, it was agreed that Vancouver Island would be included as British territory although it dipped below the forty-ninth parallel. Unclear at that time was where the boundary lay with respect to the San Juan Islands (Figure 9.5). The British claimed that the international boundary should be through the Rosario Strait, and the Americans countered that it should be the Haro Strait. The Middle Channel was proposed

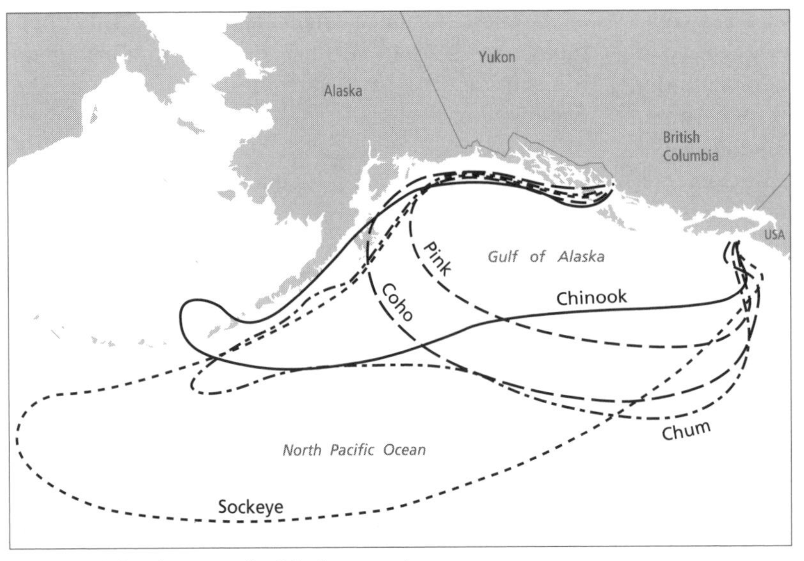

Figure 9.4 Migration routes for BC salmon species
Source: Modified with permission from Fisheries and Oceans Canada (n.d., 2-6).

as a compromise, but American insistence on the Haro Strait saw the British accede the territory. The importance of this boundary to the salmon industry was that many of the runs of salmon destined for the Fraser River come through American waters. The closures and restrictions set out in the Canadian regulations in 1889 did not apply in the United States and Americans were taking more Fraserbound salmon than were Canadians.

Commission after commission attempted to negotiate treaties with the United States. All ended in failure and frustration, with the Americans taking the lion's share of the resource. The report of the Commission of Fisheries for 1931, which ends with a threat, expresses these sentiments:

> The failure of the Senate to ratify this treaty is a matter of grave concern in Canada. For years she has sought to maintain the sockeye-salmon fishery of the Fraser System by closely restricting the operations of her fishermen and by extensive hatchery operations. Notwithstanding her continuous efforts for the last thirty years, the fishery has steadily declined. She has failed to maintain it because of lack of support and assistance on the part of the United States. Her patience is close to exhaustion. Failing favourable action on the treaty now before the Senate, it is seriously proposed that Canada abandon further efforts to maintain the sockeye-salmon fisheries of the Fraser System and, in lieu thereof, collect all the sockeye-eggs obtainable in the Fraser River basin, transfer them to her hatcheries on the Skeena and Rivers Inlet, and liberate the resulting fry in these waters, over which she has complete control. The suggestion is entirely practical, because it has been repeatedly demonstrated that adult sockeye return to spawn in the fresh waters in which they spent the first year of their life. (Canada 1931, 12)

Tough words, but no ratification of a treaty occurred until 1937, and even then it included only sockeye. An examination of Table 9.3 indicates the positive results of these negotiations.

Many of the pink salmon return through the San Juan Islands, and this species became the next target of negotiations. A treaty was signed in 1957 only after a very sophisticated seine fleet was sent off Vancouver Island, threatening to intercept the pinks before they reached Washington State waters. It took until 1985 to negotiate an agreement on all species of salmon. This agreement had an eight-year review clause attached to it and the treaty was cancelled at the 1993 review, putting the Canada-US relationship back sixty years. The inevitable fish war serves the interests of no one, particularly with respect to future salmon stocks.

Managing salmon at the international level has been only intermittently successful. The 1997 round of treaty negotiations failed, marked by the blockade of an Alaskan ferry in Prince Rupert followed by threats and counterthreats of lawsuits. The 1998 negotiations fared little better. Federal fisheries minister David Anderson announced a total ban on fishing coho salmon in Canadian waters because of the fear that some important races of coho, notably the Skeena and Thompson River runs, were

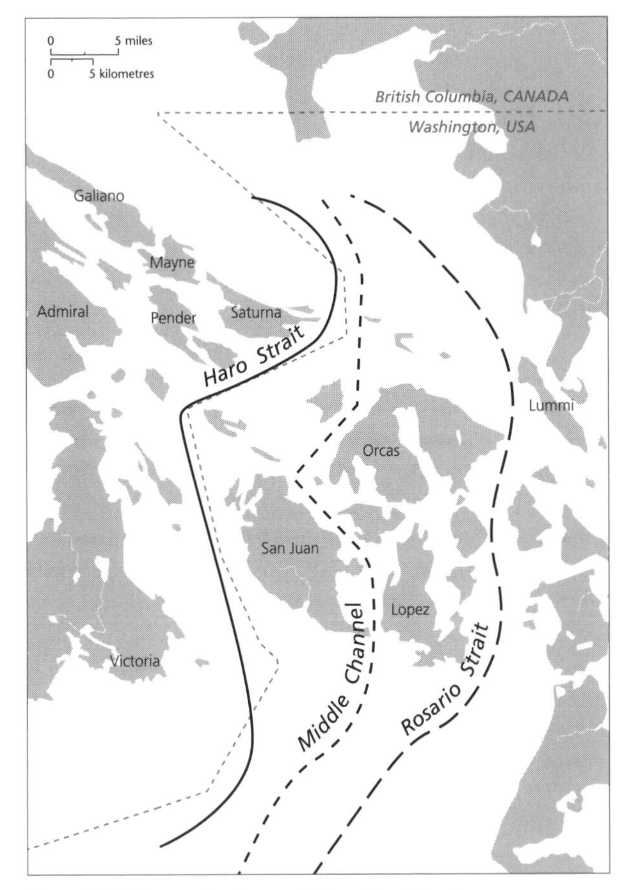

Figure 9.5 Salmon migration and international borders
Source: Modified from Lyons (1969, 214).

Table 9.3

Fraser River sockeye salmon catches for selected years

	Canada		United States		Total
	cases	%	cases	%	cases
1913	719,796	30.1	1,673,099	69.9	2,392,895
1917	148,164	26.5	411,538	73.5	559,702
1921	39,631	27.8	102,967	72.2	142,598
1925	35,385	24.0	112,023	76.0	147,408
1929	61,569	35.5	111,898	64.5	173,467
1933	52,465	29.0	128,518	71.0	180,983
1937	100,272	62.5	60,259	37.5	160,531
1941	171,290	60.8	110,605	39.2	281,895
1945	79,977	60.1	53,055	39.9	133,032
1949	96,159	54.4	80,547	45.6	176,706
1953	191,123	51.7	178,323	48.3	369,446

Source: Lyons (1969, 508).

close to extinction. Alaska issued a statement denying that the coho was threatened and stating that Alaskan fishers intended to catch their share (CBC 1998). The Pacific salmon "fish war" between Canada and the United States ended with a ten-year agreement in June 1999 (Fisheries and Oceans Canada 1999), enabling the managers of this resource to turn their attention to the many issues within their jurisdiction. A further ten-year agreement was completed in 2009. This agreement "places strong emphasis on conservation, stability of access for harvesters, and the sustainability of the Pacific salmon resource" (Fisheries and Oceans Canada 2009c).

The BC fishery also experienced many internal conflicts. By the 1950s, the main issue was how to regulate the increasing capacity of the salmon fishing fleet. The Sinclair Report of 1958 "proposed a system of restricted vessel licenses and levies on catches to dampen incentives for over-investment" (Ross 1987, 189). It was the Davis Plan of 1968 that actually resulted in the reduction of the fleet, from 6,104 vessels in 1969 to 4,707 vessels in 1980. The unexpected consequence was new investment in the remaining vessels; some were upgraded and others were replaced. The seine fleet, for example, "actually increased from 286 to 316, because the new rules allowed combining several old gillnetter licenses into a seiner license" (Meggs 1992, 195). In the end, "the capacity of the fleet is estimated to have doubled or perhaps trebled" (Ross 1987, 189).

The Pearse Commission of 1981 summarized the problems of capacity by stating that there were "too many fishermen chasing too few fish" (Pearse 1982, 2). The report touched on the sophistication and overcapacity of fishing vessels, and observed that the size of salmon being caught was getting smaller. Pearse also expressed concerns about habitat destruction and its role in eliminating races of salmon, and the lack of a sufficient information base for managers to make decisions about run openings or escapement (allowing salmon to proceed up river to spawn):

> Salmon managers face four major difficulties in developing pre-season plans. First, they seldom know

> with much confidence how many fish will enter a fishery. Second, they cannot reliably predict the time the stock will enter the fishery. Third, they do not know how many vessels will participate in a particular fishery. The highly mobile fleet in the salmon fishery responds quickly and often unpredictably to fishing opportunities along the coast. Sometimes managers refrain from planning openings for small runs because of the threat of excessive fishing effort being directed to the available stocks. Fourth, information about the stocks and their spawning requirements is so weak that the escapement targets are little better than guesses (p. 40).

Many of Pearse's "solutions" were contentious. The recommendation for more research on salmon was reasonable, but the reduction of 50 percent of the fleet through auctioning licences was extremely controversial. Equally controversial was the suggestion of tying salmon quota to vessel and gear type, which in effect would privatize the resource. In other words, commercial fishers would no longer be able to compete for as many salmon – or as much of the common property resource – as they could within an opening. Instead, a licence would entitle the fisher to a specific quota of salmon, in effect giving ownership of a portion of the resource and thus privatizing it. Pearse also recommended licences and restrictions for the sport fishery and, finally, he encouraged investment in hatcheries and fish farming. Few of his controversial recommendations were acted upon, though he had pointed out many of the problems faced by the federal managers.

FACING THE PROBLEMS TODAY

The freshwater habitat, which is crucial to both ends of the salmon's life cycle, requires much more attention than it has received. The expansion of urbanization and many forms of industry are destroying salmon spawning grounds, and this process must be reversed if the number of salmon is to increase (Glavin 1996; Clapp 1998; Werring 2007; David Suzuki Foundation 2008a). Alcan Aluminum posed a major threat with its plan to build a dam, Kemano 2, on a tributary of the Skeena River that has significant salmon runs. This project was cancelled in 1987 in favour of the Kemano Completion Project – a diversion of 87 percent of the Nechako River where "one fifth of all Fraser sockeye runs migrate" (United Fishermen and Allied Workers' Union 1993, 2). Both plans were viewed as disastrous by the United Fishermen and Allied Workers' Union, and the provincial government cancelled the project in 1995. The David Suzuki Foundation (2008b) states that "the West Coast of Vancouver Island once boasted 1,200 stocks. Now, some 718 – more than half – are extinct, at moderate risk of extinction or considered stocks of special concern. Province-wide, at least 142 salmon populations have vanished forever."

An even greater danger is present and may be the most crucial factor in the survival of this industry. The video recording *Phantom of the Ocean* (1997) documents how global warming is affecting ocean temperatures. The 7°C line is the critical mark: when ocean temperatures exceed 7°C there are no salmon. In the Pacific Northwest, ocean temperatures are warming, and this may account for the decrease of salmon in the rivers of British Columbia, the increase in Alaska, and the presence of salmon (for the first time) in the Beaufort Sea.

The Salmon Enhancement Program is designed to reverse the trend of decreasing salmon stocks, although this program has two fundamentally different directions. One strategy is to enhance streams with existing runs by cleaning up obstructions to salmon migration, constructing artificial gravel beds, fertilizing lakes to increase food, and encouraging schools and community groups to establish incubation boxes in streams. The second strategy is to construct hatcheries and raise salmon to be released into the ocean. Hatcheries are expensive to maintain, however, and they have the potential to spread diseases and reduce genetic diversity. Hatchery-raised salmon are also used to supply ocean ranches and fish farms, two threats to the commercial fishery.

Ocean ranching utilizes the freshwater habitat to raise a great number of fingerlings in hatcheries, allow them to migrate to the sea, and net them all several years later when they return to the hatchery. This eliminates the need for investment in commercial vessels, and damage to streams is of little consequence as the streams are not used for spawning. British Columbia has no ocean ranches, but Alaska, which does not allow fish farming, does practice ocean ranching on streams with few or no wild salmon.

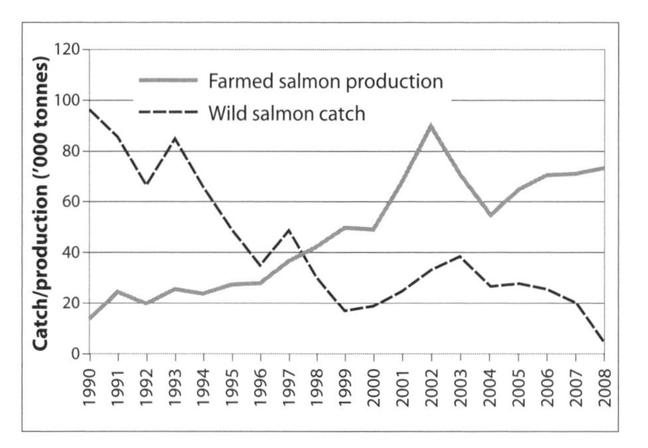

Figure 9.6 Wild and farmed salmon production, 1990-2008
Note: Includes a small amount of trout and, for 1990, steelhead.
Source: Data from Fisheries and Oceans Canada (2009d, 2010).

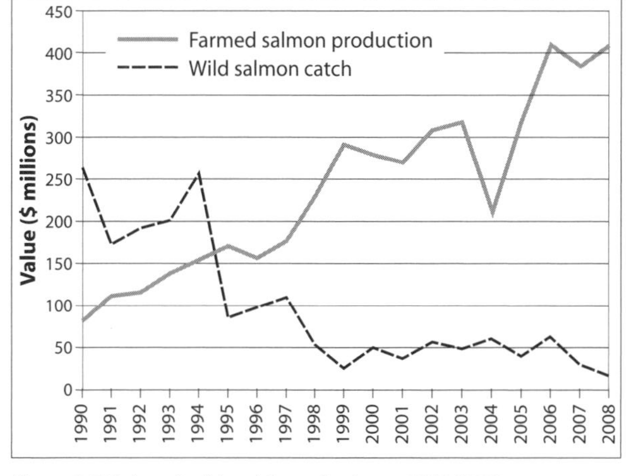

Figure 9.7 Value of wild and farmed salmon, 1990-2008
Note: Includes a small amount of trout and, for 1990, steelhead.
Source: Data from Fisheries and Oceans Canada (2009d, 2010).

The raising of salmon in netpens – nets anchored to the ocean shore in bays and inlets – reached "gold rush" proportions from the mid-to-late 1980s, much of it focused on the Sunshine Coast. Figures 9.6 and 9.7 document the very rapid growth in farmed salmon, both in tonnes produced and in dollars earned by the industry. Farmed salmon was British Columbia's leading agricultural export by 1993 (ARA Consulting Group 1994, 2-2). There were ten farms operating in 1984, 135 farms in 1989, and although there have been changes, 135 farms in 2008. Fifty companies were operating in 1989, but corporate concentration has resulted in five major companies (Marine Harvest Canada Inc., for example, owns forty-seven sites) and ten minor ones, which own five or fewer sites (BC Ministry of Environment 2009b).

Because of diseases inherently related to the stress of being caged, not all species of salmon can be farmed. Sockeye, for example, are extremely difficult to farm. The species raised in netpens are chinook, coho, and some rainbow trout, but the main species is Atlantic salmon. The preferred netpen locations are sheltered coves and inlets, and the nets are best anchored from shore. This requires a foreshore lease. The foreshore is the area between the high and low tidelines, and in British Columbia it is another common property resource, for which there is no private ownership. The proliferation of foreshore leases for fish farms has essentially privatized the foreshore and also excluded the public from many small bays and inlets. Objection was raised by recreationists and the tourism industry.

Several environmental hazards are related to this industry, including the possibility of disease transfer and genetic pollution, especially when the salmon in netpens escape and mix with wild stocks. Such escapes are well documented. Other concerns include the contamination of the aquatic environment beneath the netpens, and this has led to further regulation (see Blore 1997; D.W. Ellis and Associates 1996; Western Canada Wilderness Committee 1998; DeMont 2004; Watershed Watch Salmon Society 2010). One of the biggest issues has been the accusation that sea lice breeding in the many netpens in the Broughton Archipelago (off Port McNeill at the north end of Vancouver Island) had spread to wild pink salmon: "In 2002, 6-65 sea-lice [per fish] were found on young salmon smolts trying to migrate out of the Broughton. Of the approximately 3 million Pink salmon that migrated out of the Broughton, only approximately 148,000 returned, creating one of the most significant crashes of pink salmon ever witnessed in the Broughton Archipelago" (Greenpeace 2008).

The federal Department of Fisheries and Oceans (DFO) conducted research on the problem of sea lice and farming Atlantic salmon in the Pacific. Atlantic salmon are the preferred farmed species and account for approximately 95 percent of farmed salmon because they bring high value and are easy to raise in netpens (BC Ministry of Environment 2009c). DFO came to the conclusion that farmed Atlantic salmon pose little risk to the wild salmon industry and that they do not increase the presence of sea lice. The CBC TV program Disclosure, however, "obtained government documents that reveal DFO knew that its study of the problem was seriously flawed" (CBC 2003). In February 2003, the federal government ordered many of the fish farms in the Broughton Archipelago emptied as a precautionary measure. According to Marjorie Wonham and Martin Krkošek, writing for the David Suzuki Foundation (2007), "Sea lice have frequently killed over 80 per cent of the annual wild salmon returns to the Broughton ... salmon are approaching extinction and will collapse by 99 per cent in four years, or two salmon generations, if the infestations continue."

The commercial fishery views the fish farming industry as a threat on a number of fronts. Worldwide production of farmed salmon has increased the supply and is blamed for driving the price of all salmon down, while disease transfer to wild stocks looms as a threat to the commercial catch. Andrew Nikiforuk (1996, 103) remarks that "besides affecting price, the controversial raising of salmon in marine feedlots, complete with ocean-sterilizing pesticides and antibiotics, has devastated BC's salmon export business." Alaska does not allow fish farms to operate in its waters and the commercial fishery this side of the border would like to see the provincial government enact similar legislation.

The global increase in fish farms has been accompanied by an increased need for feed. The result has been a massive harvest of non-commercial fish to feed farmed salmon. The ratio is approximately five-to-one – in other words, "it takes about five kilograms of wild fish to create one kilogram of farmed salmon" (Boesveld 2009). Fortunately, Canada's fish feed industry has been shifting away from wild fish feed and oil (25 percent) to vegetable and poultry proteins (75 percent).

The fish farming industry has struggled to produce a product of good quality. Initial locations on the Sunshine Coast in the mid-1980s proved to be disastrous, not only because of the problems mentioned above but because the water was too warm in summer and plankton bloom suffocated the fish. Millions of farmed salmon died and were disposed of at the landfill site near Sechelt, creating a distinct odour. Few salmon farms exist on the Sunshine Coast today (Figure 9.8).

Figure 9.8 shows the numerous fish farms in the Broughton Archipelago. These farms are in close proximity to salmon-bearing streams and, consequently, pose a major threat of sea lice infestation. The location and number of fish farms permitted in any area can be cause for alarm, and it is the responsibility of managers to set limits. When fish farms are located in salt water, the federal government should be in charge. However, when fish farms were anchored to the foreshore, the federal government allowed the provincial government (back in 1988) to become the lead regulator. This decision has recently been challenged by environmental organizations, and the Supreme Court of BC ruled in February 2009 that the federal government should carry the responsibility. Importantly, the federal government also has the responsibility to protect wild salmon. A number of environmental groups have proposed a solution to the problem of fish farming – closed containment or farming on land – and they are now pressuring the federal government to adopt this position. Closed containment eliminates the problems of waste, escape, disease transfer, and sea lice. The fish farming industry is opposed, however, because of the considerable costs.

First Nations have a recognized food fishery right to salmon through the Indian Act. Early regulations prohibited First Nations from selling salmon except through the normal regulatory system of licensing. More recently, pursuit of Aboriginal rights through the courts has produced an interesting development in the fishing industry. The Supreme Court ruled that

- Aboriginal and treaty rights are capable of evolving over time and must be interpreted in a generous and liberal manner;
- governments may regulate existing Aboriginal rights only for a compelling and substantial objective such as the conservation and management of resources; and,
- after conservation goals are met, Aboriginal people must be given priority to fish for food over other user groups (BC Ministry of Aboriginal Affairs 1993, 2).

The Nisga'a Treaty, ratified in 2000, includes an allotment of salmon on the Nass River for food and ceremonial

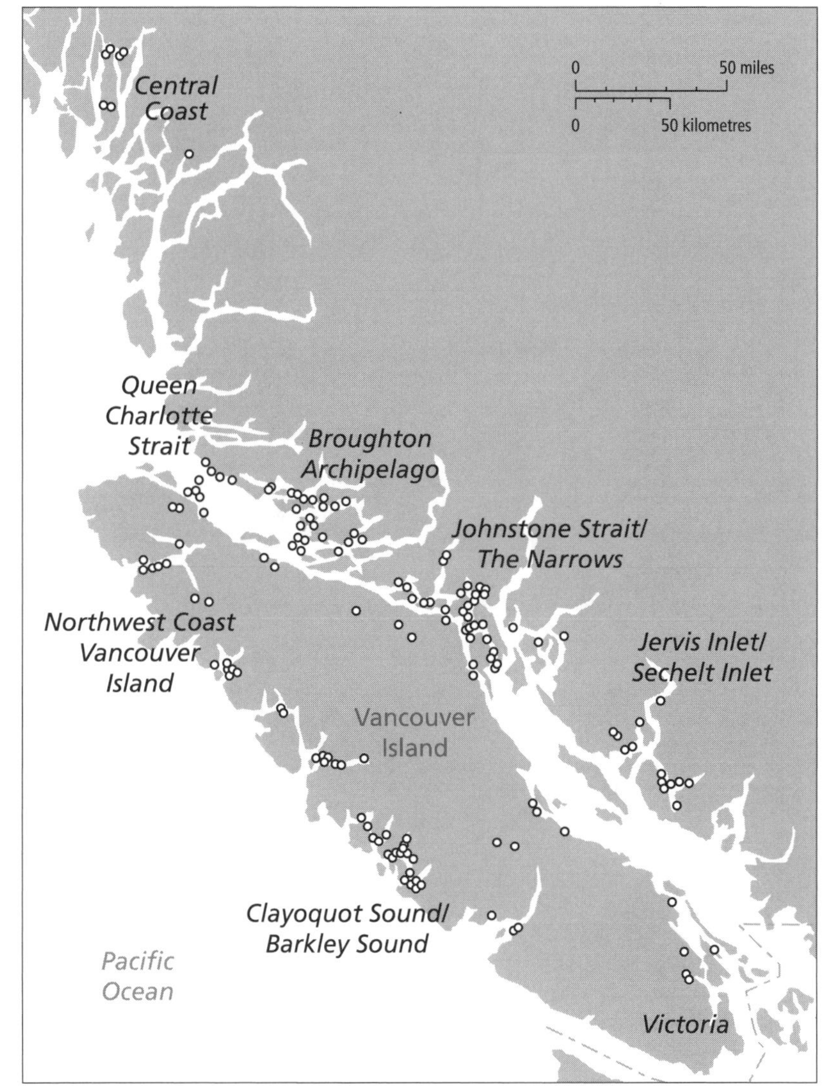

Figure 9.8 Location of salmon farms, 2010
Sources: Modified from BC Ministry of Agriculture and Lands (2010).

practices. Tangent to this Treaty is a separate Own Source Revenue Agreement that included a twenty-five year (renewable after fifteen years) Harvest Agreement for a commercial allocation amounting to 13 percent of the sockeye and 15 percent of the pink salmon on the Nass River. This allotment is on the basis of a percentage of the run to encourage future enhancement of salmon up the Nass. This agreement sets the precedent for turning salmon, a common property resource, into a **private property resource** in the sense that it will allocate salmon to a group. Basing the entitlement on a percentage, though, may supply the incentive to enhance the stocks for everyone.

The Tsawwassen Treaty, ratified in 2007, likewise includes an agreement outside the treaty for a commercial allocation of a percentage of sockeye, chum, and pink salmon on the Fraser River. A number of bands, however, have negotiated interim agreements that include an allocation of salmon (McRae and Pearse 2004, 13; Garner and Parfitt 2006). Moreover, the federal government has initiated a Native-only fishery, much to the displeasure of commercial fishers. They went to court in 2003 and had this "race-based" fishery ruled contrary to the Charter of Rights and Freedoms. The *Kapp* decision was then overturned on appeal to the Supreme Court of British Columbia in 2004 (McRae and Pearse 2004, 9). Certainly the trend is that First Nations are going to gain more salmon, whether through the courts or by treaties.

The sport fishery harvests perhaps only 4 to 10 percent of the overall provincial catch, but a major multiplier effect is attached to this industry, especially with growing tourism. The purchase of fishing lodge accommodation, charters, fishing licences, equipment, and even cases of beer accounts for the highest dollar value per fish caught. Many sport fishing interests and the communities that benefit are very concerned over the decline in salmon stocks and the outright ban on catching chinook and coho in some regions. Loss of employment due to the chinook ban in 1996 was estimated at 2,175 jobs, and the loss of revenues has been estimated at $135.7 million (ARA Consulting Group 1996, S-4).

Competition for salmon provokes frequent finger pointing at commercial interests, as the sport fishing industry asks how it can be guaranteed an allotment of the hook-and-line species. In 1997 the federal government "assigned the recreational sector priority of access to chinook and coho salmon" as well as a small share of the other species of salmon (McRae and Pearse 2004, 45). The continued poor commercial harvest of salmon has prompted major changes to the whole fishing industry. The federal and provincial governments are proposing quotas for all stakeholders in 2005, which did not please the sport fishery: "The Sport Fishing Institute, representing B.C.'s recreational fishing industry, said it appears the profitable sportfishing industry will face quotas designed to prop up the ailing commercial industry" (Meissner 2004).

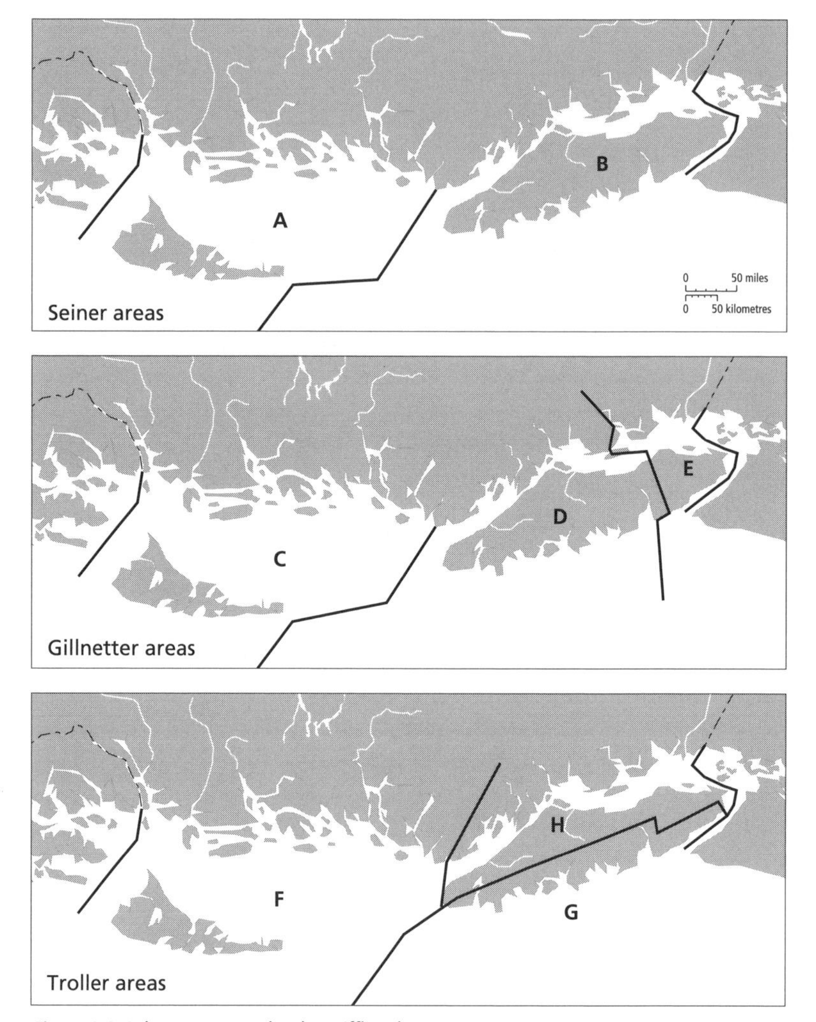

Figure 9.9 Salmon areas under the Mifflin Plan

The commercial fishery itself is not unified, as trollers, gillnetters, and seiners all compete for a share of the dwindling resource. The threat of overcapacity has been handled largely through restricting openings and closings for various runs. The Mifflin Plan, implemented in 1995, pursued the old idea of eliminating fishing vessels through a government buyback program at the same time that it introduced area licensing and the stacking of licences. The coast is divided into three areas, or zones, for gillnetters and trollers, and two zones for seiners (Figure 9.9). A fisher's licence is tied to the area he or she has traditionally fished. To fish another area, a fisher must buy out a licence from that area, thereby acquiring multiple licences – or "stacking the licence" – and replacing another fishing vessel. By 2003, over half the commercial vessels had been retired, with considerable discontent over the process (Table 9.4).

The Mifflin Plan is not without controversy. Figures 9.6 and 9.7 show that the plan was implemented at a time of record low salmon catches and low values, and that situation has only become worse. Initially, the ARA Consulting Group (1996, S-9) assessed the commercial fishing sector job loss in 1996 to be 5,625, due to the poor catch and the Mifflin Plan. When these losses are combined with a further 2,175 jobs lost in the sport fishery due to area closure and catch-and-release rules, an estimated 7,800 jobs had vanished. There was at that time, however, a $200 million federal government program for "Rebuilding the Resource" and "Helping People and Communities" (Canada 2003).

Table 9.4

Commercial salmon fishing licences by gear type, 1995-2009

	1995	2003	2009
Seine	536	260	239
Gillnet and troll	3,570	1,562	1,412
Total	4,106	2,166	1,651

Source: Fisheries and Oceans Canada, Statistical Services Unit (2009b).

Fewer licences mean that the value of each one increases. The small fishers left to fish in one area only may find it difficult to make a living each year if they cannot afford to purchase another licence in another area. Concern has been expressed, particularly in the more remote commercial fishing communities, that the existing large corporate concentration of the industry will become even more dominant because corporations can afford to purchase licences. Nikiforuk's (1996, 100) comment on the Fisheries Council of British Columbia puts corporate concentration in perspective: "Although it represents 240 processors, eight companies – including B.C. Packers Ltd., the Canadian Fishing Co., Ocean Fisheries Ltd. and J.S. McMillan Fisheries Ltd. – control the bulk of trade. Its members own most of the seine fleet, which has enough capacity to wipe out all the salmon stocks in one season, yet they still complain they can't get a regular supply of fish." Some of these corporations have more control than others; BC Packers and Canadian Fishing have more than 100 seine licences and over 140 others (Seale 1996, 21).

These concerns over corporate concentration were expressed in the mid-1990s, and conditions pertaining to ownership have actually become worse because of a shift in the commercial fishery to individual transferable quotas (ITQs). These quotas "are a fisheries management system with three common characteristics: they include quotas or defined shares of the catch, are allocated to individual fishermen or their vessels, and are transferable to some degree, allowing fishermen to buy, sell, lease and trade them" (Ecotrust Canada 2009). A report by Ecotrust Canada suggests that this system has resulted in a major increase of "armchair fishermen" – fishers who do not fish but lease out their quotas. "By way of example, in the pilot ITQ fishery for northern Chinook salmon, almost half the quota was leased from 2005 to 2007."

The federal role of gatekeeper has not been easy. The government attempts to regulate the number of salmon caught by designating run openings. Unfortunately, it does so based on inaccurate information about salmon stocks, primarily because it has little to no control over international and provincial activities that affect salmon: "We have 3,000 ecologically and genetically distinct populations of salmon on the West Coast, 1,000 spawning streams and about 150 DFO types in the field" (Nikiforuk 1996, 101). It is a daunting task to gather full information about the various salmon species, their migrations and habitats, or the number of vessels that will participate in any opening. Under this system of management, harvesters are free to take as many fish as they are able to in any opening. The federal government, which is responsible for both conserving salmon stocks and keeping the fishing industry economically viable, is in a "no win" situation.

Changes in management strategies are, however, taking place. After five years of consultation, Fisheries and Oceans Canada instituted the Wild Salmon Policy (WSP) in 2005. Conservation of wild salmon is the highest priority, and the policy establishes conservation units (CUs) for each region to determine the total allowable catch (TAC). The WSP is necessary for establishing a scientifically based (TAC), but Julie Gardner (2009), reporting for the David Suzuki Foundation and Watershed Watch, has concerns that "include a lack of finances and people to support implementation, as well as support for the monitoring efforts required for ongoing management."

A more fundamental question is whether salmon should remain a common property resource or a private property resource. The Nisga'a Treaty, as mentioned above, includes a twenty-five-year (renewable) agreement that gives them a percentage of salmon on the Nass River. This is the beginning of the privatization of salmon, and the treaty led the federal and provincial governments to commission the McRae-Pearse Report in 2004. McRae and Pearse recommended for the 2005 season a system of individual transferable quota (ITQ) licences, renewable for twenty-five years, a radical departure from the prior system. Essentially, salmon become a private property resource and allocation to the commercial fishery is made on the basis of size of vessel and history of fish catches (McRae and Pearse 2004, 58). Individual transferable quotas might be a foreign concept to the commercial salmon fishery but not to other commercial fisheries. In any event, ITQs are coming, and some salmon runs have been targeted as trial runs. One of the greatest concerns is the transferability of quota. The change has already resulted in quota leasing (at very high costs) and increasing numbers of absentee vessel owners.

OTHER COMMERCIAL FISHERIES AND LESSONS FOR THE SALMON FISHING INDUSTRY

The story of halibut is an interesting one with lessons for other commercial fisheries. It was recognized in the early twentieth century that halibut were becoming overfished, and because most of the harvest was in international waters (beyond three miles or 4.8 kilometres in those days), serious international agreements were required if this resource were to remain sustainable. In 1923 Canada and the United States negotiated an agreement that imposed restrictions on the catch. While the overall catch was limited, the individual catch was still managed by gatekeeper principles. Halibut remained a common property resource, but within international quota agreements, or total allowable catch (TAC).

By the early 1990s, approximately 400 licence holders were in competition for the TAC. The competition was intense, and vessels used dragger technology to maximize their catch in the six-day opening. This mad rush for halibut put huge quantities on the market, which in turn drove the price down. The system created the worst possible conditions. Draggers damaged the ocean floor ecosystem, the nets hauled in halibut of all ages, and this type of harvesting did not discriminate against other species of groundfish, which simply became by-catch. And, for all the investment in vessels and gear, it amounted to a short opening where excess supply depressed prices.

In 1991 DFO initiated a two-year trial catch-quota system (also referred to as individual vessel quota, or IVQ). There were no openings and closings: licence holders could fish when they preferred, which let them gauge weather and market conditions. They were limited instead to a harvest determined by DFO on the basis of the vessel size and its historical catch. In this system there is no competition with other fishers. Halibut is also no longer a common property resource; it is a private property resource whose ownership is based on each licensed vessel.

In the two-year trial period, vessels fished for 200 days, not six, to maximize market prices. Many fishers went back to long-lining technology, a system of baiting individual hooks that are weighted down along the bottom of the ocean floor, thus moving away from destructive dragger technology. Halibut fishers then voted at the end of the two years on whether to return to the gatekeeper system or retain the IVQ system: 91 percent voted to retain the catch-quota system (Alden 1996).

The question arises whether this could be applied to commercial salmon fishing. It would not be easy because salmon are anadromous and therefore more complex to regulate in terms of quota. As well, there are variations each year in stocks (also referred to as runs or races), and the allocation of salmon must take into consideration a great number of interests. In response to the decline in salmon catch, along with the twenty-five-year agreement negotiated by the Nisga'a for a percentage of salmon on the Nass River for commercial purposes, some salmon runs have been targeted for ITQs on a trial basis.

Privatization has its critics. The Cruickshank Commission (1991) warned that the small fisher would lose to large corporate interests located in large communities. Chris Newton and his colleagues (2005) have similar views on a transferable quota system and warns that these "systems suffer from severe drawbacks, including: taking fisheries benefits out of local communities, significantly weakening the regulatory authority of government, and encouraging fishermen to only keep the highest value fish." Ecotrust Canada (2009) warns that the biggest problem is the transferability of the quota, and it cites the halibut fishery as an example of the need for regulations. "In 1993, 19 percent of the halibut quota was temporarily transferred from one vessel to another during the year. (Fishermen lease quota by temporarily transferring the ownership of quota.) By 2008, the ratio skyrocketed to 106 percent of the TAC, evidence of high levels of leasing. Today, lease fees are effectively charged on almost every pound of halibut quota in BC."

The herring fishery is another fishery with a long history not only commercially but as an important food source for many First Nations. Like the salmon fishery, it has faced crises such as overcapacity of fishing technology, habitat destruction, changes to market conditions, and changes to the aquatic ecosystem.

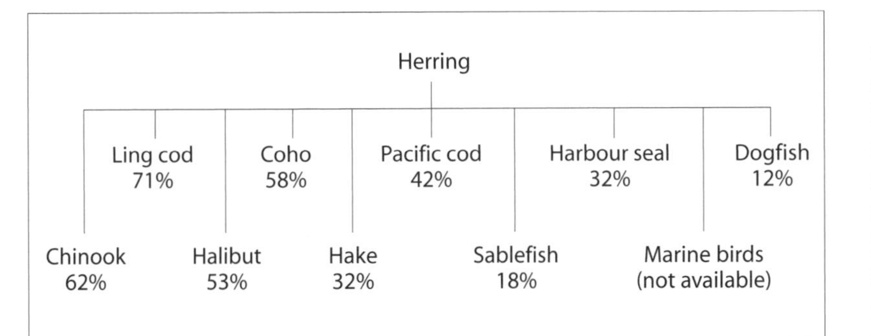

Figure 9.10 Importance of adult Pacific herring in predators' diets, west coast Vancouver Island
Source: Data from Fisheries and Oceans Canada (1998).

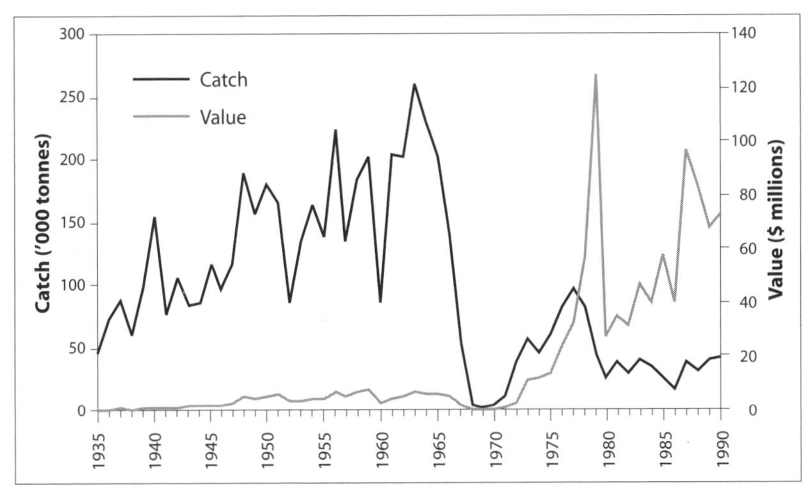

Figure 9.11 Herring catch and landed value, 1935-90
Sources: Data from Statistics Canada (1983, series N25-37); Fisheries and Oceans Canada, Statistical Services Unit (2004d).

Herring, from a food web perspective, are the most important fish species on the Pacific Northwest coast. Figure 9.10 indicates the relationship between herring and the west coast Vancouver Island predators. One significant predator, however, is not represented in the diagram: commercial fishers. From the early twentieth century to the middle of the Depression, most herring was exported to Asia after being dried and salted. Herring was also canned, especially in low salmon years. A significant change in the rules occurred in 1913 that later had profound effects on this fishery: seine boats were allowed to catch this schooling variety of fish.

From 1935 until the collapse of this fishery in 1967, huge quantities of herring were caught and turned into fish oil (for soap, shortening, margarine, and other products) or fish meal (for animal feed including cat and poultry food), a practice known as a reduction fishery. The reduction fishery characterized herring as a high bulk, low value commodity, as shown in Figure 9.11. The crash and four-year closure of this fishery resulted from a combination of adverse environmental conditions (especially warmer weather during El Niño events) and a powerful and sophisticated seine fleet.

The "new" herring fisheries that emerged in the 1970s were roe, spawn-on-kelp, and food and bait. Figure 9.11 indicates the revival of this industry, particularly with respect to the value of herring. Table 9.5 gives a more up-to-date assessment of the industry, wherein the volume can be compared to the landed value and the wholesale value. As can be observed, the roe fishery, which primarily serves the Japanese market, dominates.

Managing the herring fishery since its collapse has been a difficult task for the federal government. The complex relationships outlined in Figure 9.10 mean that other fish, including commercial species, depend upon herring, and the managers must take these relationships into consideration. Ocean warming is particularly detrimental as this increases the number of hake that then consume the herring. As well, a warmer ocean results in Pacific mackerel migrating north and eating herring and young salmon. The destruction of spawning habitat due to human habitation and pollution is another problem. For example, herring no longer spawn in Nanaimo Harbour, Ladysmith Harbour, or Pender Harbour (Fisheries and Oceans Canada 1998).

Table 9.5

Herring catch, landed value, and wholesale value, 1996-2008

Fishery	1996	1997	1998	1999	2000	2001	2002	2003	2004	2005	2006	2007	2008
Volume of herring landed ('000 tonnes)													
Spawn on kelp	0.4	0.4	0.4	0.4	0.4	0.4	0.4	0.4	0.4	0.31	0.31	0.21	0.14
Roe herring	21.8	31.2	32.9	26.5	27.2	23.9	26.3	28.8	23.1	29.1	22.3	10.5	10.5
Food and bait	0.4	0.3	0.6	0.2	0.2	0.2	0.2	0.2	1.3	1.4	0.9	1.1	0.8
Total	22.6	31.9	33.9	27.1	27.8	24.5	26.9	29.4	25.2	30.8	23.5	11.8	11.4
Landed value ($ millions)													
Spawn on kelp	22.2	17.7	8.9	7.8	10.4	10.5	10.6	9.3	5.6	3.2	4.3	6.5	3.8
Roe herring	77.2	49.0	26.0	40.4	39.0	35.5	40.0	36.0	27.9	28.4	13.5	12.8	10.9
Food and bait	0.2	0.2	1.7	0.4	0.7	0.2	0.2	0.2	0.5	0.9	0.5	0.9	0.8
Total	99.5	66.9	36.6	48.6	50.1	46.2	50.8	45.5	34.0	37.5	18.3	20.2	15.5
Wholesale value ($millions)													
Spawn on kelp	22.2	17.7	8.9	9.8	12.5	12.8	12.8	11.5	7.6	4.1	5.2	7.1	4.2
Roe herring	164.9	97.8	96.0	108.0	113.8	97.3	113.6	88.1	87.0	82.0	51.2	45.3	40.5
Food and bait	2.5	1.7	2.8	3.4	2.8	2.5	3.1	3.3	3.0	3.0	3.5	3.8	3.0
Total	189.7	117.2	107.7	122.0	129.1	112.6	129.5	102.9	97.6	89.1	59.9	56.2	47.7

Note: Landed value represents the price paid to fishers, and wholesale value represents the price after processing.
Source: BC Ministry of Agriculture, Food, and Fisheries (2004); BC Ministry of Environment (2009a).

Even worse is the large overcapacity in the fishing fleet, despite such adjustments as limited entry in 1974, area licensing in 1981, and a total allowable catch (TAC) in 1983. This last is based on 20 percent of the assessed biomass and calculated for each of the five herring zones: Strait of Georgia, west coast Vancouver Island, central coast, Queen Charlotte Islands, and Prince Rupert District. The TAC system retained the gatekeeper concept of openings, and the excess fishing capacity made overharvesting the TAC the common practice. In 1998 the concept of "pooling" was introduced, under which "only a designated number of seiners fish, but all share the catch" (McRae and Pearse, 2004, 21). This has reduced the amount of herring caught in excess of the TAC. The McRae-Pearse report suggests that this system of pooling be examined for smaller runs of salmon. It also recommended a twenty-five-year individual transferable quota for this fishery.

Trawlers, or draggers, similar to those off the east coast of Canada, harvest most of the groundfish (halibut, cod, hake) in BC waters. Another similarity to the east coast is that the majority of the trawlers were foreign fleets until the imposition of 200-mile (322 kilometre) territorial waters in 1977. By 1979, the DFO had imposed a host of management rules. These "included license limitation, the establishment of Total Allowable Catches (TACs), and imposing species/areas closures, area/time closures and vessel trip limits on groundfish of commercial importance" (Fisheries and Oceans Canada 2004). These guidelines contained no restrictions on some species of groundfish, such as Pacific cod, until 1992, and in 1996 individual vessel quotas (IVQs) were introduced. Figure 9.12 graphs the catch and value of Pacific cod. The unrestricted catch resulted in a collapse in the harvest of Pacific cod by the mid-1990s and there has been little recovery since.

Figure 9.13 shows the catch and value of all groundfish between 1990 and 2008 and illustrates some significant swings in catch, although they are mainly downward. Until recently, however, the value of groundfish has experienced a steady increase in value. One of the concerns about trawler fishing is the by-catch – the harvest of species not targeted. This fishery is also managed through ITQs and a total allowable catch.

The commercial shellfishery is classified as including all invertebrates, which covers a great number of species.

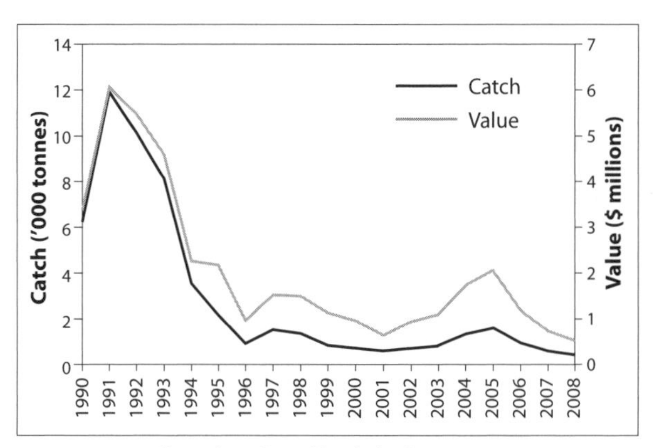

Figure 9.12 Pacific cod catch and landed value, 1990-2008
Source: Fisheries and Oceans Canada (2009d).

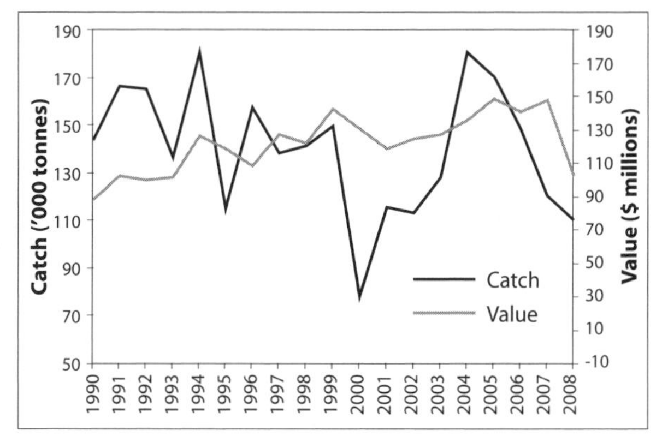

Figure 9.13 Groundfish catch and landed value, 1990-2008
Source: Fisheries and Oceans Canada (2009d).

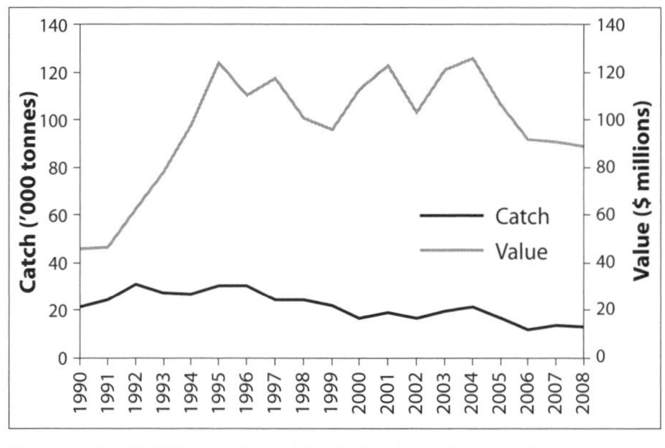

Figure 9.14 Shellfish catch and landed value, 1990-2008
Source: Fisheries and Oceans Canada (2009d).

Figure 9.14 tells the story of significant increases in value, but declining catches. This is a complex fishery to manage because of the variety of life cycles and habitats for such species as geoduck clams, Manila clams, scallops, prawns, shrimp, Dungeness crabs, sea urchins, and others. Some are caught through traps, such as prawns and crab; some are harvested by divers using high pressure hoses (e.g., geoduck clams), and still others are caught through trawling. The decline in catch noted in Figure 9.14 is related to the imposition of a variety of management techniques including closures, quotas, and trap limitations, necessitated by the concern about overharvesting (Fisheries and Oceans Canada 2003). For example, the abalone fishery has been closed since 1990, and in 2003, "northern abalone were legally listed and protected as threatened under the *Species at Risk Act* (Fisheries and Oceans Canada 2009a)." This fishery has been managed through ITQs since 1979, and Ecotrust Canada (2009) suggests that "poor monitoring led to poaching and over-fishing by licensed harvesters." Whether DFO manages fish species as a common property resource and acts as the gatekeeper or converts fish to a private property resource by issuing ITQs, the message is that strong monitoring of the aquatic environment, along with strict enforcement, is paramount. A key recommendation of the David Suzuki Foundation (2008b) is to "double the number of habitat enforcement officers, with the powers of inspection and ticketing, in the Pacific Region."

SUMMARY

All the competing interests for the salmon resource share the desire to increase stocks. To produce more salmon may well be possible, although it will take considerable effort to undo the environmental damage to the freshwater habitat in many areas. On the one hand, it is encouraging that salmon can be introduced to enhanced water systems and that their regeneration period is only a few years. On the other, all the activities that negatively influence the aquatic environment and the complexity of managing all the competing interests are discouraging. There have been some positive signs. The 2010 sockeye salmon return to the Fraser River was the highest since 1913, Canada has reached an agreement with the United States, many First Nations are advanced in their treaty negotiations, and the federal government has allocated salmon to the sport fishery along with developing the Wild Pacific Salmon Policy. Though satisfying all interests may not be possible, no interests will be satisfied if salmon stocks continue to be depleted at the current rate.

The McRae-Pearse report, *Treaties and Transition*, which was commissioned by both the federal and provincial governments, made some radical proposals slated for the 2005 fishing season. This report recommends that all commercial fish become a private property resource and that individual fishers have long-term tenure (ITQs). This report was motivated by the twenty-five-year Nisga'a agreement, and adopting ITQs is certainly the direction that more-and-more commercial fisheries, including salmon, are moving in. Nevertheless, the more serious issue is good scientific monitoring of the environment and the enforcement of regulations.

REFERENCES

Alden, E. 1996. "To Save Our Salmon, Privatize Its Future." *Vancouver Sun*, 26 July, A19.

ARA Consulting Group. 1994. *The British Columbia Farmed Salmon Industry: Regional Economic Impacts*. Vancouver: ARA Consulting Group.

–. 1996. *Fishing for Answers: Coastal Communities and the British Columbia Salmon Fishery*. Vancouver: ARA Consulting Group.

Barker, M.L. 1977. *Natural Resources of British Columbia and the Yukon*. Vancouver: Douglas, David, and Charles.

BC Ministry of Aboriginal Affairs. 1993. *Information about Landmark Court Cases*. Brochure. Victoria: Crown Publications.

Blore, S. 1997. "Betting the Farm." *Georgia Straight* (Vancouver), 10 July, 15-18.

Canada, Department of Fisheries. 1931. *Report of the Commission of Fisheries for the Year 1931*. Ottawa: Queen's Printer.

Clapp, R.A. 1998. "The Resource Cycle in Forestry and Fishing." *Canadian Geographer* 42 (2): 129-44.

Cruickshank, D. 1991. *A Commission of Inquiry into Licensing and Related Policies of the Department of Fisheries and Oceans: The Fisherman's Report*. Victoria: Mock.

D.W. Ellis and Associates. 1996. *Net Loss: The Salmon Netcage Industry in British Columbia*. Vancouver: David Suzuki Foundation.

DeMont, J. 2004. *A Stain upon the Sea*. Madeira Park: Harbour Publishing.

Fisheries and Oceans Canada. 1998. *Sustaining Marine Resources: Pacific Herring Fish Stocks*. SOE Bulletin No. 98-2 (Winter).

–. n.d. *The Incredible Salmonids*. Brochure. Ottawa: Ministry of Supply and Services Canada.

Forester, E., and A. Forester. 1975. *British Columbia's Commercial Fishing History*. Saanichton, BC: Hancock.

Glavin, T. 1996. *Dead Reckoning: Confronting the Crisis in Pacific Fisheries*. Vancouver: Greystone.

Hume, S. 1997. "Carelessness, Greed Nearly Destroyed Salmon Run." *Vancouver Sun*, 7 July, B3.

Lyons, C.P. 1969. *Salmon: Our Heritage*. Vancouver: Mitchell Press.

McRae, R.M., and P.H. Pearse. 2004. *Treaties and Transition: Towards a Sustainable Fishery on Canada's Pacific Coast*. Report for Ministry of Fisheries and Oceans and Government of British Columbia. Ottawa.

Meggs, G. 1992. *Salmon: The Decline of the British Columbia Fishery*. Vancouver: Douglas and McIntyre.

Nikiforuk, A. 1996. "The Empty Net Syndrome." *Canadian Business* (October): 99-109.

Pearse, P. 1982. *Turning the Tide: A New Policy for Canada's Pacific Fisheries*. Vancouver: Commission on Pacific Fisheries Policy.

Roos, J.F. 1991. *Restoring Fraser River Salmon*. Vancouver: Pacific Salmon Commission.

Ross, W.M. 1987. "Fisheries." In *British Columbia: Its Resources and People*, ed. C.N. Forward, 179-96. Western Geographical Series vol. 22. Victoria: University of Victoria.

Seale, R. 1996. "Fishing for Answers." *Nature Canada* (Autumn): 20-4.

Stacey, D. 1982. *Sockeye and Tinplate*. Victoria: Royal BC Museum.

United Fishermen and Allied Workers' Union. 1993. *A River Run Dry: Alcan Built a Dam and No River Ran*. Vancouver.

Western Canada Wilderness Committee. 1998. Wilderness Committee Educational Report 17 (2): 1-4.

FILMS

Raincoast Storylines Ltd. in association with the Canadian Broadcasting Corporation. 1997. *Phantom of the Ocean: The Mystery of Global Climate Change*. Kelowna: Filmwest Associates (distributor).

INTERNET

Boesveld S. 2009. "Fish Farms Put Squeeze on Marine Resources." *Globe and Mail*, 8 September. www.theglobeand

mail.com/life/food-and-wine/fish-farms-put-squeeze-on-marine-resources/article1279812/.

BC Ministry of Agriculture and Lands. 2010 "Fish Farming on and around Vancouver Island and Coastal British Columbia." Map. www.agf.gov.bc.ca/fisheries/Finfish/cabinet/marine_fishfarms.pdf.

BC Ministry of Agriculture, Food, and Fisheries. 2004."BC Herring Landings, 1996-2003." www.agf.gov.bc.ca/fish_stats/Herring.htm.

BC Ministry of Environment. 2009a. "BC Herring Production." www.env.gov.bc.ca/omfd/fishstats/graphs-tables/herring.html.

–. 2009b. "Farm Sites." www.al.gov.bc.ca/fisheries/licences/cabinet/detailed_mw_farm_sites.pdf.

–. 2009c. "Salmon Aquaculture in British Columbia." www.env.gov.bc.ca/omfd/fishstats/aqua/salmon.html.

BC Salmon Marketing Council. 2004. "Go Wild." Home page. www.bcsalmon.ca.

BC Stats. 2009. "Quick Facts about British Columbia: 2009 Edition." www.bcstats.gov.bc.ca/data/qf.pdf.

CBC. 1998. Almanac. "No Fish – No Future." 22 May. www.radio.cbc.ca/news/fish/980522_b.html.

–. 2003. Disclosure. "Fish Farm Flap." 4 February. www.cbc.ca/disclosure/archives/030204_salmon/introduction_print.html.

David Suzuki Foundation. 2008a. "David Suzuki: Fishing for Salmon Answers." www.straight.com/article-141011/david-suzuki-fishing-salmon-answers.

–. 2008b. An Upstream Battle: Declines in 10 Pacific Salmon Stocks and Solutions for Their Survival. www.davidsuzuki.org/publications/downloads/2008/DSF_UpstreamBattle.pdf.

Ecotrust Canada. 2009. "A Cautionary Tale about ITQ Fisheries." www.ecotrust.ca/fisheries/cautionarytale.

Fisheries and Oceans Canada. 1999. "A History of the Pacific Salmon Treaty." Backgrounder B-HQ-99-29(113). June. www.dfo-mpo.gc.ca.

–. 2003. "Invertebrates." *Fish Stocks of the Pacific Coast: Online Book*. www.pac.dfo-mpo.gc.ca.

–. 2004. "Groundfish." *Fish Stocks of the Pacific Coast: Online Book*. www.pac.dfo-mpo.gc.ca.

–. 2009a. "Abalone: Pacific Region." www.pac.dfo-mpo.gc.ca/fm-gp/commercial/shellfish-mollusques/abalone-ormeau/index-eng.htm.

–. 2009b. *Analysis of Commercial Fishing Licence, Quota, and Vessel Values*. www.pac.dfo-mpo.gc.ca/fm-gp/picfi-ipcip/docs/analys-mar-31-2009.pdf.

–. 2009c. " Canada and the United States Renew Treaty to Conserve Pacific Salmon Stocks and Ensure Long-Term Sustainability of Pacific Salmon Fishery." www.dfo-mpo.gc.ca/media/npress-communique/2009/pr01-eng.htm.

– 2009d. "Commercial Fisheries: Landings – Seafisheries." www.dfo-mpo.gc.ca/stats/commercial/sea-maritimes-eng.htm.

–. 2010. "Aquaculture: Production Quantities and Values." www.dfo-mpo.gc.ca/stats/aqua/aqua08-eng.htm.

Gardner, J. 2009. "Knowledge Integration in Salmon Conservation and Sustainability Planning." David Suzuki Foundation and Watershed Watch. www.watershed-watch.org/.../files/WSP-Strategy4Report-Gardner_2009.pdf.

Garner, K., and B. Parfitt. 2006. "First Nations, Salmon Fisheries and the Rising Importance of Conservation." www.fish.bc.ca/files/First%20Nations%20Salmon%20Fisheries.pdf.

Greenpeace. 2008. "Threats: Aquaculture." www.greenpeace.org/canada/en/campaigns/Seafood/Resources/Fact-sheets/Threats-Aquaculture/.

Meissner, D. 2004. "Report Rules on B.C. Salmon Catches." Cnews. 5 May. cnews.canoe.ca/CNEWS/Canada/2004/05/05/448332-cp.html.

Newton, C., O. Langer, M. Weinstein, and P. Copes. 2005. "Who Is the Salmon For? Privatising Salmon Won't Help B.C. Communities, or Fish." In Canadian Council of Professional Fish Harvesters *Current Issues Bulletin* January 13: 5-7. www.ccpfh-ccpp.org/cgi-bin%5Cfiles%5C050117,Current-Issues-Bulletin,E.pdf.

Sierra Legal Defence Fund. 1998. "New NAFTA Challenge Filed and Canada Failing to Enforce Fisheries Act against Polluting Mines in B.C." www.sierralegal.org/.

Watershed Watch. 2010. "Watching Out for BC's Wild Salmon." www.watershed-watch.org/.

Werring, J. 2007. "High and Dry: An Investigation of Salmon-Habitat Destruction in British Columbia." David Suzuki Foundation. www.davidsuzuki.org/publications/reports/2007/high-and-dry-an-investigation-of-salmon-habitat-destruction-in-british-columbia/.

West Coast Vancouver Island Aquatic Management Board. n.d. "Fisheries Overview by Gear Type." www.westcoastaquatic.ca/fisheries_overview.htm.

Wonham, M. and Krkošek, M, for the David Suzuki Foundation. 2007. "Fish Farms Drive Wild Salmon toward Local Extinction." www.davidsuzuki.org/publications/downloads/2007/SEA_LICE_BROCHURE_FINAL.pdf.

10

Metal Mining

The Opening and Closing of Mines

Metal mining has the reputation of having opened up British Columbia to non-Native settlement and development. It began with the discovery of gold in the lower reaches of the Fraser River in 1858 and subsequent discoveries up the Fraser and into the Cariboo by the 1860s. The region had been held by the British mainly through Hudson's Bay Company fur trade posts with a few hundred Europeans, and it faced a huge challenge when approximately 30,000 miners descended on the landscape (Chapter 4). The various mining booms have had a powerful and permanent impact on the geography of the area. Because minerals are a nonrenewable resource, ore eventually runs out and mines close. A community that has depended solely on a mine for its economic well-being often becomes a ghost town, or must struggle to reinvent itself.

This chapter examines the metal mining industry of British Columbia. Mining of energy minerals such as coal, natural gas, and oil is discussed in the following chapter. The focus here is on gold, silver, copper, and other valuable minerals throughout the province. Understanding the physical landscape, and especially its geology, is important to the discovery of minerals, but a host of other factors influence the economic viability of any mineral going into production. Technology is very important to all aspects of mining, from mineral exploration to the end uses of metals. Technological innovations can affect the production costs of and supply and demand for the final product. The economics of mining is also affected by access to often isolated mine sites, by political decisions, by the cost of energy and other inputs, and, more recently, by environmental considerations. The world market price reflects the overall supply and demand for any mineral.

This chapter begins with a historical overview of the production of metals in British Columbia and moves on to a more specific examination of mines that have been into and out of production since 1990. Examining the

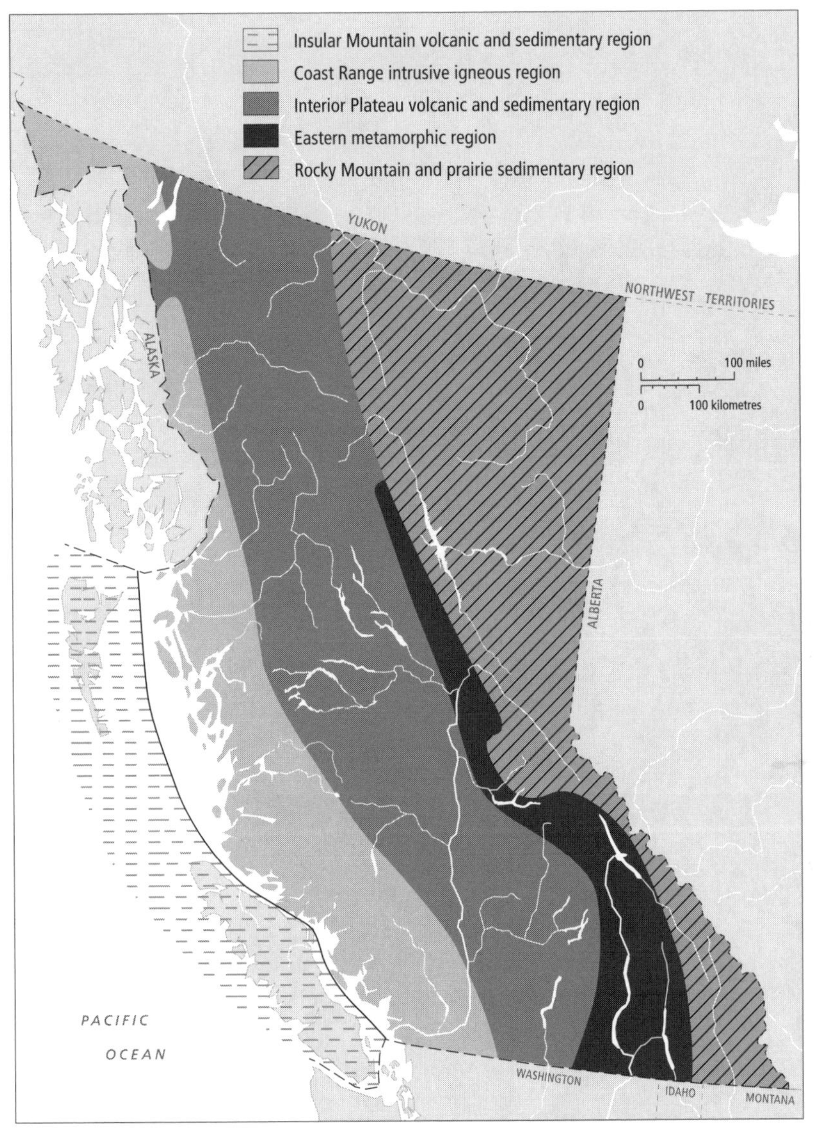

Figure 10.1 Geologic regions of British Columbia
Sources: Data from Rider (1978), 13; Geological Survey of Canada (1970, 30).

fundamental question of why mines open and close reveals numerous interrelated factors. Finally, it is suggested that few new metal mines will open, putting the future of this industry into question.

FACTORS INFLUENCING METAL MINING

The complex geology of British Columbia, including the ages of different rock structures and the tectonic forces responsible for their formation, is explained in greater detail in Chapter 2. Figure 10.1 divides the province into five general geological regions according to their mineralization characteristics.

The zone farthest west is the Insular Mountain volcanic and sedimentary region, encompassing Vancouver Island and Haida Gwaii. Geologically, this region is mainly a mix of intrusive and extrusive igneous rock with sedimentary layers. Some gold has been discovered and considerable quantities of coal mined, but the most important metal in this region has been copper. The Coast Mountains are referred to as the Coast Range intrusive igneous region. In this mainly granitic rock formation, deposits of gold, copper, and iron ore have been mined. The central portion of British Columbia, much of it referred to as the Interior Plateau, was created largely by massive lava flows. This is the Interior Plateau volcanic and sedimentary region and it has high mineralization, with major deposits of copper, molybdenum, gold, mercury, and asbestos. Next to it is the eastern metamorphic region, made up of very old and very hard rock. It is also rich in minerals. In the Kootenay region, especially, silver, lead, zinc, copper, and even gold have been mined. The most easterly region, the Rocky Mountain and prairie sedimentary region, is, as the name suggests, made up of sedimentary rock. As in Alberta, deposits of coal, oil, and natural gas have been developed there.

Some minerals were mined prior to the gold rush: coal on the north coast and Vancouver Island and some gold in Haida Gwaii. Coal was mined for some time, but these gold finds were small, short-lived ventures. The subsequent history of metal mining in the province and the relative importance of each metal can be seen in Table 10.1. The table compares only the most important metals mined in the province. As can be seen, gold dominated the industry until a silver boom in the Kootenays late in the nineteenth century. New mining techniques and new demands then led to the development of copper, lead, and zinc mining. By the 1960s, some iron ore was being mined on Texada Island in the Gulf of Georgia. The big discovery, though, was molybdenum, which hardens steel when used as an alloy. More recently, new metals such as cadmium, platinum, tungsten, and palladium are being developed, thus increasing the percentage of other metals.

Table 10.1

Value of mineral production as a percentage of total metal mined, 1860-2009

	Gold	Silver	Copper	Lead	Zinc	Molybdenum[a]	Iron	Other
1860	100	–	–	–	–	–	–	–
1870	100	–	–	–	–	–	–	–
1880	100	–	–	–	–	–	–	–
1890	83	17	–	–	–	–	–	–
1900	42	20	14	23	–	–	–	–
1910	44	10	35	10	2	–	–	–
1920	13	16	40	14	16	–	–	–
1930	8	11	28	31	17	–	–	–
1939	41	8	13	22	15	–	–	–
1950	10	6	7	37	40	–	–	–
1960	6	6	7	31	41	–	9	–
1970	2	3	42	12	15	19	7	–
1980	13	11	48	4	4	20	–	–
1990	15	8	56	1	8	6	–	–
1997	17	7	47	3	16	6	–	–
1998	20	8	46	2	16	7	–	2
1999	28	11	35	3	17	5	–	1
2000	21	9	47	2	16	4	–	2
2001	23	10	49	2	11	5	–	<0.3
2002	27	12	48	<0.2	4	2	–	8
2003	28	11	46	<0.4	5	4	–	6
2004	13	6	34	0.1	2	27	–	17
2005	9	4	42	0.1	3	21	–	21
2006	9	4	59	0.1	3	13	–	12
2007	9	4	55	0.2	3	15	–	14
2008	9	2	53	0.2	2	16	–	18
2009	15	2	66	0.2	3	9	–	5

a Molybdenum, because of confidentiality, has been estimated since 2003.

Sources: BC Ministry of Energy and Mines (2007); Natural Resources Canada (2010); BC Stats (2009).

The Ministry of Energy, Mines and Petroleum Resources often combines all mining activities in its statistics – metals, industrial minerals (e.g., sand and gravel), structural minerals (e.g., asbestos and gypsum), and energy

Table 10.2

Value of mineral production in British Columbia, 1998-2009

	1998	1999	2000	2001	2002	2003	2004	2005	2006	2007	2008	2009[a]
	($ millions)											
Metal Mining												
Copper	681.4	418.6	733.0	687.8	612.1	588.8	910.4	1,289.3	2,191.7	1,877.8	1,613.1	1213.4
Gold	302.5	332.3	334.3	317.3	340.0	354.5	349.4	288.7	343.7	289.4	278.5	269.6
Silver	120.8	124.7	140.1	132.8	152.4	142.4	172.7	128.9	149.5	119.4	66.2	39.9
Zinc	231.4	201.7	243.7	153.4	83.1	57.5	61.3	82.4	119.2	105.0	70.3	50.4
Lead	24.1	30.0	31.6	25.7	4.4	4.1	2.9	4.3	3.8	6.9	7.3	4.1
Molybdenum	98.8	59.6	62.7	73.8	28.6[b]	48.5[b]	706.2[b]	659.4[b]	484.0[b]	504.6[b]	486.6[b]	216.8[b]
Other	24.8	16.0	25.7	3.6	101.5	135.1	458.8	648.3	440.3	488.1	554.3	34.6
Total	1,483.7	1,183.0	1,571.2	1,394.5	1,288.0	1,282.4	2,661.7	3,101.3	3,732.2	3,391.2	3,076.3	1828.8
Non-metals												
Aggregates[c]	208.1	218.5	223.9	216.6	230.5	238.2	238.5	278.1	273.7	347.2	378.2	305.6
Industrial minerals	245.5	246.4	284.3	296.4	310.2	339.9	355.0	364.4	363.2	424.1	695.7	283.1
Total	453.6	464.9	508.2	513.0	540.7	578.1	593.5	642.5	636.9	771.3	1,073.9	588.7
Fuel minerals												
Coal	956.0	796.9	812.1	959.3	1,034.9	1,000.0	1,190.9	2,229.6	2,105.3	1,949.0	3,738.0	3,316.5
Oil	303.3	374.8	665.7	554.3	550.8	544.2	612.0	666.7	1,013.0	989.0	1,215.0	719.0
Natural gas	1,174.4	1,655.8	3,826.4	4,832.2	3,468.2	5,366.5	5,834.9	7,774.2	5,858.0	5,723.0	7,501.0	3,410.0
By-products[d]	82.2	132.3	276.0	246.2	232.5	273.9	337.5	455.6	179.0	–	–	
Total	2,515.8	2,959.9	5,580.3	6,592.0	5,286.3	7,184.6	7,975.3	11,126.1	9,155.3	8,661.0	12,454.0	7,445.5

a Preliminary estimates
b Estimate based on production times average price for the year.
c Sand, gravel, and stone.
d Included in natural gas values for 2007-9.
Sources: BC Ministry of Energy and Mines (2007); BC Stats (2010a); BC Ministry of Finance (2010).

minerals (e.g., coal) – as Table 10.2 shows. Revenues from metal mining underwent considerable swings from 1998 to 2009, but the most significant difference is in the revenues from the fuel minerals (coal, oil, and natural gas).

British Columbia has a long history in mining and a potentially productive future, as there is no lack of minerals. Being rich in minerals and making production of them economical, however, are two different things. Figure 10.2 categorizes the various stages of metal mining, from exploration to end use. All have to be considered in the industry. Technological innovation plays a role at each stage. The search for gold, for example, was accomplished in the nineteenth century with the simple technology of a shovel, a gold pan, and plenty of hard work. This was placer mining, which extracted gold in a pure form from existing or old stream beds. Next came sluices, dredges, and hydraulics, more costly technologies that allowed the exploitation of even greater gold deposits. Not all gold is in a pure form, though; most minerals or ores are "bound" to other minerals and require the accompanying rock to be smashed and crushed in order to extract the metal. This is **lode mining**, and it requires more sophisticated technologies with much higher costs than placer mining. Figure 10.3 puts into perspective the amount of gold produced by both methods.

Today's prospectors, searching for metal-bearing outcroppings armed with a rock hammer and a knowledge of geology, supplement their search with techniques ranging from remote sensing to diamond hole drilling.

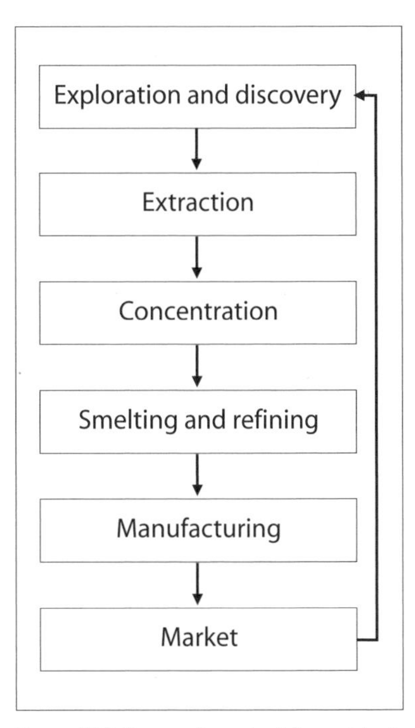

Figure 10.2 Stages of metal mining production

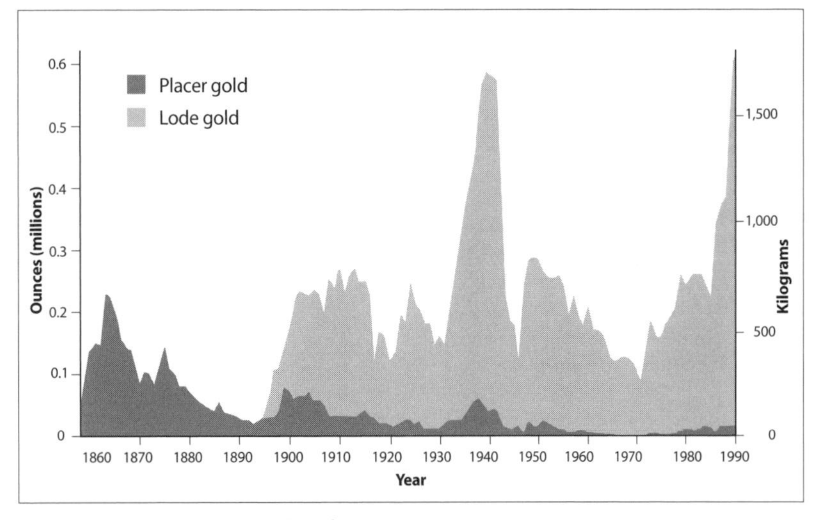

Figure 10.3 Placer and lode gold production, 1860-1990
Source: British Columbia, Geological Survey Branch (1990, 20).

Assessing the quality and quantity of any ore body is crucial to production. The development of processing technologies can change the quality/quantity equation considerably. Copper, for example, was economically viable only at 10 percent grade or higher before the twentieth century. By 1900 new techniques and new demands – such as copper wire for the distribution of electricity – made 2 percent copper ore viable, and by the 1940s, 1 percent was considered viable. These percentages are very significant. A 10 percent copper ore body, for example, would result in 100 kilograms of pure copper being refined from 1,000 kilograms of copper ore. When the percentage drops to 1 percent, only 10 kilograms of copper will become available and 990 kilograms will remain as wastes, or tailings. The 1960s development of the open-pit method of mining, using huge earth-moving equipment, was able to operate with 0.5 percent copper, and 0.4 percent may be feasible today (Barker 1977, 28; Ross 1987, 163). As open-pit technologies replaced underground mining, low grade ores could be economically processed only if there were huge quantities.

Changes in technology reduced the labour required as mining made the transition from a labour-intensive to a capital-intensive industry. Hard rock mining with a pick and shovel and manual loading of a rail cart is an occupation of the past. Today's open-pit miners operate enormous trucks, drills, and shovels after assessing the lode by computer. With far fewer underground mines and no more old rock drills, the workplace is safer. (The drills were known as "widow makers" because most drillers inhaled the dust into their lungs and died from silicosis.) There are, however, far fewer workers per tonne of minerals produced today. Figure 10.4 reveals the change in metal mining employment for British Columbia between 1980 and 2007. The other factor, as can be observed in the graph and in Table 10.2, is that employment increases with higher metal values, leading to more exploration and, potentially, the opening of new mines.

Low grade ores are crushed and concentrated, and most are exported at this stage. Smelters refine or smelt these concentrates into a pure state, but the Trail smelter has not processed ore from British Columbia since the Sullivan Mine closed in 2001. There is a second smelter at Kitimat, but it is used for aluminum and therefore operates on an entirely different basis. The key factor for the production of aluminum is inexpensive electrical energy. The raw material required is bauxite, which is not mined in British

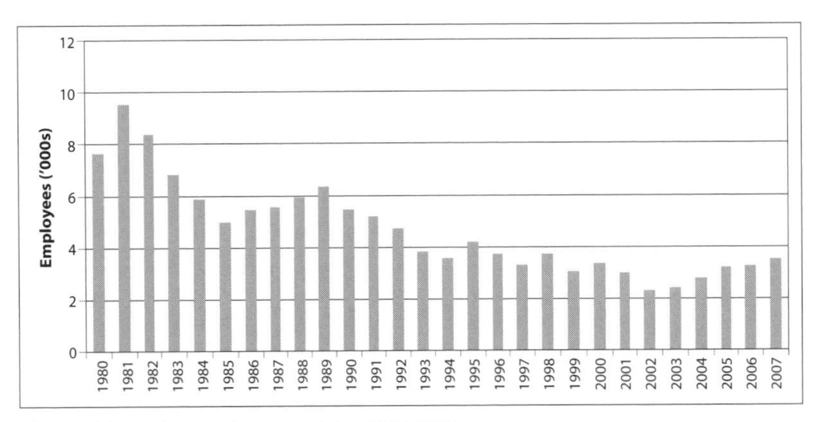

Figure 10.4 Employment in metal mining, 1980-2007
Sources: BC Stats (2008); Statistics Canada (2004, table 281-0024).

Columbia but must be imported from tropical regions such as the West Indies or Australia.

The last two stages of the diagram in Figure 10.2 – manufacturing and market – refer to the end uses of the mineral being mined. Demand for particular end uses influences the world market price, and price is central to economic viability. World market price is also the link between the beginning and the end categories in Figure 10.2, as market demand spurs on new exploration and discovery. Both manufacturing and market are affected by changes in technology. One needs only to consider some of the traditional uses of copper, such as electrical wire and pipes for plumbing. Fibre optics made of silica-derived glass threads and plastic pipe made from petroleum compete with these traditional uses or markets today.

The technological development and market demand for molybdenum, or "moly," is a different story. Molybdenum was discovered and used to make armour-piercing shells during the Second World War. With the Korean War, Vietnam War, and Cold War, a high demand was maintained for military uses and even for the "star wars" scenarios of the aerospace industry. This initial type of demand put molybdenum in the category of a "war-based" mineral. British Columbia is the only Canadian producer and one of the leading molybdenum producers in the world. With the end of the Cold War, the demand for this mineral fell off, and so did the world market price. More recently, demand has increased again because molybdenum is used in stainless steel products, in alloy steels for oil and gas pipelines and for drilling for hydrocarbons, in the aircraft and automobile industries, and in a myriad of other ways (e.g., as a lubricant and as a catalyst in a number of chemical compounds).

Technological aspects of the processes of mineral production from discovery to market are not the only factors influencing costs in this industry. British Columbia has a vertical and rugged topography, which often makes access by rail, boat, or road a significant cost. Nelson, Trail, Greenwood, and a number of other communities in the Kootenays had smelters in the 1890s and early 1900s, but several factors were necessary to make them economically viable: "A smelter needs three things: a satisfactory process of separation, an assured supply of ore, and reasonably priced fuel. In these early days, neither the metallurgical process nor the quality of coke was satisfactory" (Taylor 1978, 132). Pioneer smelters appeared along the main line of the Canadian Pacific Railway, but not surprisingly, most of them failed. A widely distributed infrastructure of railways, mainly owned by the CPR, was constructed throughout the Kootenays to link the region east to southern Alberta, south to the United States, north to the CPR main line, and west to the Okanagan and Vancouver. With transportation in place, going from exploration and discovery to production was much easier.

Politics was very much a part of the decision to provide rail systems to the Kootenays, and politics is still a significant factor in this industry. The provincial government, as the manager of this resource, has played a mainly facilitative role in mining development by building infrastructure such as railways, roads, and hydroelectric developments, as well as by providing direct subsidies, Crown grants, tax incentives, and other beneficial legislation. Nevertheless, the mining industry has also perceived politics as playing a negative role.

Bill 31, passed in 1973, illustrates the relationship between demand, world market price, competition, profits, and taxation. Copper is often referred to as a war-based mineral because one of the big demands for it is in the manufacture of shell casings for bullets. The acceleration of the Vietnam War in the late 1960s and early '70s increased the demand for copper. On the supply side, in the United States, a major strike in the copper mines lasted for months, while in Chile, a major producer of copper, the government decided to nationalize foreign-run mines. Large exports from the United States and Chile were thus interrupted, driving the world market price for copper to over $1.40 per pound. For BC copper producers this was a windfall. The newly elected NDP government thought that the people of the province should also benefit from the windfall, and Bill 31 was introduced as a "super tax" on these "excess" profits. The industry screamed that the taxes were unfair. The real irony was that by the time the government got around to implementing the tax, world conditions had changed and the price of copper had plummeted to thirty-seven cents per pound. The strikes were over, Chile was producing copper, as were other countries such as Zaire and Australia, and the Vietnam War was winding down. Oversupply and underdemand created bankrupt conditions almost overnight, and there was nothing left to tax. It was the government of British Columbia that took the blame, though, for the downturn in the mining economy.

Politics around the issues of taxation, decreased employment, and insufficient government support for the mining industry continue. The September 1993 issue of the *Mining in BC Newsletter* remarks that "government is largely responsible for the fact that exploration is declining, and for the fact that other provinces and countries are luring B.C. mining companies to invest and create jobs in their jurisdictions" (Mining Association of British Columbia 1993, 1). The Liberal government has responded to the noncompetitive nature of the industry by reducing corporate income tax rates and eliminating some capital taxes on mining companies. As well, "The government has also committed itself to improving the certainty of access to resources by adopting a 'go or no-go' designation for exploration and development" (Savings and Credit Unions of British Columbia 2002). Have these measures satisfied the mining industry? It would appear not, for the *Vancouver Sun* (2010) states: "Government red tape has strangled exploration and mine development. Delays caused by a tangle of often overlapping regulations are the principal reasons no major new mines have opened in BC in 13 years."

Another concern (which is also political) has been the increase in parks and protected areas in the province, which in effect shuts the mining industry out in these areas. The Muskwa-Kechika Management Area Act (1998, amended 2002) is designed to take into account the many values of and interests in 6.3 million hectares of this remote area in northern British Columbia, allowing resource development in certain regions. The mining industry walked out of the planning process before the act was finalized, believing that the act would discourage investment. In response, the provincial government instituted a "two-zone system for mineral exploration and mining" that "identifies lands that are closed to mineral development and those that are open to mineral exploration and mining, subject to applicable legislation" (Integrated Land Management Bureau, 2010).

Since the 1970s, the environmental costs associated with mining have become more important in costs of operation. The mining of low-grade ores results in waste materials, or tailings. Legislation now requires mines to deal with waste materials effectively, an added cost to factor into production. Tailings are often high in acid, and plans must be implemented to prevent the contamination of ground- and surface-water sources (Environmental Mining Council of British Columbia 2006). This issue was so controversial for the proposed copper-rich Windy Craggy Mine of the Tatshenshini River area of northwestern British Columbia that the proposal was rejected, and the region was declared a park in 1993. The imposition of these provincial and federal regulations is viewed by the industry as an obstacle to acquiring permits for new mines – and as another level of politics.

The contaminated water in tailing ponds is another source of pollution, as is the discharge of wastes from smelters to the air, water, and land. Mining also provokes potential land-use conflicts. In response to these environmental concerns, the politics of regulation has come into play. Again, as the example of the Muskwa-Kechika Management Area illustrates, government is viewed by this industry as a stumbling block. Still another consideration

Table 10.3

World market prices for selected minerals and years, 1971-2009

	Gold	Silver	Copper	Lead	Zinc	Molybdenum
	(US$/oz)		(US$/lb)			
1971	35	1.56	0.38[a]	0.12	9.13	1.39
1974	166[a]	4.98[a]	0.70	0.16[a]	0.29[a]	1.65[a]
1980	708	23.95	0.94	0.32	0.27	9.63
1985	434	8.40	0.72	0.13	0.38	3.55
1990	469	5.60	1.13	0.30	0.67	2.68
1995	384	5.21	1.33	0.28	0.53	7.42
2000	280	5.00	0.82	0.21	0.56	2.51
2001	271	4.39	0.72	0.22	0.44	2.31
2002	310	4.60	0.71	0.21	0.35	3.59
2003	363	4.88	0.81	0.23	0.39	5.21
2004	409	6.66	1.30	0.40	0.48	15.92
2005	445	7.32	1.67	0.44	0.63	31.05
2006	604	11.55	3.07	0.59	1.49	24.46
2007	697	13.38	3.24	1.17	1.47	30.22
2008	872	14.90	3.15	0.94	0.85	28.78
2009[b]	1,025	15.00	2.45	0.80	0.80	12.00

a The average price has been used since 1970 for copper and 1974 for gold, silver, lead, zinc, and molybdenum.
b Estimated
Sources: BC Ministry of Energy and Mines (2007); British Columbia Ministry of Finance (2009); Global Infomine (2010).

Table 10.4

Production and value of gold and copper for selected years, 1995-2009

	Gold		Copper	
	Quantity (g)	Value (US$)	Quantity (kg)	Value (US$)
1995	19,925,161	327,600,129	278,929,540	1,119,163,080
2000	25,129,245	334,294,340	270,688,178	733,023,586
2001	23,614,119	317,373,759	275,245,104	687,837,515
2004	20,408,114	349,448,136	244,134,848	910,387,848
2006	15,594,001	338,842,048	292,275,925	2,259,292,900
2008	9,322,863	278,511,209	217,571,608	1,613,075,902
2009[a]	7,666,011	269,552,279	210,077,009	1,213,404,804

a Preliminary estimates
Sources: BC Ministry of Energy and Mines (2010); Natural Resources Canada (2010).

in the exploration-to-extraction phase of mining is the recognition that these minerals, for the most part, are on First Nations land. With so few treaties completed, mining companies are obligated to negotiate with First Nations. Scott Simpson (2009) suggests that "red tape threatens to strangle BC's mining industry."

Government red tape, the creation of parks that exclude mining, and the increased costs associated with environmental protection are viewed by the metal industry as major challenges to new mines opening. These "costs," however, are relatively certain hurdles to overcome; far less certain are world market prices for metals. World market price fluctuates, in some cases dramatically, to reflect the many changes in supply and demand. Precious metals such as gold, which have few practical uses, are not as influenced by the over- and under supply and demand, and changes in end use, that metals such as copper have. Gold has a very long history as a currency, and even paper money was once based on gold. A gold standard had been widely adopted by most countries by 1900 but had to be abandoned in the 1930s during the Depression; even so, the price of gold remained fixed until 1970. Table 10.3 shows the dramatic increase in the price when the market was allowed to establish its own level. With the relatively high world market price today for a metal that is one of the heaviest of all the elements, it is no wonder that most new mining ventures in British Columbia over the past two decades have been gold mines.

Table 10.3 also shows the fluctuating world market prices for other metals mined in British Columbia. As stated earlier, copper is our most important metal, and unfortunately, its price is among the most volatile. Table 10.3 provides the average yearly price, but not "spot" prices such as the high US$1.47 per pound reached in 1995 that triggered expectations of continued high prices. In June 1996, a $2.6 billion trading scandal in Japan pushed copper prices down (Savings and Credit Unions of British Columbia 1996, 10). The continued slump and crisis in Asian markets kept the price of copper low until the mid-2000s, when it reached the highest prices known, sparking interest in the opening of new mines.

The price per pound of many minerals has rarely kept pace with inflation and, more recently, the Canadian exchange rate (to the American dollar) has decreased the value of metals, although prices rose dramatically for all metals between 2004 and the recession of 2008

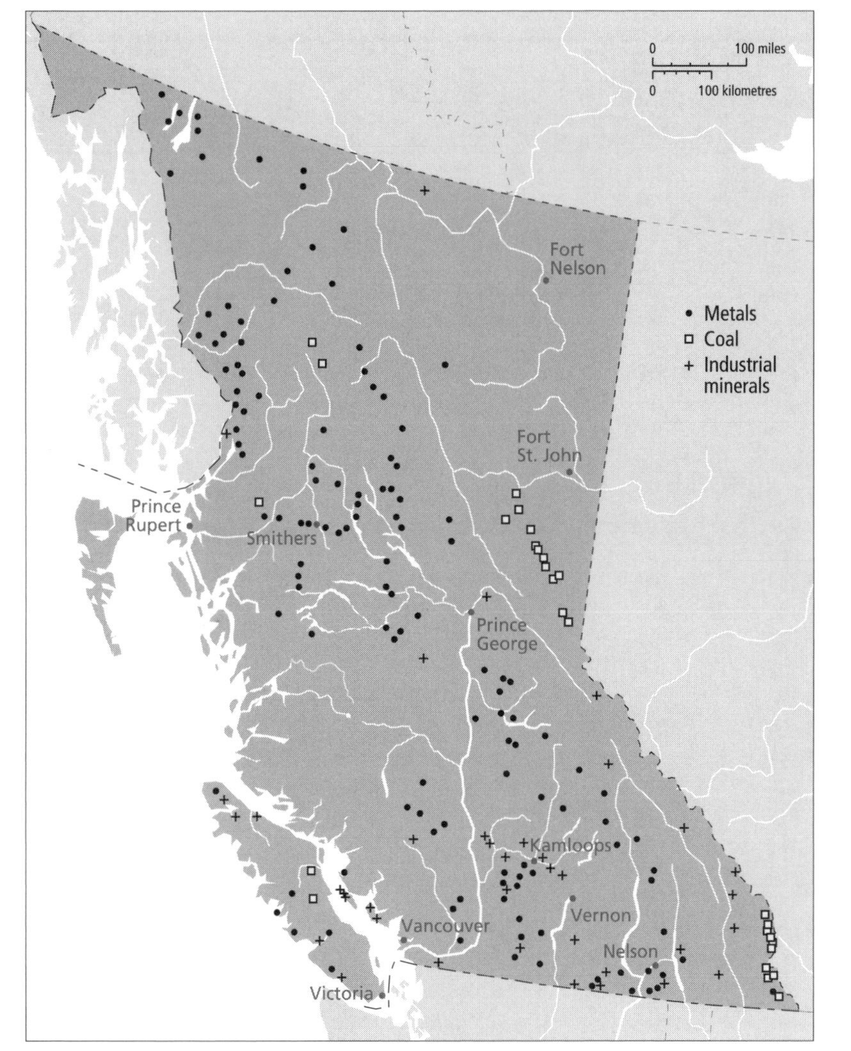

Figure 10.5 Potential mines, 2008
Source: BC Ministry of Energy, Mines, and Petroleum Resources (2009).

(e.g., copper peaked at over US$4/lb in July 2008 but fell to US$1.35/lb by January 2009). Individual mines are, therefore, in the position of maintaining the same efforts for production, often with the same costs, but with smaller returns. This is evident in Table 10.4, which shows the quantity and value of gold and copper from 1995 to 2009. Comparing the figures for 1995 and 2001 for both metals, it is evident that, while production increased for gold and declined only slightly for copper, values dropped, and precipitously for copper. The rebound in values for both gold and copper by 2008 represents vastly improved conditions for the industry, but the recession at the end of 2008 reduced the value of copper considerably.

THE OPENING AND CLOSING OF MINES IN BRITISH COLUMBIA

To this point, the discussion has included an overview of the history and many influences on the development of the metal mining industry. Figure 10.5 shows prospected mineral deposits throughout British Columbia, demonstrating that the province is rich in minerals. Such richness might lead one to expect a great number of mines in the future. Few will materialize, however, as new mines face high capital costs of production, stringent environmental regulations, and transportation costs from remote sites. Table 10.5 lists operating mines from 1990 through to 2008, indicating when individual mines have opened and closed. Figure 10.6 shows their locations, an important factor in the analysis of the industry.

The list of producing metal mines is not numerous; there were twenty-one metal mines at some stage of operation in 1990 (including those temporarily shut down or in the process of closing), but only six in operation in 2003. When market prices rebounded, a number of mines reopened, and two new mines came onstream – Lexington-Grenoble and Max Molybdenum. There is a wide variation among mines in tonnes milled – the volume of minerals that have gone through the concentrator and been exported – one measure of the size of the mining operation. Only a few of the producing mines process just one mineral, even though mines such as Island Copper or Highland Valley

Table 10.5

Production, employment, and operation of metal mines, 1990-2008

Mine	Principal minerals	Ore milled ('000 tonnes)	Average employment	1990	1991	1992	1993	1994	1995	1996	1997	1998	1999	2000	2001	2002	2003	2004	2005	2006	2007	2008
Afton	Cu, Au, Ag	2,655	208	X	S			R	X	X	C											
Beaverdell	Pb, Ag, Zn	36	34	X	C																	
Bell	Cu, Au, Ag	5,423	303	X	X	C																
Blackdome	Au, Ag	73	102	X	C																	
Brenda	Cu, Au, Mo, Ag	4,282	190	C																		
Endako	Mo	9,706	196	X	X	X	X	X	X	X	X	X	X	X	X	X	X	X	X	X	X	X
Equity Silver	Cu, Au, Ag	3,146	205	X	X	X	X	C														
Eskay Creek	Au, Ag	115	150					O	X	X	X	X	X	X	X	X	X	X	X	X	X	C
Gibraltar	Cu, Au, Mo, Ag	12,635	301	X	X	X	S	R	X	X	X	S						R	X	X	X	X
Golden Bear	Au, Ag	70	115	X	X	X	X	S			R	X	X	X	C							
Highland Valley	Cu, Au, Mo, Ag	49,030	1,230	X	X	X	X	X	X	X	X	X	SR	X	X	X	X	X	X	X	X	X
Huckleberry	Cu, Au, Mo, Ag	7,000	214								O	X	X	X	X	X	X	X	X	X	X	X
Island Copper	Cu, Au, Mo, Ag	18,361	549	X	X	X	X	X	C													
Johnny Mountain	Au, Ag	87	56	S			RS															
Kemess South	Cu, Au	18,633	343									O	X	X	X	X	X	X	X	X	X	X
Lawyers	Au, Ag	185	145	X	X	S																
Lexington-Grenoble	Cu, Au	72	n/a																			O
Max Molybdenum	Mo, W, Pb, Zn, Cu	72	n/a																		O	X
Mount Polley	Cu, Au	6,570	n/a								O	X	X	X	S				R	X	X	X
Myra Falls	Cu, Zn, Au, Ag,	1,460	379	X	X	X	X	X	X	X	X	S	R	X	S	R	X	X	X	X	X	SR
Nickel Plate	Au, Ag	1,141	173	X	X	X	X	X	X	C												
Premier	Au, Ag	736	166	X	X	X	X	X	X	S												
Quesnel River	Au	438	80						O	X	X	S									R	X
Samatosum	Cu, Au, Pb, Ag, Zn	169	92	X	C																	
Shasta	Au, Ag	64	28	X	C																	
Silvana	Pb, Ag, Zn	32	38	X	X	S																
Similkameen	Cu, Au, Ag	6,677	329	X	X	S	R	X	X	S												
Snip	Au, Ag	155	134		O	X	X	X	X	X	X	X	X	X	C							
Sullivan	Pb, Ag, Zn	400	379	X	X	X	X	X	X	X	X	X	X	X	C							
Table Mountain	Au	33	35					O	X	X	X	X	S							R	S	

Note: Au = Gold; Ag = Silver; Cu = Copper; Mo = Molybdenum; Pb = Lead; W = Tungsten; Zn = Zinc; O = Opened; S = Shutdown; R = Re-opened; X = Operating; C = Closed.
Sources: BC Ministry of Energy and Mines (2007); BC Stats (2010b).

are often referred to as copper mines. The average employment for each mine also varies considerably. The number of employees is related to the tonnes milled and also to mining processes. Notice, for example, the tonnes produced at Island Copper and at Myra Falls, both on Vancouver Island (Table 10.5). The higher number of employees in relation to output at Myra Falls represents the difference between the underground operation at this mine and the open-pit process at Island Copper.

Mines go out of production, either temporarily or permanently, for various reasons. Metals are nonrenewable resources, and most permanent closures have occurred because of exhaustion of the ore (e.g., Island Copper, Brenda mines, and Sullivan Mine). Of the mines operating in 2008, both the Kemess South and Huckleberry mines are close to the end of their lifespan. Temporary shutdowns are related to various factors, such as strikes and lockouts, environmental issues, a slump in

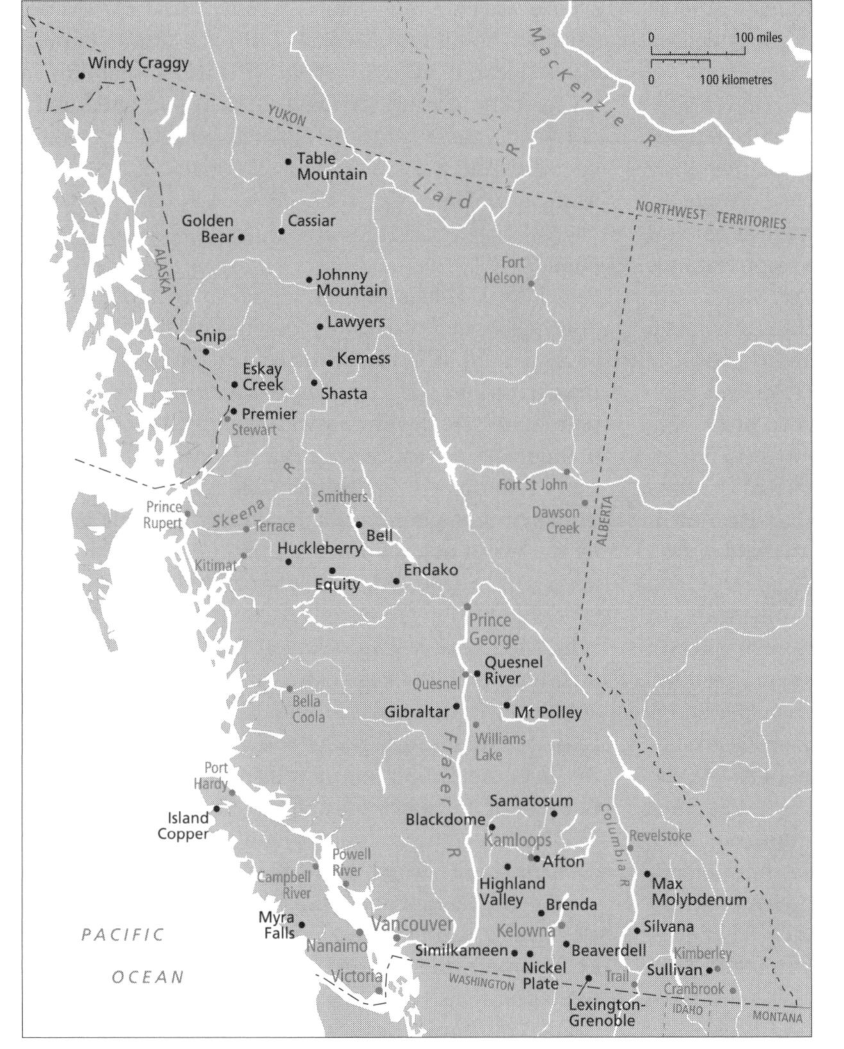

Figure 10.6 Metal mines, 1990-2008
Source: BC Ministry of Energy, Mines, and Petroleum Resources (2009)

the world market price of the metal, or decisions made by large mining corporations involving a combination of these reasons. The Similco Mine at Princeton, for example, like many copper mines in British Columbia, is owned by a large multinational corporation. Head office made the decision to shut the mine down in 1993 because the price of copper was declining with little likelihood of higher returns. Prices did rebound and the mine reopened in 1994, but low prices in 1996 resulted in another closure. Table 10.5 indicates a similar story for a number of mining operations.

As stated earlier, the collapse of the Asian markets for copper depressed the price of that metal from the mid-1990s to the early 2000s; as a result, the Gibraltar Mine shut down in February 1998, Myra Falls in December 1998, and Highland Valley in May 1999, not because they had exhausted their copper but because the price of copper was not high enough to make a profit (Savings and Credit Unions of British Columbia 1999). The price of copper did not rebound until 2004, but the price of gold increased, resulting in the reopening of Myra Falls and Highland Valley; Gibraltar reopened in 2004. Afton, at Kamloops, has also closed down for several periods, again because of the world market copper price. Afton is unique in British Columbia because it is designed to smelt the particular type of copper ore in the Kamloops area, creating other cost considerations. The longest-running mine in British Columbia, the Sullivan Mine (opened in 1900), employed 2,122 miners in 1949, but this had been reduced to 660 by 1999. Its closure in 2001 has forced the town to turn to tourism and forestry to maintain employment (CBC 2000). Record high prices for metals since 2004 have led to renewed interest and investment, and it is likely that several other mines will open between 2010 and 2012. Copper Mountain (south of Princeton) is the revival of an existing mine and New Afton (Kamloops) is a new underground gold-silver-copper mine at the old Afton mine site. Three new mines are close to opening: Mount Milligan (copper and gold, near Mackenzie), Red Chris (copper and gold, south of Dease Lake), and Prosperity (copper and gold, southwest of Williams Lake). It should be noted that there are serious issues with respect to environmental assessments for the latter two mines because their tailings will be dumped into fish-bearing lakes.

One of the strengths of BC mines is that most produce more than one metal. British Columbia is a copper producing region of Canada, but copper, as previously seen, has a very volatile world market price. Fortunately, these large copper mines also produce gold, silver, and in some cases, molybdenum. Three new copper mines opened in 1997-98 (Huckleberry, Kemess South, and Mount Polley) on the expectation of rebounding metal prices (which occurred in 2004), but all three have other valuable metals associated with their ores, such as gold. Huckleberry and Kemess South remain as producing mines, but Mount Polley closed in 2001 and did not reopen until the price of copper rebounded in 2004. When a mine has only one metal, such as Endako with molybdenum, it is considerably more vulnerable to world market volatility.

The story of Cassiar (see Figure 10.6) is an example of what happens when an isolated company town relies on one resource and the demand for that resource declines. The mine at Cassiar produced asbestos, which is an industrial mineral rather than a metal. Since asbestos has been recognized as carcinogenic, there has been less and less demand for it. Although plenty of asbestos remains in the deposit, the low world market price does not make it economically viable. In 1993 the Cassiar mine closed and another ghost town was added to the list in British Columbia. Other mining communities – such as Granisle, which is in a relatively isolated location on Stuart Lake (see Figure 10.6) – have not been reduced to ghost town status, but they are certainly less viable with only 350 residents.

A significant number of the mines listed in Table 10.5 primarily produce gold, or gold and silver. They are relatively small and tend to start up, shut down, and start up again. This on-and-off activity is related to the high price of gold and the ability of a small mining company to offer shares and raise money for short runs of production.

The closure of a mine is usually related to economic considerations, but the province also has a history of mines being closed and their communities dying because of natural hazards. In 1965, a landslide at the Granduc Mine near Stewart killed twenty-six workers and shut the mine down for a considerable length of time. The copper smelter community of Anyox, northeast of Prince Rupert, was reduced to ashes by a forest fire in 1939. Sandon, in the West Kootenays, was wiped out by a flood, and Britannia Beach, just north of Vancouver, was hit by both a landslide and a debris torrent. Many mines are at risk from natural hazards.

SUMMARY

British Columbia is rich in mineralization and has had a long history of metal mines opening and closing, adding to the pattern of boom and bust (and the numerous ghost towns) throughout the province. A host of economic and political factors, beginning with the economic recession of the 1980s, has caused the industry to decline. The primary metal mined in British Columbia is copper, although on the global scale this supply has little impact on the world market price. All the metals mined in the province are exported, usually as concentrates, and British Columbia therefore has little control over the demand side. The industry is very capital intensive, making any operation high in fixed costs even though revenues are volatile. The exception in metals is gold. Although its value fluctuates, it is nonetheless consistently so high that even small amounts of gold are often worth mining. Because of the high costs and high risks involved, metal mining in British Columbia is dominated by multinational corporations, which are frequently foreign owned.

When all factors are considered, the pattern of mine closures exceeding mine openings is not surprising. Three new copper mines opened in 1997-98 in expectation that the high prices of copper and other metals in 1995-96 would continue. These were the first copper mines to open in British Columbia in nearly twenty years. Unfortunately, the unpredictable global economic climate caused most metal prices to decline from late 1996 to 2004, forcing one of the mines to close until copper prices revived. Record high metal prices between 2004 and 2008, and rebounding prices since then, have resulted in the reopening of closed mines and the opening of new, mainly copper-gold mines. Metal mining in British Columbia peaked in the 1960s and '70s, with major investments in copper and molybdenum mining, especially. Many of these large mines had exhausted their deposits by the 1990s. The consequent mine closures, and overall decline of the industry, have brought tragic levels of unemployment to several communities. The boom in metal prices since 2004 has ignited new investment and another round of mine openings that has led to more jobs in construction and production. Also on the positive side, the vast experience gained in mining has proven to be an exportable commodity for British Columbia.

REFERENCES

Barker, M.L. 1977. *Natural Resources of British Columbia and the Yukon*. Vancouver: Douglas, David, and Charles.

Mining Association of British Columbia. 1993. *Mining in BC Newsletter* 6, 8 (September): 1-4.

Nutt, R. 1998. "Gibraltar First Casualty of Weak Copper Prices." *Vancouver Sun*, 13 March, D1.

Ross, W.M. 1987. "Mining." In *British Columbia: Its Resources and People*, ed. C.N. Forward, 162-76. Western Geographical Series vol. 22. Victoria: University of Victoria.

Savings and Credit Unions of British Columbia. 1996. *Economic Analysis of British Columbia* 16 (9): n.p.

–. 1999. *Economic Analysis of British Columbia* 19 (3): n.p.

–. 2002. *Economic Analysis of British Columbia* 22 (4): n.p.

Taylor, G.W. 1978. *Mining: A History of Mining in British Columbia*. Saanichton, BC: Hancock.

INTERNET

BC Ministry of Energy and Mines. 2010. "Production and Values: Metals." www.empr.gov.bc.ca/mining/MineralStatistics/Pages/Default.aspx.

BC Ministry of Energy, Mines, and Petroleum Resources. Operating Mines and Selected Major Exploration Projects in British Columbia. 2008. Map. www.empr.gov.bc.ca/mining/geoscience/publicationscatalogue/openfiles/2009/pages/2009-1.aspx.

BC Ministry of Finance. 2009. *2009 British Columbia Financial and Economic Review*. 69th ed. www.fin.gov.bc.ca/tbs/F&Ereview09.pdf.

BC Ministry of Finance. 2010. *2010 British Columbia Financial and Economic Review*. 70th ed. www.fin.gov.bc.ca/tbs/F&Ereview10.pdf.

BC Stats. 2004. "The Economy, Mining." In Quick Facts about British Columbia, 7-8. www.bcstats.gov.bc.ca.

–. 2008. "Metal Mining Employment." www.empr.gov.bc.ca/Mining/MineralStatistics/MineralEconomySnapshot/1980toPresent/Employment/Pages/default.aspx.

–. 2009. "All Minerals Production and Values: 1998 to 2008." www.empr.gov.bc.ca/Mining/MineralStatistics/MineralEconomySnapshot/1992toPresent/ProductionandValues/Pages/default.aspx.

–. 2010a. "British Columbia Mineral Production." www.empr.gov.bc.ca/Mining/MineralStatistics/MineralEconomySnapshot/1992toPresent/ProductionandValues/Pages/default.aspx.

–. 2010b. "Mine Operating Status, 2010." www.em.gov.bc.ca/mining/miningstats/12mineopstatusmetcoal.htm.

CBC. 2000. "Sullivan Mine Closure." 1 September. www.cbc.ca/news/story/2000/09/01/bc_mine000901.html.

Environmental Mining Council of British Columbia. 2006. "EMCBC Mining and the Environment Primer: Acid Mine Drainage." www.miningwatch.ca/en/emcbc-mining-and-environment-primer-acid-mine-drainage.

Global Infomine. 2010. "Commodity Prices." www.infomine.com/commodities/.

Integrated Land Management Bureau. 2010. "Muskwa-Kechika Management Area: Background Information – Mineral Exploration and Mine Development in Muskwa-Kechika." archive.ilmb.gov.bc.ca/slrp/lrmp/fortstjohn/muskwa/plan/background/mining.html.

Natural Resources Canada. 2010. "Mineral Production." mmsd.mms.nrcan.gc.ca/stat-stat/prod-prod/ann-ann-eng.aspx.

Rider, J.M. 1978. "Geology, Landforms, and Surficial Materials." In *The Soil Landscapes of British Columbia*, ed. K.W.G. Valentine, P.N. Sprout, T.E. Baker, and L.M. Lavkulich. BC Ministry of Environment: Soils. www.env.gov.bc.ca/soils/landscape/1.3geology.html.

Simpson, S. 2009. "Red Tape Threatens to Strangle BC Mining Industry." *Vancouver Sun*, September. www.vancouversun.com/tape+threatens+strangle+mining+industry/1964450/story.html.

Statistics Canada. 2004. "Employment, Unadjusted for Seasonal Variation, by Type of Employee for Selected Industries, Annual." CANSIM Table 281-0024. www.statcan.gc.ca/.

Vancouver Sun. 2010. "Lift the Regulatory Burden on BC's Mining Industry." 26 January. www2.canada.com/vancouversun/news/editorial/story.html?id=fec39196-a7e5-41c1-b9a8-527562de8d17&p=2.

11

Energy

Supply and Demand

Energy resources in British Columbia have come in many forms, as animal power, wood, and water wheels gave way to coal, electricity, oil, and natural gas. Demand, both locally and for export, has changed considerably over time, and technology has played a lead role in the development of new energy sources for old and new uses. Within British Columbia are spatial patterns of supply and demand; not all energy resources are spread uniformly throughout the province, nor are the demands for energy (Chapman 1960).

This chapter describes the overall historical and spatial development of energy in British Columbia and then examines in turn coal, oil, natural gas, and electricity – the energy sources on which we rely most. Each of these energy sources is reviewed in terms of past to present development, and included in the review are some of the main issues arising from use of these forms of energy.

Amory Lovins (1977) challenges our dependence on traditional, centralized energy by rethinking energy strategies for the future. In Lovins's view, we must focus on decentralized, renewable forms of energy technologies and explore the demand side of energy. What role can individuals play in using and saving energy? How will British Columbia fulfill its demands for energy in the future? This view, which translates into "green energy" options, is particularly relevant because Canada has signed the Kyoto Protocol and thus committed to reducing greenhouse gases. A controversial side to this direction has been the debate over the production of green energy by independent power producers (IPPs) rather than by the provincial government via its Crown corporation BC Hydro.

HISTORICAL OVERVIEW OF ENERGY DEVELOPMENT

British Columbia's early energy needs were met by wind for sailing ships and wood for fuel. By the 1830s, coal was being mined on Vancouver Island and added to the energy options. The gold rush brought a great number of people to the region and signalled the beginning of many industries to follow, as well as a need for ever greater amounts of energy. Horses, oxen, donkeys, and people performed a fair amount of the work. Flumes moved logs and water wheels ran mills and mining operations, while wood and coal were burned for heating, cooking, and running the steam engine. As railways became the main mode of transport and the steam engine became essential in industry, the demand for coal steadily increased.

Table 11.1 reflects the changing demand for energy in the province. As it shows, from the 1920s to the 1960s, oil, or petroleum, became increasingly important. It was used for transportation, heating, creating electricity, smelting, and, eventually, myriad petrochemical products from paint to lipstick. A significant technological change occurred with the wide adoption of the diesel engine, which replaced the coal-driven steam engine and reduced the demand for coal. British Columbia's growing dependence on oil, most of which has come from Alberta since the 1950s, is little different from the rest of Canada and the industrialized world. Nonetheless, the energy crisis of the 1970s caused a major reassessment of British Columbia's energy supply and demand, particularly with respect to dependence on oil. Natural gas, electricity, and hog fuel (wood wastes and black liquor from pulp operations) became increasingly important sources of energy and alternatives to oil.

The rising price of energy caused other developments not evident in the statistics. Solar energy and wood stoves became popular at the individual, residential level. BC Hydro, a Crown corporation, investigated the potential of tidal power, geothermal production, nuclear power,

Table 11.1

Energy sources, 1925-2008

Year	Coal	Wood	Oil	Natural gas	Electricity
	(% use)				
1925	52	16	24	–	8
1935	30	18	30	–	22
1945	31	12	29	–	28
1955	14	13	65	–	8
1965	2	8	60	15	15
1975	1	7	54	23	15
1996	1	16a	36	29	18
1999	1	–	45	33	22
2002	<1	–	42	35	23
2007	1	18	36	24	21
2008	1	17	36	25	20

a Hog fuel or wood wastes.

Note: Not all years add up to 100 percent because of rounding.

Sources: Barker (1977); Statistics Canada (2007, table 128-0002; 2010a, table 2-12).

coalfired thermal power, and extensive hydroelectric megaprojects. The federal and provincial governments began to explore new ways to acquire oil in order to reduce dependence on Alberta's supplies. Attention was also given to conservation measures.

The demand for energy in British Columbia and supplying this demand with BC energy resources are two different components of the energy question. As electricity was required, dams were constructed to produce hydroelectricity. Supply of electricity met demand until the major development of both the Columbia and Peace River dams in the 1960s and '70s, at which time a surplus was produced and the province was able to export hydroelectricity. Oil was imported prior to its discovery in the Peace River region in the 1950s. Unfortunately, there has never been enough BC oil to meet the demand, and Alberta oil is still needed to supplement it. On the positive side, natural gas discoveries have produced a surplus of this energy source. Coal made a major production recovery in the late 1960s, but the demand (shown as only 1 percent in Table 11.1) was not from the domestic market. Japan, and later Korea, needed coking coal for the steel industry and thermal coal for electricity, thus reviving coal production in British Columbia during that decade. Table 11.2 presents both production and consumption of the main energy sources in the province from 2000 to 2008. Coal and natural gas continue to be major exports. As for electricity production, consumption has caught up to production and for some years has exceeded it, and oil production meets about 20 percent of the provincial demand.

Energy demand can be further broken down into various sectors (Table 11.3). Industry consumes the greatest amount of energy, followed by the transportation sector, and then residential and commercial use. These requirements can be fulfilled by many types of energy, although transportation demands are met almost exclusively by petroleum.

Another aspect of the industry is the spatial differences between where energy is produced and where it is consumed. The largest demand is in the Lower Mainland, which has the greatest concentration of population. Industrial, commercial, residential, and transportation demands are met with a range of energy sources, few of which come from this region. Pipelines transport oil and natural gas, and transmission lines bring electricity.

Table 11.2

Energy production and consumption, 2000-8

Year		Coal (tonnes/year)	Natural gas (10^9 m^3)	Crude oil (10^6 m^3)	Electricity (GWh)
2000	Production	25.7	26.6	3.2	68,241
	Consumption	0.5	7.1	11.1	63,582
2001	Production	27.0	29.9	3.2	57,332
	Consumption	1.0	6.5	10.9	61,079
2002	Production	24.4	32.4	3.3	64,945
	Consumption	0.4	5.7	11.1	62,636
2003	Production	23.1	30.8	2.8	63,051
	Consumption	0.4	5.8	11.3	60,221
2004	Production	27.3	36.0	2.7	60,496
	Consumption	0.6	6.0	12.1	61,966
2005	Production	26.7	33.9	2.5	67,811
	Consumption	0.5	6.3	11.7	65,790
2006	Production	23.1	30.8	2.4	62,021
	Consumption	0.5	6.2	11.4	68,561
2007	Production	25.7	33.2	1.4	72,212
	Consumption	0.6	6.4	11.6	69,252
2008[a]	Production	26.6	32.8	1.3	65,824
	Consumption	0.6	6.3	11.7	68,007

a Estimated for natural gas and oil
Sources: BC Ministry of Finance (2009); Statistics Canada (2010c, table 126-0001).

Table 11.3

Energy consumption by sector, 1978-2007

	1978	1986	1990	1996	2000	2002	2005	2007
	(%)							
Residential	15.5	15.0	14.8	14.0	13.7	14.0	13.8	13.9
Commercial	10.0	10.0	11.6	12.0	11.0	12.7	10.9	10.5
Transportation	26.0	25.0	31.1	35.0	31.8	31.8	33.1	32.2
Industrial	48.5	50.0	42.5	39.0	43.5	43.5	42.2	43.4
Total	100.0	100.0	100.0	100.0	100.0	100.0	100.0	100.0

Note: Some rows do not add up to 100 percent because of rounding.
Sources: Sewell (1987); BC Ministry of Finance and Corporate Relations (1987, 95; 1991, 69; 1997, 88); Nyboer, Bennett, and Muncaster (2004, 35); Natural Resources Canada (2009a).

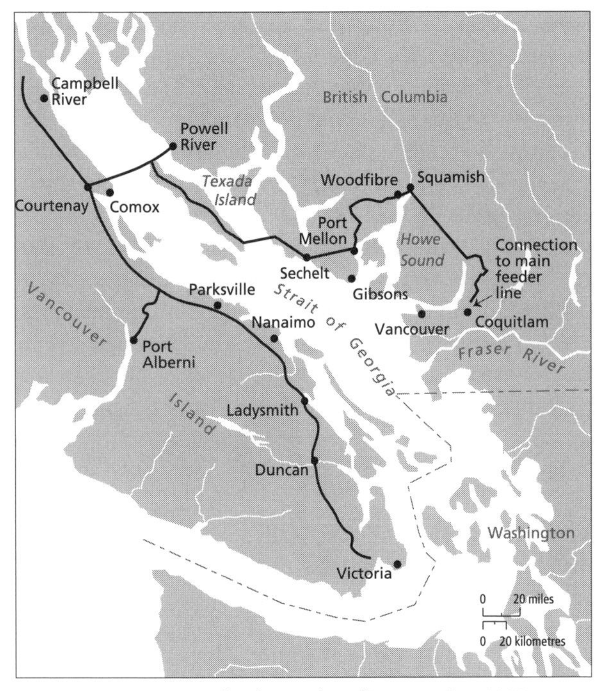

Figure 11.1 Vancouver Island natural gas line extension, 1995
Source: Modified from Pacific Coast Energy Corporation

Vancouver Island does not have large river systems suitable for generating hydroelectricity, and until the 1990s, natural gas was not available there. Essentially, 70 percent of the island's electricity comes from the mainland via two corridors: an undersea southern corridor built in the 1940s and replaced in 2007 and a northern corridor referred to as the Cheekeye-Dunsmuir transmission line (built in the 1980s). A natural gas pipeline from the mainland was built in the 1990s to supply gas to the island as well as to pulp mills and communities in Howe Sound and the Sunshine Coast (Figure 11.1).

A human and environmental price is paid by any region in which energy is generated. In regions where dams are constructed, for example, valleys are turned into reservoirs. Transportation of this energy via transmission lines and pipelines has a further environmental cost (Wilson 1978). The electrical demand in the Vancouver Island/Lower Mainland region is over 70 percent of the provincial total, but the region produces only 12 percent of the electricity (BC Transmission Corporation 2009). The region thus gains the benefit of electrical production while paying little of the environmental cost. For the development of energy sources in the future, it is important to assess all the costs and benefits for individual regions and attempt to equalize costs and benefits across the province.

COAL

Coal is a fossil fuel. It is graded according to the amount of carbon it contains because that determines its ability to burn and produce heat (Figure 11.2). Lignite coal has the least amount of carbon and gives off relatively low heat values. Bituminous coal has an increased carbon/heat ratio, and anthracite coal has the highest rating. British Columbia has all three grades of coal, and the known deposits of coal can be seen in Figure 11.3.

Coal was one of the first resources to be mined in British Columbia. Initial discoveries in the 1830s led to some

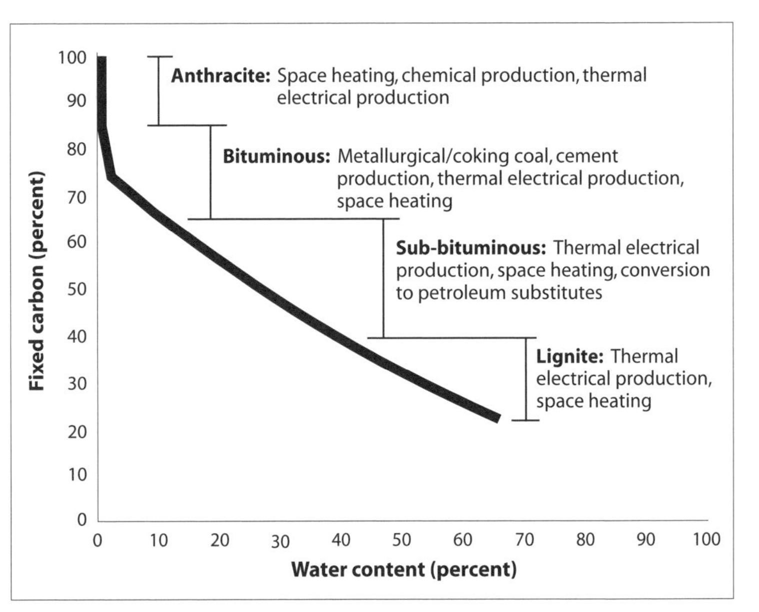

Figure 11.2 Grades of coal
Source: Data from Smith (1989, 13); Enger et al. (1983, 169).

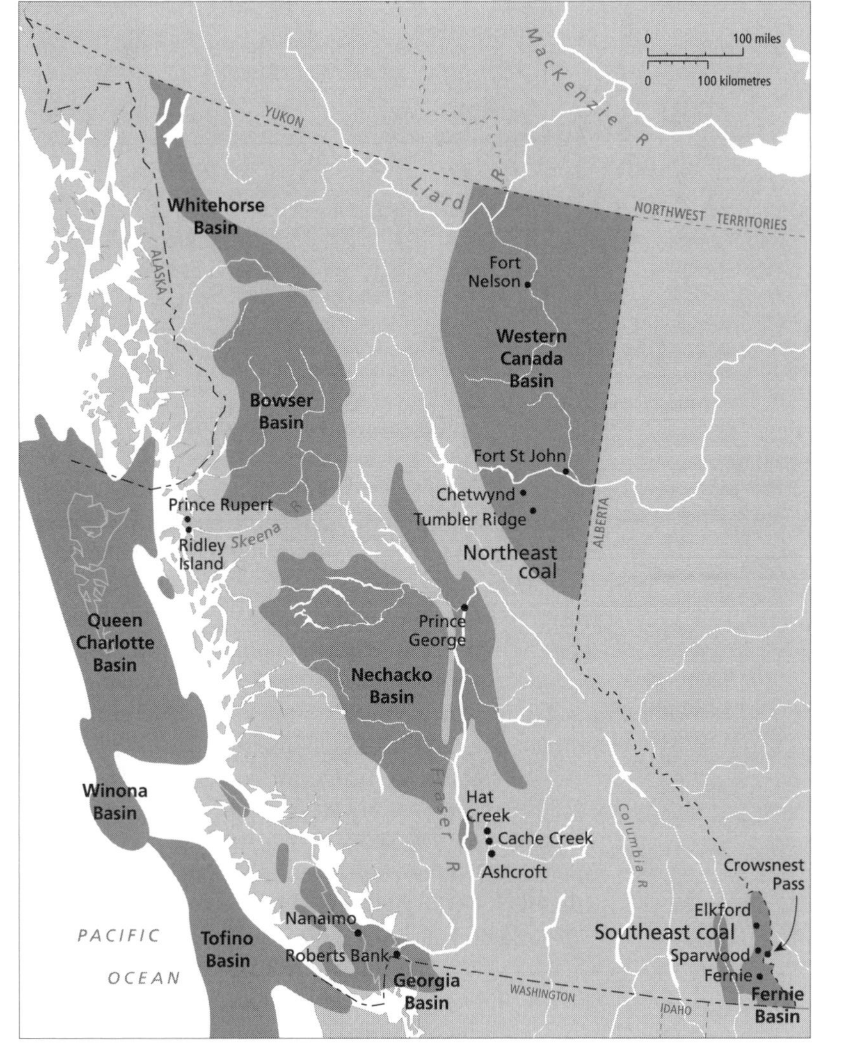

Figure 11.3 Sedimentary basins containing coal, natural gas, and oil
Sources: Modified from Sewell (1987, 230); BC Ministry of Energy and Mines (2004).

mining at the north end of Vancouver Island and near present-day Prince Rupert. Substantial discoveries and developments in Nanaimo, Cumberland, and other central island locations followed (Seager 1996). As the demand for coal increased, discoveries were made in the East Kootenays and production got under way. Petroleum had replaced coal by the 1950s and many coal mines closed, but coal production had revived by the end of the 1960s, almost exclusively for export to Japan. The industry took off, with huge multinational corporate investments in capital-intensive, open-pit mining and concentrating of coal. Old mining communities such as Natal and Michel near the Crowsnest Pass were phased out and the new communities of Sparwood and Elkford were built to house the coal miners. With government help, dedicated rail lines were built from the Crowsnest Pass region to the new, large coal terminal of Roberts Bank. Coal became a major export commodity to Japan and later Korea. Figure 11.4 reveals these changes in coal production over the past century. Table 11.4 indicates the continued strong ties to the Japanese and Korean markets, although some diversification of British Columbia's coal exports has occurred and China has now become an important customer.

By the 1960s, demand for metallurgical, (or coking), coal, along with minor quantities of thermal coal, was responsible for new investment. The primary use of thermal coal is to create steam to turn turbines for electrical production, and small amounts of thermal coal are also exported to Japan and Korea. Metallurgical coal is used almost exclusively in the iron and steel industry, which requires high grade bituminous coal. British Columbia's bituminous deposits in the East Kootenays, known as southeast coal, were targeted for both these end uses.

The story of northeast coal production is much more controversial, as decisions made for this region at national and provincial political levels were based on international events and influences. Figure 11.5 outlines the projected and real demands of the Japanese steel industry in the 1970s and '80s. The wildly exaggerated projected increase translated into a similarly exaggerated demand for BC

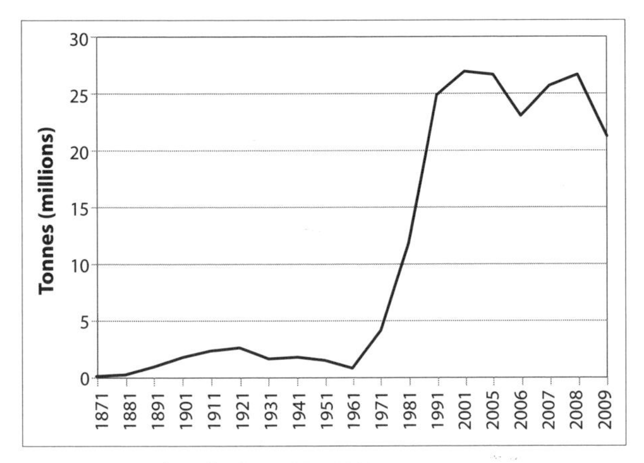

Figure 11.4 Coal production, 1871–2009
Sources: BC Ministry of Energy and Mines and Petroleum Resources (2008); Statistics Canada (2010b, table 135-0002).

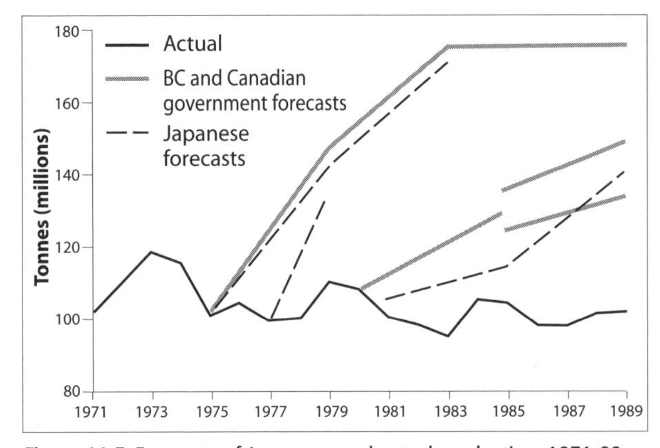

Figure 11.5 Forecasts of Japanese crude-steel production, 1971-89
Source: Modified from Halvorson (1990, 319.)

Table 11.4

Main destinations of coal exports, 2007

Region	Tonnes
Japan	8,219,319
South Korea	5,040,161
Germany	1,691,687
Brazil	1,544,854
United Kingdom	1,267,716

Source: BC Stats (2008).

coal. The energy crisis over oil began in 1973, causing global reconsideration of energy use. In Japan, almost all raw materials for production were imported, including oil as the energy base. As the price of oil and oil-based transportation increased, Japan's steel production became less competitive, and even though it projected major increases they did not materialize. The governments of British Columbia and Canada produced their own exaggerated projections, which were accompanied by foreign investment for plant construction, operation, and equipment, and contracts for coal. The two levels of government invested an enormous amount of money in infrastructure for the northeast coal region: for the new community of Tumbler Ridge, the coal port facility at Ridley Island, the double tracking of the CN Rail line to Prince Rupert, and a new electric rail line between Tumbler Ridge and Chetwynd.

The Quintette and Bullmoose mines of northeast coal had just gone into operation when the recession of the 1980s began. Consequently, they experienced financial troubles from the beginning, particularly when the Japanese renegotiated the coal contract down from over $110 per tonne in 1981 to $82.40 per tonne in 1998 and also reduced their annual volumes imported. British Columbia was caught in global oversupply with a demand that had stabilized. The Quintette Mine closed in 1999 because of low coal prices and the Bullmoose Mine in 2003 because the coal was exhausted. A rebound in coal prices in 2004 has resulted in new metallurgical mines – the Trend (2005) and Wolverine (2006) mines near Tumbler Ridge and the Brule Mine (2007) near Chetwynd. The energy crisis of the 1970s caused BC Hydro to consider the use of coal to produce electricity. Hat Creek Coal, a mine with huge reserves of low grade lignite coal, was acquired by BC Hydro. This deposit, along with coal from the East Kootenays and some from Vancouver Island, was proposed for thermal electrical production. None of these projects were initiated, for reasons to be reviewed later in the chapter.

Burning coal to produce electricity results in significant greenhouse gas emissions. As well as coming in various grades, coal contains varying amounts of sulphur, which is also released into the atmosphere when burned and contributes to acid rain. The residents of the Hat Creek/Cache Creek/Ashcroft area expressed great concern over the impact of acid rain.

Coal has been a major energy export commodity for British Columbia (see Table 10.2). The province has plenty of it, but so do many other regions of the world. Until recently, this abundance has kept the world market price at relatively low levels. The economic slump of the Japanese and Korean markets had a serious impact on northeast coal production and, with the closure of both mines, the population of Tumbler Ridge felt threatened until coal prices rebounded and new mines in the area opened. Coal, whether used here (very little) or in other parts of the world (considerable), does contribute to global climate change and acid precipitation. Research and development is therefore necessary to develop more environmentally friendly ways to burn coal, or we will need to develop greener alternatives.

Coalbed methane (CBM) is another form of energy related to coal, but it is actually methane gas or natural gas. All coal beds contain methane, and now that the price of natural gas is relatively high, CBM is an energy source worth pursuing, and the provincial government is encouraging its development. According to the Ministry of Energy, Mines and Petroleum Resources (2007a), the methane has few impurities, although below 2,000 metres, the methane "cannot be extracted at economic rates." The extraction process is somewhat complex because the gas permeates the coal. When a well is drilled, groundwater must first be pumped out to decrease the pressure in the coal and thus release the methane. This means that initially water is extracted and, over time, methane becomes the main substance from the wells drilled. The water extracted must be treated or re-injected to deep aquifers to ensure that it is not contaminated.

CBM extraction is being encouraged in the Peace River region, the Kootenays, the south central interior, and especially on Vancouver Island, where supply options are limited and its discovery would relieve the pressure of increased energy demand. To date, mainly experimental wells have been drilled in the province; however, some commercial production has begun in the Hudson Hope area of the Peace River. In 2010, the Dogwood Initiative published the "Citizen's Guide to Coalbed Methane" (2010), which warns the public that this energy source, like conventional natural gas, only adds greenhouse gases when burned and leads to climate change. Moreover, CBM results in an industrial landscape (more roads, pipelines, power lines, compressor stations) because CBM requires far more wells than conventional natural gas; as well, in US regions where there has been a great deal of CBM extraction, there have been lawsuits over contaminated groundwater.

A grave concern is that there are "no legislated baseline testing requirements" on groundwater production in British Columbia, "making it impossible to determine whether the quality of water has changed as a result of CBM development" (Dogwood Initiative 2010). Other issues include the flaring off of gas in the initial phase of CBM production (which can lead to many health issues for humans and animals) and the lack of consultation with First Nations. For example, the Tahltan First Nation has protested Shell Canada's attempts to explore for CBM in the Sacred Headwaters since the company was awarded tenure in 2004" (Dogwood Initiative 2010). The province declared a moratorium on exploration in 2008 to assess the implications of CBM exploration and possible production. Opposition to mining operations in the remote Flathead Valley of the east Kootenays also resulted in a moratorium in 2010.

OIL

Of all the sedimentary basins shown in Figure 11.3, only the Peace River region in the northeastern part of the province has produced oil and natural gas. Both of these valuable energy sources are found at depths between 1,000 and 3,000 metres and both are the product of organic materials created through photosynthesis millions of years ago. This organic material was covered by sediments; pressure and heat converted it into petroleum or natural gas.

Until 1954, British Columbia's oil demands were met with imported oil shipped from California to the oil refineries of Burrard Inlet. This changed with the discovery of oil in the Peace River region and the construction of the TransMountain Pipeline, which brought both BC and Alberta oil to refineries in the Lower Mainland (Figure 11.6) New refineries were constructed in Prince George and Kamloops to serve the needs of the interior and, later, one at Taylor in the Peace River region. The need for and dependence on oil continues today.

The energy crisis of the 1970s saw the price of oil climb from US$3.01 per barrel in July 1973 to over US$11.00 per

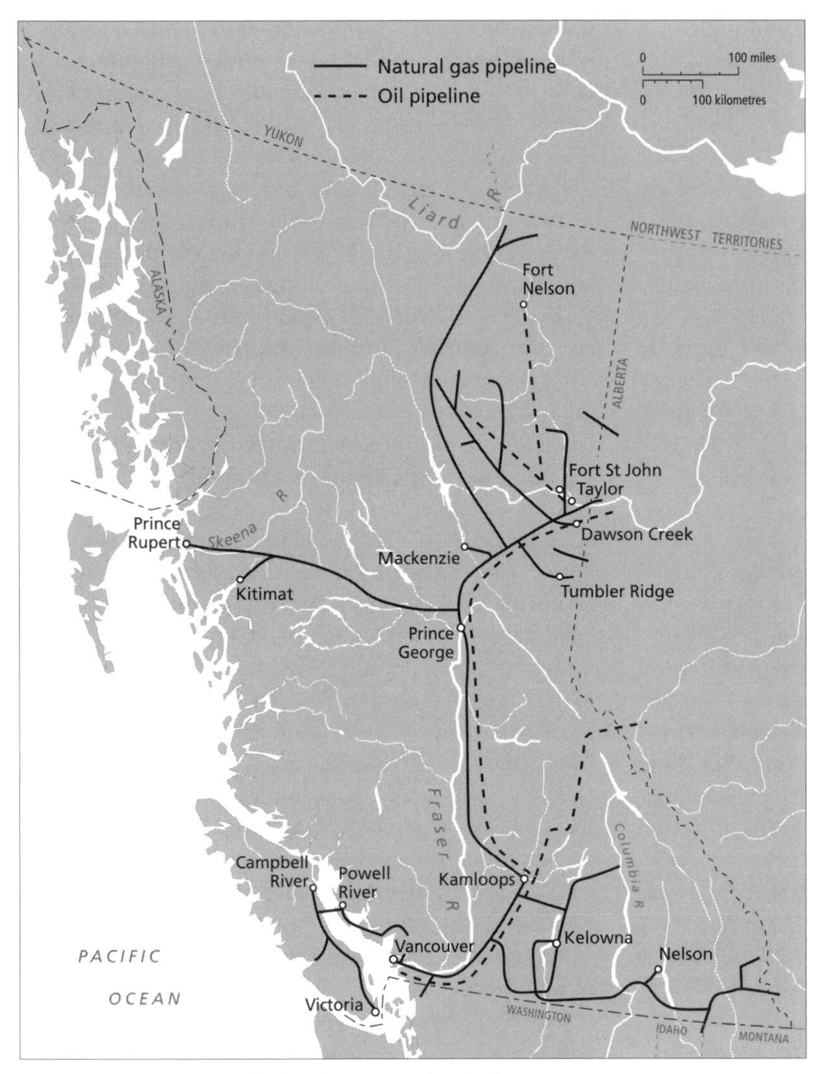

Figure 11.6 Natural gas and oil pipelines in British Columbia
Source: Modified from British Columbia, Ministry of Finance and Corporate Relations (1998, 87).

barrel by January 1974. By the late 1970s, oil was close to US$40.00 per barrel, as shown in Figure 11.7. This represented a significant increase in cost to everything that had an oil component, whether gasoline or any product made from crude petroleum such as plastics, fertilizers, and paints. An atmosphere of crisis was manufactured over cost and the fear of global oil shortages. Regions without oil, or with insufficient oil, worried about their balance of payments. For Alberta, the crisis produced a windfall in investment and development of the traditional oil patch as well as technologies to produce oil from tar sands.

British Columbia received some of this attention with investments in the Peace River/northeast region and prospects of offshore oil. Various companies staked out their territories, mainly between Haida Gwaii and the mainland, but offshore prospects met with two obstacles. First, the government of British Columbia claimed the inland waters and any oil revenues derived from them (Figure 11.8). This claim challenges the terms of Confederation, which gave the federal government sole jurisdiction over oceans. The second, and more serious issue, is the prospect of drilling for oil in the most seismically sensitive area of Canada. An earthquake could rupture an oil well, with serious consequences for the aquatic environment. A series of moratoria have prevented the drilling process. Meanwhile, the price of oil descended rapidly in the 1980s, returned to the $40/barrel level in 2004, and then soared over $140/barrel in July 2008. The recession then brought demand and prices down.

The 1970s energy crisis brought another response to prospective oil shortages that involved British Columbia. Alaska proposed to move its oil reserves by tanker either to the Cherry Point refinery in Washington State's Puget Sound, which is connected to the Burrard refineries in Vancouver by pipeline, or to the port of Kitimat, where a pipeline would then be built to the Lower Mainland. In either case, the risk of transporting oil in tankers along the west coast of British Columbia was met with a great deal of anxiety by its residents. Neither of these schemes materialized, and the oil spill of the *Exxon Valdez* in 1989

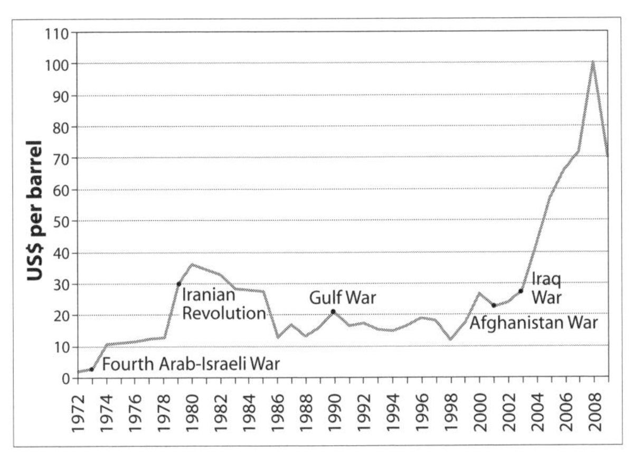

Figure 11.7 Oil prices, 1972-2009
Sources: BC Ministry of Finance (2009); Global InfoMine (2010); British Petroleum (2010).

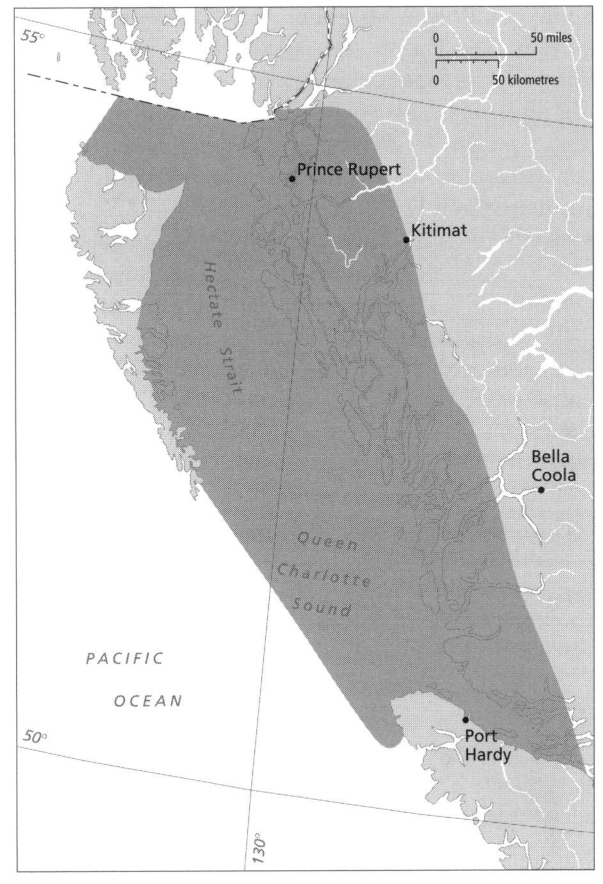

Figure 11.8 Offshore oil claim by BC government
Source: Modified from Whiteley (1982, E5).

off the coast of Alaska, along with the explosion of a deep-sea oil rig and massive oil spill in the Gulf of Mexico in April 2010, serves as a real reminder of the catastrophic level of disaster that can be caused by oil spills. And the threat continues. The Enbridge Gateway Pipeline Project is a megaproject proposal that includes two pipelines (crude oil in one and condensate – thick oil plus chemicals – in the other) to be constructed between the tar sands in northern Alberta and Kitimat on BC's West Coast. If constructed, it would result in some two hundred oil tankers per year carrying crude oil or condensate to markets in Asia and the United States. The fear of an oil spill, especially since the Gulf of Mexico Disaster, has resulted in opposition from many, including "over 150 First Nations, businesses, organizations and prominent Canadians" (Goodwin 2010).

The Enbridge Project has raised many environmental issues in addition to the prospect of coastal oil spills. The proposed pipelines are in excess of 1,1000 kilometres, and they "would require over 1,000 stream and river crossings" (SkeenaWild Conservation Trust 2008), raising fears of pipeline ruptures and their effect on water and land. Moreover, much of the land to be crossed in British Columbia is First Nations land, where there is considerable oppositon from nations such as the Wet'suwet'en and the Carrier Sekani Tribal Council. The West Coast Environmental Law Society (2009) advises that the pipeline should not be advanced until a great number of environmental questions have been assessed, including increasing tar sands oil, increased greenhouse gases, and increased tanker traffic.

Although the energy crisis resulted in some shifting to other energy sources, oil is still in the highest demand in the province. British Columbia produces approximately 20 percent of the provincial demand and imports the rest from Alberta. This lack of oil self-sufficiency is another incentive to develop and use made-in-British Columbia energy sources.

Since the late 1990s, Alberta has captured much of the refining capacity for western Canada, and only two refineries remain open in British Columbia – the Husky refinery in Prince George and the Chevron refinery in Burnaby (Ostergaard 2002, 222). On the environmental side, pipeline rupture is a concern. Tragically, a rupture occurred on 1 August 2000 near Chetwynd, spilling at least a million litres of oil into the Pine River, closing down that community's water supply, and harming fish and fish habitat. Pembina Pipeline was fined $200,000 to restore the damaged habitat (Environment Canada 2004). A more significant question is whether this thirty-five-year old pipeline will be safe in the future.

Of long-term concern is Canada's signature on the Kyoto Protocol and commitment to cut down on greenhouse gases, which translates into discouraging the use of petroleum. The discouragement is occurring, but in an unwelcome form – higher oil prices. The Iraq and Afghanistan wars, increasing demand, and the threat of curtailed supply drove the price of oil to an unprecedented US$140 per barrel in mid-2008, but the recession of that year adjusted the price downward significantly. The inflationary implications, however, were similar to those of the 1970s energy crisis, namely, the cost of transportation and therefore of all deliveries depends on the price of oil. In turn, everything that is delivered must increase in price to offset the higher price of fuel, as must all products with a petroleum base.

There is an interesting link between the tough economic times faced by the forest and fishing industries and the oil and gas industry, which is booming. The high price of oil and natural gas, along with the provincial government's desire to become self-sufficient in oil, has renewed interest in overturning the long-standing moratorium on offshore drilling for these hydrocarbon sources. This would provide jobs for coastal communities suffering from forestry and fishing layoffs. The Geological Survey of Canada suggests that some 734 billion cubic metres of natural gas and 1.56 billion cubic metres of crude oil are contained in the Queen Charlotte Basin alone, and that other offshore basins contain natural gas predominantly (Canadian Centre for Energy Information 2004, 13).

These reserves represent an enormous amount of revenue, and the Liberal government has created its own scientific team to investigate the removal of the moratorium, initially targeting 2010 as the removal date. A BC scientific and technical review conducted in 2002 "concluded that although there are some information gaps, there are no scientific or technical barriers to offshore exploration and development" (Canadian Centre for Energy Information 2004, 13). A report by the federal government discusses these "gaps," however, and suggests that if the moratorium is to be lifted a number of environmental concerns must be addressed, including "at-risk species and habitats, sea-floor stability, currents, tides, waves, earthquake risk and oil-spill potential" (p. 14). The high risk of oil spills also led the federal report to recommend limiting tanker traffic. More recently, the BC Energy Plan (2007) "re-affirmed its commitment to offshore oil and gas exploration and development, its request to Canada to lift the federal moratorium and reiterated that the provincial moratorium will be lifted at the same time" (BC Ministry of Energy, Mines and Petroleum Resources 2007b). Environmental groups are lining up to oppose this development, citing noise and vibrations from seismic testing and the possibility of large and small oil spills (David Suzuki Foundation 2010). Since the Gulf of Mexico spill, the federal government, which has jurisdiction over offshore drilling, has stated: "The moratorium on offshore oil development in BC won't be lifted any time soon" (Pynn 2010).

Plenty of work must be done simply to ensure that environmental concerns about offshore drilling are taken into account by truly independent assessors. Another consideration is Aboriginal fishing rights and title to the ocean, which are part of the ongoing treaty negotiations. Revenue sharing may be the largest hurdle, and a crucial one; the federal government maintains jurisdiction and ownership offshore, and the provincial government is operating under the expectation of suitable revenue-sharing arrangements, as are First Nations. These, however, have yet to be negotiated.

NATURAL GAS

For a resource that was discovered by accident, natural gas has become a great success story. In the process of drilling for oil in the 1950s, energy companies discovered natural gas in considerable quantities in the northeast region. Sufficient gas was found to build the Westcoast Transmission (now Westcoast Energy) Pipeline in 1957, connecting the Fort Nelson/Fort St. John area with the Lower Mainland and the Pacific northwestern region of the United States. From this main line, branches were built to Kitimat and Prince Rupert, and from Kamloops into the Okanagan and the Kootenays (Figure 11.6). Connections have also been made to Alberta's supply of natural gas. The most recent expansion of natural gas distribution, in the early 1990s, was from the Vancouver area to Squamish and the Sunshine Coast and across to Vancouver Island (Figure 11.1). This expansion connected natural gas lines to the coastal pulp mills in these areas and provided some communities and other industries with this energy option.

Having an abundance of natural gas and a shortage of oil, British Columbia has promoted the use of natural gas over oil products, including converting automobiles to this fuel. More recently, the substitution of natural gas for oil has been encouraged on environmental grounds since natural gas burns more cleanly and gives off fewer greenhouse emissions (British Columbia Energy Council 1994). Natural gas can also be used for a variety of non-energy products, such as chemicals, paints, and plastics. In addition, it is "the primary feedstock for hydrogen fuel cell technology, which may soon begin to displace the internal combustion engine in both motor vehicle and remote thermal-electric generation applications" (Ostergaard 2002, 226).

Natural gas cogeneration plants – which produce electricity plus steam – are common in Alberta because of its surplus of natural gas and lack of major river systems to dam for hydroelectric power. British Columbia produces most of its electricity by damming river systems and only 11 percent through natural gas cogeneration. The oldest cogeneration station is on Burrard Inlet near Vancouver. With 950 megawatts (MW), it is used mainly as a backup system. BC Hydro runs others, including two relatively small (46 MW) plants at Prince Rupert and Fort Nelson. Two cogeneration plants are run by private corporations: a 250 MW facility built at Campbell River on Vancouver Island in 1999 and a 120 MW plant built at Taylor in 1990. Approximately 2,600 megawatts were planned by BC Hydro in the mid-1990s on the assumption that natural gas cogeneration has fewer drawbacks than large hydroelectric dams (British Columbia Energy Council 1994). Since then, however, Canada's signing of the Kyoto Protocol in 2002, the "crisis" of Vancouver Island's energy supply, and the uncertain, but higher, natural gas prices have ignited divergent views with respect to energy use overall and natural gas specifically.

The Liberal government's 2007 BC Energy Plan supported renewable energy sources for electricity. In 2008, the government implemented several follow-up policies: a carbon tax on all fossil fuels, including natural gas; the Greenhouse Gas Reduction (Cap and Trade) Act, to reduce greenhouse gases; and an Environmental Management Act amendment that states that "electricity generation facilities are required to have 'net zero' greenhouse gas emissions" (BC Ministry of Energy, Mines and Petroleum Resources 2009). These new directions have resulted in the provincial government's ordering of "BC Hydro to yank the aging Burrard Thermal generating plant off its roster of baseline electricity sources" (Simpson 2009).

Vancouver Island does not currently have enough electricity to supply its own needs, a deficit that will only increase with the population. Most of its electrical energy has come via five undersea cables from the mainland, which had to be replaced by 2007. BC Hydro's original plan was to build a second natural gas pipeline through its subsidiary, Georgia Strait Crossing Pipeline Ltd., known as the GSX Canada pipeline project. This line was initially to be linked to a cogeneration plant at Port Alberni, but after this location was rejected in public hearings, Duke Point near Nanaimo was selected in 2003. Many environmental groups and others were opposed to Vancouver Island meeting its future energy needs by burning natural gas, thus creating more greenhouse gases, which is contrary to the Kyoto Protocol. With the crisis of replacement imminent, BC Hydro has constructed new undersea cables.

The price of natural gas (Figure 11.9) and the revenues from this energy source (Table 11.5) are an encouragement for the provincial government to extract even more in the future, including, as stated above, offshore gas. As well, there is a proposal for yet another pipeline to the Kitimat area, where liquefied natural gas (a process that cools the gas and compresses it) would be exported to Asia. The potential date for this operation is 2013 (Ebner 2009).

There are a number of environmental and health risks associated with natural gas. Certain gas wells produce sour gas, gas with hydrogen sulphide in it, which can be deadly. Of course, people are not supposed to be exposed to this gas, which is heavier than air, but a sour gas well

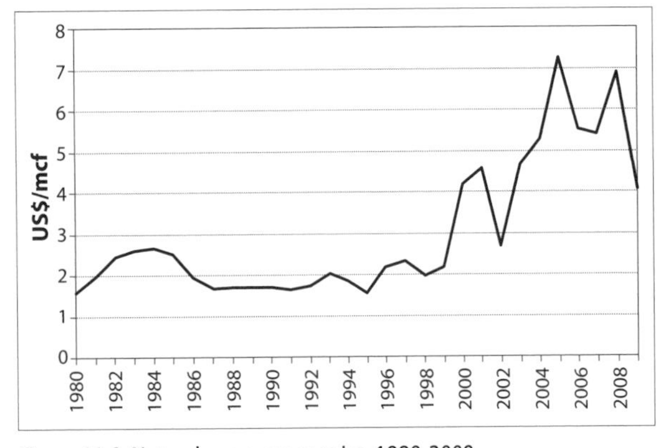

Figure 11.9 Natural gas average price, 1980-2009
Note: mcf = thousand cubic feet.
Source: BC Ministry of Finance (2009); British Petroleum (2010).

Table 11.5

Natural gas production, value, and overall natural gas revenues, 1998-2009

	Production (billion m³)	Value ($ millions)	Revenues ($ millions)
1998	20.1	1,154.0	379.6
1999	20.7	1,577.0	587.3
2000	21.4	3,826.0	1,306.2
2001	25.2	4,834.0	1,249.0
2002	27.5	3,458.0	836.0
2003	25.5	5,396.0	1,056.0
2004	32.0	5,827.0	1,230.0
2005	32.8	7,821.0	1,439.0
2006	32.8	5,956.0	1,912.0
2007	32.6	5,723.0	1,207.0
2008	33.4	7,501.0	1,321.0
2009	33.0	3,294.0	1,314.0

Sources: BC Ministry of Energy and Mines (2004a); BC Ministry of Finance (2009, 2010).

did explode near Pouce Coupe in the Peace River region on 22 November 2009, releasing 30,000 cubic metres of sour gas. The community organized its own evacuation (Arsenault 2010). EnCana, the company responsible for the well, has been the target of sabotage by someone (or more than one) who has sent a letter to the local paper complaining "about the production of 'crazy' natural gas, a possible reference to toxic sour gas that is produced in the area and shipped by EnCana." There have been six attacks on EnCana property since 2008 (Rynor 2009). As of 2010, no one has been arrested, even though the company has posted a million dollar reward.

ELECTRICITY

To create electricity, turbines are rotated to produce a current, which is then transported via transmission lines. There are, however, many ways to rotate turbines. Coal, oil, natural gas, wood (hog fuel), or even garbage (solid waste) can produce thermal electricity. These fuels are burned to boil water, the water produces steam, and the steam, in turn, rotates the turbines. Nuclear power performs the same function: atoms of uranium are split (in a controlled environment) to generate the energy needed to create the steam. Steam, or geothermal energy, also occurs in nature, especially in seismically active regions of the world. Countries such as Iceland and New Zealand have used these sources extensively to produce electricity. Wind, waves, and ocean currents can also turn turbines, and the sun shining on photovoltaic cells (e.g., those in a solar calculator) can produce electricity. Each of these means of generating electricity results in different efficiencies and costs (including to the environment), but the most efficient process is by moving water through a turbine. The many large river systems in British Columbia, combined with high elevations, make enormous hydroelectric generation possible. To develop hydroelectricity megaprojects, though, dams for large reservoirs must be constructed to regulate the major natural fluctuations in discharge rates throughout the year. Dams impound immense volumes of water, which can be released at controlled rates to maintain an even production of electricity.

The first electricity production in British Columbia was by steam at the Moodyville sawmill in North Vancouver in 1882 (BC Hydro 1979, 6), and the first hydroelectricity production was in Nelson in 1896 (Sewell 1987, 234). From these humble beginnings, demand increased slowly throughout the province until the 1950s. The ensuing megaproject era involved the construction of large dams and a major increase in the supply of electricity (Figure 11.10). Table 11.6 provides an overview that includes differences between hydro and thermal generation imports, exports, and provincial consumption.

The escalation in electrical energy production was fuelled by the expansion of the mining and forest industries after the Second World War. The megaproject era for British Columbia began with the Kemano 1 project, built in the 1950s. The Kenny Dam reversed a portion of the flow of the Nechako River system via a tunnel through the Coast Mountains to the Kemano generating station. This project was developed and built by Alcan to produce aluminum at Kitimat (Figure 11.11).

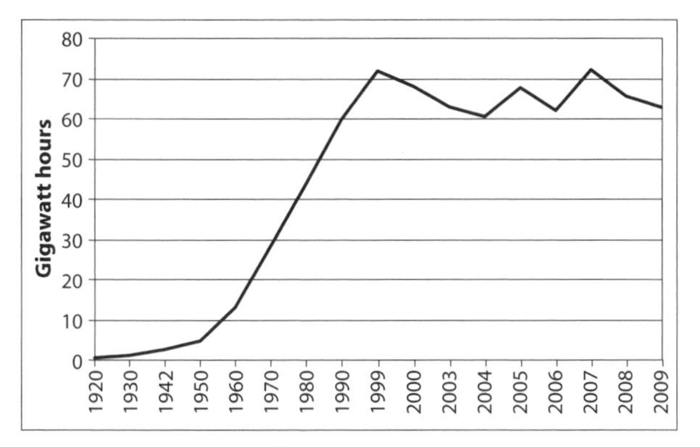

Figure 11.10 Generation of electricity, 1920-2009
Sources: Barker (1977, 54); BC Ministry of Finance (2009, 2010).

Table 11.6

Supply and consumption of electricity in British Columbia, 1994-2009

	Supply (GWh)					Consumption (GWh)		
	Net provincial generation							
	Hydro	Thermal	Total	Imports	Total supply	Exports	Total consumption	Net exports
1994	53,979	7,036	61,015	7,836	68,851	9,541	59,311	1,705
1995	49,814	8,192	58,006	6,385	64,391	3,972	60,419	(2,413)
1996	67,329	4,436	71,765	3,289	75,053	10,390	64,664	7,101
1997	61,772	5,189	66,961	4,316	71,278	12,114	59,163	7,798
1998	60,849	6,861	67,710	5,056	72,766	10,619	62,147	5,563
1999	61,588	6,457	68,045	6,807	74,852	12,529	62,323	5,722
2000	59,754	8,487	68,241	6,039	74,280	10,689	63,582	4,659
2001	48,338	8,994	57,332	10,154	67,486	6,408	61,079	(3,747)
2002	58,627	6,318	64,945	5,769	70,714	8,078	62,636	2,309
2003	56,689	6,362	63,051	7,071	70,0122	9,901	60,221	2,515
2004	53,281	7,214	60,496	8,261	68,757	6,791	61,966	(1,470)
2005	60,605	7,207	67,811	7,226	75,037	9,247	65,790	2,021
2006	54,772	7,249	62,021	12,695	74,716	6,155	68,561	(6,540)
2007	64,738	7,473	72,212	8,027	80,239	10,987	69,252	2,960
2008	58,774	7,082	65,856	12,027	77,883	9,844	68,038	(2,183)
2009	55,872	6,334	62,206	11,585	73,790	7,950	65,841	(3,635)

Note: Numbers in brackets indicate years in which British Columbia imported coal-produced electricity from Alberta.
Sources: BC Ministry of Finance (2009, 2010).

By the 1950s, the provincial government realized that electrical energy production was a crucial component to industrial expansion, especially for the interior of the province. In 1956, for example, the provincial government granted a vast area of north central British Columbia to Swedish investor Axel Wenner-Gren for northern development. The region extended from Summit Lake, north of Prince George, to Lower Post on the Yukon border. Wenner-Gren's promises of pulp mills, mines, and a monorail never materialized, but both he and the provincial government recognized the large amount of hydroelectricity that could be produced by the Peace River.

Both the BC and the US governments at that time also realized that building dams on the Columbia River would provide the dual benefits of electricity and flood control. Dams hold back the peak flows, thus reducing floods, and the impounded water is then used to raise low flows in winter and thus maintain steady levels of electrical production. Direct political involvement occurred in 1961, when the provincial government took over the private utility BC Electric, created the Crown corporation BC Hydro, and launched the "two rivers policy." Major dams were designed and built for the Peace and Columbia Rivers. Through the Columbia River Treaty, Americans paid for three dams to be constructed in British Columbia (Keenleyside, Duncan, and Mica) and received the benefit of hydroelectricity and flood control from the Columbia River, while BC Hydro supplied British Columbians with electricity from the Peace.

The damming of the Columbia River resulted in massive reservoirs, loss of plant and animal species, and the relocation of some 2,000 residents (Wilson 1978). Recent research suggests with the aid of hindsight that the people in the Kootenays paid the price for the Columbia

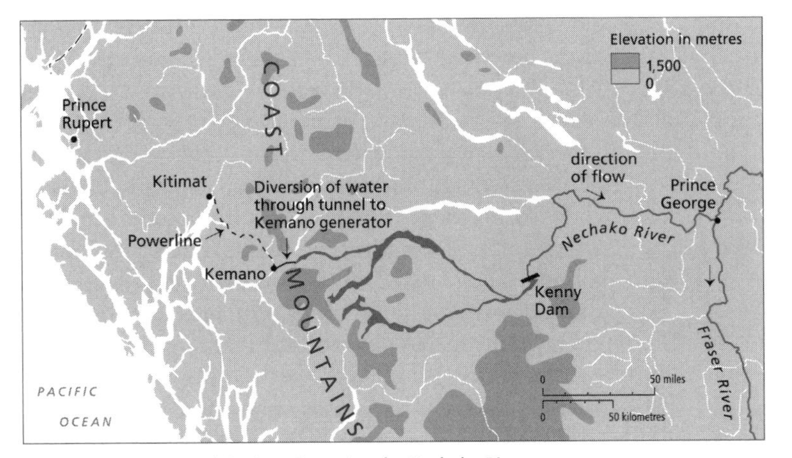

Figure 11.11 Kemano 1 Project: Damming the Nechako River
Source: Data from Sewell (1987, 231).

By the mid-1970s, the accuracy of BC Hydro projections was a matter of debate, particularly after the energy crisis of 1973. The Crown corporation did not anticipate the forest industry's reaction to increased energy costs, namely, to use hog fuel to generate large amounts of electricity for its own purposes. Similarly, BC Hydro did not account for conservation measures requiring less energy use. Figure 11.12 illustrates the differences in projected demand according to BC Hydro, the BC Energy Commission, and environmental groups over a twenty-year period. Like compound interest, it took only a few years before a significant difference appeared. One can observe the large gap in projections by 1996, particularly between the sharp increase expected by BC Hydro and the 1 to 2 percent per year increase foreseen by environmental groups. The implications of these projections can be seen in the list of projects planned by BC Hydro in 1976 (Table 11.7).

To fulfill the exaggerated demand forecast by BC Hydro, CANDU nuclear reactors, coal-fired thermal generators, and a great number of dams were on the drawing board:

River dams. According to Toller and Nemetz (1997, 27), "Hydro dams have adversely affected the ecology, economy, and social fabric of the Kootenays." Similar developments and adverse effects on First Nations communities have been documented following the building of the Kemano Dam on the Nechako River and the WAC Bennett Dam on the Peace River.

Another controversial hydroelectric proposal in the 1960s was the Moran Dam, approximately twenty-five kilometres north of Lillooet on the Fraser River. The benefits would be hydroelectric power and flood control (Hardwick 1962, 63). The costs would be mainly to the races of salmon no longer able to spawn in the upper half of the Fraser River basin. In this instance, fishing interests outweighed energy interests and the dam was not constructed.

There seemed to be no end to the investment, however, nor to the need for electrical energy, with the rapid expansion of the forest and mining industries in the 1960s. Through BC Hydro, the provincial government used electrical energy as an industrial strategy. Large dams require a minimum of ten years' lead time before they produce electricity, and it is therefore essential to anticipate electrical demand accurately (Marmorek 1981). In other words, huge amounts of money need to be invested at the start to build a dam, with the realization that it will take at least ten years to move from planning, through construction, to power generation.

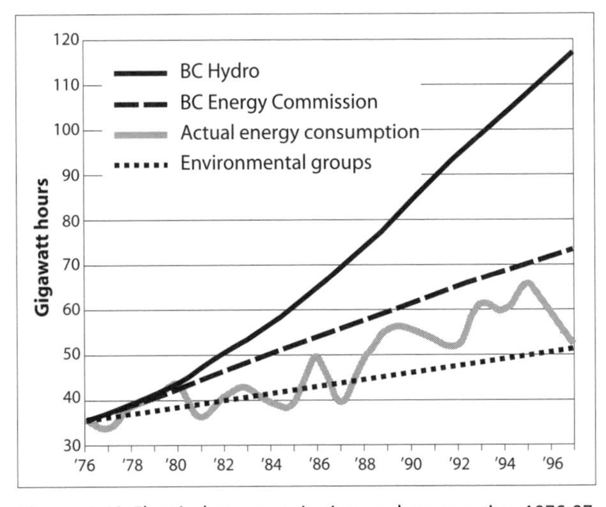

Figure 11.12 Electrical energy projections and consumption, 1976-97
Sources: Data from British Columbia Energy Commission (1976, 25); Farrow (1977, 36); British Columbia, Ministry of Economic Development (1987, 64); British Columbia, Ministry of Finance and Corporate Relations (1993, 252); (1998, 283).

Table 11.7

Major proposed electrical energy projects, 1976

Projects	At site capacity (Mw)	Average annual energy (millions of Kwh)	Earliest in-service date	Completion date
Hydroelectric				
Site One	700	3,150	1979	1980
Seven Mile	700	3,150	1980	1979
Revelstoke	2,700	7,970	1981	1984
McGregor Diversion	0	3,110	1985	n.c.
Kootenay Diversion	0	875	1984	n.c.
Kemano 2 (Completion)	1,200	6,480	1983	n.c.
Peace Site C	900	5,010	1984	n.c.
Peace Site E	750	3,920	1984	n.c.
Murphy Creek	400	2,100	1983	n.c.
Duncan Bay (Vancouver Island)	150	270	1980	n.c.
Kokish (Vancouver Island)	160	255	1980	n.c.
Elaho	500	1,710	1983	n.c.
Homathko	1,450	7,550	1984	n.c.
Skeena (Cutoff Mountain)	1,080	6,370	1987	n.c.
Iskut-Stikine	3,025	19,500	1986	n.c.
Liard River	3,800	23,600	1987	n.c.
Yukon-Taku	3,700	22,800	1990	n.c.
Upstream Fraser	5,800	29,850	1987	n.c.
Downstream Fraser	2,420	15,510	1985	n.c.
Thermal projects				
Hat Creek (proven)	2,500	16,500	1982	n.c.
Hat Creek (possible)	2,500	16,500	1985	n.c.
East Kootenay	1,800	12,000	1983	n.c.
Comox (Vancouver Island)	550	3,500	1981	n.c.
CANDU nuclear plants (two)	1,200	8,400	1985	n.c.

Note: n.c. = not completed
Sources: British Columbia Energy Commission (1976, 72); British Columbia, Ministry of Finance and Corporate Relations (1998, 87).

all to produce electricity for an industrial demand based on a past performance that had little relationship to the future. The recession of the 1980s saw the shutdown of mines and mills and a further reduction of electrical energy demand. Only a few of the projects on the list were ever built, the largest being the Revelstoke Dam. Figure 11.13 shows the main dams and transmission lines now in the province compared to the projected energy schemes of the 1970s.

BC Hydro projections also raised the question of the environmental cost of progress and the impact of mega-projects. Many of the proposed projects were vehemently opposed by the public. Concerns over nuclear wastes, low level radiation, and possible contamination meant few citizens wanted a nuclear power plant in their backyard. Coal-fired thermal plants raised fears of acid rain and increased greenhouse gases. Damming the Fraser River and destroying the salmon runs was considered a sacrilege by many. Impounding water to create huge reservoirs behind hydroelectric dams also causes enormous environmental problems and affects many people and their way of life. The proposed Site C Dam, for example, next to Fort St. John on the Peace River (approved for development on 9 April 2010), will, if it passes environmental reviews, flood some of the finest agricultural land in the province.

The plan to divert the McGregor River into the Parsnip River would have joined two entirely different ecosystems – one that flows to the Pacific and one that flows to the Arctic. The Parsnip/Peace Rivers have different parasites and organisms from the McGregor/Fraser system, and it was feared that the two might find ways to invade one another, with devastating consequences to fish stocks. Alcan, at Kitimat, had plans to expand its aluminum production that required increased hydroelectricity. Kemano 2 was a plan to dam the Nanika River, a tributary of the Skeena, and to reverse its flow into the Ootsa reservoir, thus increasing the capacity of the existing hydroelectric system. This controversial project was cancelled in 1987 in favour of the Kemano Completion Project. That equally controversial proposal included drilling a second tunnel through the Coast Mountains and draining far more water from the Nechako (Christensen 1995). This project was halfway completed when the provincial government cancelled it in 1995; Alcan

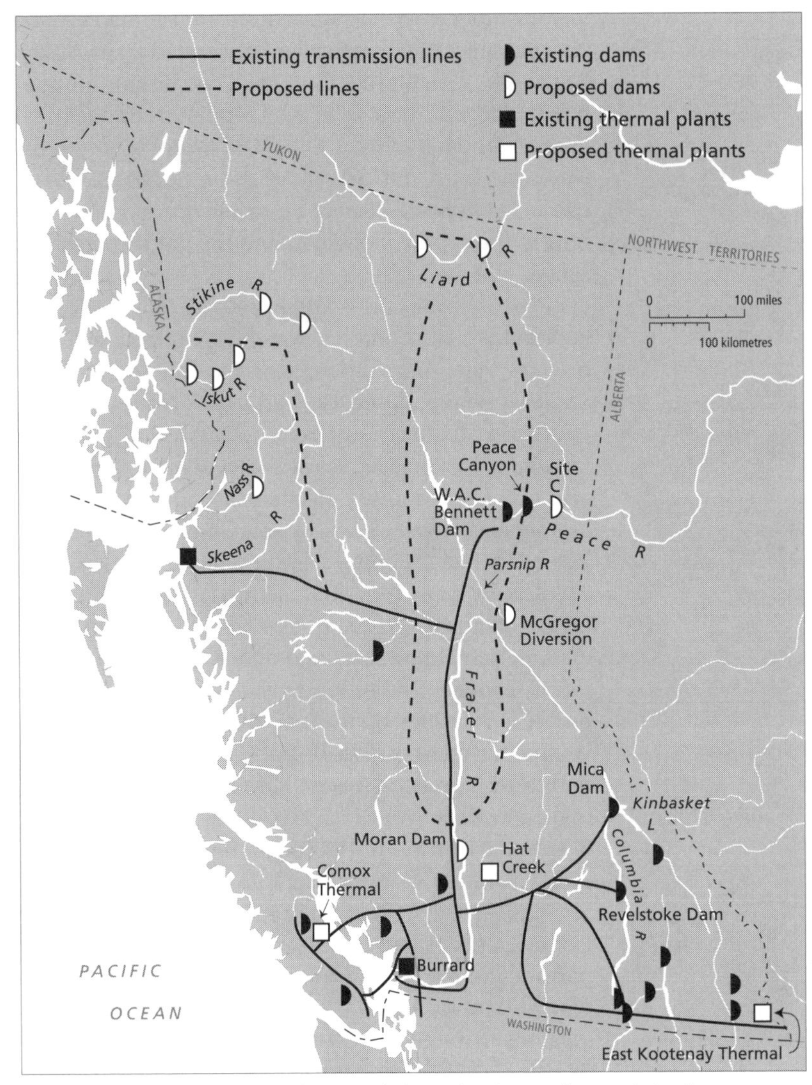

Figure 11.13 Main developed and proposed electrical projects and transmission lines
Sources: Data from Sewell (1987, 231); BC Hydro (1994, 3).

and the province settled out of court in 1997 with "replacement power for a proposed smelter expansion" (Ostergaard 2002, 218).

Transmission lines, which would have been built from northern dams to distribute power to southern consumers, raised other environmental concerns (Figure 11.13). If these dams had been constructed, a great swath of forests would have been cleared for the transmission lines to run from one end of the province to the other. Uninsulated lines have a corona discharge, from escaping ions, and research suggests that many health risks are related to this hazard (see Brodeur 1993; Young 1974).

Many environmental concerns therefore need to be kept in mind when considering where future gigawatt hours of electricity will come from. The regulatory process has also changed since the megaproject era of the 1960s and '70s. Legislation such as the Fish Protection Act (1997) and Species at Risk Act (2002) makes the damming of rivers much more difficult. Other factors have to be considered. First Nations must be consulted and agreements reached for further production and transmission. BC Hydro, a Crown corporation, must comply with provincial government policies and mandates, one of which states that, with the exception of Site C, BC Hydro can increase gigawatt hours only by upgrading existing facilities, not building new ones. BC Hydro also takes orders from the BC Utility Commission (BCUC), which guides (accepts or rejects) BC Hydro's Long Term Acquisition Plan, along with any rate increases.

One thing is certain: British Columbia is growing in population, and there is a corresponding increase in the need for electrical energy. Figure 11.10 clearly shows that production peaked in the mid-1990s and that variations since then have been mainly related to water availability. Table 11.6 is more revealing, for it shows both increasing demand and increasing imports from Alberta and the United States. BC Hydro states that "electricity demand is expected to grow by approximately 20 to 40 per cent (from 59,000 GWh/yr to 78,000 GWh/yr) over the next 20 years" (BC Hydro 2010a).

Fulfilling this need will be a major challenge for BC Hydro. Through its Energy Plan, BC Hydro intends to increase electrical energy production until it is energy self-sufficient by 2016, even under conditions of low water flows. Moreover, the "new" energy will come from renewable (green energy) sources and, if the electricity is produced by either natural gas or coal, then it must have zero greenhouse gas emissions (BC Hydro 2010a).

BC's Clean Energy Act was put into law in June 2010, setting further goals and ojectives for BC Hydro's Energy Plan. Site C has been given the green light (subject to environmental review), and electricity, including energy produced through the Natural Resources Pulp and Paper Green Transformation Funding Program (a federal subsidy for pulp-and-paper mills to produce electricity through hog fuel), will be purchased from IPPS. The passage of this legislation is a clear indication that the province is going to be a significant exporter of clean energy through BC Hyro's subsidiary Powerex. There are other changes such as smart meters that encourage individuals to conserve and produce excess energy that BC Hydro will put into the grid, and there is a special $5 million fund to encourage First Nations to become involved in producing clean energy (BC Ministry of Energy, Mines, and Petroleum Resources 2010). The Act also modifies the role of BCUC in that it no longer has power to comment on energy purchase agreements (energy from IPPs) or on Site C or upgrades to existing turbines. But BCUC will continue to be the approving body for BC Hydro rates. The Act sets out a bold direction and, as expected, there has been criticism, mainly that BCUC is no longer a watchdog over BC Hydro's production and acquisition of future energy. The follow-up conern is that the increase in power production will result in higher rates for BC consumers (Hunter 2010).

The Site C Dam will certainly increase electrical supply without adding to greenhouse gases (about 4,800 GWh) by its production, but many other environmental, cultural, and economic factors (including greenhouse gases from flooded land) need to be considered, including the twelve-year lead time necessary for a dam of that magnitude. Natural gas, coalbed methane cogeneration, and the production of electricity from the province's huge coal reserves are other alternatives, but these all produce greenhouse gases and, unless they can reach zero emissions, are not viable options. Nuclear power has also been ruled out by the provincial government. Additional generators could, however, be added to the existing Mica (1,805 MW) and Revelstoke (1,980 MW) dams, along with upgrades to their old generators, with few environmental consequences. BC Hydro's Power Smart program, a program to conserve energy ("demand supply measures"), also intends to save 3,000 gigawatt hours by 2012 and over 11,000 by 2027 (BC Utilities Commission 2009, 118).

Because BC Hydro is a Crown corporation, electrical energy production is influenced by political decisions and subject to controversy. The Liberal government split BC Hydro into two in 2003: BC Hydro continues to generate and distribute electricity to consumers, while British Columbia Transmission Corporation (BCTC) manages, maintains, and operates the transmission grid (BC Hydro 2010b). This change alone created public concern that the original BC Hydro was being dismantled and privatized, leaving customers to face potentially escalating electricity rates with no government controls. The BC Clean Energy Act has reversed the 2003 decision, and the BCTC has been merged back into BC Hydro.

Table 11.6 shows that BC Hydro has increasingly had to import electricity. But is this the whole story in terms of electrical generation and what appears to be a escalating shortfall? The assumption is that only BC Hydro (and not the provincial government) is responsible for electrical production. Powerex is a wholly owned subsidiary of BC Hydro that is responsible for buying and selling power (along with natural gas and other products and services). It has access not only to the energy that BC Hydro produces but also to all the other private producers of electricity (e.g., existing IPP production, Alcan at Kitimat, Tech Cominco at Trail, and Fortis at Kootenay Lake). As well, Powerex has access to 1,170 megawatts of capacity from the Columbia River Treaty (referred to as the "Canadian entitlement to downstream benefits"). Another Crown corporation, the Columbia Power Corporation, has formed a joint venture with the Columbia Basin Trust (Citizens of the Kootenays) to develop electrical energy. Because these corporations are not considered part of BC Hydro, it only seems like British Columbians are short of electricity. Meanwhile, environmental groups are demanding that future electricity be supplied by renewable energy sources that have a minimal impact on the environment and that energy consumption be reduced in the future. In other words, the estimated demand increase of 20 to 40 percent over the next twenty years should be reduced drastically.

BC Hydro's service plan for 2010-11 and 2012-13 does factor in significant conservation measures. For new power, however, "BC Hydro will continue to purchase power from Independent Power Producers that use clean or renewable resources such as wind, water, biomass, waste heat and geothermal using competitive call and other acquisition processes" (BC Hydro 2010a). BC Hydro will also refurbish and upgrade existing facilities to produce more electricity. As of 2009, IPPs provide 8,374 GWh of energy (14 percent of total domestic electricty requirements) to the system (BC Hydro 2010a). On 11 March 2010, nineteen more IPPs (fourteen run-of-river and five wind proposals) were approved under BC Hydro's Clean Power Call. If they are all built, the province will gain 2,400 GWh of energy (BC Hydro 2010a). However, buying power from IPPs in general and from those IPPs that use run-of-river in particular has led to heated debate. There are concerns regarding insufficient environmental assessment, not only of the project itself but also of transmission lines and road-building activities. The plan has also received a number of other criticisms, including that these are private and not public (i.e., BC Hydro) ventures, that IPPs are being given the rights to river and stream systems, that the energy they produce will increase the cost of electricity, and that they are exporting the power they generate to the United States.

In defence of IPPs, the Independent Power Producers Association of BC (IPPBC) points out that all of the run-of-river projects, regardless of size, are subject to numerous environmental reviews (fifty-two permits are required), many of the projects are relatively small (under ten megawatts), and the IPP bears the financial risk of production. The IPP leases (not owns) the water system for the length of the contract (usually twenty to thirty years), and the electricity contract is with BC Hydro. Powerex is the marketing arm of BC Hydro, which imports and exports energy (i.e., IPPs do not export power). Most of these projects are on Crown lands where no treaties have been negotiated and, therefore, an IPP must be negotiated with First Nations, which is easier for a company than the provincial government to do. For the provincial government to propose a development on First Nations territory, it must either negotiate a treaty or an interim measures agreement, and both are complex. The cost of electricity produced through IPPs is higher than the electricity produced by the large heritage dams built many years ago, but new infrastructure to produce electricity will increase costs, regardless of who builds the project (private or public). Through the Clean Energy Act, Site C (a BC Hydro Project, that is, a public project) has been given the go ahead, and its costs will be in the billions. Site C would be the first new major hydroelectric dam to be built since the 1970s; completion of this project would make apparent the difference in cost per gigawat hour between private and public infrastructure. In anticipation of these higher costs, BC Hydro has applied to the BCUC for approval of a 6.11 percent increase in electricity rates.

RETHINKING ENERGY OPTIONS

Most of the energy used in British Columbia is supplied by large oil and gas corporations or Crown corporations such as BC Hydro. There is often little thought given to types of energy or where it comes from, the most important concern being the cost.

Amory Lovins (1977) observed the energy directions being promoted in the 1970s by governments and industries as they reacted to the energy crisis. He used the term **hard energy path** to describe the typical forecasting that BC Hydro, and many other utilities in the world, were engaged in at that time. He saw that the perceived energy crisis of the period created strategies that centralized energy sources and focused on how to find more energy. From this supply-side perspective, he also saw great increases in costs for energy because the rivers most easy to dam had already been dammed and the most accessible oil and gas had already been exploited. Future energy supplies, according to Lovins, would rely on nuclear and coal-fired generators, and costs would increase because of the inherent environmental problems attached to both of these nonrenewable energy sources. The hard energy path is implied in Table 11.7, with BC Hydro projections of a need for CANDU reactors, coal-fired thermal plants, and many more hydroelectric dams.

An alternative to the hard energy path is the **soft energy path**, which focuses on decentralized forms of energy, soft technologies, conservation of energy, and recognizing the link between lifestyle and energy use. The term "decentralized" implies that individuals – citizens, local government, corporations – begin to make decisions about the types of energy they use, including alternative forms of energy. Lovins (1977, 38-39) uses five principles to define the characteristics of the technologies along this path:

1 *Use renewable technologies.* Make use of the sun, wind, wood, wastes, and water.
2 *Use diverse technologies.* Avoid thinking in terms of "all or nothing" regarding energy use. If solar energy can be utilized to produce 10 percent of heating, then this represents a 10 percent reduction in conventional sources. Use many different energy sources, and renewable ones where possible.
3 *Use simple technologies.* Build and maintain renewable systems, such as a solar-heated hot water system that makes use of understandable technology and is relatively simple to construct and operate. A solar-heated system is not necessarily unsophisticated, but it does not depend on highly specialized, costly technical expertise.
4 *Match energy quality to end use needs.* Recognize how various energy sources are utilized to perform jobs, or end uses. A simplified classification, using heat, divides end uses or tasks into three categories:

    - high grade tasks: above 148°C (300°F)
    - medium grade tasks: approximately 148°C (300°F)
    - low grade tasks: below 148°C (300°F)

    Combustion in automobiles, smelters, and many electrical motors, for example, is a high grade task requiring high grade energy sources such as oil and natural gas. Conversely, hot water, space heating, and air conditioning require energy below 148°C. Many renewable energy sources such as solar and wood can be employed for these demands, reserving high grade energy for high grade tasks.
5 *Match energy in scale and geographic distribution to end use needs.* The energy demands of the individual household, the neighbourhood, the community, and the region are matched with the renewable energy sources that exist in the region to fulfill the demands. Windy areas can utilize windmills, for example, while those having considerable amounts of sunshine use solar technologies.

The soft energy path shifts attention to the demand side of energy, or how to make energy go further. The need for conservation of energy recognizes that British Columbians and Canadians in general are among the highest per capita consumers of energy in the world (Natural Resources Canada 2009b). Lovins (2008) refers to many of the conservation measures employed as "technical fixes" and outlines the many ways that US corporations and institutions have cut back on energy needs and saved vast amounts of money. BC Hydro's Power Smart program, for example, has promoted the need to save energy and describes ways to do so. It suggests saving energy by using insulation and energy efficient machinery, redesigning buildings, cogenerating energy, and many other techniques that make resources go further with no change to lifestyle. The program also has cash incentives to promote energy savings. A change in habits – riding a bike, walking, carpooling, using public transit, turning down the thermostat, or turning off the air conditioning – will result in far greater savings of energy.

New green projects are occurring as BC Hydro approves wind, water (run-of-river), and biomass projects (related to pulp mill production and the pelletization of mountain pine beetle wood). These are important directions to reduce the carbon footprint and prevent climate change. The David Suzuki Foundation (2010) has a host of publications that encourage policies by all levels of government to deal with climate change, including green, renewable energy options and energy conservation. The provincial government through the BC Energy Plan (2009) has established a complementary policy that tackles the production of greenhouse gases through a carbon tax, a cap-and-trade program, and policies of zero carbon emissions for electricity production. Yet there is still plenty of resistance by environmental groups that want to stop run-of-river projects and others who argue that windmills are not attractive. Breaking away from a hydrocarbon dependent way of life will not be easy, but it is necessary.

SUMMARY

Energy resources have been abundant in British Columbia and important to its development. Prompted by technological change and new demands for energy, the province has provided not only for domestic energy needs but also for an export market. The exception is oil, which is imported from Alberta.

Along with the spatial distribution of the many energy resources, world events and politics have shaped the geography of energy in British Columbia. Unprecedented industrial expansion in the 1950s, '60s, and '70s required a major increase in energy production and created an expectation of never-ending growth. The provincial government moved from a policy of facilitating and encouraging energy projects to one of becoming directly involved in megaprojects to increase electrical production.

With the energy crisis of the early 1970s, there was a move away from oil and toward natural gas and electricity. Increased environmental awareness provoked a skeptical reaction to proposed major electrical generation projects, and the recession of the 1980s brought a new reality to energy development. The forest industry implemented a decentralized source of energy (hog fuel), changes to the building code improved insulation standards, and energy efficiency was rewarded. BC Hydro's Power Smart program stressed an overall emphasis on conservation.

Hybrid vehicles are becoming more common, reducing the dependence on petroleum, but they increase the need for electricity. Electricity demands are increasing and are related directly to a growing population, economic expansion, and the increased use of electronic devices. Environmental organizations such as the David Suzuki Foundation send a clear message: the use of fossil fuels with emissions that contribute to greenhouse gases and acid rain must be diminished. The production of electricity must come from green energy sources that are more sustainable and have less impact on the environment. The provincial government through the BC Energy Plan has enacted policies that head in this direction. The lessons of Amory Lovins remind us that individuals can take responsibility for supplying their own forms of renewable energy (i.e., solar power), implement energy conservation, and recognize that energy use is a lifestyle choice.

REFERENCES

Barker, M.L. 1977. *Natural Resources of British Columbia and the Yukon.* Vancouver: Douglas, David and Charles.

BC Hydro. 1979. *Power Perspectives '79.* Vancouver: BC Hydro.

–. 1994. *Making the Connection: The BC Hydro Electric System and How It Is Operated.* Vancouver: BC Hydro.

BC Ministry of Economic Development. 1987. *British Columbia Facts and Statistics.* Victoria: Crown Publications.

BC Ministry of Finance and Corporate Relations. 1987. *1987 British Columbia Economic and Statistical Review.* Victoria: Crown Publications.

–. 1991. *1991 British Columbia Economic and Statistical Review.* Victoria: Crown Publications.

–. 1993. *1993 British Columbia Economic and Statistical Review.* Victoria: Crown Publications.

–. 1997. *1997 British Columbia Economic and Statistical Review.* Victoria: Crown Publications.

–. 1998. *1998 British Columbia Financial and Economic Review.* Victoria: Crown Publications.

British Columbia Energy Commission. 1976. *British Columbia's Energy Outlook 1976-1991.* Victoria: The Commission.

British Columbia Energy Council. 1994. *Planning Today for Tomorrow's Energy.* Vancouver: The Council.

BC Utilities Commission. 2009. *BC Utilities Commission Decision on BC Hydro's Application for Approval of the 2008 Long-Term Acquisition Plan.* 27 July.

Brodeur, P. 1993. *The Great Power-Line Coverup: How the Utilities and the Government Are Trying to Hide Cancer Hazards Posed by Electromagnetic Fields.* Boston: Little, Brown.

Chapman, J.D. 1960, "The Geography of Energy: An Emerging Field." In *Occasional Papers in Geography,* 31-40. Canadian Association of Geographers, BC Division, no. 1. Vancouver: Tantalus.

Christensen, B. 1995. *Too Good to Be True: Alcan's Kemano Completion Project.* Vancouver: Talon.

Enger, E.D., J.R. Kormelink, B.F. Smith, and R.J. Smith. 1983. *Environmental Science: The Study of Relationships.* Dubuque, IA: Wm. C. Brown.

Farrow, M. 1977. "Hydro Energy Growth 'Suspect.'" *Vancouver Sun,* 3 March, 36.

Halvorson, H. 1990. "The British Columbia Coal Industry." In *The Pacific Rim: Investment, Development and Trade,* 2nd ed., ed. P.N. Nemetz, 310-22. Vancouver: UBC Press.

Hardwick, W.G. 1962. "The Moran Dam and Availability of Cultivated Land in Interior British Columbia." In *Occasional*

Papers in Geography, 62-68. Canadian Association of Geographers, BC Division, no. 3. Vancouver: Tantalus.
Lovins, A.B. 1977. *Soft Energy Paths*. New York: Harper.
Marmorek, J. 1981. *Over a Barrel: A Guide to the Canadian Energy Crisis*. Toronto: Doubleday.
Ostergaard, P. 2002. "Energy Resources in BC's Central Interior." *Western Geography* (Canadian Association of Geographers) 12: 216-29.
Pacific Coast Energy Corporation. 1995. *Vancouver Island Natural Gas Line Extension: 1995*. Brochure. Vancouver: The Corporation.
Seager, A. 1996. "The Resource Economy, 1871-1921." In *The Pacific Province: A History of British Columbia*, ed. H.J.M. Johnston, 205-52. Vancouver: Douglas and McIntyre.
Sewell, W.R.D. 1987. "Energy." In *British Columbia: Its Resources and People*, ed. C.N. Forward, 226-56. Western Geographical Series vol. 22. Victoria: University of Victoria.
Smith, G.G. 1989. *Coal Resources of Canada*. Geological Survey of Canada Paper 89-4. Hull, QC: Ministry of Supply and Services Canada.
Toller, S., and P.N. Nemetz. 1997. "Assessing the Impact of Hydro Development: A Case Study of the Columbia River Basin in British Columbia." *BC Studies* 114 (Summer): 5-30.
Whiteley, D. 1982. "Offshore Claim Negotiations Less Friendly." *Vancouver Sun*, 10 September, E5.
Wilson, J.W. 1978. "Electric Power Development in British Columbia: A Case of Metropolitan Dominance?" In *Vancouver: Western Metropolis*, ed. L.J. Evenden, 79-93. Western Geographical Series vol. 16. Victoria: University of Victoria.
Young, L.B. 1974. *Power over People*. New York: Oxford.

INTERNET

Arsenault, C. 2010. "Sour Gas Line Explosion Leaves Bad Taste for Northern BC Residents." Rabble.ca, 11 February. www.rabble.ca/news/2010/02/sour-gas-line-explosion-leaves-bad-taste-northern-bc-residents.
BC Hydro. 2010a. "Service Plan 2010/11–2012/13." www.bchydro.com/etc/medialib/internet/documents/info/pdf/service_plan_2010_11_2012_13.Par.0001.File.service_plan_2010_11_2012_13.pdf.
–. 2010b. "Transmission System." www.bchydro.com/about/our_system/transmission_system.html.
BC Ministry of Energy and Mines. 2004a. "All Minerals Production and Values: 1998-2003." www.em.gov.bc.ca/mining/miningstats.
–. 2004b. "Sedimentary Basins in British Columbia." Oil and Gas – Resource. www.em.gov.bc.ca/subwebs/oilandgas/resource/resource.htm.
BC Ministry of Energy, Mines and Petroleum Resources. 2007a. "Coalbed Methane in BC." www.bcstats.gov.bc.ca/pubs/exp/exp0804.pdf.
–. 2007b. "Offshore Oil and Gas in BC: A Chronology." www.empr.gov.bc.ca/OG/offshoreoilandgas/OffshoreOilandGasinBC/Pages/AChronologyofActivity.aspx.
–. 2008. "British Columbia Coal Sold and Used." www.em.gov.bc.ca/mining/miningstats/31coalsold.htm.
–. 2009. "Energy Plan Report." www.energyplan.gov.bc.ca/.
–. 2010. "New Act Powers BC Forward with Clean Energy and Jobs." www2.news.gov.bc.ca/news_releases_2009-2013/2010PREM0090-000483.htm.
BC Ministry of Finance. 2009. *British Columbia Financial and Economic Review 2009*. 69th ed. www.fin.gov.bc.ca/tbs/F&Ereview09.pdf.
BC Ministry of Finance. 2010. *British Columbia Financial and Economic Review 2010*. 69th ed. www.fin.gov.bc.ca/tbs/F&Ereview10.pdf.
BC Stats. 2008. "Exports – April 2008." www.bcstats.gov.bc.ca/pubs/exp/exp0804.pdf.
BC Transmission Corporation. 2009. "Long-Term Electricity Transmission Inquiry." www.bchydro.com/planning_regulatory/long_term_electricity_transmission_inquiry.html.
British Petroleum. 2010. "Statistical Review of World Energy 2010." www.bp.com/productlanding.do?categoryId=6929&contentId=7044622.
Canadian Association of Petroleum Producers. 2004. BC Oil Industry Statistics, 1997-2004. Table. www.capp.ca/raw.asp?NOSTAT=YES&dt=NTV&e=PDF&dn=34089.
Canadian Centre for Energy Information. 2004. "Canada's Evolving Offshore Oil and Gas Industry." March. www.centreforenergy.com.
David Suzuki Foundation. 2010. "Protecting Canada from an Oil Spill." www.davidsuzuki.org/issues/oceans/science/marine-planning-and-conservation/protecting-canada-from-an-oil-spill/.
Dogwood Initiative. 2010. "Citizen's Guide to Coalbed Methane." www.dogwoodinitiative.org/publications/reports/citizens-guide-to-coalbed-methane-in-british-columbia.
Ebner, D. 2009. "BC Emerges as Natural Gas Player." *Globe and Mail*, 10 August. www.theglobeandmail.com/report-on-business/industry-news/energy-and-resources/bc-emerges-as-natural-gas-player/article1247206/.
Environment Canada. 2004. "Successful Investigation Results in Guilty Plea and a $200,000 Financial Penalty." News release, 22 April. www.ec.gc.ca/media_archive/press/2004/040422_n_e.htm.

Global InfoMine. 2010. "Brent Crude Oil." www.infomine.com/investment/charts.aspx?mv=1&f=f&r=2y&c=cbrent_crude_oil.xusd.ubarrel#chart.
Goodwin, S. 2010. "No Pipeline, No Tankers, No Problem." www.greenpeace.org/canada/en/Blog/no-pipeline-no-tankers-no-problem/blog/11988.
Hunter, J. 2010. "BC Liberals' Clean Energy Act Opposed from Within." *Globe and Mail,* 1 June. www.theglobeandmail.com/news/national/british-columbia/bc-liberals-clean-energy-act-opposed-from-within/article1587409/.
Lovins, A.M. 2008. "Special Report: Energy Efficiency – The Case for Efficiency." Forbes.com. www.forbes.com/2008/07/03/energy-efficiency-biz-energy_cx_al_0707efficiency_lovins.html.
Natural Resources Canada. 2009a. "Comprehensive Energy Use Database." www.oee.nrcan.gc.ca/corporate/statistics/neud/dpa/trends_tran_bct.cfm.
–. 2009b. "Energy Policy." www.nrcan.gc.ca/eneene/polpol/index-eng.php.
Nyboer, J., M. Bennett, and K. Muncaster. 2004. "Energy Consumption and Supply in British Columbia: A Summary Review." Report prepared for the British Columbia Ministry of Energy and Mines. www.cieedac.sfu.ca.
Pynn, L. 2010. "B.C. Offshore Drilling Moratorium Stays: Prentice." *Vancouver Sun,* 21 May. www.globaltvbc.com/world/offshore+drilling+moratorium+stays+prentice/3058241/story.html.
Rynor, B. 2009. "Letter Warns of More Attacks on BC Pipelines." Canada.com, 16 July. www.canada.com/news/Pipeline+bomber+actions+incredibly+risky+Terrorism+expert/1797395/story.html?id=1797395.
–. 2009. "BC Government Orders Burrard Thermal Write-off." *Vancouver Sun,* 29 October. www.vancouversun.com/technology/government+orders+Burrard+Thermal+writeoff/2159484/story.html.
SkeenaWild Conservation Trust. 2008. "Enbridge." skeenawild.org/conservation-issues/enbridge/.
Statistics Canada. 2007. "Supply and Demand of Primary and Secondary Energy in Terajoules." CANSIM Table 128-0002. www.statcan.gc.ca/
–. 2010a. "Primary and Secondary Energy, Terajoules – British Columbia." Catalogue No. 057-003-XWE, Table 2-12. www.statcan.gc.ca/pub/57-003-x/2008000/t041-eng.htm.
–. 2010b. "Production and Exports of Coal." CANSIM Table 135-0002. www.statcan.gc.ca/.
–. 2010c. "Supply and Disposition of Crude Oil and Equivalent." CANSIM Table 126-0001. www.statcan.gc.ca/.
West Coast Environmental Law Society. 2009. "Enbridge Gateway Pipeline Project: Draft Joint Review Panel Agreement and Terms of Reference." wcel.org/sites/default/files/file-downloads/Letter%20Regarding%20Legal%20Concerns%20with%20the%20Federal%20Government's%20Process%20for%20the%20Enbridge%20Northern%20Gateway%20Pipeline.pdf.

Agriculture

The Land and What Is Produced

12

British Columbia has a mountainous landscape and therefore lacks agricultural land for commercial production. Pockets of agricultural land have been developed, and this resource has been important to the settlement and development of many parts of the province. A host of physical conditions influence the range and types of agricultural commodities that the land is capable of producing. Many agricultural decisions are based not solely on these physical parameters but on human considerations such as the economic viability of producing a commodity. The range of viable agricultural products has changed as development of new seeds and products, farming methods and technologies, transportation routes and modes, political decision making, and consumer tastes have shaped and reshaped the agricultural landscape of British Columbia.

The most important ingredient in farming is land, and in many ways it is a nonrenewable resource because it is so difficult to create "good" soil. Over time in some locations of the province agricultural land has been converted to other uses. An already limited agricultural land resource has shrunk, and along with it the capability to produce food for the future. To protect agricultural land, the BC government in 1973 established agricultural land reserves with a strict set of rules on their use. This tactic was unique in Canada and much of the world, but it is not without its critics.

From its beginning, farming in British Columbia has been in transition. As conditions change so do the products and methods of farming; today, approximately 200 agricultural commodities are produced. The combination of traditional and new, exotic products has created regional specialization and considerable hope for future growth and prosperity. The 1991-96 census shows that British Columbia was one of the few provinces in Canada to increase its number of farms, leading all other provinces; however, between 1996 and 2006 the number of farms decreased (Table 12.1) in the province and across Canada. Still, British Columbia has more farms than Manitoba as well as all the Maritime provinces combined.

PHYSICAL CONDITIONS

There is very little good agricultural land in British Columbia. Figure 12.1 shows the pockets of primary farmland, or agricultural land reserves, along with larger areas of range lands and the few areas of potential farmland in the north. These lands make up the total of British Columbia's arable land: approximately 3 percent of the land base is farmland and another 30 percent has some agricultural capability (BC Stats 2004, 8).

The soils of British Columbia are influenced by a combination of factors such as climate, vegetation, elevation, slope, and surficial geology, resulting in a variety of soils and agricultural capabilities (Valentine et al. 1978; and see Chapter 2). The Canada Land Inventory uses a seven-point scale to grade soils in terms of agricultural capabilities, modified by descriptive subclasses giving details of crop limitations (Environment Canada 1972). Table 12.2 outlines the seven soil classifications and twelve subclass characteristics (limitations to farming) as well as assessing the area of land and agricultural reserve land in each of the seven classes. Typically, a farm

Table 12.1

Census farms in Canada, 1991-2006

Province	1991 farms	1996 farms	% change	2001 farms	% change	2006 farms	% change
Newfoundland	725	731	0.8	645	-11.8	558	-13.5
Prince Edward Island	2,361	2,200	-6.8	1,845	-16.1	1,700	-7.9
Nova Scotia	3,980	4,021	1.0	3,920	-2.5	3,795	-3.2
New Brunswick	3,252	3,206	-1.4	3,030	-5.5	2,776	-8.4
Quebec	38,076	35,716	-6.2	32,140	-10.0	30,675	-4.6
Ontario	68,633	67,118	-2.2	59,730	-11.0	57,211	-4.2
Manitoba	25,706	24,341	-5.3	21,070	-13.4	19,054	-9.6
Saskatchewan	60,840	56,979	-6.3	50,595	-11.2	44,329	-12.4
Alberta	57,245	58,990	3.0	53,655	-9.0	49,431	-7.9
British Columbia	*19,225*	*21,653*	*12.6*	*20,290*	*-6.3*	*19,844*	*-2.2*
Canada	280,043	274,955	-1.8	246,925	-10.2	229,373	-7.1

Note: The term "census farm" refers to a farm, ranch, or other agricultural operation that produces agricultural products for sale.
Sources: Statistics Canada (2001b, 2006).

includes sloping land, land beside a stream, boggy land, and so forth, and will therefore have a number of designations, as can be seen in the following example:

$$\underset{\text{I}}{5^{6}_{\text{W}}} - 4^{4}_{\text{W}}\,(4^{6}_{\text{W}} - \underset{\text{I}}{3^{4}_{\text{W}}})$$

The farmland thus described is in a wet area and has two designations. Both class 5 and class 4 lands are in an unimproved state. The small 6 above the W indicates that 60 percent of this land is class 5 and the land has excess water, which is the subclass W. The other 40 percent of the land (hence the small 4) is class 4; however, it has excess water and is subject to flooding or inundation (the I subclass). The bracketed classification, or second designation, is interesting because it suggests that this land could be improved if the moisture problems were addressed through installing a drainage system or some other technique. The improved land would then be 60 percent class 4 land and 40 percent class 3 land, even though excess moisture and some flooding might still occur. The classification system is dynamic in that it recognizes the potential for land improvement. Perhaps the most important value of the Canada Land Inventory is as a scientific and objective means of assessing agricultural land.

Another major constraint to agriculture is climate. The combination of prevailing westerly winds bringing relatively mild moist conditions in winter, significant latitudinal variation influencing incoming solar radiation, and high ranges of mountains affecting air masses and moisture means that both temperature and precipitation vary greatly throughout the province (see Chapter 2). Figure 12.2 maps the average number of frost-free days per year throughout the province, revealing significant differences from region to region. High elevations and northerly latitudes subject much of the province to frosts even in summer, with fewer than fifty frost-free days a year. The relatively warm Pacific Ocean combined with westerly flow of winds gives a distinct advantage to the coast. The southern interior river valleys of the province also have

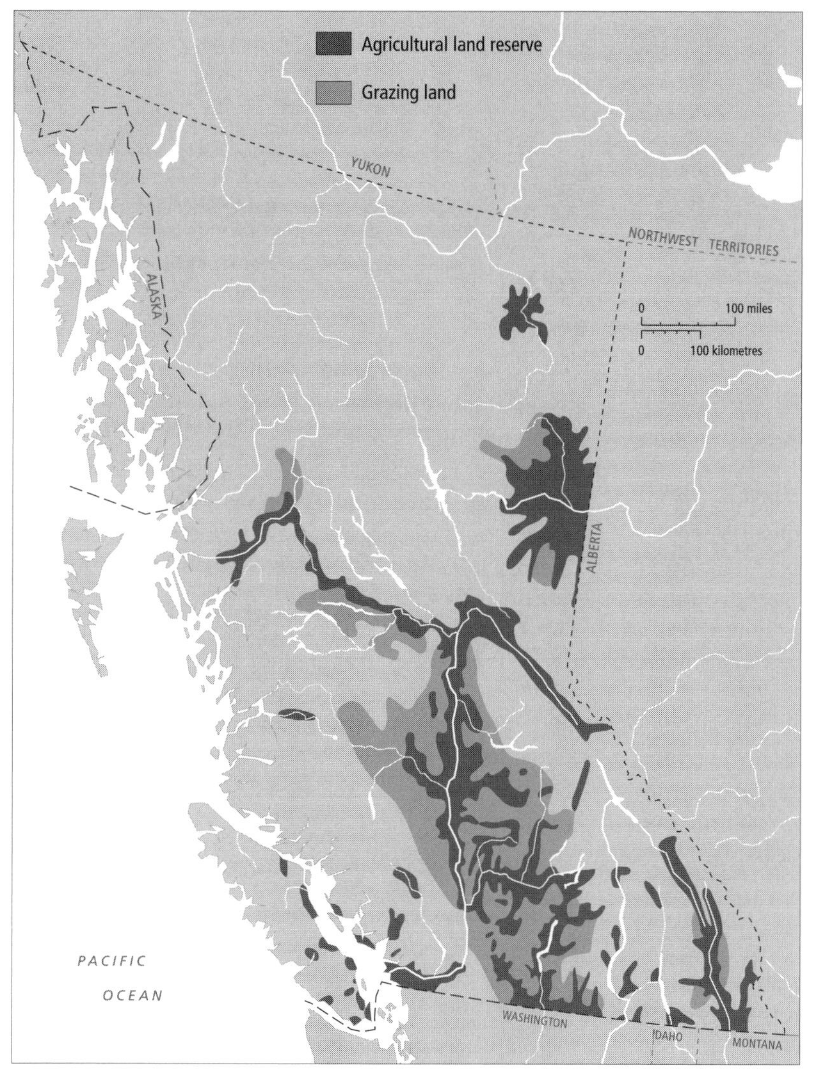

Figure 12.1 Farming areas in British Columbia
Source: Modified from Dalichow (1972, 40); Farley (1979, 81).

Table 12.2

Soil classification, area, and subclasses

Soil class	Canada Land Inventory characteristics	British Columbia		Agricultural land reserve	
		Area (ha)	% of total land	Area (ha)	% of total ALR
1	Widest range of crops	69,948	0.2	56,655	1.2
2	Moderate limits	397,688	1.3	302,163	6.4
3	Moderately severe limits	999,778	3.3	717,637	15.2
4	Severe limits	2,131,867	7.1	1,454,159	30.8
5	Only permanent pasture/forage	6,138,294	20.5	1,482,487	31.4
6	Natural grazing	5,357,900	17.9	448,523	9.5
7	No capability	14,900,513	49.7	259,671	5.5
Total		29,995,988	100.0	4,721,295	100.0

Subclass characteristics

C	Adverse climate	I	Inundation (flooding)	R	Bedrock near the surface
D	Undesirable soil structure	M	Moisture deficiency (droughtiness)	T	Topography (slope)
E	Erosion	N	Salts	W	Excess water
F	Low fertility	P	Stoniness	X	Combination of soil factors

Note: Totals do not always equal 100 percent, due to rounding.
Sources: British Columbia, Environment and Land Use Committee Secretariat (1976); Berry (1988, 25).

a considerable number of frost-free days. The number of frostfree days is crucial to planting and harvesting crops.

Another frequently used statistic for calculating quality of agricultural land is growing degree days. When the temperature is above 6°C plant growth is active. (Some research suggests 5°C.) Each day that the temperature is above 6°C is recorded as a growing degree day, and a yearly total of degrees above that benchmark is calculated (Matthews and Morrow 1985, 38).

Table 12.3 provides a regional summary of physical characteristics, including potential agricultural limitations. Kelowna, Cranbrook, and Kamloops, for example, are all in southern interior river valleys with little precipitation. The hot, dry, desert-like conditions of summer make irrigation systems a requirement for producing agricultural products. Because of its higher elevation, Cranbrook's agricultural conditions are further limited by considerably fewer frost-free days than either Kelowna or Kamloops experience, but its southerly latitude results in more growing degree days than Kamloops. Fort St. John has more frost-free days than does Cranbrook and more precipitation in the important summer growing season, but because of its northern latitude, it has fewer growing degree days.

All these physical characteristics are the foundations on which an agricultural industry is built. Human factors are also complex because farmers must attempt to make a living. The physical constraints of the land are just one element with which a farmer contends. Others are the economic and political climate for agricultural production and the ability to adapt to it.

THE DEVELOPMENT OF AGRICULTURE IN BRITISH COLUMBIA

The above discussion was based on the development of a commercial agricultural industry. Nevertheless, the land has sustained people throughout the ages, and food has been abundant ever since the end of the last ice age. First Nations practised fishing, hunting, and gathering to support themselves, but agriculture was unknown in British Columbia (Dalichow 1972, 23; also see Chapter 5). The First Nations lived off the resources of the land without ever planting crops or domesticating animals.

The first Europeans in what is now British Columbia resided in fur trade forts and practised agriculture initially

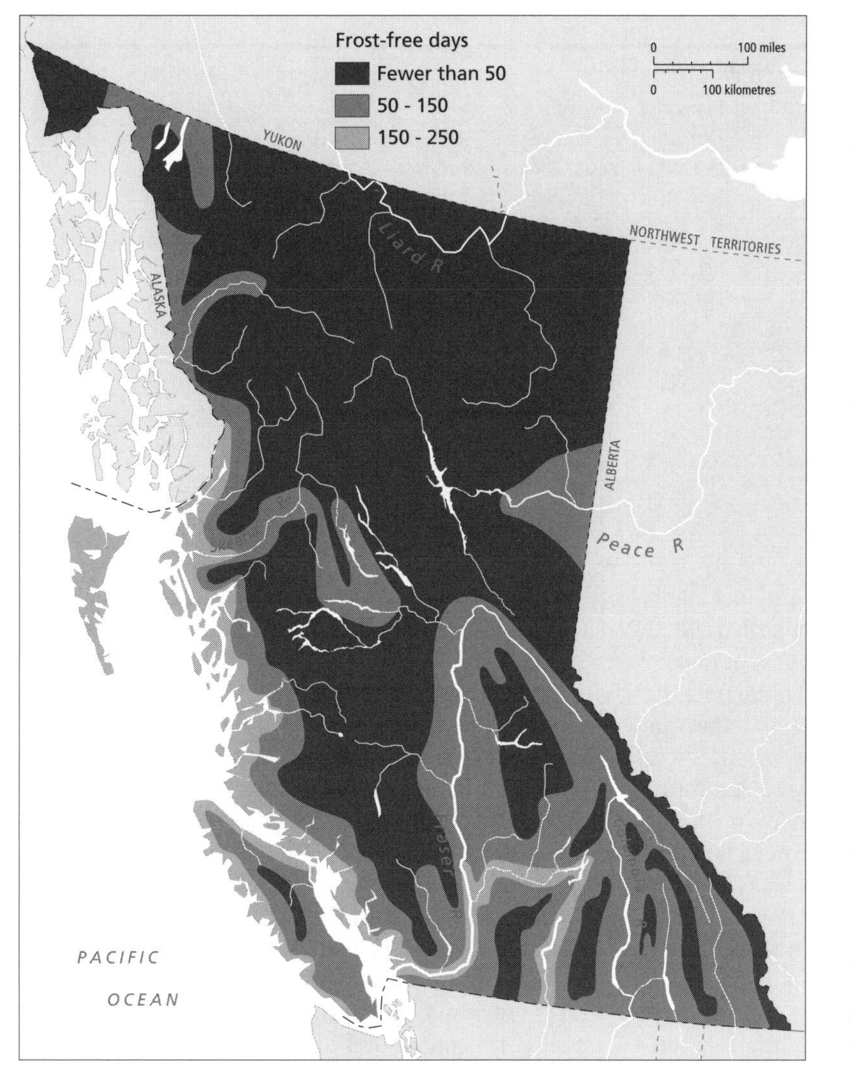

Figure 12.2 Frost-free days in British Columbia
Source: Modified from BC Ministry of Agriculture, Food, and Fisheries (2004).

for subsistence, growing vegetables, planting a few fruit trees, and sometimes caring for cows and chickens. Early churches similarly planted gardens on their land. Horses were raised in the southern interior for transporting furs (see Chapter 4). By the 1840s, the Hudson's Bay Company was attempting some commercial agriculture as well as promoting agricultural settlement. Neither was particularly successful. Perhaps the most important aspect of these limited and isolated agricultural endeavours is that they built a base of information for the wave of settlers who came shortly after. Wherever there was a Hudson's Bay Company fort or a church, there was information about the soil, which crops would grow and which would not, the limitations of water and climate, and the basics for a commercial agricultural industry.

The gold rush of 1858 brought a wave of gold seekers along with many who sought to earn a living providing food for the miners. Farms spread throughout southeastern Vancouver Island, the Fraser Valley, and, when gold was discovered there, north into the Cariboo. Ranchers followed gold settlement as well. Cattle drives, often originating in the United States, used the Okanagan-to-Kamloops route and the Thompson River system to bring meat to the Cariboo communities. Ranches also became popular in these areas. There were neither rules nor much concern about the number of cattle grazing on the natural grasses – mainly bunch grass – in these dry interior valleys. Serious overgrazing resulted. The newly built Cariboo Road, though, was important in facilitating the expansion of the agricultural industry. More wheat was grown, flour mills were built, the number of dairy farms increased, and more vegetables and other agricultural produce were grown to supply the remote interior.

The early colonial governments of Vancouver Island and British Columbia had a land policy that encouraged settlement on 160-acre parcels through a process of preemption, especially after 1861 (Harris 1997; Little 1996). Preemption was a system whereby "an occupant who improved a claim to the value of ten shillings ($2) per acre

Table 12.3

Physical characteristics and potential limitations for agriculture

Agricultural area	Recording station	Altitude (m)	Average annual precipitation (mm)	Average summer precipitation (mm)	Frost-free days	Growing degree days (above 6°C)[a]	Potential limitations
Vancouver Island	Victoria	19	883	81	284	1,920	Climate, soil
Lower Mainland	Vancouver	4	1,199	150	229	2,085	Hydrography, vegetation
Okanagan	Kelowna	430	381	71	146	1,914	Climate, hydrography
Kootenay	Cranbrook	926	401	102	82	1,686	Climate, soil
South central interior	Kamloops	345	279	84	176	2,309	Climate, hydrography
North central interior	Prince George	579	554	137	81	1,492	Climate, soil
North coast	Terrace	55	1,161	157	132	1,562	Vegetation
Peace River	Fort St John	695	466	180	123	1,312	Climate

a Growing-degree days are calculated by adding together the number of degrees above 5°C the temperature reaches each day over the course of the year.
Source: Modified from Dalichow (1972, 18, 44); Environment Canada (2010).

would receive a certificate of improvement, making the land eligible for sale or mortgaging" (Little 1996, 78). By this policy it was hoped that Europeans would be encouraged to settle; however, both Harris and Little note that many preemptions were for land speculation rather than agricultural development.

The decline of the gold rush by the late 1860s brought a decline in agriculture. Many farmers abandoned their land and others became subsistence farmers. Ranchers fared somewhat better because ranching was connected to distant markets by cattle drives (Dalichow 1972, 33).

Transportation became a major influence in commercial agriculture. The industry increased only marginally until the building of the Canadian Pacific Railway in the 1880s. The thousands of railway workers represented a market for beef and other agricultural produce, and the completed line provided connections to the Prairies and central Canada, while the port provided a connection to the world. The mining boom of the late 1880s and 1890s in the Kootenays resulted in more railway expansion, more people, and more demand for agricultural products. As transportation provided access to British Columbia's agricultural land and to markets, demand for agricultural land led to more land speculation and claims of the free 160-acre homestead available after 1875, just as it had across the Prairies (Demeritt 1995-96).

Farms and ranches became widespread in the southern portion of the province. The Kootenays, Okanagan, South Thompson River, Lower Fraser Valley, and southern Vancouver Island were popular areas. The 1891 total population was 98,173, of which some 22,000 were farmers on 6,500 farms (BC Ministry of Agriculture and Food n.d.). Agriculture was becoming an important industry in terms of employment, commodity export, and settlement. The mild winter climate, combined with an abundance of inexpensive fuel wood for energy, saw the development of a greenhouse industry in the Lower Mainland and southern Vancouver Island. Greenhouse vegetables could be transported easily to a Prairie market. The rapid growth of the Vancouver region provided its own market for agricultural products, and the Fraser Valley developed specializations in dairy products and market gardening.

Further railway construction, and even the speculation of railways to come, opened up other agricultural areas. Rumours that the Grand Trunk Pacific (absorbed into the Canadian National Railway, or CNR, in 1922) would terminate on the north end of Vancouver Island enticed a number of settlers, mostly Scandinavian, into Quatsino Sound, Cape Scott, Sointula, and farther up the coast to Bella Coola. The ultimate destination for the CNR was Prince Rupert, however, thus opening up the more northerly regions such as the Bulkley Valley for farming. The Peace River block of agricultural land became accessible on the Alberta side with the expansion of a railway from Edmonton by 1918. Extension of this railway to Dawson Creek by 1930 made the BC Peace River region somewhat accessible, but it was the extension of the Pacific Great Eastern (later the British Columbia Railway and today CN) by 1958 that connected the region to Vancouver and provided much greater ease of access. In the Okanagan, Lower Mainland, and southern Vancouver Island, local railways were also built to facilitate agricultural production.

Table 12.4

Farms, farm area, average farm size, population, and urbanization, 1901-2006

	Farms	Area (ha)	Average farm size (ha)	BC population	% urban
1901	6,501	606,285	93.3	178,657	50.5
1911	16,958	1,028,700	60.7	392,480	51.9
1921	21,973	1,158,705	52.7	524,582	47.2
1931	26,079	1,434,510	55.0	694,263	43.1
1941	26,394	1,633,770	61.9	817,861	54.2
1951	26,406	1,904,310	72.1	1,165,210	52.8
1961	19,934	1,825,335	91.6	1,629,082	72.6
1971	18,400	2,358,315	128.2	2,184,621	75.7
1981	20,012	2,467,330	123.3	2,744,465	78.0
1991	19,225	2,392,300	124.4	3,282,061	80.4
2001	20,290	2,587,118	127.5	4,078,447	82.0
2006	19,844	2,835,458	143.0	4,113,487	85.0

Sources: Statistics Canada (1983, 2006).

Table 12.5

Farm size and gross receipts, 2006

Farm size (ha)	Number of farms	% of total farms	Annual gross receipts	Number of farms	% of total farms
Under 4	5,335	26.9	Under $10,000	9,466	47.7
4-28	7,251	36.5	$10,000-$24,999	3,194	16.1
29-97	3,574	18.0	$25,000-$24,999	1,936	9.8
98-307	1,968	9.9	$25,000-$99,999	3,629	18.3
308-648	833	4.2	$100,000-$249,000	1,536	7.7
649-1,425	528	2.7	$250,000-$499,999	889	4.5
Over 1,425	355	1.8	$500,000 and over	1,130	5.7

Source: Statistics Canada (2006).

Over the years, many economic conditions changed. Small family farms and ranches became less competitive, giving way to large producers and corporate farms and ranches. New, capital-intensive technology – agribusiness – was taking over by the 1960s. Table 12.4 shows the early increases in the number of farms and in the overall population of the province, then the decline in the number of farms from the 1950s to the early 1970s as farms grew larger and the population of the province mushroomed. More and more people in this period were living and working in urban environments. With the creation of agricultural land reserves in the early 1970s, however, along with encouragement to farm, the number of farms has levelled off. When farms are classified by size and income, however, it becomes apparent that the increase has been primarily in small farms (Table 12.5). Over 63 percent of BC farms are 28 hectares (69 acres) or less in area, and nearly three-quarters show gross sales of under $25,000. Most farmers in British Columbia today farm part time and have other employment.

Far fewer British Columbians are employed in agriculture today than historically. Just under 1.5 percent of the workforce is employed on farms, in greenhouses, nurseries, and other horticultural operations, in veterinary offices, hatcheries, grooming, and other agriculture related services (BC Ministry of Finance 2009). Even so, the revenues from agriculture have steadily increased, although the concerns over bovine spongiform encephalopathy (BSE), or mad cow disease, lowered revenues from livestock during the early 2000s (Figure 12.3). The variety of crops has also increased considerably since the 1970s.

As Chapter 1 described, time-space convergence occurs when changing transportation technologies shrink the time it takes for people, goods, or information to travel between places. After the CPR was built across Canada, for example, the trip from the east coast to the west took seven days, as opposed to weeks travelling via the river systems and trails or sailing around South America. In a sense, the railway shrank the distance, bringing Halifax and Vancouver considerably closer together. The initial time-space convergence accomplished by wagon roads

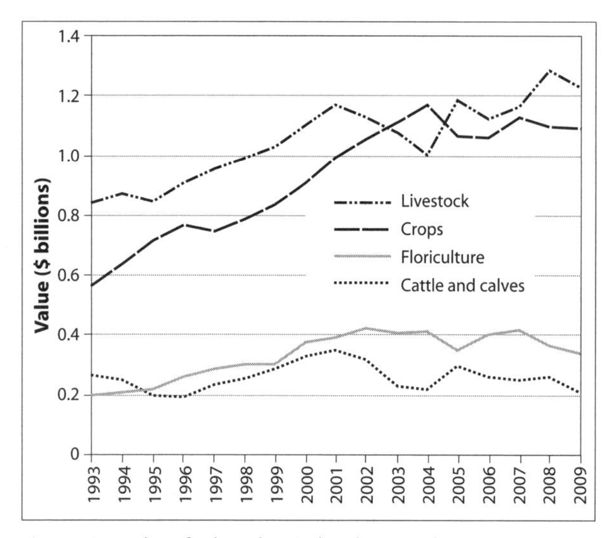

Figure 12.3 Value of selected agricultural commodities, 1993- 2009
Source: Data from Statistics Canada (2010).

and railways was important to the expansion of agricultural settlements and agricultural production in British Columbia. The more recent time-space convergence of highway systems and refrigerated trucks, container ships, and jet air cargo carriers, however, has not always been so beneficial to agriculture in this province. These transportation technologies brought global competition. Vegetables and fruit grown in California, Mexico, China, and elsewhere are now common commodities at BC supermarkets.

Global competition has also been spurred by genetic engineering, large corporate investment and control of produce, changing consumer expectations and demands, and free-trade politics – all of which have made agriculture in British Columbia a highly competitive industry. Both commodities and markets have changed drastically, and the challenge is often to develop new agricultural products and find new places to sell them.

The global movement of agricultural products and outside competition was initially met with provincial legislation in 1937. The Natural Products Marketing Act and subsequent acts created marketing boards that controlled the supply of agricultural produce coming into Canada, production within provinces, and trade between them. Marketing boards were intended to prevent overproduction leading to product price decline and farm bankruptcy. They are still in existence today. A milk quota, for example, was established in 1956, setting a limit on the number of gallons of fluid milk a farmer could legally produce. Because it is perishable, fluid milk is destined mainly for the local market. Industrial milk includes dairy products such as butter, cheese, yogourt, ice cream, and powdered milk. These have a much longer shelf life, can be transported greater distances, and are subject to a federal quota system. British Columbia has a small part in the national quota of industrial milk. Quebec has claim to 47 percent of the market, Ontario has 31 percent, and British Columbia holds a paltry 5.8 percent (BC Stats 2008). The system has resulted in a fluctuation in the number of cows but considerably fewer dairy farms, as can be seen in Table 12.6.

Table 12.6

Number of cows and dairy farms in British Columbia, 1941-2006

	Cows	Dairy farms		Cows	Dairy farms
1941	62,402	6,109	1986	75,005	2,601
1951	51,819	6,526	1991	74,919	2,038
1961	62,402	4,075	1996	82,008	1,644
1971	80,485	5,285	2001	71,401	1,044
1976	81,506	4,818	2006	72,756	812
1981	89,279	3,695			

Sources: Dalichow (1972, 53); Statistics Canada (2001a).

Marketing boards exist for producers of milk and dairy products, vegetables, fruit, wheat, and poultry. The last category includes chickens, turkeys, and eggs. Control of the supply side of agriculture, along with transportation subsidies for products such as wheat, have produced a rather artificial agricultural economy. The Free Trade Agreement (FTA) of 1989, the North American Free Trade Agreement (NAFTA) of 1994, and the global meetings of the World Trade Organization (WTO) have produced global economic pressures that seriously challenge our controlled system of agricultural production (BC Stats 2008).

AGRICULTURAL LAND RESERVES

Agricultural land can be improved through drainage, adding nutrients to the soil, and a number of other measures designed to fit local conditions. The process is often slow and expensive, and in this sense, agricultural land is a nonrenewable resource. In other words, it is difficult to make more of it. Agricultural land is the base for producing agricultural products now and for the future, and it is therefore important to assess the value of being able to feed the regional population versus depending on imported food. Once land is lost to other uses, options to produce more food are also lost (Rees 1993).

Population in British Columbia has steadily increased, albeit not evenly throughout the province. The Lower Fraser Valley, southern Vancouver Island, and the Okanagan have experienced some of the greatest population growth in the province since the Second World War. The trend has been accompanied by land development for transportation, industry, commercial functions, and housing. It has also meant the loss of agricultural land to construction. By the late 1960s and early 1970s, an estimated average of over 4,000 hectares of arable land per year was being converted to other uses (Farley 1979, 80). These losses were felt in the high growth regions such as the Lower Fraser Valley and Okanagan. The same land use conversion process was occurring in other regions, but more slowly. The newly elected NDP government in 1972 took the political action of "freezing" all agricultural land, and the next year passed the Land Commission Act, designed to preserve all lands assessed or zoned as farmland from class 1 to class 4 under the Canada Land Inventory. Initially, some 4.56 million hectares (11.4 million acres) of land came under the designation of agricultural land reserve (ALR).

There was considerable opposition to the creation of the agricultural land reserves. Developers, land speculators, local governments, and even farmers resisted reserving land for agricultural practices alone and appealed to have their land excluded. The Agricultural Land Commission (ALC) was the body responsible for appeals, but its mandate was to base decisions on the quality of the soil. As noted earlier, good soil can be scientifically and objectively defined through the Canada Land Inventory assessment.

The NDP were voted out in 1975, and the Social Credit were returned to government. By this time, the farmers and the public were largely in favour of the land reserve system, so the Land Commission Act was kept on the books. Changes were made to the appeal process in 1977,

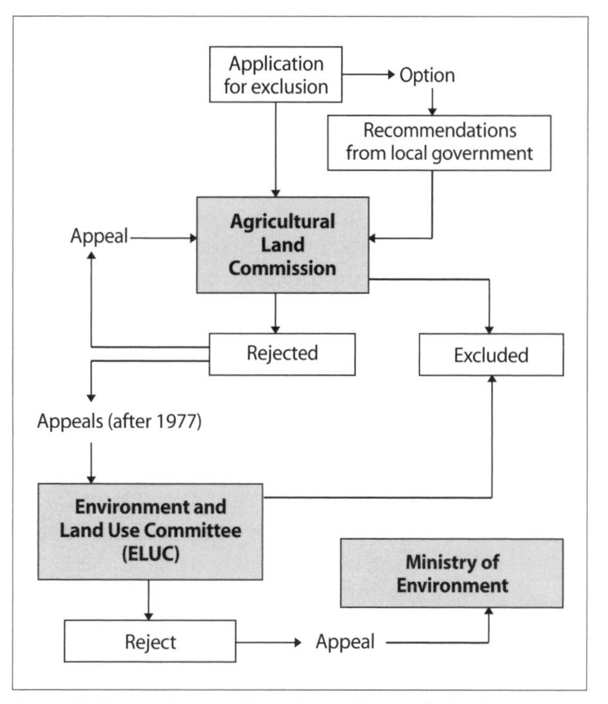

Figure 12.4 Appeal process for exclusion from agricultural land reserves
Source: Modified from Canada Works Project (1978, 5-6).

making it easier but much more political to exclude land from the agricultural reserve. The new appeal process allowed appeals to go to the Environment and Land Use Committee (ELUC), and its mandate put priority on "best use" for the land in question. If an appeal to ELUC was unsuccessful, an applicant could appeal directly to the minister of environment (Figure 12.4). Many appeals resulted in ugly confrontations and headlines in the major newspapers. The 1989 decision by the provincial government to allow golf courses on agricultural land set off another round of charges of political manipulation and favouritism and the recognition that some of the best agricultural land was being lost.

By 1991 the NDP was back in government and rescinded the process that allowed appeals to ELUC as well as to the Ministry of Environment. The negative publicity and confrontations when individuals attempted to exclude their land from the reserve between 1975 and 1991 resulted in minimal exclusion of agricultural land. NDP amendments to the Agricultural Land Commission Act in 1998 allowed land to be excluded or subdivided in the ALR for special reasons or for "provincial interests." Since the Liberal government has been in power, further amendments (in 2002) made exclusion for provincial interests more transparent, but it opened the door to exclusion for "community need." A number of communities have since had ALR land excluded for urban expansion. The 2002 amendments also included the decentralization of the Commission's decision-making authority to six agricultural regions. There is concern "that these small, local panels have led to discrepancies in decision-making between regions and may place undue stress on panellists from supporters of ALR exclusion applications" (Smart Growth n.d.).

There is still opposition to the ALR. Diane Katz (2009), in a report for the Fraser Institute, suggests that ALR land is responsible for the high cost of housing in the Lower Mainland, because it deprives individual owners from being able to make a market for their land. The ALR designation has made agricultural land prohibitively expensive. She concludes that "the ALR is not only harmful but unnecessary; human ingenuity and market forces are fully capable of meeting the food demands of British Columbia's growing population through increased productivity and efficiency" (p. 37). For those in favour of the ALR, the signing of the Tsawwassen Treaty, which saw the removal of over 200 hectares of agricultural land from the reserve (likely for industrial development), represents another means of losing agricultural land. Another loss of some of the best (Class 1) ALR land will occur in the Peace River Valley if the Site C Hydroelectric Dam is built.

Table 12.7 outlines the regional distribution of ALRs, including the changes each region has experienced since 1974. All areas have had more land excluded than included, with the exception of northern British Columbia and the Lower Mainland (now referred to as the South Coast). It is interesting to note that the northern region contains one-half of the agricultural land in the province. It is also interesting that the Vancouver Island, and Okanagan-Shuswap regions, which are extremely important agriculturally, have the highest exclusions but contain a very small land base.

British Columbia has taken one of the strongest stands of any province in Canada, and indeed most countries of the world, in protecting farmland. This bodes well for agricultural land and its production possibilities now and in the future. Some farmers have expressed concern, however, that agricultural land reserves save the land but not necessarily the farmer. Several programs have been implemented to address farm income, including crop insurance (now production insurance) and a joint provincial-federal net income stabilization account program. Still, the farmer must contend with all the vagaries of the physical elements and the markets, along with marketing boards and global demands to remove agricultural subsidies. Perhaps it is no wonder that there are so many part-time farmers in British Columbia.

Table 12.7

Changes in area of agricultural land reserve by region, 1974-2008

	1974 land (ha)	2008 land (ha)	% change	% of ALR
Vancouver Island	130,161.9	95,214	-26.8	2
Lower Mainland	181,821.8	190,428	4.7	4
Okanagan-Shuswap	257,246.9	238,035	-7.5	5
Kootenays	399,109.3	380,856	-4.6	8
Central Interior	1,494,210.6	1,475,818	-1.2	31
Northern British Columbia	2,258,744.8	2,380,352	5.4	50
Total	4,721,295.3	4,760,703	0.8	100

Source: BC Agricultural Land Commission (2009).

AGRICULTURAL COMMODITIES AND REGIONAL SPECIALIZATION

The geography of agriculture in British Columbia has been subject to many influences. The rugged, mountainous landscape, limited amounts of soil in classes 1 to 5, and extensive variations in climatic conditions have meant that less than 4 percent of the land base is available for agriculture. On the human side of the scale, historical patterns of settlement, transportation and technology developments, and a host of economic and political changes have resulted in general agricultural specializations with regional associations.

Table 12.8 gives a breakdown, by value, of the main categories of the agricultural commodities produced from 2001 to 2008. The value of crop receipts has increased the

Table 12.8

Agricultural receipts by commodity, 2001-8

	2001	2002	2003	2004	2005	2006	2007	2008
				($'000s)				
Crops								
Floriculture and nursery	359,964	387,288	406,650	412,185	350,635	400,521	420,456	446,109
Greenhouse vegetables	163,835	180,249	226,204	254,223	231,200	232,803	213,900	207,247
Other Vegetables and potatoes	172,097	157,718	163,595	163,102	170,442	126,907	176,108	237,844
Berries and Grapes	90,023	106,798	129,455	160,367	152,718	146,796	169,877	172,249
Treefruit	63,939	59,898	69,250	69,723	70,316	83,230	82,407	76,647
Other crops	120,322	125,068	119,294	109,052	92,933	69,931	91,890	92,253
Total crops	970,180	1,017,019	1,114,448	1,168,652	1,068,244	1,060,188	1,154,638	1,232,349
Livestock								
Dairy products	364,478	368,630	393,102	395,105	410,992	405,445	434,428	482,281
Cattle and calves	347,722	317,399	228,290	220,725	299,178	263,651	249,267	259,670
Poultry and eggs	328,596	328,088	339,649	274,845	363,760	358,880	382,557	440,367
Other livestock	129,965	111,920	113,204	112,959	110,049	95,307	95,227	98,756
Total livestock	1,170,761	1,126,037	1,074,245	1,003,634	1,183,979	1,123,283	1,161,479	1,281,074

Source: Statistics Canada (2010).

greatest amount since 2001. Floriculture and nurseries, including sod farms (Lower Mainland and southern Vancouver Island), have been very successful and increased in value each year except 2005. Vegetables, especially greenhouse vegetables (also Lower Mainland and southern Vancouver Island), peaked in 2005 and have since fallen off. Berries and grapes (Lower Mainland, Vancouver Island, and Okanagan) have experienced a much greater increase in value than tree fruits (Okanagan) over the last eight years.

The production of livestock and livestock products has increased overall, but 2003 was a disastrous year because the BSE incident closed the United States border to cattle and calves (central interior and north). The following year, the avian flu outbreak in the Fraser Valley resulted in the destruction of some 16 million birds. This was devastating for poultry producers, who lost an estimated $40 million, although there was some compensation for farmers from the government (Canadian Press 2004). Dairying (southern Vancouver Island, Lower Mainland, and Okanagan-Shuswap regions) has been a success story because, with the exception of a minor setback in 2006, the industry has maintained increases each year. Concentrations have emerged in other areas: honey, grain, and canola production in the Peace River region; ginseng in the southern interior; organic farming (especially in the Lower Mainland, southern Vancouver Island, and the Okanagan); and various other specialty crops throughout the province.

Through the census, British Columbia is divided into eight development regions (See Figure 7.10, p. 125). Each has advantages and disadvantages in producing agricultural commodities.

The Vancouver Island/Coast Region

The Vancouver Island/coast region is accessible mainly by ferry, making transportation of agricultural products onto and off the island (and the remote coastal mainland) expensive. With a population nearing three-quarters of a million, much of it concentrated on the southeastern end of the island, there is a substantial market as well as sufficient agricultural land for dairying, sheep and cattle ranching, growing vegetables, fruit, and a variety of specialty crops such as kiwi fruit, organic vegetables, wasabi (Japanese horseradish), and raising livestock such as ostrich and fallow deer. The southern end of the region has the mildest winter temperatures and the greatest number of growing days in the province, which allows the 2,855 mostly small farms to produce a wide range of crops. Greenhouse production of flowers and vegetables has become important, as has raising grapes for an increasing farm-gate winery industry. Dairy products are the highest value agricultural product in the region, but the industry serves "less than half of the local demand" and "only about 25% of the produce consumed in the region is grown locally" (Grow BC n.d., 13). Consequently, there is still plenty of opportunity to increase local production.

The Mainland/Southwest Region

Nearly 2.5 million people, or 60 percent of British Columbia's population, live in this region, which includes the Sunshine Coast Regional District, the Lower Fraser Valley, and the north region to Lillooet. The region "produces over 70% of BC's dairy products, berries, vegetables, poultry, eggs, pork, greenhouse vegetables, mushrooms, floriculture and nursery products" (Grow BC n.d., 15). Many specialty vegetable crops are produced to serve an ethnically diverse population. Mild weather, plenty of frost-free days, good soil, a large population (local demand), proximity to the United States (an export market), and excellent transportation infrastructure have resulted in this region's 5,410 farms generating over half the province's agricultural revenues. Proximity to the United States has had both positive and negative implications. It is a huge market for BC products; however, as with softwood lumber, there have been trade wars. In 2001, the successful export of greenhouse tomatoes to the United States resulted in a unilateral 33.95 percent antidumping duty being levied against Canadian producers. The World Trade Organization ruled in 2002 that American farmers had not been affected by the export of tomatoes, and the duty was dropped. Still, these actions highlight the unpredictable nature of resource exports. Another concern has been over crossborder shopping. Lower Mainland residents can easily purchase considerably cheaper milk, cheese, and other dairy products (CTV 2010).

Tourism is directly related to agriculture in the form of farm tours and bed and breakfasts, and the region is the leader in the food- and beverage-processing sector with annual shipments valued at about $2.8 billion (about 87 percent of the total provincial value of food and beverage shipments) (Grow BC n.d., 17). There are, however, local issues and conflicts. Because the region has the largest population base and agricultural sector, there is constant pressure to convert the ALR lands for purposes other than agriculture. There are also water issues, especially in the Lower Fraser Valley, where aquifers are involved and there is concern about agricultural runoff. Noise and smells are other causes for concern, although the Right to Farm Act has attempted to mitigate these issues. The success of agriculture in this region, and the shortage of agricultural land, has made the region's land the highest priced in the province.

The Thompson/Okanagan Region

In the Thompson/Okanagan region, farmers in the Thompson and Nicola valleys rely mainly on cattle ranching, while those in the Okanagan are major producers of tree fruits and, more recently, grape-wine. The southern interior has traditionally been cattle country and remains so, primarily with cow and calf operations. The finishing-off stage of cattle ranching usually involves feeding cattle grain in feedlots, most of which are located in Alberta, although there is an increasing number of local feedlots. There are several issues in this industry, including BSE and public concern over the consumption of red meat. Fortunately, the Free Trade Agreement offset the downward trend in local beef production by giving Canadian ranchers greater access to the US market. On the other hand, reduced red meat consumption has encouraged the production of white meats with lower fat content, including chicken, turkey, and pork, as well as fish products and organic vegetables.

The Okanagan is famous for its orchards. It produces apples ("over half of Canada's apple exports" [Grow BC n.d., 18]), pears, and 96 percent of British Columbia's soft fruits, such as peaches, apricots, and cherries (p. 19). There has always been stiff competition in the fruit industry from US producers, who have access to cheaper labour. To overcome this disadvantage, farmers have reduced the number of old, labour-intensive McIntosh and Delicious apple trees and switched to the newly developed Gala, Fuji, and Ambrosia apples. These varieties are planted thirty to sixty centimetres apart with 8,000 to 12,000 stems per hectare. They receive water and fertilizer by drip irrigation and grow only two metres tall. With

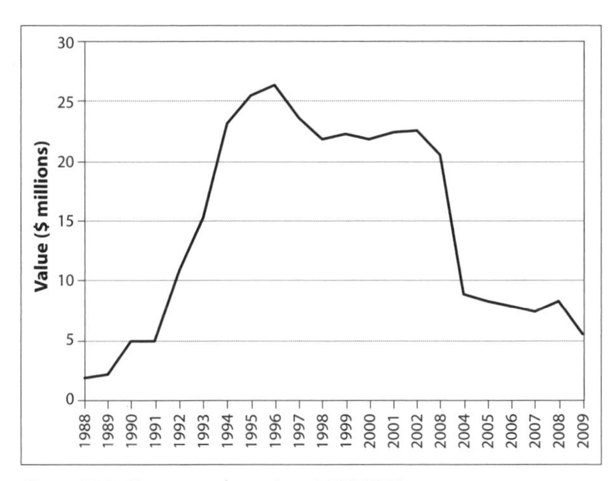

Figure 12.5 Ginseng cash receipts, 1988-2009
Source: Statistics Canada (2010).

no need for ladders to pick the fruit, the labour hours are reduced. These new apple varieties are in high demand by both farmers and consumers and have allowed apple production to remain competitive. High technology has reached most forms of agriculture.

The Okanagan region produces dairy products and vegetables, and two new specialty crops have appeared in recent years: wine grapes and ginseng. Ginseng production, mainly for the Asian market, took off in the late 1980s and reached peak value in 1996 (Figure 12.5). Ginseng takes four years to mature and is easy to recognize because it requires a black shade cloth to protect the crop from intense summer sun. Figure 12.5 contains a lesson for farmers who plan to jump into exotic crops such as ginseng. When production increased in the mid-1990s, supply outdistanced demand and prices fell, collapsing much of the industry by 2004.

Grape producers were one of the groups hardest hit by the Free Trade Agreement, which dropped protection for locally grown grapes. Approximately 50 percent of the Okanagan grape growers went out of business within several years. Fortunately, the provincial government allowed producers to import grape vines from around the world, and it introduced new regulations for producing wine from locally grown grapes. Two new categories of licence were introduced in the 1980s: the Farm Gate Winery licence (minimum 10 acres of land, 100 percent your own grapes, and a maximum of 5,000 gallons) and the Estate Winery licence (minimum 20 acres of land, can purchase up to 50 percent of grapes, and a maximum of 30,000 gallons). These licences allow individual farmers to sell directly from their farms. Today, however, the distinctions no longer exist, and wine growers must apply to the government for a licence to produce wine from their grape farms. The success of this industry is based on new varieties of vines and a new marketing system, the Vintner's Quality Association (VQA) labelling system, which guarantees a quality wine from 100 percent BC grapes. Many award-winning wines have been produced. These wines have helped to build the reputation of the Okanagan region and have tied it firmly to the tourism industry. The Okanagan produces most of the premium grapes in the province, making it the centre for the expanding industry.

The Okanagan-Similkameen district used to be characterized by roadside fruit and vegetable stands. They still exist but the concept of direct marketing has added to the attraction. Fresh fruits and vegetables are available seasonally, but so too are many value-added products such as juices, jams, jellies, fruit leather, and pies and pastries.

Because the entire Thompson-Okanagan region, with its 5,700 farms, is arid, expensive irrigation systems are necessary for crops to grow. The availability of water presents a particularly large challenge to farmers in the southern Okanagan. Much like the Lower Fraser Valley, the Okanagan is a popular region that is growing in population, putting pressure on ALR lands adjacent to urban areas and raising the price of agricultural land.

The Kootenay Region

Agriculture is carried out mainly in the lowlands of the Columbia and Kootenay rivers of the Kootenay region. This is a mountainous region with only 1,349 farms. Livestock and forage crops are the main commodities, but the region's moderate climate allows for apple, poultry, honey, and many vegetable crops and Christmas tree production. With relatively few people in the region and fairly high transportation costs, commercial agriculture for export faces a number of economic constraints. On the other hand, the higher cost of transportation works in the favour of local producers, and there is room for expansion.

The Cariboo/Central Region

The Cariboo/central region includes much of the grassland and open forests of the Interior Plateau, which is ideal for grazing cattle. Compared to other regions, there are considerably fewer frost-free days in this more northerly region; within the region, however, there are considerable variations (e.g., more frost-free days in the river valleys than on upland benches). The region has been the centre for cattle ranching since the Cariboo gold rush of the early 1860s. The whole Cariboo/central region has been overgrazed and, in response, range laws have been enacted, and ranchers have turned to fodder crops to overwinter their stock (Wood 1987, 150). Today, the region is still devoted to ranching, but its 1,781 farms have also diversified into dairy and lamb production, along with potatoes, cabbages, turnips, cauliflower, and carrots. Exotic commodities such as fallow deer, bison, and ostrich are also produced. On the food-processing side, the provincial government authorized a class "A" slaughter house in 2009 (it will likely be in operation by 2011) that "will increase local meat and value-added sales of beef, bison, fallow deer, hogs, lamb and ostrich" (Grow BC n.d., 12). The region has fairly easy access to the Alberta market for agricultural products, and the price of land is relatively low.

The North Coast Region

The north coast region is wet and heavily treed, with the exception of inland areas such as Terrace. It has a relatively short growing season because of the climate. Few people live in this area (57,663 as of 2006), and there are only 134 farms. Most of the existing farms are in the fertile river valleys, and the creation of new farms requires clearing the forests, which makes it costly. Because ranching is the main revenue source in this region, forage crops are needed for winter feed. Farmers also produce vegetables and some fruit, mostly for the domestic market. Much of the region is isolated and transportation to distant markets adds to its disadvantages, which support a local market.

The Bulkley/Nechako Region

Even fewer people live in the Bulkley/Nechako region (39,352 as of 2006), but it has 886 farms. Again, ranching is the primary agricultural occupation, along with dairying and the produciton of some cereal crops. Most farms are located in the fertile river valleys in the southern portion of the region. The region has a short growing season, which raises costs for feeding cattle in the winter. Transportation costs are also high, and the need to clear forest land also raises costs. Conversely, farmland is considerably less expensive than in the Lower Mainland. "For example, the average cost of land in the Vanderhoof area is about 10 to 15 times lower than land near Abbotsford" (Grow BC n.d., 6).

The Peace River/Northeast Region

The Peace River/northeast region is prairie, and its agriculture is dominated by grain. It has 1,729 farms and produces 77 percent of the province's grain crop, as well as canola and cattle. Grain farmers in this region face the same economic pressures as farmers on the Prairies, namely, the phasing out of low-volume grain elevators (e.g., Pouce Coupe, Taylor, and Buick) in favour of few high-volume ones (e.g., Fort St. John and Dawson Creek). This increases transportation costs. In addition, low world market prices persist because of global subsidies for wheat. Of course, farmers always face the vagaries of the weather. Figure 12.6, which traces world wheat prices from 2000 to 2009, shows a significant boost in prices in 2007-8. This price increase was influenced by a number of related events, including drought in Australia (which

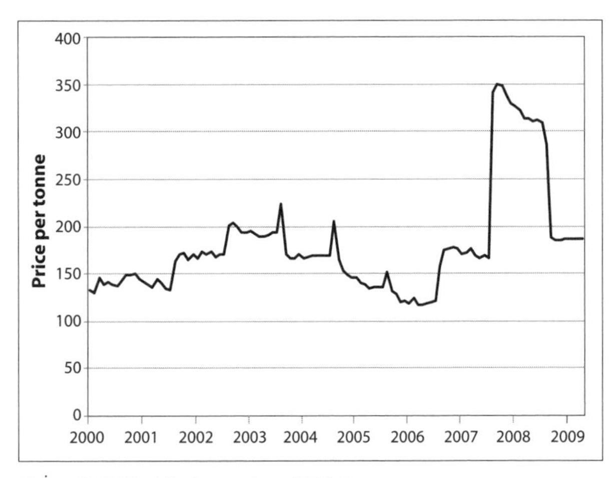

Figure 12.7 World wheat prices, 2000-9
Source: Statistics Canada (2009).

reduced supply) and US subsidies for biofuel production. The decision to produce biofuels resulted in the switch from wheat to corn, thus reducing supply and elevating the price of wheat. The collapse of the US economy and the global recession of 2008 also ended high wheat prices.

As in most of northern BC, this region is also cattle country, but "90% of calves leave the region to be finished elsewhere, particularly in Alberta" (Grow BC n.d., 9). Given the region's large grain harvest, it is a potentially good place for feedlots. The exotic products grown in the region include bison, and the milder climate adjacent to the Peace River allows a variety of vegetables to be grown for the local market. Forage crops are significant, as are honey farms. Although the growing season is short, the daylight hours in the summer months are long. For new farmers starting up, the cost of land is lower than in the rest of British Columbia. This region also has the largest block of Class 1, 2, and 3 land – the best land in the province.

SUMMARY

Farming is a balancing act between the many physical and human factors that affect the economic viability of producing agricultural commodities. Good agricultural land with few limitations is relatively rare in this province and confined mainly to valley bottoms. Some of this land is in the southern and western portions of the province, which have the physical advantage of a long growing season and mild winters, allowing for a greater range of agricultural commodities. The area also has the majority of the population, providing a ready market for products.

Transportation and time-space convergence have always played a key role in agriculture, bringing the advantage of ready markets and the disadvantage of competition. Competition from outside British Columbia and overproduction from inside resulted in political decisions to control imports of agricultural products and in marketing boards to control the internal supply. Political decisions at the federal level – FTA, NAFTA, and GATT (now WTO) rulings – have introduced international and global competition and broken many of the safeguards of protectionism. In this process of change, old farming methods and technologies have given way to new ones and to new market opportunities, both local and international.

Political decisions in British Columbia have given serious consideration to the preservation of agricultural land, creating agricultural land reserves. Farming and nonfarming uses for land still compete, particularly in the more heavily populated areas, but little agricultural land has been lost and Right to Farm legislation has reduced the conflicts.

As the "rules" of farming have changed, the number of agricultural commodities has increased, and marketing has kept pace. The development of the wine industry and VQA labelling is a good example of these changes. Another development is direct marketing, which is popular in the Okanagan. Many farmers are establishing roadside outlets to sell their produce, along with value-added products such as preserves, dried fruits and vegetables, and even pastries.

Tastes have changed. Many people are concerned about their diet and about how and where the food they eat is produced. Production of white meat, fish, and organic vegetables has expanded to meet this demand. Although consumption of red meat has declined, the opening of the US market to Canadian producers following the Free Trade Agreement (1989) has maintained red meat production in the province.

There are approximately 200 agricultural commodities produced in this province, and some are exotic plants and animals. Llamas, snub-nosed pigs, and herbs such as echinacea are being produced commercially, exemplifying the possibilities in farming today. Overall, this industry has developed with some remarkable regional strengths, and the number of farms is stable. With British Columbia still importing over half its food, there is plenty of opportunity for the agricultural industry to expand.

REFERENCES

BC Environment and Land Use Committee Secretariat. 1976. *Agricultural Land Capability in British Columbia*. Victoria: The Secretariat.

Berry, J. 1988. *Agriculture and Food in British Columbia*. Ottawa: Agriculture Canada.

Canada Works Project. 1978. *Preserving Agricultural Land*. New Westminster, BC: Douglas College.

Dalichow, F. 1972. *Agricultural Geography of British Columbia*. Vancouver: Versatile.

Demeritt, D. 1995-96. "Visions of Agriculture in British Columbia." *BC Studies* 108 (Winter): 29-60.
Environment Canada. 1972. *The Canada Land Inventory: Soil Capability Classification for Agriculture,* Report no. 2. Ottawa: Ministry of the Environment.
Farley, A.L. 1979. *Atlas of British Columbia: People, Environment and Resource Use.* Vancouver: University of British Columbia Press.
Harris, C. 1997. *The Resettlement of British Columbia: Essays on Colonialism and Geographical Change.* Vancouver: UBC Press.
Little, J.I. 1996. "The Foundations of Government." In *The Pacific Province: A History of British Columbia,* ed. H.J.M. Johnston, 68-96. Vancouver: Douglas and McIntyre.
Matthews, G.J., and R. Morrow, Jr. 1985. *Canada and the World: An Atlas Resource.* Toronto: Prentice-Hall.
Rees, W.E. 1993. *Why Preserve Agricultural Land?* Vancouver: UBC School of Community and Regional Planning.
Wood, C.J.B. 1987. "Agriculture." In *British Columbia: Its Resources and People,* ed. C.N. Forward, 138-61. Western Geographical Series, Vol. 22. Victoria: University of Victoria.

INTERNET

BC Agricultural Land Commission. 2009. "ALR Statistics." www.alc.gov.bc.ca/alr/stats/Statistics_TOC.htm.
BC Ministry of Agriculture, Food, and Fisheries. 2004. "Fast Stats 2004." www.agf.gov.bc.ca/stats/faststats/brochure 2004.pdf.
–. n.d. "History of BC Agriculture in British Columbia." www.agf.gov.bc.ca/aboutind/history.htm.
BC Ministry of Finance. *2009 Financial and Economic Review.* www.fin.gov.bc.ca/tbs/F&Ereview09.pdf.
BC Stats. 2004. "Quick Facts: The Economy." www.bcstats.gov.bc.ca/data/qf.pdf.
–. 2008. "Milking the System: Is Canada's Supply Management System an Impediment to Free Trade?" www.bcstats.gov.bc.ca/pubs/exp/exp0808.pdf.
Canadian Press. 2004. "Avian Flu Stamped Out in B.C.'s Fraser Valley." CTV News. 27 May. www.ctv.ca/.
CTV. 2010. "As HST Kicks In, West Coasters Head South to Shop." 2 July. www.ctvbc.ctv.ca/.
Grow BC. n.d. "The Regions." www.aitc.ca/bc/uploads/growbc/5_regions.pdf.
Katz, D. 2009. *The BC Agricultural Land Reserve: A Critical Assessment.* The Fraser Institute. www.fraserinstitute.org/research-news/display.aspx?id=13485.
Smart Growth BC. n.d. "Agricultural Land Commission." www.smartgrowth.bc.ca/Default.aspx?tabid=199.
Statistics Canada. 1983. Historical Statistics of Canada. Catalogue 11-516-X1E. Series M12-33. Farm holdings, area of land in farm holdings, census data, Canada and by province 1871-1971. www.statcan.gc.ca/pub/11-516-x/sectionm/4057754-eng.htm.
–. 2001a. Census of Agriculture. "Cattle and Calves: Dairy Cows." www.statcan.gc.ca/.
–. 2001b. Census of Agriculture. "Farms Classified by Total Farm Area, 2001, and Gross Farm Receipts, 2000." www.statcan.gc.ca/.
–. 2006. "Census of Agriculture, Farm Data and Farm Operator Data." www.statcan.gc.ca/.
–. 2009. "Farm Product Prices, Crops and Livestock." CANSIM Table 002-0043. www.statcan.gc.ca/.
– 2010. "Farm Cash Receipts, Annual (Dollars)." CANSIM Table 002-0001. www.statcan.gc.ca/.
Valentine, K.W.G., P.N. Sprout, T.E. Baker, and L.M. Lavkulich, eds. 1978. *The Soil Landscapes of British Columbia.* BC Ministry of Environment: Soils. www.env.gov.bc.ca/soils/landscape/index.html.

Water

An Essential Resource

A map of British Columbia quickly reveals a province with many rivers and lakes. British Columbia has "access to more than one-third of Canada's run-off, which itself amounts to almost one-tenth of the world total" (Sewell 1987, 199). The province is fortunate to have such an abundance of freshwater supplies. The basic physical conditions of ground- and surface water are seen in the hydrologic cycle. Seasonal variation also plays a role, especially on the discharge rates of the river systems. These physical parameters determine water availability, including surpluses and deficits in different regions.

Water is both a renewable and a mobile resource that has many uses in society: drinking, transportation, irrigation, recreation, hydroelectricity, fish habitat, commercial and industrial uses, as a coolant, to dilute effluents, and as a place to discharge wastes. These can be categorized into uses that are compatible and those that are not. Consequently, the various levels of government that manage this resource often have difficult decisions to make. Over time more population, new industries, and new demands for water have placed some of our sources in jeopardy. In Walkerton, Ontario, where the groundwater (the main water supply for the community) was contaminated with E. coli bacteria in May 2000, seven citizens died and over 2,000 became ill. This tragic situation was a wakeup call across Canada for how water is managed, and British Columbia has many water use conflicts that need to be addressed.

PHYSICAL PROPERTIES OF WATER

With one-third of Canada's supply of fresh water, it appears that British Columbia has an abundance of this resource. Yet the water is not equally distributed, its quantity fluctuates seasonally, and its quality can vary.

The hydrologic cycle is the first step in understanding the physical parameters of water (Figure 13.1). Extreme climate conditions throughout British Columbia are responsible for a good deal of the spatial variation of water (Chapter 2). Westerlies bring moist, relatively warm air masses from the Pacific, which are then forced to rise over the mountain chains of Vancouver Island and the Coast Mountains. As the air rises, it cools, condenses,

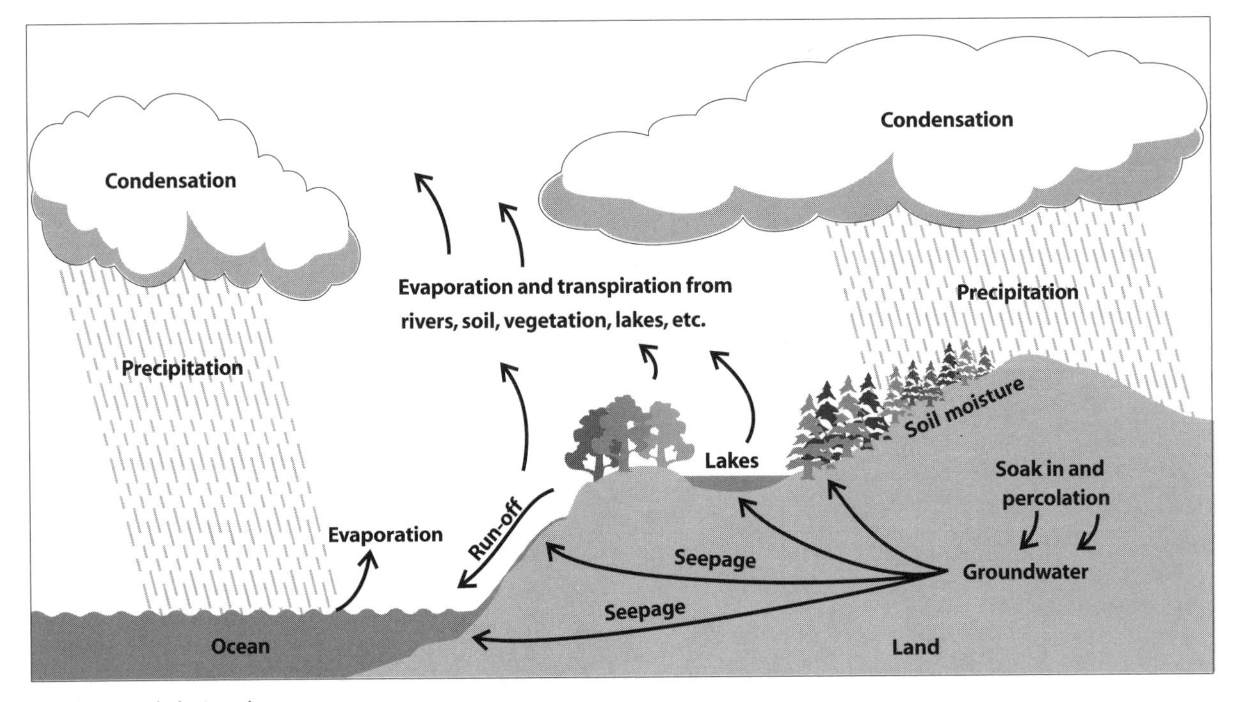

Figure 13.1 Hydrologic cycle

Table 13.1

Drainage area, length, and discharge of major rivers

River	Drainage area (km^2)	Length (km)	Average annual discharge (m^3/sec)
Fraser	231,312	1,368.5	3,600
Liard	143,418	483.0	1,300
Peace	128,451	281.7	1,400
Columbia	103,004	750.3	2,800
Skeena	54,488	579.6	1,000
Stikine	50,362	539.3	1,000
Nass	20,839	378.3	900

Source: Farley (1979, 38).

and releases moisture on the western slopes of the mountains. This process, as mentioned earlier, is called the orographic effect. The farther inland and the greater the number of intervening mountain chains, the less influence is sustained from these Pacific air masses. Other climatic influences also play a role in varying the pattern of precipitation and account for record high rainfall in regions such as the west coast of Vancouver Island, in contrast with the extremely arid South Thompson and Okanagan regions.

Evaporation and transpiration – moisture given off to the air by plants – are important variables in the hydrologic cycle and act as the basis of the orographic effect. Once precipitation occurs over the land, surface water flows with gravity, meeting and merging with other streams in the watershed and often forming a large river. The watershed is the total geographic area where surface water is collected for any stream or river system. It is important to recognize that the Fraser River watershed is the largest in the province, with over 230,000 square kilometres – approximately one-quarter of British Columbia (Table 13.1). Other large river systems include the Columbia, Skeena, Nass, and Stikine, all of which flow to the Pacific. The Peace and Liard River systems flow into the Mackenzie and thus to the Arctic. Vancouver Island, Haida Gwaii, and much of coastal British Columbia have numerous rivers and streams but much smaller watersheds.

Figure 13.1 also shows groundwater as part of the hydrologic cycle, and it is an important source of fresh water. Approximately 25 percent of British Columbians rely on groundwater as a source of water (Douglas 2008). Groundwater results from surface water percolating through the permeable soil of the earth to find its way to aquifers and eventually back to the ocean, completing the cycle.

Seasonal cycles influence this model in many ways. Transpiration is considerably higher in summer, when more vegetation is growing, and incoming solar radiation,

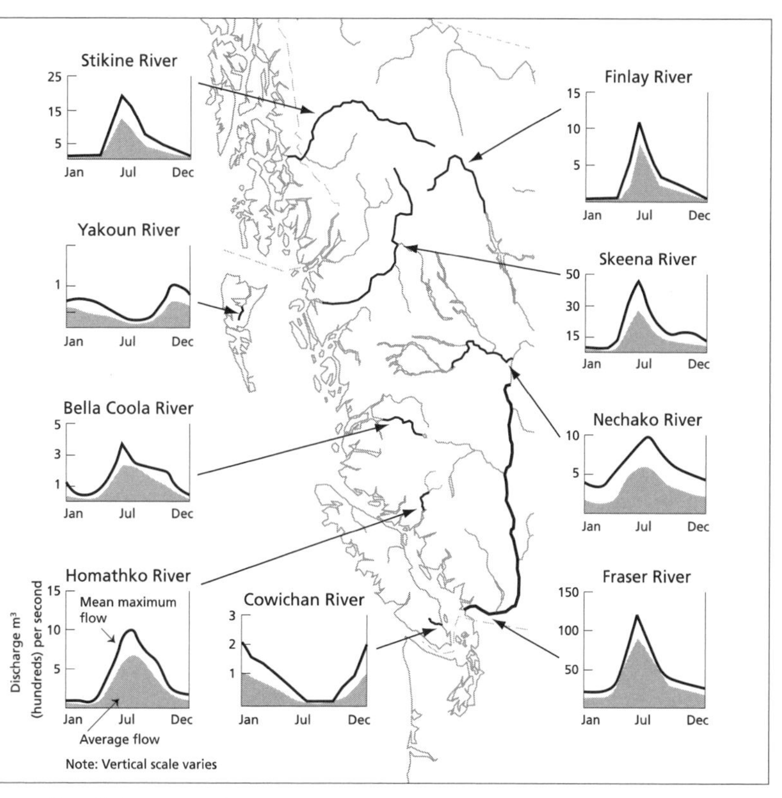

Figure 13.2 Discharge rates for selected rivers
Source: Modified from Farley (1979, 39).

or insolation, is much more intense. The main seasonal influences on surface run-off are the amount of precipitation, the accumulation of snowpack, and spring weather conditions that affect the rate of thaw. Seasonal variability of river systems is measured as the rate of discharge (Table 13.1, Figure 13.2). Essentially, two patterns emerge for British Columbia's rivers. First, the large rivers drain the interior, running either to the Pacific or to the Arctic. They build up water in the form of snow and discharge these stored volumes in the spring. Second, in coastal areas, smaller river systems have considerably less water stored as snow, and their discharge peaks at the time of the greatest amount of rainfall. The coastal and interior rivers therefore peak at different times of the year, yet both types of system raise the concern of flooding (see Chapter 3).

Greenhouse gases, which result in climate change (see Chapter 2), are having a significant influence on discharge rates. Glaciers, which are a source of water for many rivers, are receding as average temperatures increase. Moreover, the double or even triple jeopardy is that hotter weather results in reduced precipitation and increased evapotranspiration, thus reducing available water supplies even more. For already drought-prone regions such as the south Okanagan, reduced water supplies will bring many challenges.

One way to reduce the risk of flooding is changing the seasonal pattern of discharge rates, which can be accomplished by building a dam. A dam impounds water, creating a reservoir of stored water and thus creating the ability to reduce the run-off at peak times and increase the flow at low water times. Despite these benefits, there are disadvantages to damming and impounding water.

Evapotranspiration ratios, with all their lines and labels, appear rather complex at first glance, although each line in itself is fairly straightforward (Figure 13.3). The precipitation line for a coastal region such as Vancouver is relatively high in winter and much lower in summer (Figure 13.3a). Conversely, the line of potential evapotranspiration, which graphs the theoretical relationship between the growing season, incoming solar radiation, and the ability for moisture to be given off to the atmosphere, has the reverse curve (Figure 13.3b). This line shows the volume of water that could be evaporated and transpired if water were available. The centre of Figure 13.3c shows, below the line of potential evapotranspiration, the line of actual evapotranspiration. Typically this is a considerable amount in the late spring and early summer, but by August there is very little moisture left and the gap between actual and potential evapotranspiration widens. By combining the graphs of annual precipitation, potential evapotranspiration, and actual evaporation, the periods of water surplus and deficit can be assessed (Figure 13.3c).

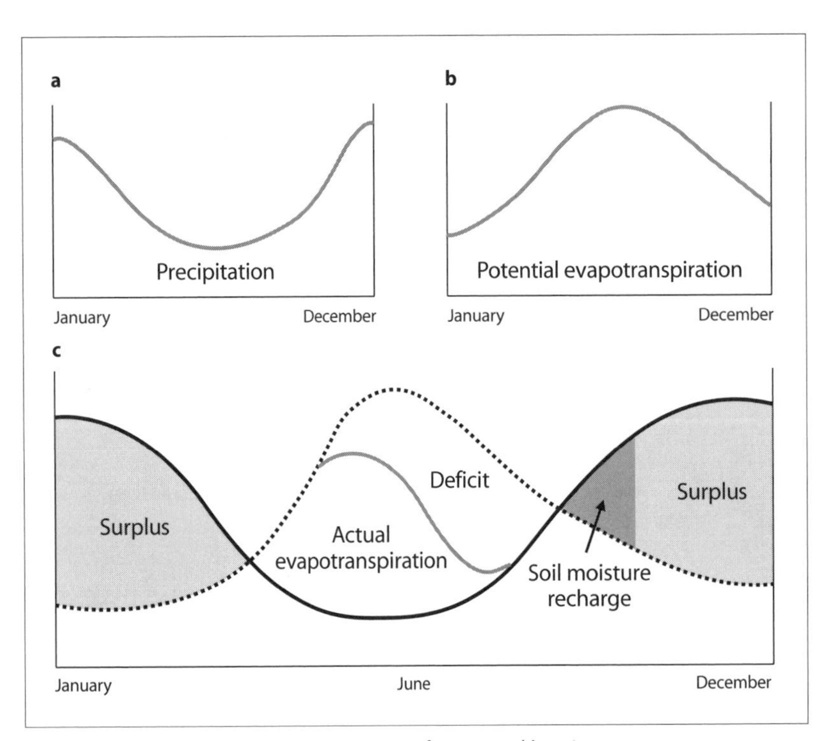

Figure 13.3 Theoretical evapotranspiration ratio for a coastal location

Following the hypothetical model of Figure 13.3c for each month of the year, it is evident that the January/February rainy season has a surplus of water, and one can infer the need for storm drains at that time. As precipitation decreases, plants grow, and as solar energy increases, greater evaporation and transpiration occur. By June, the potential for evapotranspiration outstrips available moisture and a deficit occurs. This deficit increases throughout the summer, requiring the irrigation of farms and the watering of gardens. By autumn, the leaves are falling, evapotranspiration has subsided with the decrease in incoming solar radiation, and precipitation is increasing. Soil moisture recharge is the term to describe the state of the soil at this time. Using the analogy of the soil as a sponge, the summer months cause the sponge to dry out, whereas the increase in precipitation and decrease in evapotranspiration in the fall allow the sponge to absorb, or recharge, with moisture. By late fall, the soil is saturated and any further precipitation cannot be absorbed; it is surplus.

The evapotranspiration ratios for Pemberton, Kamloops, Vancouver, and North Vancouver illustrate very different water budgets and concerns about water supply (Figure 13.4). These physical characteristics are important to keep in mind when considering how water is used as a resource.

WATER AS A RESOURCE: USING IT AND ABUSING IT

Understanding the quality, quantity, and timing of flows for any water system is crucial. Water quality has to be assessed in terms of the many uses of the resource, whether for fish habitat, swimming, industrial purposes, or drinking. "The Guidelines for Canadian Drinking Water Quality" (Federal-Provincial-Territorial Committee on Health and the Environment 2010) establish a standard for potable water, or water that is safe to drink. It should be noted that these standards can change, usually because of new information about toxicity levels of various elements in the water system. Arsenic, for example, often from groundwater supplies, was once believed to be safe at 0.05 milligrams per litre, but that figure was reduced to half (0.025 milligrams per litre) in 2006 and, as of 2010, it is 0.010 milligrams per litre.

Quantity is also important for any water source. Having sufficient water for all users during the year may be difficult, particularly when the source is limited. If a water supply is required for domestic uses including fire fighting, for example, as well as for industry, irrigation, and fish habitat, some uses may have to be curbed. Some regions of British Columbia are growing rapidly, and an increased population means more demand for water. The question of quantity may limit residential and industrial growth.

The third factor, timing of flows, recognizes that river and stream systems have variable discharge rates. Unfortunately, water use also tends to vary seasonally, and the highest use periods are opposite to the highest supply periods. A typical example is the need in most communities to put watering restrictions in place for gardens and lawns during the summer months. This is precisely the period when most stream systems, as Figure 13.2 shows, are at their lowest discharge rate. Human activities such as logging in watersheds can also affect the timing of flows. Clear-cut logging practices may result in much faster run-off, with even less water being available during the dry season. By impounding water and releasing it in the low water season, dams also change the timing of flows.

The combination of these factors has already reached a critical level in other regions of Canada (e.g., Alberta) and the world (e.g., Australia). Chris Wood (2008) documents the problem of North American water shortages in *Dry Spring*. The Bow River watershed (a sub-basin of the Saskatchewan River), which runs through Calgary, Alberta, may serve as an example for managing British Columbia's dry regions. In the Bow River Basin, the fundamental issue is the allocation of water. Water licences have been the means to secure a supply of water for farmers, municipalities, and so on. Water licences, though, have been tied to land uses, based on seniority (i.e., first come, first served), and largely taken for granted. The Bow River Basin has reached capacity in terms of demand, and there is great concern that global warming will reduce the availability of water in the future. As a consequence, the Bow is now a closed basin; no more water licences will be issued. The Alberta government has also separated water licences from land uses and allows them (or at least a portion) to be sold. New developments that will require water (e.g., shopping centres, golf courses, and housing developments) must now purchase water from an existing licence holder.

This predicament of having insufficient water in a river system for present and future users raises others questions

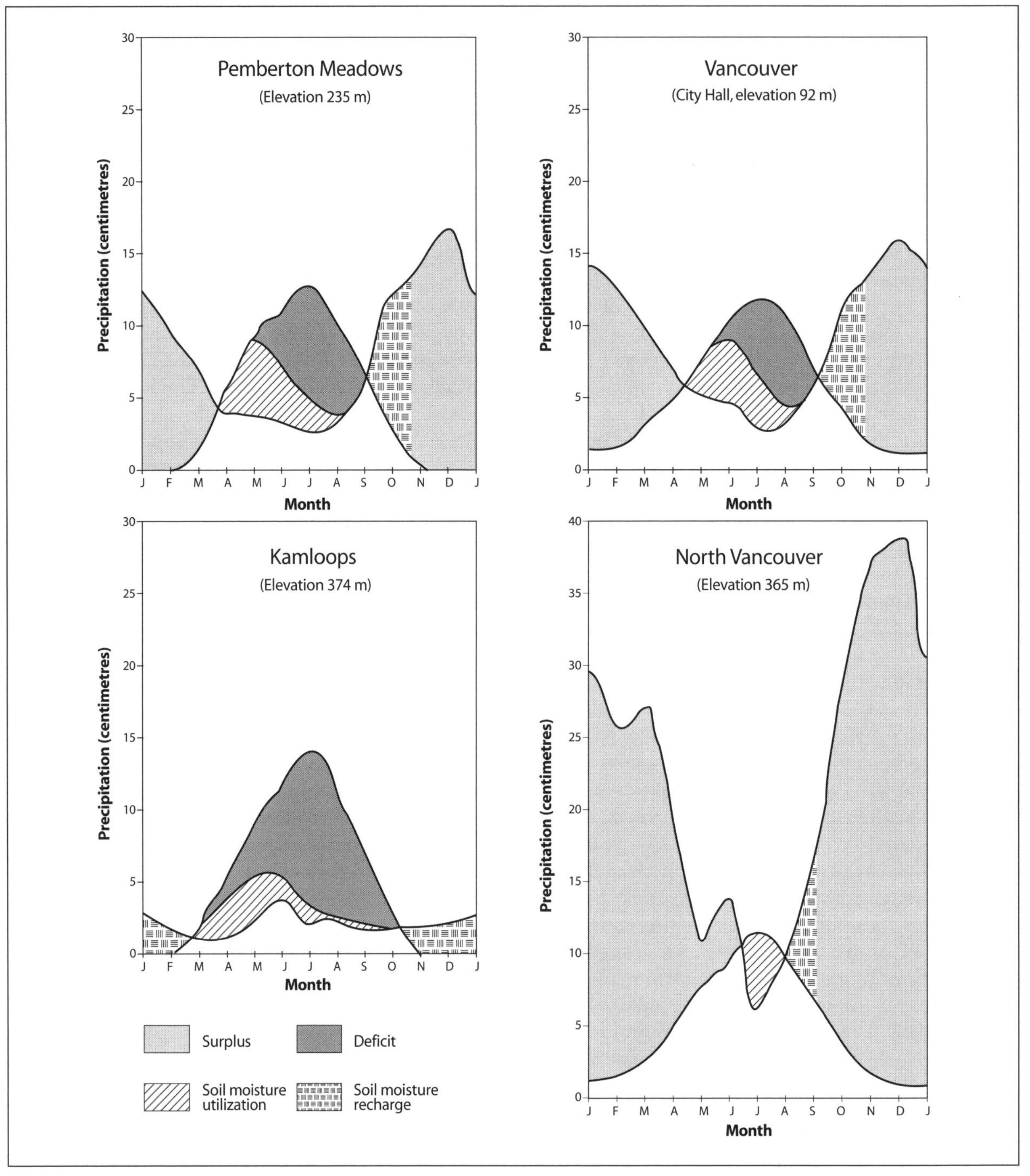

Figure 13.4 Evapotranspiration ratios for selected communities
Source: Canada, Department of Environment (1972).

Table 13.2

Categories of water use

Withdrawal	Consumed	In situ	In stream	Inadvertent
Water is withdrawn, altered, and returned. e.g., pulpmills	Water is withdrawn and not put back directly e.g., agriculture	Water is used in its natural state e.g., recreation	Water is used while remaining in river systems e.g., hydro dams	Water quality is affected negatively and accidentally e.g., leachate

Sources: Foster and Sewell (1981); Sewell (1987).

such as, can the supply of water to a river system be increased? As discussed above, dams and reservoirs can change the timing of flows, but these structures have environmental consequences when reservoirs flood valleys and change the discharge rate. Are diversions from other watersheds possible? Again, these actions have environmental implications and, in some instances, they are potentially catastrophic. In the case of Alberta's Bow River, this option does not even exist. Other concerns include the health of the river system itself. How much water must be retained in the watershed to ensure a viable ecosystem of aquatic plants and animals? On the human side, how does a government address the historical exclusion of First Nations from acquiring water licences?

Table 13.2 categorizes the major uses of water, all of which affect water quality, quantity, and timing of flows. The withdrawal of water is increasing to meet the multiple needs of a province with a growing population. Domestic, commercial, and industrial uses, ranging from car washes to the enormous quantities of water required by pulp mills, require prodigious daily withdrawals of water. This water is used and then returned in an altered state. In some cases, effluents have been added and, in others, the water temperature has been raised.

There are many uses for which water is not only withdrawn but also consumed. The largest consuming industry of water is agriculture. Irrigation removes water from surface or groundwater sources and does not return it to the same source. Of course, it still remains part of the hydrologic cycle, as this water then evaporates and transpires into the atmosphere. The resource is also consumed as potable water for drinking, in commercial products such as beverages, and in many industrial processes. For some municipalities and rural areas, limited potable water can restrict population growth.

In situ uses of water include all the values we have for water systems in their natural state. Recreational activities such as swimming, boating, and whitewater rafting, to name a few, are important to local populations and to the tourism industry. The river systems leading to the Pacific are essential to the salmon fishing industry because freshwater lakes and streams provide the habitat for the beginning and the end of the salmon life cycle. A number of freshwater species of fish and animals, which are important to the recreation and tourism industry, require water in its natural, unaltered state.

The major in-stream use of water in British Columbia is the generation of hydroelectricity. The many rivers and streams permit a great deal of production, although most of our present hydroelectricity comes from dams on the Peace and Columbia River systems. Once a dam is built, the environment is considerably changed. The discharge rate of the river is modified below the dam, reducing high peak flows of the spring and increasing low discharge volumes of summer and fall. This may dry up downstream wetlands and deltas and thus affect the habitat of fish and wildlife. The W.A.C. Bennett Dam had just this impact on the Peace-Athabasca delta. Impounding water behind a dam creates a reservoir, in some cases so large that it affects the climate of the region. Williston Lake, the result of the W.A.C. Bennett Dam, is the largest lake in British Columbia and modifies the climate of the surrounding area. Furthermore, dams and lakes/reservoirs can conflict with other resource uses such as fishing, forestry, hunting, trapping, and farming.

Dams do not change the overall amount of water discharged in a river system unless one river system is diverted into another. The Kenny Dam on the Nechako River, as discussed earlier, diverts the flow of water from the Fraser River system and redirects it through a tunnel in the Coast Mountains to the Kemano hydroelectric plant, which supplies the aluminum plant at Kitimat (see Figure 11.12, p. 202). This dam, referred to as the Kemano 1 project, was not built without controversy. It reduced salmon stocks on the Nechako, displaced people (the Cheslatta Band, particularly) without adequate compensation, and reduced the amount of water discharged to the rest of the Fraser, again affecting salmon habitat.

Other water diversion schemes have been proposed, such as the diversion of the McGregor River into the Parsnip River/Williston Lake/Peace River system and of the North Thompson River into the Columbia River system. Both these plans would increase the volume of water available to existing hydroelectric generators and thereby increase the production of electricity at a more economical rate. Again, these plans raise some serious environmental concerns.

By far the most ambitious and controversial proposal has been the North American Water and Power Alliance (NAWAPA), wherein diversions were proposed from Alaska, the Yukon, and British Columbia to quench the thirst of California, the American southwest, and Mexico (Figure 13.5). This massive scheme would have required the construction of 240 reservoirs, 112 irrigation systems, and 17 navigation channels (Foster and Sewell 1981, 31). The economic, physical, and environmental impacts would have been enormous for this megaproject of the early 1960s. This grand plan was put on the shelf, but smaller diversion projects involving part of the NAWAPA plan were carried out. Most notable was the Columbia River Treaty (1961), by which a number of dams were constructed on the BC side of the border to provide Americans with hydroelectric energy and flood control.

More recently, the Free Trade Agreement of 1989 and the North American Free Trade Agreement of 1994 raised concerns over the export of fresh water and further water diversion schemes. The video recording *Captured Rain: American Thirst, Canadian Water* (Raincoast Storylines 2000) documents an American company's lawsuit against the Canadian government over its refusal to allow the export of water from British Columbia. The argument from the American side is that water, under the North American Free Trade Agreement, is a tradable commodity like oil or gas and not subject to the political decisions of provincial or federal governments. No decision has occurred, but the implications challenge the essence of sovereignty.

Figure 13.5 North American Water and Power Alliance Plan (NAWAPA)
Source: Modified from Foster and Sewell (1981, 33).

The last category shown in Table 13.2, inadvertent, includes all the changes to water systems that alter quality. Historically, river systems were a powerful influence for settlement because they provided a ready source of drinking water, transportation, irrigation, and waste disposal. As communities grew and their agricultural and industrial base expanded, liquid wastes, solid wastes, industrial effluent, toxic wastes, farming chemicals, and so forth were increasingly discharged into the ground- and surface water systems. Road building, clear-cut forestry practices, mining activities, overgrazing, and other activities within the watersheds have also had a negative impact on water systems, often changing the timing of flows.

The most damaging effect of all has been an increase in the amount of sediment in stream systems. Sediment can be devastating to fish habitat, as it settles on the stream bottom and smothers the fish eggs there. Sediment is also the main cause of turbidity, one of the major problems of potable water. Turbidity is the result of the fine sediments in water systems that are almost impossible to filter. Turbid water then flows through the water mains of communities, settling in and coating the pipes. This environment can then become home for undesirable bacterial microorganisms. To combat these bacteria, most communities add chlorine to their water systems. Ever increasing bottled water sales are evidence of water quality concerns.

Surface water systems have gone through an inordinate number of changes as a result of combined human and natural activity, reducing the options for water use along those systems from source to mouth. Groundwater is somewhat more protected, but even here leachates from agricultural activities – fertilizers especially – have rendered some supplies unusable without prior boiling. The level of nitrates is known to exceed the Canadian drinking water standard (10 milligram/litre) in areas such as Langley/Brookswood, Hopington, Abbotsford/Sumas, Osoyoos, Spallumcheen, and Grand Forks, a state of affairs normally caused by inadequate or badly managed septic systems, animal feedlots, manure storage facilities, and overuse of nitrogen-rich fertilizers (BC Ministry of Health 2000).

Sewage, liquid wastes that are flushed into our water systems, is another controversial issue for most communities. Processes to purify sewage include primary, secondary, and tertiary treatment. Each stage removes greater amounts of materials and chemicals before the processed waste reenters the water system, but each is costly to implement. Few communities opt for tertiary treatment. Several reports have been issued on the state of sewage discharge in major Canadian cities, outlining specific concerns. The first *National Sewage Report Card* (Bonner and McDade 1994) included Victoria (the Capital Regional District) and Vancouver in its study. Victoria received a failing grade because the city has almost no sewage treatment prior to discharge into the Juan de Fuca Strait; Vancouver fared little better, with a D minus. Victoria was finally ordered by the provincial government in 2006 to build a treatment plant. As of 2010, the plant is still at the report stage, while some 129 million litres of raw sewage is pumped into the Strait each day (*Times Colonist* 2010). The 2010 Olympics was a catalyst for turning some of Vancouver's sewage into energy recovery. A treatment plant (which cost $30 million) built at False Creek passes the sewage through heat pumps, which, in turn, heat homes and water for approximately 16,000 residents (Bellett 2010). The sewage is then sent on to the treatment plant at Iona. Herein lies the real problem of sewage for the Vancouver area: the two main treatment plants (Iona in Richmond and Lion's Gate in West Vancouver) are only primary treatment plants. They "dump the equivalent of 780 million litres (or 312 Olympic size swimming pools) of sewage effluent into the Strait of Georgia every day" (Sauve 2008). The third treatment facility (at Annacis Island in Delta) is no better. There are plans to upgrade the plants by 2020.

Figure 13.6 lists communities that achieved secondary or better wastewater treatment by 2004. Environment Canada stated that 65.4 percent of the population had secondary or better treatment in 2006 (Environment Canada 2010). Still, the above discussion on minimal sewage treatment for the two major population regions in the province (Metro Vancouver and the Capital Regional District) recognizes that it will take considerable time before improvements to waste water occur.

Figure 13.7 gives an ecosystem perspective of water use and abuse. Each community and resource activity affects the next in line downstream, often resulting in huge costs for filtration and purification. The concept of **biomagnification** suggests that very small amounts of toxic waste

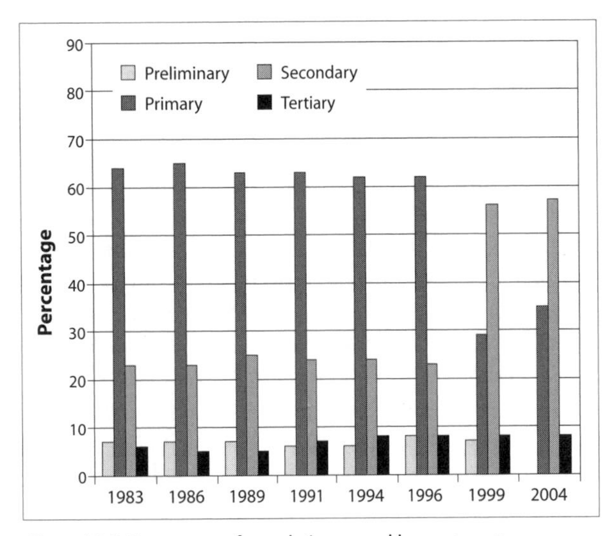

Figure 13.6 Percentage of population served by wastewater treatments, 1983-2004
Source: BC Ministry of Environment (2007).

are ingested at the lowest level of the food chain; as the next level up consumes the lowest level organisms, the toxicity builds. With people at the top of the food chain, we must take great care over the food we eat. One can also question whether the Canadian Drinking Water Quality Standards should tolerate any level of toxic chemicals in our drinking water.

FOCUS ON POTABLE WATER

As stated above, most of the drinking water in British Columbia comes from surface sources – rivers, lakes, and streams – that are abundant throughout the province, although some 1 million residents get their water from wells that tap into groundwater aquifers. However, these water sources, both surface and groundwater, are within watersheds that include many activities detrimental to water quality. The words of Marq de Villiers (2000, 29) should be kept in mind: "Water can be polluted, abused, and mistreated, but it is neither created nor destroyed, it only migrates." The focus here is the activities that affect water quality and quantity and the political policies in place to manage this critical resource.

The multibarrier approach to safe drinking water has been adopted by all levels of government in Canada. It has five components:

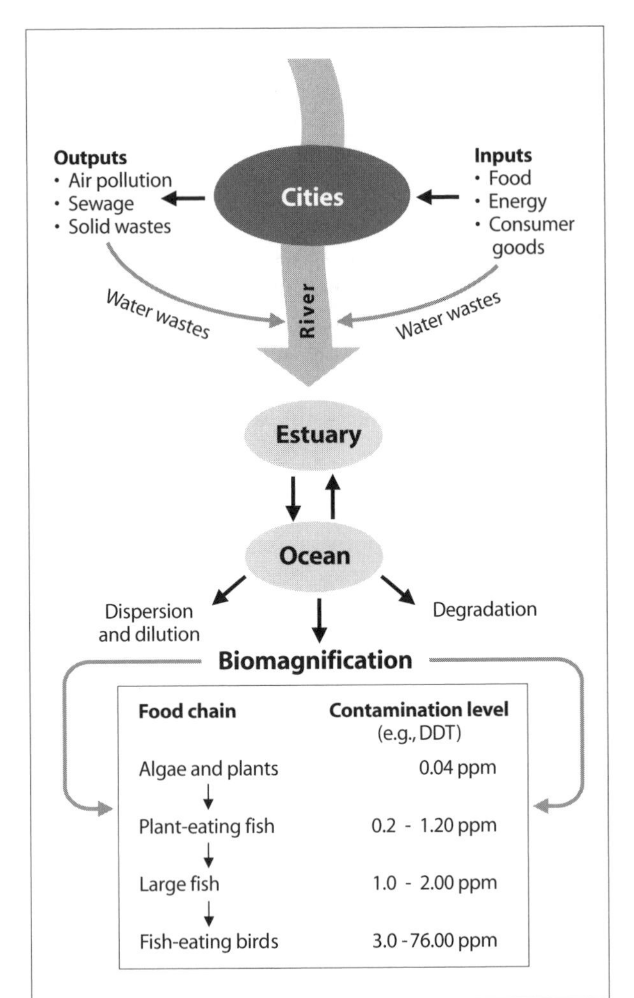

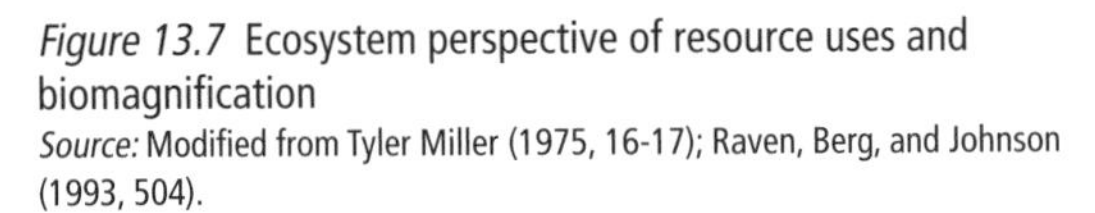

Figure 13.7 Ecosystem perspective of resource uses and biomagnification
Source: Modified from Tyler Miller (1975, 16-17); Raven, Berg, and Johnson (1993, 504).

1 Source water protection: whether surface or groundwater sources, watersheds must be protected against contamination.
2 Treatment: there are a number of options (and costs) to ensure safe drinking water (e.g., chlorination, filtration, ultraviolet light).
3 Distribution: some waterlines extend hundreds of kilometres and therefore require insystem treatment such as maintaining chlorination levels.

4 Monitoring: there is a need for constant and vigilant monitoring of water quality by competent managers.
5 Response plan: accidents do occur either by human error or through natural disasters, and it is critical to have an emergency plan in place to respond quickly to the conditions.

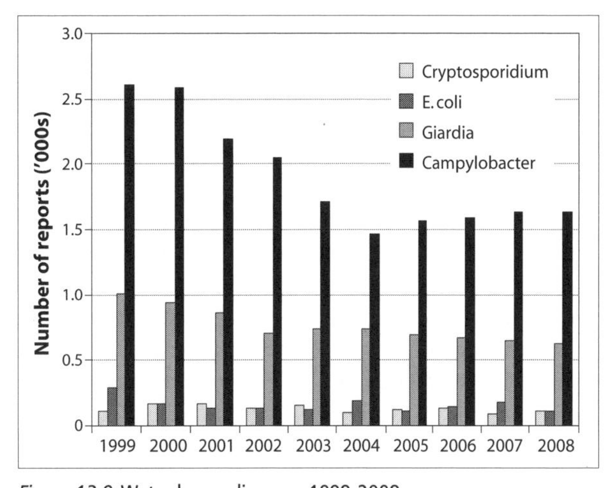

Figure 13.8 Water-borne diseases, 1999-2008
Source: BC Centre for Disease Control (2009).

Managing Surface Sources

One of the challenges in managing surface water in the province is assessing the range of acidity (low pH) to alkalinity (high pH). The Canadian Drinking Water Standard allows a pH range between 6.5 and 8.5. Those communities on the high end of the scale (mainly in the interior) experience "hard" water, which does not lather easily and coats kettles, dishwashers, and pots with deposits of calcium and magnesium. Coastal water is often described as "soft," as this lower pH water lathers easily, but its acidity dissolves not only minerals but also copper pipes. The Vancouver water supply, for example, was considerably below 6.5 until 1999 when measures were taken to raise the pH to 6.7. There are costs related to the pH of water, but of greater concern to most citizens are the incidence of diseases carried by water.

There have been many cases of contamination by disease-causing organisms in British Columbia. Figure 13.8 highlights a number of common water-borne diseases that persist in the province. A report by the Sierra Legal Defence Fund (Christensen and Parfitt 2003) linked many of these diseases to activities within watersheds such as "logging, mining activities, road building, cattle grazing, and garbage dumping. Such developments can increase sediment loading in creeks – always a risk to water supplies because sediment can mask bacterial, viral or parasitic agents" (p. 9). Boil-water advisories issued by the provincial government are another signal that water quality has deteriorated. There were 540 boil-water advisories in 2008 alone in this province (CBC 2008). With respect to these advisories, Bob Patrick and Reid Kreutzwiser (2003, 2) suggest, however, that "this trend reflects, in part, an increased emphasis on water quality testing and reporting after the BC Safe Drinking Water Regulations took effect in October 1992, as opposed to a sudden deterioration in water quality."

Source water protection is neither simple nor inexpensive. The water supply of about two-thirds of BC's population comes from watersheds under the local control (privately owned) of the Greater Vancouver Regional District (GVRD) and the Capital Regional District (CRD). This is a positive step in the direction of source water protection, for it means restrictions to public access and control over activities within these watersheds. The rest of the province is supplied by "surface water sources in over 700 community watersheds" (Patrick 2004, 6), and these watersheds are within provincial jurisdiction, mainly Crown forestland. In other words, communities have rights to the water, but not to the watershed. The difficulty for the provincial government is weighing all the interests in a watershed. For example, putting the brakes on mining, forestry, and other land uses or putting a stop to development in or near community water sources often requires compensation. Since the provincial government has encouraged these activities, it is reluctant to reverse its policies. The alternative (to source water protection) is to focus on systems to clean, purify, and produce potable water (treatment), but "relying solely on water treatment technology is not sufficient to ensure public health" (Patrick and Kreutzwiser 2003, 4). Moreover, if the activities leading to contaminated water continue, water treatment will become a constant local government cost.

Kelowna is a case in point. In 1996 the community faced a serious outbreak of cryptosporidium, likely brought about by abundant animal manure from farms surrounding Okanagan Lake, a main source of Kelowna's water. The city initially spent approximately $300,000 to flush out its entire water system with heavily chlorinated water, but the by-products of chlorination include chloroform, which is believed to cause cancer (Hogue 2001). Kelowna then spent some $44 million on treatment improvements. Local governments view this approach as treating the symptom, not the cause and, worse, local taxpayers bear the costs and responsibility for providing safe drinking water. On a provincial scale, the BC Auditor General's 1998-99 report states: "For the approximately 100 municipalities outside Victoria and Vancouver that use unfiltered surface water, we estimate the capital cost of installing filtration would be about $700 million and the extra cost of financing, operating and maintaining the new treatment plants would be about $30 million a year."

The provincial government in 2002 passed the Drinking Water Protection Act, and by May 2003 the Drinking Water Protection Regulations were implemented. There were, however, considerable criticisms of these measures. As Bob Patrick (2004, 7) pointed out: "The word 'watershed' does not appear in the definition section of the Act," and, while "drinking water source" is defined, "this definition does not recognize the watershed as integral to a source of water supply." The Sierra Legal Defence Fund was even more critical, giving the Act a failing grade for source water protection because it does not deal with activities in the watershed: "logging, cattle-grazing, mining and other land uses with the potential to harm drinking water and human health will not be subject to scrutiny" (Christensen and Parfitt 2003, 18). Furthermore, it criticized the list of potentially harmful waterborne contaminants because it does not inlude giardia and cryptosporidium, even though these parasites have caused many outbreaks of waterborne disease in British Columbia (p. 14). More recently, the province has published new water policies in a report titled "Living Water Smart" (BC Ministry of Environment 2008). The report is fairly comprehensive, although most policies will not be implemented until 2012 or later. For example, some groundwater sources will have regulations by 2012, water use in British Columbia will be 33 percent more efficient by 2020, and the quality of drinking water in all Aboriginal communities will meet the same provincial standards applied across British Columbia by 2015.

Many have offered criticism and advice regarding the proposed changes, including the fact that "the Plan omits critical details about implementation, time lines, and funding for policy and legal changes. Also, the contentious issue of water governance is not resolved, so outstanding questions of the appropriate division of responsibilities for water management remain" (Nowlan 2008). Most communities (over 4,500 water systems in the province), along with environmental groups, are particularly interested in policies that ensure source water protection.

Managing Groundwater Sources

Groundwater, like surface water, can become contaminated. Naturally occurring minerals such as arsenic in bedrock can contaminate well water. The communities of 100 Mile House, Bowen Island, Burns Lake, Chase, Kamloops, Quesnel, the Sunshine Coast, Vanderhoof, Vernon, and Williams Lake have wells in their vicinities where arsenic levels exceed the drinking water guidelines. Because wells and the aquifers they draw from are part of a watershed, human activities here can also contaminate drinking water. Walkerton, Ontario, is the most tragic Canadian example to date. In May 2000, "After heavy rainfall washed bacteria from cattle manure into the town well, which for years was known to be vulnerable to contamination, residents began to experience bloody diarrhea, vomiting, cramps, and fever – all symptoms of E. coli" (Shin 2010). The disaster resulted in seven deaths and 2,300 cases of illness from E. coli and campylobacter poisoning. Figure 13.9 illustrates a well's relationship to the watershed, drawing attention to the need for wellhead protection to prevent a Walkerton mishap.

The BC government introduced new groundwater regulations in July 2004 "establishing standards to ensure wells are properly drilled, sealed, maintained and closed. Included in the regulations are qualification requirements for well drillers and well pump installers and the creation of a provincial registry for those qualified persons. The regulations also require wells to be floodproofed so runoff contamination cannot occur during

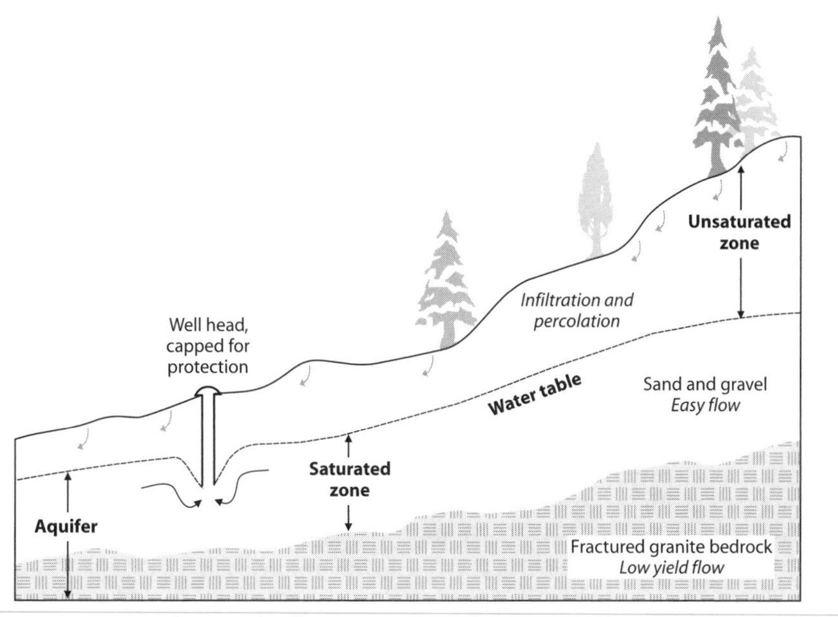

Figure 13.9 Groundwater and wells

flooding or heavy rains" (BC Ministry of Water, Land and Air Protection 2010). These are important regulations that address wellhead concerns, but nothing in the regulations deals with users of the watershed who may contaminate the groundwater.

The provincial government has been assessing groundwater conditions and, out of the 815 aquifers mapped, "64 are rated as heavily developed" and "28 are rated as highly vulnerable to contamination and 53 have documented quality concerns, such as salty water, nitrates, and arsenic" (BC Ministry of Environment 2007). The new water Act mentioned above will regulate groundwater use in priority areas and large groundwater withdrawals by 2012. The legislation has been criticized on the grounds that all groundwater use (not just priority areas and large groundwater sources) should require a licence and be included in the regulations (Nowlan 2008). Ecojustice's (2009) advice to the province is to "treat water as one interconnected resource by requiring water management plans to evaluate both groundwater and surface water systems and the linkages between them." Again, the most important means of safeguarding water, whether from a well or the surface, is to incorporate source water protection for watersheds.

Managing Water: Other Considerations

Whether water comes from a surface supply or a well, it is essential to view water within the context of the entire watershed. As the Sierra Legal Defence Fund and others argue, the cost of source water protection – managing the watershed to prevent contamination – is lower than the costs associated with water treatment technologies (Christensen and Parfitt, 2003; Patrick and Kreutzwiser 2003). For example, the estimated cost of the Walkerton disaster is $155 million (Shin 2010); source water protection measures would have cost a fraction of this amount. New York City realized this: "Faced with $8 billion in capital cost and $300 million in annual operating and maintenance cost for high-tech water treatment, New York City opted to invest in source water protection. Thus far, the source protection option has cost New Yorkers less than $1 billion, and has contributed to greater awareness of water quality issues between all stakeholders" (Patrick and Kreutzwiser 2003, 5).

The Walkerton catastrophe pointed out other problems; namely, inadequate training of water personnel led to substandard monitoring and failure to inform the public. British Columbia's Drinking Water Protection Act (2002) and Drinking Water Protection Regulations (2003) include the necessity for water operators to be certified and for the public to be kept informed about water quality. The new water Act will ensure a report on the state of our water by 2012 and every five years thereafter.

Water quality is of major importance, but quantity is also a concern for many communities. Canada is one of the highest consumers of water on a per capita basis (second only to the United States), and in regions such as the Okanagan-Similkameen – where the population increased from approximately 25,000 in 1971 to over

83,000 in 2010 and is predicted to be 92,000 by 2036 (BC Stats 2010) – water use is under close scrutiny (Statistics Canada 2003). The challenge is considerable, considering that "Okanagan residents use 675 litres of water per person, per day – year round, on their residential properties. This is more than twice the Canadian average (329 litres), and much higher than that of other countries. Yet, the Okanagan has the lowest per person water availability in Canada" (Okanagan Basin Water Board 2010). The promise in the new water Act is that 50 percent of new municipal water needs will be acquired through conservation by 2020.

Awareness of water use and new attitudes are essential. Some communities have implemented water meters as a means of forcing high water users to reduce their consumption or pay more; other communities have encouraged low flush toilets by offering a cash incentive to phase out older, high water consumption toilets. Communities such as Kamloops have also encouraged xeriscaping – planting native vegetation that requires little or no watering. Sun Rivers, a private subdivision on Kamloops Indian Band land, has a unique private water system wherein each house has two water lines. One delivers potable water from a high-tech water treatment system to sinks and showers, while the second line delivers untreated water for toilets and outside taps used to water lawns and gardens. All of these initiatives are important, particularly as the need for water increases with population growth. An even greater challenge may be presented by global warming, as water-short communities become desperate for water.

SUMMARY

The historical view of water resources regarded them as inexhaustible, as having little or no value in economic decisions, and often as a means of disposing of wastes. Our overuse and abuse of water has brought home the reality that it not only has a value but is essential to our economic and physical well-being. With water's many uses for many interest groups, this mobile, renewable resource is not easy to manage.

The federal government has the responsibility for oceans and for any changes to freshwater systems crossing international boundaries. It also has interests and obligations with respect to water systems: flood control, salmon habitat, hatcheries, Aboriginal issues, and so forth. The provincial government manages many aspects of the freshwater systems as well and is responsible for regulating withdrawals, consumption, and quality of water, mainly through the Ministry of Environment and the Ministry of Health. Ministry responsibilities overlap with those of other managers, such as the Ministry of Forests and Range, and local governments as purveyors of water for their communities. Water issues are complex, because they involve the quantity, quality, and timing of flows of surface water and quantity and quality of groundwater and because both surface and groundwater are components of watersheds. Growing populations, particularly in water-short regions of the province, and global warming complicate the issue further.

The Walkerton, Ontario, tragedy has caused all governments in Canada to examine the supply of water. Within British Columbia, the Drinking Water Protection Act was passed in 2002, with Drinking Water Protection Regulations in 2003 and Groundwater Protection Regulations in 2004. These were important steps in recognizing the importance of drinking water, setting out directions for wellhead protection as well as for certification of water system operators and informing the public on water quality. The Act and regulations, however, failed to take into account the activities in watersheds that cause contamination. A new water Act promises improvement to source protection, limits on water licences of forty years, and emphasis on water conservation. What is required is a Quality of Water Act that will give priority to the municipal/regional districts' interests in water quantity, quality, and timing of flows and will restrict forestry, cattle grazing, mining, and other watershed activities from affecting this fundamental resource. As population and the uses for water increase, the need for planning and action becomes more urgent.

REFERENCES

Bonner, M., and G. McDade. 1994. *The National Sewage Report Card*. Vancouver: Sierra Legal Defence Fund.

Canada, Department of Environment. 1972. *Canada Water Yearbook*. Ottawa: The Ministry.

Canada, Ministry of National Health and Welfare. 1993. Guidelines for Canadian Drinking Water Quality. 5th ed. Ottawa: The Ministry.

Christensen, R., and B. Parfitt. 2003. *Watered Down: A Report on Waterborne Disease Outbreaks in British Columbia 1980-2002.* Vancouver: Sierra Legal Defence Fund.

de Villiers, M. 2000. *Water.* Toronto: Stoddart.

Farley, A.L. 1979. *Atlas of British Columbia: People, Environment, and Resource Use.* Vancouver: UBC Press.

Foster, H.D., and W.R.D. Sewell. 1981. *Water: The Emerging Crisis in Canada.* Toronto: James Lorimer.

Patrick, R. 2004. "Source Water Protection in British Columbia: A Perspective from Political Ecology." Paper presented at the annual meeting of Canadian Association of Geographers, Moncton, NB.

Patrick, R., and R. Kreutzwiser. 2003. "Who's Guarding the Well? The Status of Drinking Water Source Protection in British Columbia." Paper presented at the annual meeting of Canadian Association of Geographers, Victoria, BC.

Raven, P.H., L.H. Berg, and G.B. Johnson. 1993. *Environment.* Orlando, FL: Saunders College Publishing.

Sewell, W.R.D. 1987. "Water Resources." In *British Columbia: Its Resources and People,* ed. C.N. Forward, 198-225. Western Geographical Series vol. 22. Victoria: University of Victoria.

Tyler Miller Jr., G. 1975. *Living in the Environment: Concepts, Problems, and Alternatives.* Belmont, CA: Wadsworth.

Wood, C. 2008. *Dry Spring: The Coming Water Crisis for North America.* Vancouver: Raincoast.

FILMS

Raincoast Storylines. 2000. *Captured Rain: American Thirst, Canadian Water.* Director: Jerry Thompson. Co-Producers: Bette Thompson and Terence McKeown.

INTERNET

BC Auditor General. 1999. "Protecting Drinking-Water Sources." Report 5. June. www.bcauditor.com/pubs/subject/environment?page=1.

BC Centre for Disease Control. 2009. *2008 British Columbia Annual Summary of Reportable Diseases.* www.bccdc.ca/NR/rdonlyres/59BFCFBB-933D-4337-9305-E3E5FF30D272/0/EPI_Report_CDAnnual2008_20091202.pdf.

BC Ministry of Environment. 2007. "Environmental Trends in British Columbia: 2007." www.env.gov.bc.ca/soe/indicators/01_population_economic/technical_paper/population_economic_activity.pdf.

–. 2008. "Living Water Smart." www.livingwatersmart.ca/.

BC Ministry of Health. 2000. "HealthLink BC File No. 05, February 2000: Nitrate Contamination in Well Water." www.healthlinkbc.ca/healthfiles/hfile05.stm.

–. 2010. "BC's Ground Water Protection Regulation." www.env.gov.bc.ca/wsd/plan_protect_sustain/groundwater/gw_regulation/GWPR_private_well_owners.pdf.

BC Stats. 2010. "Population Projections." www.bcstats.gov.bc.ca/data/pop/pop/dynamic/PopulationStatistics/.

Bellett, G. 2010. "Vancouver Sewage-to-Heat Neighbourhood Energy Centre Goes Live." *Vancouver Sun.* www.vancouversun.com/business/Vancouver+sewage+heat+neighbourhood+energy+centre+goes+live/2442089/story.html.

CBC. 2008. "1,775 Boil-Water Advisories in Canada Require Action: Report. www.cbc.ca/health/story/2008/04/07/boil-advisory.html#ixzz0lHkPjQ4D.

Douglas, T. 2008. "Groundwater in British Columbia: Management for Fish and People." *Streamline Watershed Management Bulletin* 11, 2. www.forrex.org/streamline/ISS37/streamline_vol11_no2_art4.pdf.

Ecojustice. 2009. "Backgrounder: Statement of Expectations on Reform of the BC Water Act from BC Nongovernmental Organizations." www.ecojustice.ca/media-centre/media-backgrounder/backgrounder-statement-of-expectations-on-reform-of-the-bc-water-act-from-bc-nongovernmental-organizations.

Environment Canada. 2010 "Municipal Water and Wastewater Survey." www.ec.gc.ca/eau-water/default.asp?lang=En&n=ED7C2D33-1.

Federal-Provincial-Territorial Committee on Health and the Environment. 2010. "Guidelines for Canadian Drinking Water Quality." 6th ed. www.hc-sc.gc.ca/ewh-semt/pubs/water-eau/sum_guide-res_recom/intro-eng.php.

Hogue, E. 2001. "Chloroform and Canser." pubs.acs.org/cen/topstory/7944/7944notw4.html.

Nowlan, L. 2008. *Smarter Water Laws: The Key to Living Water Smart – BC's Newest Water Plan.* www.watergovernance.ca/PDF/PoWGSmarterWaterLaws.pdf.

Okanagan Basin Water Board. 2010. "Okanagan Water Supply and Demand Project." www.obwb.ca/fileadmin/docs/100326_faq.pdf.

Sauve, K. 2008. "Sewage Treatment a Mess for New Mayor. *Thethunderbird.com.* thethunderbird.ca/2008/11/13/sewage-treatment-a-mess-for-new-mayor/.

Shin, M. 2010. "Thirsty for Answers: The Tenth Anniversary of the Walkerton Water Tragedy Is in May: Has Canada Learned Its Lesson?" www.insidethebottle.org/thirsty-answers-tenth-anniversary-walkerton-water-tragedy-may-has-canada-learned-its-lesson.

Statistics Canada. 2003. "Human Activity and the Environment: Annual Statistics." *The Daily*, 3 December. www.statcan.ca/Daily/English/031203/d031203a.htm.

Times Colonist. 2010. "Every Day Victoria Dumps 129 Million Litres of Raw Sewage into the Juan de Fuca Strait." www2.canada.com/victoriatimescolonist/features/sewage/index.html.

14

Tourism

A New and Dynamic Industry

Tourism throughout the world has escalated at a remarkable pace over the past fifty years and brought with it a great deal of revenue and opportunity (Table 14.1). In fact, by the year 2000, tourism was considered the leading employer and leading industry in the world (Nickerson and Kerr 2001). In 2008 it was estimated that travel and tourism generated US$974 billion in revenues per year and was responsible for over 79 million jobs globally (World Tourism Organization 2010). Yet there was a 4 percent decline in tourism in 2009 because of concerns over the H1N1 flu pandemic and the global recession. In addition, volatile oil prices have increased the price of transportation and security threats have resulted in border hassles that discourage many.

Looking back, this province took little notice of the economic importance of tourism until a major recession hit Canada (and most of the world) from 1981 to 1986. Although the recession seriously affected traditional resource-based industries, tourism not only maintained its share of economic activity but increased it. Canada designated 1984 the year of tourism, a declaration that symbolized the growing importance of the industry.

Expo 86 became the catalyst for a whole new awareness of tourism's potential in British Columbia. Tourists came from all over the world and from the rest of Canada to participate at the fair and to experience the BC landscape. The exposition achieved two major objectives: spreading the word on the many tourist attractions in the province, and having tourists return with their friends and relatives. With Expo 86, tourism reached a new level in terms of the number of visits and revenues, and these gains were not only maintained but also expanded over the following years.

Table 14.1

International tourism estimates, globally, 1950-2009

	Total international visits (millions)	Receipts from international tourism (US$ billions)
1950	25.3	2.1
1960	69.3	6.9
1970	165.8	17.9
1980	287.8	103.5
1990	455.9	261.0
1994	531.4	335.8
2004	763.0	622.8
2006	846.0	733.0
2009	880.0	946.0

Source: World Tourism Organization (2010).

Defining tourists and tourism is difficult; the industry is complex, as it involves both private business and all levels of government. Like other industries in British Columbia, tourism is affected by events ranging from the local (e.g., the closure of salmon fishing areas, avian flu, forest fires, contaminated water) to the global (e.g., terrorism, global recession, SARS, H1N1, volatile oil prices, and a volatile Canadian dollar). Adding to the complexity is the conflict over land use between tourism and the traditional resource extraction industries. The need for planning is essential, as this new and expanding industry requires a great deal of coordination between sectors of government and business.

Tourism has been important to the economic growth and diversification of British Columbia, especially as traditional resources have been depleted. This dynamic service industry is a significant industry in the province. By 2008, visitors were spending $13.8 billion in the province, tourism and tourism-related jobs employed 131,000 British Columbians, and there were nearly 18,000 businesses involved in the sector (Tourism British Columbia 2009). Nevertheless, the benefits of tourism have not been spread evenly over the eight tourist regions of the province.

TOURISM AS A RESOURCE INDUSTRY

Tourism as a resource industry is considerably different from the traditional extractive resources of forestry, mining, and fishing. Wood, minerals, and fish are harvested, often undergo some form of processing, and then for the most part are exported from British Columbia to gain income for the industry and the province. Tourism reverses this process by enticing tourists to come into the regions of the province.

British Columbia has a great variety of tourist attractions. The rugged topography and wilderness of both the coastal and the interior landscapes and the many recreational activities associated with them throughout the year are strong drawing cards. The combination of topography and climatic diversity enhances recreational

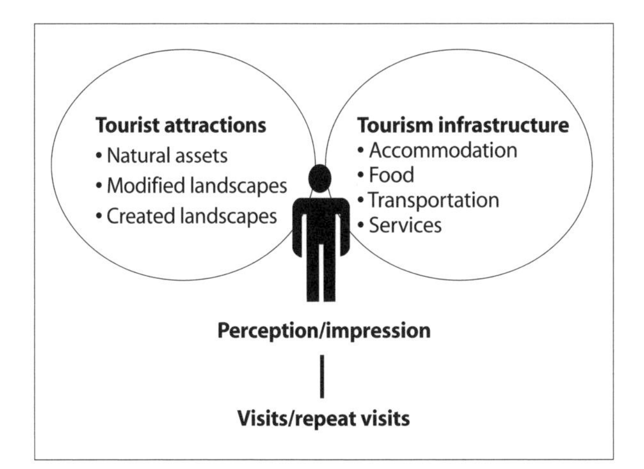

Figure 14.1 Relation of tourism industry to resources

opportunities and gives rise to a mix of unique flora and fauna species that attracts tourists. As well as the natural features, which are frequently promoted as "supernatural BC," many arts and cultural resources including First Nations sites, heritage sites, and sports events attract visitors. The industry also includes facilities and hospitality – not just how tourists are treated but the whole tourist infrastructure of accommodations and eating establishments. Figure 14.1 outlines many of the factors that attract tourists to a location.

This model categorizes tourist attractions into natural assets, modified landscapes, and created landscapes, each of which can attract tourists. British Columbia's many natural attractions include scenic ocean landscapes, wildlife, and the rugged Cordilleran alpine landscape, to name a few. The modified landscape describes natural settings that have been developed for human use, such as swimming beaches and ski slopes. Created landscapes, or as Philip Dearden (1983, 78) calls them, "anthropocentric attractions," are created or built environments such as Butchart Gardens, the Royal BC Museum, rodeos, music and arts festivals, First Nations centres, and amusement parks. Complementing these attractions is the essential tourism infrastructure: where tourists sleep and eat, how they travel, and how they acquire services. The combination of all these experiences leaves an impression on the tourist, and this impression, or perception, is essential to return visits, which generate continued income and economic well-being for the region and the province as a whole.

Collecting data on tourism is difficult because tourists are difficult to define. For example, to define a tourist one has to decide the distance a person must travel to qualify as one. How long does a person have to stay away from home, and at what point in an extensive stay does one cease to be a tourist? Is there a link between tourism and travelling because of work or business? Many individuals and government bodies have used definitions in an attempt to quantify the number of tourists and understand their impact on any region (see Mathieson and Wall 1982; Cooper et al. 1993; BC Stats 2009a; Halseth 2002). The World Tourism Organization defines a tourist as "someone who travels to and stays in a place outside their usual environment for not more than one consecutive year for leisure, business and other purposes not related to the exercise of an activity remunerated within the place visited" (BC Stats 2009a). Peter Murphy (1987, 402) uses precise distances and makes the definition specific to British Columbia, describing as tourism, "the travel, the activities, and services used by any British Columbia resident beyond a 40 kilometre (25 mile) radius from home for the purposes of personal enjoyment and the travel, the activities, and services used by nonresidents who enter the province for any reason other than work. It is recognized that elements of tourism are often involved when people travel for business purposes."

Although it is difficult and somewhat arbitrary to keep statistics about distances travelled, this formulation succeeds in identifying the range of places tourists come from, and there is a recognition that "people visiting their vacation homes or cottages are considered to be tourists" (BC Stats 2003). The acknowledgment that work and tourism may be related is especially relevant as more and more business is done in convention centres at locations such as Whistler, where tourist activities are part of the business package. Setting a time limit of one year for international tourists and six months for internal tourists may also be arbitrary, but it does recognize the number of tourists who see British Columbia as a safe, relaxing, and enjoyable destination for extended periods (BC Stats 2004, 1).

Tourism and recreation are closely related to each other and to the concept of leisure, since people require free

Table 14.2

Tourism categories

Adventure tourism		Cultural tourism, including arts, heritage, and sports	Industrial tourism
Auto travel	Kayaking	Butchart Gardens	Ginseng factory tours
Back country skiing	Mountain biking	Cultural displays (e.g., Doukhobor village, Sun Yat Sen Gardens)	Hydroelectric, mine, fish farm, sawmill, pulpmill, and farm tours
Bike touring	Rafting	Events and festivals (e.g., Expo, Oktoberfest, wine, music, and arts festivals, Chemainus murals, Barkerville)	Mining, forestry, agricultural museums
Bird/wildlife viewing	Rock climbing	First Nations sites (e.g., petroglyphs, totems, museums, interpretive centres)	Wine and beer tours
Boat travel	Sailing	Royal BC Museum, UBC Museum of Anthropology, municipal museums	
Canoeing	Scuba diving	Sports events (e.g., 2010 Winter Olympics, Commonwealth Games)	
Cruise ship travel	Spelunking		
Downhill skiing/boarding	Trail/horseback riding		
Fishing	Train travel		
Hiking	Wind surfing		
Hunting			

time to engage in either recreation or tourism. Table 14.2 outlines a range of tourism activities. Assessing tourism is rather complex, as it involves collecting information on a host of interrelated components, among them travel from communities, destination(s), activities, facilities used, and the creation of a positive impression on tourists in order to attract greater numbers in the future. Studying recreation, on the other hand, involves a somewhat narrower focus, namely, recording people engaged in recreational activities or events in particular locations. In British Columbia, recreational activities are often categorized in terms of either geographic location or infrastructure requirements, such as indoor versus outdoor settings. Marine-based recreation, for example, encompasses the many activities, from scuba diving to swimming to wind surfing, that take place in the diverse marine environment. The provincial government makes no distinction between adventure tourism, nature-based tourism, or commercial recreation; they are all defined as "outdoor or recreational activities provided on a fee-for-service basis, with a focus on experiences associated with the natural environment" (Integrated Land Management Bureau 2010). Greg Halseth (2002, 134) makes "a distinction between tourism as involving 'visitors' to a region and recreation as involving activities engaged in primarily by 'local residents.'"

Tourism is a multifaceted industry that relies on all levels of government and involves a multitude of private interests. The federal and provincial governments are fundamental to the industry. They develop and maintain transportation systems (highways, airports, ferry systems, and ports) and provincial and national parks, and establish the policies and laws that affect the private sector. In 2008, for example, there were 972 provincial parks and protected areas (totalling over 13.5 million hectares) that include 340 campgrounds, 11,000 campsites, 263 day-use areas, and 6,000 kilometres of hiking trails (BC Stats 2009b). The provincial government created a Crown corporation called Tourism British Columbia to manage tourism in 1997, raising revenues through a hotel room tax and spending the money to promote the development and growth of the industry (BC Ministry of Finance and Corporate Relations 1998, 229).

Local governments also play a crucial role in the tourism industry because they are responsible for zoning much of the tourist infrastructure and using the local tax base to build facilities for both residents and tourists. Municipalities and regional districts sponsor the development of a great number of facilities and activities that attract tourists: historical sites, museums, theatres, entertainment centres, parks, recreation complexes, and so on. Tourism has added a degree of much needed economic

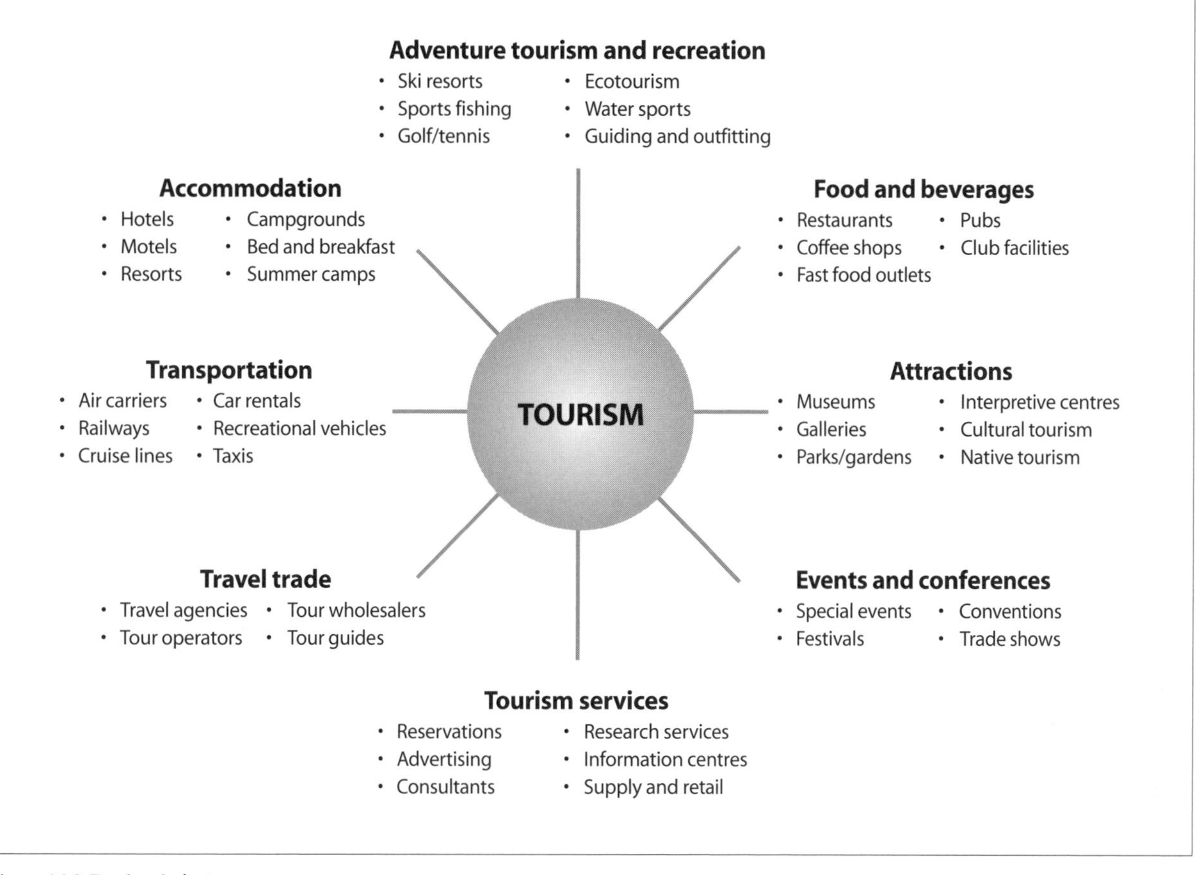

Figure 14.2 Tourism industry sectors
Source: Modified from Canadian Regional Forecasting (1991, 4).

diversity to communities that depend largely on one resource, such as forests or minerals, providing another way for people to make a living. For some towns, such as Chemainus and Tofino, the conversion from a forest-based economy to a tourist-based one has been profound.

The private sector contains the full range of tourist businesses, from large multinational corporations to many small, family-oriented businesses. Figure 14.2 gives an overview of the many components of the private sector, which generates the most tourism employment and investment. Tourism generated nearly a billion dollars ($937 million) in tax revenues for the provincial government in 2008. The industry's export earnings are second only to wood products, and there is an expectation that tourism will require some 84,000 new workers over the next decade (Tourism British Columbia 2009).

Coordination and planning are essential among the many levels of governments and thousands of private tourist-related businesses. A major area of conflict has been the traditional resource harvesting of fish, minerals, and forests. Overharvesting of salmon has resulted in sport fishing closures and restrictions on fishing for certain salmon species. The consequent loss of employment and revenues to fishing lodges, fishing charters, restaurants, and coastal communities generally are obvious. Less obvious though still serious are conflicts over forest clear-cutting and mining operations. The provincial government's move to increase parks and protected areas is one sign that it is attempting to preserve the natural setting and to enhance tourist values.

The issue of capacity also generates conflict and demonstrates the need for planning. Tourism is often referred to as a "clean" industry in the sense that tourists come to look or to enjoy recreation, spending money in the process and then leaving. This view of tourism – that it does not alter the environment – is not entirely accurate. Tourism and tourist-related facilities and activities do change the landscape, and in some regions it may be better to establish thresholds for tourism. When a community begins to erect many tourist attractions – such as billboards, amusement parks, water slides, and restaurants – local residents may not be content with the noise, excess traffic, or the aesthetics of the billboard landscape. Similarly, as a region attracts too many tourists, causing delays on roads and ferries, its regular users may become annoyed. Tourist overuse of provincial parks, marine parks, scuba diving sites, and the many other destinations in the public domain can have a negative impact on the environment, making tourism less than a clean industry.

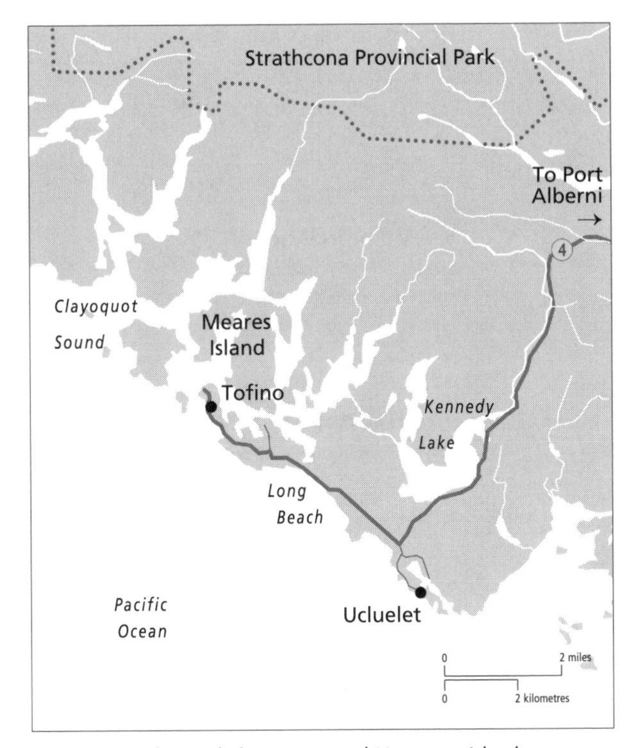

Figure 14.3 Tofino and Clayoquot Sound, Vancouver Island

TOFINO: THE EVOLUTION OF A TOURIST LANDSCAPE

The transition from a resource extraction-based landscape to a tourist-based landscape is not without conflicts. Tofino and the adjacent Clayoquot Sound exemplify many of the resulting struggles and changes to the landscape (Figure 14.3).

Table 14.3 summarizes the many human influences on the area from pre-European times to the 1990s. The trajectory shows first the transition from traditional First Nations use of the land to the imposition of new landscape values based on resource extraction. The Nuu'chah'nulth were deterritorialized and their land was taken over by fur traders, missionaries, prospectors, fishers, foresters, and settlers. Throughout the fairly long period from exploration to corporate hegemony, the Tofino/Clayoquot region was portrayed as a rather exotic wilderness, appealing to the adventurous traveller. Both travellers and tourists became considerably more numerous once forest roads, built by the 1970s, gave access to Long Beach. The real struggle over use and development of this landscape evolved in the 1980s and 1990s with improved forest technology of clear-cut logging and the need by the industry for these old-growth trees.

The institutional structure and history of logging interests, including the establishment of long-term forest tenure for large corporate entities such as MacMillan Bloedel and Fletcher Challenge, came into direct confrontation with a host of new values. A showdown flared between those who wished to clear-cut Meares Island and First Nations groups making Aboriginal title claims, along with environmental groups that attempted to stop logging, some by spiking trees. A court injunction in 1985 stopped logging on Meares Island. For the rest of Clayoquot Sound, the battle to preserve environmental and tourist values waged on through a 1991 change in provincial government and a number of government sponsored round table discussions attempting to reach consensus among competing land uses. By 1993, Clayoquot had become a global environmental icon, made popular by celebrities such as Robert Kennedy Jr. and international rock groups such as Midnight Oil from

Table 14.3

Overview of factors influencing the Clayoquot landscape

	To 1770s	1770s-1820s	1820s-1880s	1880s-1920s	1920s-1980s	1980s-1990s
Phase	Traditional precontact	Exploration, fur trade	Strategic control of trade	Settlement of frontier	Corporate hegemony	Environmental amenity
Authors	Elders, shamans chiefs	Explorers, naval officers, fur traders	Traders, missionaries, government agents, travellers	Land developers, government agents, travellers	Corporations, government public relations, travel writers	Environmentalists, tourism operators, Natives
Landscape images	Mythic, totemic images	Native life, landscape resources	Frontier posts, wild coast, scenic beauty, mining	Land clearance, settlements, mining, canneries	Canneries, "working forests,"[a] fish farms, scenic beauty	Scenic beauty, clear-cuts, wildlife
Landscape impact	Use of forest resources, shoreline fish traps, seasonal settlements	Trading posts, cutting of spars, extirpation of sea otters	Fortified villages, missions, trading posts, destruction of fish traps	Land clearance, settlements, mines, canneries	Clear-cuts, logging roads, sawmills, mines	Viewscape protection, new forest practices, parks
Institutional arrangements	Defence of hereditary ownership	British defence of contested trade hegemony	Timber grants, Indian reserves, pre-emptions, mining claims	Pre-emptions, resource sales, transportation infrastructure	Timber licences, tree farms, integrated resource management	Landscape management, parks, ecosystem management, co-management

a Selective logging and small clear-cuts to designated areas for large corporations under tenures such as tree farm licences.
Source: Modified from White (1999, 79). Used with permission.

Australia. Over 800 protesters were arrested in the summer of 1993. The largest mass arrest in Canadian history not only provided publicity but also gave a new signification to the landscape: "This was a place for which ordinary people were prepared to go to jail. It was also giving Canada a bad reputation internationally ('Brazil of the North'), leading to forest product boycotts and international condemnation" (White 1999, 210).

Brian White (1999) documents well the various steps involved in the transition from an industrial extractive landscape to one recognized for its many touristic values. By 1999, a new forest company, Iisaak – a joint venture between the Nuu'chah'nulth and MacMillan Bloedel (now Weyerhaeuser) – was created and the major environmental groups, the Western Canada Wilderness Committee and Greenpeace, had agreed to a logging plan for the Clayoquot. In 2005 "the Central Region First Nations took over 100% ownership of Iisaak when they purchased the 49% share from Weyerhaeuser making Iisaak a 100% privately owned First Nations forest company" and owner of Tree Farm License 57 (Iisaak 2009). Harvesting is still conducted through the guidelines established by the Cayoquot Sound Science Panel.

Tofino's economy has diversified considerably from dependence on commercial fishing and logging to major tourist attraction. Tofino has become the tourism gateway to the spectacular setting of Long Beach and Clayoquot Sound, as well as providing whale watching, storm watching, kayaking, and surfing.

TOURISM IN BRITISH COLUMBIA

How many tourist visits occur each year? From where do tourists originate? To which regions or locations do they go? What attractions and facilities do they use? How much revenue is generated? These questions are of interest from the perspective of geography. Table 14.4 compares overnight tourist numbers and revenues, along with the origin of visitors, from 1997 to 2009. During this period, overall revenues increased each year, but the number of tourists varied because of a number of factors. The 9/11 terrorist attacks in the United States caused many to fear flying, and many consider increased border security a hassle. Table 14.4 shows that the number of US visitors to British Columbia is declining. Pandemics – severe acute respiratory syndrome (SARS) in 2003, avian flu (H5N1) in 2004-5, and swine flu (H1N1) in 2009-10 – have likewise caused great concern and reduced the number of visitations. The value of the Canadian dollar (Fig. 14.4) reaching parity with the US dollar made visits to Canada less affordable, and volatile oil prices (see Figure 11.7, p. 197) have increased the price of transportation (sometimes greatly). There are other unpredictable global conditions, such as the major collapse of the economy in late 2008 and its slow recovery, that are affecting people's income and ability to take holidays. Even natural hazards such as the BC fire storms in 2003 and 2009 and the volcanic explosions in Iceland in 2010 can intervene and negatively affect the number of people travelling to the province. The tourism industry also faces heavy taxation burdens and competition from other regions.

Table 14.4

Tourism indicators, 1997-2009

	Tourist revenue and volume		Visitor entries ('000s)			
	Revenue ($ millions)	Visitors ('000s)	United States	Europe	Asia	BC tourists ('000s)
1997	8,471	21,313	5,893	449	920	10,654
1998	8,731	21,753	6,549	449	771	10,654
1999	9,080	22,053	6,862	485	842	–
2000	9,449	22,491	7,006	498	893	10,761
2001	9,582	22,381	6,895	466	864	10,761
2002	9,720	22,571	6,591	412	882	10,869
2003	9,799	21,870	6,137	410	702	10,884
2004	10,712	22,450	6,039	458	716	11,075
2005	11,463	22,886	5,751	494	724	11,418
2006	12,382	23,122	5,380	493	717	11,700
2007	13,251	–	5,062	519	699	–
2008	13,802	–	4,476	514	673	–
2009	12,706	–	4,295	471	575	–

Source: BC Ministry of Tourism, Culture and the Arts (2010).

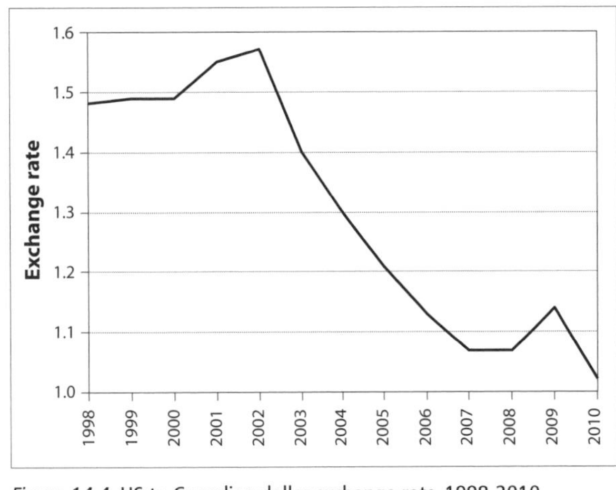

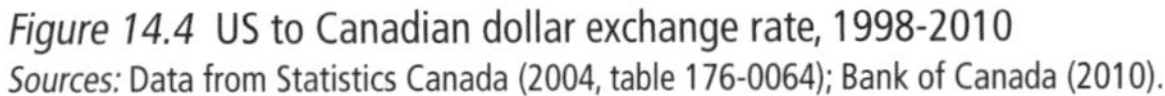
Figure 14.4 US to Canadian dollar exchange rate, 1998-2010
Sources: Data from Statistics Canada (2004, table 176-0064); Bank of Canada (2010).

On the positive side, BC has much appeal, including major events such as the 2010 Winter Olympics, which attracted hundreds of thousands of people. The province is largely an uninhabited region with an amazing variety of landscapes. It is a politically stable and safe environment for tourists, and it has well-developed tourist infrastructure, including transportation systems. Table 14.4 shows how these factors have influenced international tourism to the province. Despite the common perception that tourists come from outside the province, the table clearly demonstrates the importance of British Columbian tourists. The main tourist market for communities developing tourist and recreation facilities may be from adjacent communities.

Indeed, local and global economic uncertainty have affected British Columbia's tourism industry. Tourism relies on people having leisure time, income to spend, and comfort in travelling to a region. Because of the series of recessions and depressions since the 1980s, "the restructuring of work and employment within the North American economy now often means lower wages and the need by more and more people to work at more than one job" (Halseth 2002, 136). The provincial government also recognizes tourism "as a discretionary expenditure.

Table 14.5

Room revenue by tourist region, 2000-9

	Vancouver Island/coast	Mainland/ southwest	Thompson/ Okanagan	Kootenay	Cariboo	North coast	Nechako	Northeast	Total
					($)				
2000	245,309	846,849	183,893	63,793	43,571	20,609	8,443	29,617	1,442,085
2001	258,235	854,936	190,350	64,108	42,553	19,342	8,807	35,653	1,473,983
2002	268,451	850,333	205,413	71,712	44,514	19,016	8,843	37,904	1,506,186
2003	272,990	810,416	213,447	72,167	43,064	19,364	8,929	46,266	1,486,643
2004	295,245	863,420	229,551	76,957	46,198	20,104	8,811	51,889	1,592,176
2005	309,134	910,093	236,479	81,467	49,272	22,070	9,649	63,400	1,681,564
2006	323,895	973,879	259,128	88,989	54,701	23,127	9,791	74,099	1,807,609
2007	351,617	1,044,766	293,097	105,026	61,892	27,682	11,246	68,633	1,963,959
2008	340,450	1,056,622	298,525	104,155	62,202	27,376	10,891	77,237	1,977,458
2009	308,355	923,334	267,230	91,947	54,281	23,857	9,821	67,700	1,746,525

Source: BC Ministry of Tourism, Culture and the Arts (2010).

When the economy is weak and people have less money or feel uncertain about the future, tourism activities are usually among the first to be cut from their budgets" (BC Ministry of Management Services 2003). The decline in tourist revenues is a stark reminder that British Columbia is very much a part of economic and political developments both globally and locally.

Keeping track of the number of tourists coming to British Columbia and the revenue they generate is obviously important to the province and to the industry as a whole. Each region attempts to attract a share of tourists and their dollars, but as Table 14.5 shows, these are not distributed equally.

Tourism Regions

Tourism British Columbia divides the province into eight regions. The configuration has changed since 1997 to correspond to the provincial government's eight development regions, which are an amalgam of the twenty-eight regional districts within the province (see Figure 7.10, p. 125). This change facilitates the gathering of tourist data such as overnight stays by tourists at hotels, motels, fishing lodges, and other accommodations.

The Vancouver Island/coast region includes the whole of Vancouver Island and the area of the central coast from Powell River to Bella Coola. It comprises eight regional districts: Capital, Cowichan Valley, Nanaimo, Alberni-Clayoquot, Comox-Strathcona, Mount Waddington, Powell River, and Central Coast. This region fairly consistently increased its tourism revenues between 2000 and 2007, but the recession and uncertainty reduced its revenues for 2008 and 2009. This region is second to the Mainland/Southwest region with 18 to 20 percent of provincial room revenue (see Table 14.5). The Insular Mountains of Vancouver Island and Coast Mountains of the central coast, deep fjords, magnificent land and marine-based flora and fauna, and mild winter temperatures are the distinguishing features and major regional strengths for the tourism industry.

The south end of Vancouver Island, specifically Victoria, attracts the greatest number of tourists. Victoria wears the image and charm of a British past, and has been a favourite tourist destination for a long time. In 1991 the capital city also had the distinction of being the most popular city in Canada and eighth in a survey of favourite cities in the world (Florence, Italy, ranked first) (BC Ministry of Small Business, Tourism, and Culture 1991, 9). In 2003 *Condé Nast Traveler*'s Reader's Choice Award ranked Victoria as the "Best City in the Americas" and Vancouver Island as the "Top North American island"

(City of Victoria n.d.). A travel survey in 2007 by the Conference Board of Canada ranked Victoria third, after Calgary and Vancouver, as the most attractive city in Canada (*Times Colonist* 2007).

Recreational opportunities abound on Vancouver Island and the Gulf Islands, with sport fishing, whale watching, surfing, kayaking, skiing, bungee jumping, spelunking, camping, and hiking as some of the favourite activities. Scuba diving has become an important tourist attraction for this region and other coastal areas of the province; a recent study designated British Columbia a "top diving destination in the world" (Tourism British Columbia 2004). Pacific Rim National Park is one of the popular tourist destinations, attracting a million people a year to the west coast of the island (VancouverIsland.com 2010). This park contains three popular regions – Long Beach, Broken Group Islands, and the historical Ship Wreck Trail, also known as the West Coast Trail – accounting for some 60 percent of tourism to the region. The popularity of hikes such as the park's West Coast Trail is clear from the reservation system that the provincial government had to implement to keep the number of hikers at a manageable level. Winter sports enthusiasts are attracted to Mount Washington (near Comox). The north end of the island, which has its own spectacular attractions from Cape Scott hiking to marine activities, receives considerably fewer tourists than the better-connected south end. The central coast region is even more remote, but that is its appeal to many tourists, who travel the coast by water.

A contentious issue that affects Vancouver Island tourism primarily is salmon sport fishing. The sport fishing industry brings in the highest return per fish caught but is allotted the fewest fish. Some areas of coastal British Columbia have had moratoria on catching chinook – the largest and most sought-after species – while the commercial fleet has not had the same curtailment. For example, ARA Consulting Group (1996, S.4) estimates that the 1996 closure on chinook meant losses of 2,175 jobs and $135.7 million in revenues to the sport fishing industry. The Canada-United States agreement of June 1999 introduced some better means of managing salmon for tourism and other interests in this resource, but salmon stocks remain weak for many runs, and intermittent bans on coho, chinook, and sockeye continue. The McRae-Pearse Commission (2004) recognized the value of salmon and other species of fish to the tourist industry and has made strong recommendations for fish to be set aside for sport fishing. The more recent federal government's Wild Salmon Policy stresses conservation of salmon stocks, and fishing bans will continue.

An increasing population of both tourists and residents in the southern half of the island led to a new Vancouver Island Highway that accommodates a greater volume of traffic north beyond Nanaimo and Campbell River. Transportation to Vancouver Island is mainly via the provincial government's ferry system, which connects at Victoria, Nanaimo, Comox, and Port Hardy. The Victoria and Nanaimo crossings are extremely popular and frequently marked by the frustration of waiting for one or two sailings during the tourist season and holiday weekends. The coastal community of Powell River on the mainland is accessed by ferry from Comox as well as from the Sunshine Coast. Farther up the coastal mainland, only Bella Coola is accessible by road. The island is well accessed by regular airline service; WestJet serves Victoria and Comox, and Air Canada flies into Victoria International Airport (YYJ), Nanaimo, Comox, and Campbell River. Many other small carriers operate regular service to other Vancouver Island and coastal communities.

Table 14.5 reveals the dominance of the Mainland/southwest region in the tourism industry. It is not a large geographic region (see Figure 7.10, p. 125), consisting of only four regional districts – Greater Vancouver, Fraser Valley, Sunshine Coast, and Squamish-Lillooet – but it contains over 50 percent of the population and garners over 50 percent of the tourism room revenue in the province. The Mainland/southwest has a great range of tourist attractions. Many ocean-related opportunities exist on the Sunshine Coast, which is accessed by ferry, while the Squamish-Lillooet Regional District is graced with spectacular mountains, lakes, and valleys where year-round tourism-recreational opportunities abound. Whistler, the internationally famous resort and convention community (and companion community to Vancouver for hosting the 2010 Winter Olympics) is adjacent to Garibaldi Provincial Park and is a major tourist destination for the region. The multi-million dollar highway upgrade

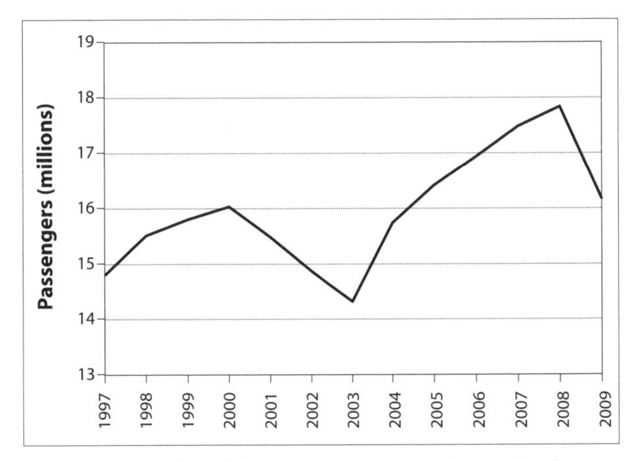

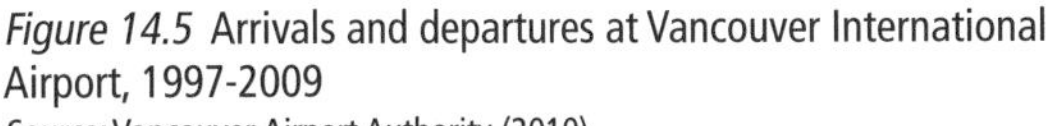
Figure 14.5 Arrivals and departures at Vancouver International Airport, 1997-2009
Source: Vancouver Airport Authority (2010).

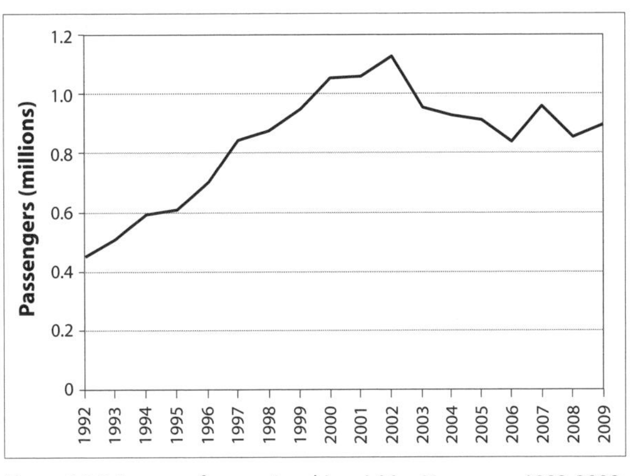

Figure 14.6 Revenue from cruise ships visiting Vancouver, 1992-2009
Source: Port Metro Vancouver (2009).

for the Olympics left a valuable transportation legacy for Squamish, Whistler, and Pemberton.

The Greater Vancouver and Fraser Valley Regional Districts also have parks, mountains, rivers, lakes, and salt water, which attract tourism-related activities. Most important, however, this large urban complex offers tourist options that smaller communities cannot, such as professional sports events, cultural activities, and the Pacific National Exposition, along with a wide range of shopping, entertainment, culinary, and other services. At the same time, large cities such as Vancouver face the challenges presented by panhandlers, homeless people, and those affected by alcohol and drug abuse – all in proximity to the core of the city and such popular tourist attractions as Gastown and Chinatown.

Vancouver, by far the largest city in British Columbia, is the focal point of many transportation networks, including highways, railways, and ferries. Its international airport and cruise ship facilities are some of the most important transportation nodes for the industry. Indeed, these links to the "outside" account for the decline in revenues in 2002-3 and 2009 for this region. For example, Figure 14.5 shows that there was a correlation between decreasing numbers of tourists arriving via Vancouver International Airport (YVR) and room declines in 2002-3 and 2009. The reduced number of airline passengers is in part due to the threat of terrorism, SARS, and a major recession but also due to higher ticket prices based on the increased cost of security, the rise in fuel prices, and the strength of the Canadian dollar.

Cruise ship tourism – destination Alaska – has also experienced a reduction in passenger revenues (Figure 14.6). The port of Vancouver's domination has been challenged by Seattle, which built cruise ship facilities in 1999. The Seattle terminal started out slowly with 120,000 passenger visits in 2000. With 875,433 passengers in 2009, however, it has a similar number and pattern of passengers as Vancouver (Port of Seattle 2010). The *Tourism Sector Monitor* also suggests that in the post-9/11 era, American passengers feel more comfortable beginning and ending a cruise in their own country. Again, terrorism fears play a significant role in the international tourism industry. Nevertheless, the demand for cruise ship travel is likely to increase as the North American population ages, living longer and with considerable disposable income.

The successful bid for the 2010 Winter Olympics had the same dynamic influence on tourism as did Expo 86. Huge numbers of people visited the region. Also similar to Expo, the Winter Olympics left a legacy of sports facilities, such as Richmond's skating oval, that will increase tourist visits long into the future. Of equal importance to the future of tourism is the legacy of transportation infrastructure, such as the extension of the SkyTrain (Canada Line) to the airport and the upgrading and rebuilding of the twisty road to Whistler. Other new and expanded facilities include hotels and other accommodation, a new convention centre, and a host of tourism-recreation services.

The Thompson-Okanagan region is made up of five regional districts – Thompson-Nicola, Columbia-Shuswap, Okanagan-Similkameen, Central Okanagan, and North Okanagan – and is home to nearly half a million British Columbia residents. Table 14.5 indicates that this region steadily increased room revenues until 2009. Kamloops is the centre for the northern portion of the region; because the Trans-Canada Highway, Coquihalla Highway, and Yellowhead Highway pass through Kamloops, it is known as the gateway to Alberta. From Kamloops one can head either east to Revelstoke and Banff or Calgary, or northeast up the North Thompson River via the Yellowhead Highway to Mount Robson and on to Jasper or Edmonton. Both the Canadian Pacific Railway and Canadian National Railway lines also pass through Kamloops, which has a regional airport as well. Tourism for this part of the Thompson-Okanagan region relies on these major transportation systems, although there are many attractions: hiking, heli-skiing, downhill skiing (at, e.g., Sun Rivers near Kamloops, Revelstoke Mountain, and Kicking Horse near Golden), fishing, boating, whitewater rafting, swimming, and sightseeing. Some specific sites are the Revelstoke Dam, Craigellachie (where the last spike of the CPR was driven), the world-famous run of sockeye salmon heading for Adams Lake, and Helmcken Falls in Wells Grey Park. As well, events such as the Merritt Mountain Music Festival (cancelled for 2010, but reopening in 2011) attract thousands of visitors each year. The challenge for tourism operators is to make these attractions a destination, since most tourists use the transportation systems to pass through the area.

South of Kamloops and the Thompson lies the Okanagan, which is famous for its many orchards with a great variety of fruit and for its many lakes, which attract water recreationists. More recently, its reputation has been enhanced by the production of high quality wines. Wine tours and wine festivals are now a major attraction in this region. Sport fishing, biking or hiking the Kettle Valley Line, many fine golf courses, casinos, and the Silver Star, Big White, and Apex ski resorts have helped this region become a year-round tourist attraction. The Okanagan-Similkameen region has easy access to the large Vancouver population via the Coquihalla Highway and a regional airport at Kelowna. Within the Okanagan, the main highway that links communities has been improved by a number of bypasses. Even so, the volume of summer tourist traffic still results in delays. Another struggle for the Thompson-Okanagan region has been the invasion of Eurasian milfoil into the many lakes, spoiling the beaches for swimming. The summer of 2003 brought an unexpected hazard – wildfires. Residents who lost homes and possessions felt the greatest tragedy, but the closure of the backwoods, the burning of many historic Kettle Valley Line trestles (rebuilt by 2005), and general fear of being caught in a forest fire were detrimental to the tourist industry. The firestorm of 2009 generated more fires than in 2003 but did not destroy as many homes. It was a reminder that, with global warming, wildfires may become more common for this region.

The Kootenay region, which is made up of Kootenay Boundary, Central, and East Kootenay Regional Districts, gains less than 5 percent of tourism room revenues in the province. The region is mountainous, with relatively narrow valleys and long lakes formed by numerous hydroelectric dams. The region has a history of mining, forestry, and farming. Because these traditional industries are in decline, tourism is viewed by many of the region's 150,000 residents as the "new" means of employment. Older ski locations such as Red Mountain at Rossland already have an international reputation, and with newer ski facilities at Fernie, Panorama (near Invermere), and Kimberley, nearly 1.5 million skiers were attracted to the Kootenay region in the 2003-4 season (Land and Water British Columbia 2004). The Bavarian theme at Kimberley, a new casino at Cranbrook, and the Victorian architecture of Nelson are other attractions. The outdoors is a spectacular setting for year-round tourism and recreation activities, and sightseeing, swimming, golfing, kayaking, mountain biking, fishing, hunting, and hiking are just some of the attractions. Hot springs, a Doukhobor village museum, and ghost towns also attract tourists. The main drawback to Kootenay country is access: the region has Air Canada service to Castlegar and Cranbrook (Canadian Rockies International Airport), but the region is not on a major arterial highway and is considerably distant from major urban centres in British Columbia, Alberta, and the United States.

The Cariboo is a large tourist region although it includes only two regional districts – Cariboo and Fraser-Fort George. This region was the site of the famous Cariboo gold rush in the 1860s. The historic Cariboo wagon road has been widened into Highway 97 and links the various Mile Houses, which were stagecoach stops. The houses also mark the distance from Lillooet to goldfields and communities such as Barkerville. This region largely relies on employment from forestry, which has been in decline, as reflected in a population drop from nearly 200,000 in 1996 to just over 160,000 in 2009. It is also a region where tourism room revenues fluctuate, as can be observed from Table 14.5.

Prince George is the major service centre for the Cariboo and is well connected by road, rail, and regular airline service. This region is also known as cattle country, and tourists are attracted by the western culture of guest ranches, cattle drives, and rodeos. Cross-country and downhill skiing, fishing in the many freshwater lakes, hunting, and a scenic landscape are also important attractions. The historical site of Barkerville brings the gold rush era to life, engaging tourists in gold panning, live theatre, and life in the 1860s. Close by, Wells, another mining community, has a casino and the popular Bowron Lake Provincial Park, where tourists can complete a circular tour by canoe in seven to ten days (but you must register ahead of time). Forestry activities are still very important to the economy of this region, however, and interfere with some tourism. Another drawback for tourism is the Cariboo's relative isolation and distance from major population centres. The main highway from Vancouver leads tourists through this region to Prince George and to other destinations.

The Nechako tourist region is a somewhat awkward configuration that attaches the Bulkley-Nechako Regional District to the remote Stikine region. This region has the smallest population, at only 39,502 in 2009 (down from 44,751 in 1996), and the least amount of tourist room revenue (see Table 14.5). The largest community in the Bulkley-Nechako Regional District is Smithers (population just over 5,000), and there are no incorporated communities at all in Stikine. There is, however, plenty of wilderness. Rugged mountains, pristine lakes, and rapidly flowing rivers all permit a range of outdoor activities: fishing, hiking, wildlife viewing, hunting, cross-country skiing, canoeing, and rafting. The few communities – Vanderhoof, Burns Lake, Houston, and Smithers – are on the highway between Prince George and Prince Rupert. The Stikine region to the north is best described as isolated, which is both an advantage and a disadvantage for tourism. For those tourists seeking a wilderness experience far from any inhabited location, this area offers a diversity of spectacular landscapes such as the Tatshenshini-Alsek Provincial Park, Spatsizi Plateau Wilderness Park, and a portion of the recently created Muskwa-Kechika Management Area. The difficulty is one of access. The Stewart-Cassiar highway, leading north from Hazelton, connects to the Alaska Highway near Watson Lake in the Yukon, making this area a considerable driving distance from southern British Columbia.

The fairly large and remote north coast tourist region is composed of the Kitimat-Stikine and Skeena-Queen Charlotte Regional Districts. Tourism room revenues are not large in comparison to other tourist regions, but they have increased considerably since 2000 (see Table 14.5). Prince Rupert and Kitimat are the principal coastal communities, and Terrace is the main inland city. Both Prince Rupert and Terrace are served by Air Canada, and these centres are linked by highway to Prince George and the rest of the province. Haida Gwaii is connected by regular ferry service over the Hecate Strait; this is a shallow, and at times treacherous body of water, but tourists can buy an "I survived the Hecate Strait" T-shirt and wear it with pride. There is also regular air service to Sandspit. Farther up the coastal mainland at the end of Observatory Inlet is Stewart, accessible via the Stewart-Cassiar Highway, which passes the picturesque Bear Glacier. Prince Rupert is often a stop for Alaskan cruise ships, and a $1.6 million investment in cruise ship port facilities in 2003 comes with the expectation that Prince Rupert will have the potential to attract 140 vessels and more than 250,000 passengers annually (Western Economic Diversification Canada 2010). These numbers have not materialized, but there were 54,867 passengers in 2009, which added some $7 million to the local economy (Cruise BC

2009). This centre is also on the regular ferry route to Port Hardy on northern Vancouver Island, via the Inside Passage.

The north coast region offers a blend of both coastal and interior-oriented activities. First Nations are a big part of this region and the tourist experience. The Haida of Haida Gwaii and coastal First Nations – the Tsimshian Gitxsan, Nisga'a, and Heiltsuk – have traditional villages, totems, and arts and crafts displays, and an upgraded road has been built to access the territory of the Nisga'a, settled by the first modern treaty. Coastal tourist attractions include whale watching, sports fishing, kayaking, and visiting a historic salmon cannery, while the variety of wilderness experiences inland includes grizzly bear tours and other wildlife viewing.

The final tourist region, northeast British Columbia, is large in area but sparsely populated. In 2009 nearly 68,000 people lived in two regional districts: Peace River and Northern Rockies. This region includes a northern portion of the Rockies, the eastern foothills of the Rockies, and the start of the Prairies. Dawson Creek, Fort St. John, and Fort Nelson, the main urban centres of the northeast, have relied on resources such as oil and natural gas, forestry, hydroelectricity, and wheat. The prairie landscape here is similar to Alberta's. The W.A.C. Bennett Dam is on the Peace River, and not far downstream is the Peace Canyon Dam, which houses a museum that displays a history of the region including the discovery of dinosaur footprints. To the south is the community of Tumbler Ridge, which was created in response to the opening of two coal mines. Both mines closed by 2003 and, like others before it, this community turned to the retirement industry and tourism, offering accommodation, skiing, golfing, hiking, and, more recently, the viewing of dinosaur footprints. Fortunately, rebounding coal prices have resulted in two new coal mines – Trend (2005) and Wolverine (2006) – opening near Tumbler Ridge. A third coal mine, Brule, opened in nearby Chetwynd in 2007.

This northern region of the province contains a great deal of remote wilderness, including the Muskwa-Kechika Management Area, where ecotourism abounds with river boating and rafting, wildlife viewing and, in winter, snowmobiling, downhill and cross-country skiing, snowshoeing, and dog sled racing. Hunting, fishing, hiking, and sightseeing also attract tourists to the region, but the main incentive for tourists is the Alaska Highway, of which Dawson Creek is Mile 0. This highway attracts a great number of tourists – mainly in the summer months – travelling to Yukon and Alaska and enjoying the wilderness or a stop at the popular Liard hot springs.

From this brief overview, it can be seen that all areas of the province have touristic values and that tourism has become an important economic component to many local economies. The mainland/southwest region dominates in terms of tourist numbers and overall revenues, but in other regions tourism is increasingly being recognized as an important diversification from traditional resource-based industries.

SUMMARY

In many ways, Expo 86 was the catalyst for the most rapidly developing industry in this province, in conjunction with the equally rapid expansion of global tourism through the 1980s and '90s. British Columbia offers many attractions for tourists: a great variety of saltwater and freshwater landscapes, spectacular alpine vistas, and an array of unique flora and fauna. As more and more people live in urban environments, the wilderness experience of British Columbia becomes increasingly desirable. Convention centres have been developed throughout the province to accommodate the internationalization of business activities that have a clear tourism dimension. All regions and communities in the province are involved in tourism, but not equally.

Tourism is a labour-intensive service industry. For the foreseeable future, tourism employment will increase, although many of the positions will pay only minimum wage. In this knowledge-based industry, information and organizational skills are essential commodities. The BC tourism industry is tied to global economic conditions, however, and these are marked by uncertainty. The terrorist activities of 2001, SARS and wildfires in 2003, wildfires in 2009, avian flu in 2004, H1N1 flu in 2009-10, serious recession since 2008, rising US-Canadian-dollar exchange rates, less leisure time, less income, increased competition, increased air fares, increased gasoline prices – all of these conditions tend to reduce spending on tourist activities. As well, the industry must work with all levels of government because governments provide the infrastructure it needs. Hosting the 2010 Olympics was a major public expenditure that attracted thousands of tourists. But to offset the costs of hosting an international event of this magnitude, British Columbia will need to attract future tourists and continue to foster interest in the province.

Tourism also relies on the quality of the environment. Since the landscape is part of the product being sold to tourists, land-use conflicts are inevitable, particularly as the industry intensifies. Traditional industries have an impact on forests, fish, and water quality that can directly affect tourism, and they therefore need to be managed. Conversely, tourism can also have a negative impact on the landscape and on the way of life for residents of British Columbia. It, too, needs to be managed.

REFERENCES

ARA Consulting Group. 1996. *Fishing for Answers: Coastal Communities and the British Columbia Salmon Fishery.* Vancouver: ARA Consulting Group.

BC Ministry of Finance and Corporate Relations. 1998. *1998 British Columbia Financial and Economic Review.* Victoria: The Ministry.

BC Ministry of Small Business, Tourism, and Culture. 1991. *Urban Tourism in British Columbia.* Victoria: The Ministry.

Cooper, C., F. Fletcher, D. Gilbert, and S. Wanhill. 1993. *Tourism: Principles and Practice.* Essex: Longman.

Dearden, P. 1983. "Tourism and the Resource Base." In *Tourism in Canada: Selected Issues and Options,* ed. P.E. Murphy, 75-93. Western Geographical Series vol. 21. Victoria: University of Victoria.

Halseth, G. 2002. "A Regional Geography of Tourism and Recreation in New Caledonia." *Western Geography* (Canadian Association of Geographers) 12: 130-62.

Mathieson, A., and G. Wall. 1982. *Tourism: Economic, Physical and Social Impacts.* Essex: Longman.

McRae, R.M., and P.H. Pearse. 2004. *Treaties and Transition: Towards a Sustainable Fishery on Canada's Pacific Coast.* Report for Ministry of Fisheries and Oceans and Government of British Columbia. Ottawa.

Murphy, P.E. 1987. "Tourism." In *British Columbia: Its Resources and People,* ed. C.N. Forward, 401-30. Western Geographical Series vol. 22. Victoria: University of Victoria.

Nickerson, N.P., and P. Kerr. 2001. *Snapshots: An Introduction to Tourism.* Toronto: Prentice-Hall.

White, B. 1999. "Authorizing the Tourism Landscape of Clayoquot Sound." PhD diss., Simon Fraser University.

INTERNET

Bank of Canada. 2010. "Exchange Rates." www.bankofcanada.ca/en/rates/exchange_avg_pdf.html.

BC Stats. 2003. "BC's Tourism Sector Struggles through 2001." *Business Indicators,* January. www.bcstats.gov.bc.ca/pubs/bcbi/bcbi0301.pdf.

–. 2009a. "Measuring the Size of British Columbia's Tourism Sector." www.bcstats.gov.bc.ca/data/bus_stat/busind/tourism/tour_meth_09.pdf.

–. 2009b. *Quick Facts about British Columbia: 2009 Edition.* www.bcstats.gov.bc.ca/data/qf.pdf.

BC Ministry of Management Services. 2003. "BC's Tourism Sector Struggles through 2001." BC Stats. *Business Indicators,* January. www.bcstats.gov.bc.ca/pubs/bcbi/bcbi0301.pdf.

BC Ministry of Tourism, Culture and the Arts. 2010. "Industry Performance: Tourism Indicators." www.tca.gov.bc.ca/research/IndustryPerformance/TourismIndicators.htm.

City of Victoria. n.d. "Quality of Life." www.victoria.ca/common/pdfs/profiles_city_qlty.pdf.

Cruise BC. 2009. "BC Cruise Contributes $1.6 Billion to Economy in 2009." www.cruisebc.ca/pdfs/CBCNews-Fall09.pdf.

Iisaak. 2009. "History of Iisaak Forest Resources." iisaak.com/ourHistory.html.

Integrated Land Management Bureau. 2010. "Adventure Tourism." www.ilmb.gov.bc.ca/adventure_tourism/index.html.

Land and Water British Columbia. 2004. "Banner Year for Kootenay Ski Resorts." www.biglines.com/articles/banner-year-kootenay-resorts.

Port Metro Vancouver. 2009. "Revenue Passenger Performance." www.portmetrovancouver.com/Libraries/ABOUT_Facts_Stats/oct09_cruise.sflb.ashx.

Port of Seattle. 2010. "Cruise Seattle: 2010 Fact Sheet." www.portseattle.org/downloads/seaport/Cruise_Facts_20100208.pdf.

Statistics Canada. 2004. "Foreign Exchange Rate in Canadian Dollars, Computed Quarterly Average." CANSIM Table 176-0064. www.statcan.gc.ca/.

Times Colonist. 2007. "Victoria Gets an 'A' for Attractive." 13 December. www.canada.com/victoriatimescolonist/news/story.html?id=fcf26fc6-e109-4832-813c-2572f94f711b&k=40310.

Tourism British Columbia. 2004. "Study Uncovers Positive Findings for British Columbia's Diving Industry." News release, 15 July. www.tourism.bc.ca.

–. 2009. *The Value of Tourism in British Columbia.* www.tca.gov.bc.ca/research/IndustryPerformance/pdfs/tourism_indicators/Value_of_Tourism_in_British_Columbia.pdf.

Vancouver Airport Authority. 2010. "2009 Annual and Sustainability Report." www.yvr.ca/ar/2009/pdfs/2009-Annual-and-Sustainability-Report.pdf.

VancouverIsland.com. 2010. "Pacific Rim National Park, West Coast Vancouver Island, BC." www.vancouverisland.com/parks/?id=404.

Western Economic Diversification Canada. 2010. "Success Stories: Port of Prince Rupert Terminal." www.wd.gc.ca/eng/11602.asp.

World Tourism Organization. 2010. "Facts and Figures." www.unwto.org/index.php.

Single-Resource Communities

Fragile Settlements

The arrival in British Columbia of Europeans, Asians, and other non-Natives was largely motivated by the expectation of wealth through resource exploitation. There was also money to be made in building the infrastructure needed to export these commodities and in constructing resource-based communities. Many communities were based on the harvesting or processing of a single resource such as fish, forests, minerals, agricultural products, and much more recently, tourism. These **single-resource communities** took a number of forms and their populations varied, reflecting the economic and social values of the day. Haphazard mining communities such as Emory Creek, Boston Bar, and Barkerville were constructed almost overnight in response to the gold rush of the late 1850s and early 1860s. A silver boom in the Kootenays in the 1890s resulted in similar disorganized, hastily constructed mining communities such as Silverton and Sandon. These communities were made up largely of individual miners and small companies operating with relatively simple technologies.

Company towns were also part of the early settlement landscape of British Columbia. These single-resource communities were built and controlled by large corporations that owned the mine, smelter, mill, or cannery. Coastal pulp mills such as Powell River, Port Mellon, and Woodfibre were all company towns that appeared in the early 1900s. Mines and smelters also required large investments, and many of them were the basis for company towns (e.g., Britannia Beach, Cumberland, and Trail). A company town was distinguished from other single-resource communities in that it was built by the company, which "exercised ownership and control over land, housing, recreation, plant, and all assets and facilities of the community including hotels, hospitals, golf courses, churches and movie theaters" (Bradbury 1979, 54). These communities were also not regulated by the Municipal Act.

Single-resource communities, like the province as a whole, were subject to a host of influences over time: changing demands for resource commodities, technological developments, political decisions affecting resource development, and unpredictable global events. Nonetheless, these developments have not had a uniform effect either throughout the province or on the development of single-resource communities. Figure 15.1 illustrates the typical pattern of development for these communities. The fate of some was sealed from the start, as the resource on which they were based was nonrenewable or not in great supply, technologies changed, or world economic price declined. Many single-resource communities became ghost towns; some lost their original resource but reinvented themselves through the development of other resources, while others grew and expanded their economic base until they were no longer classified as single-resource communities. Still others, such as Powell River and Trail, have remained economically viable as mature, incorporated single-resource communities.

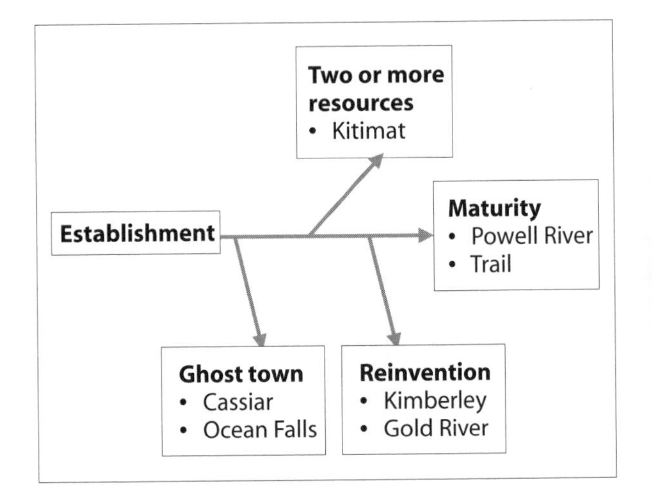

Figure 15.1 Development of single-resource communities

Lucas (1971, 12) found in his study of single-resource communities throughout Canada that they differed considerably based on the resource and on structure of employment: "Agricultural communities, for example, are made up of independent capitalists, and the concerns and patterns of life are quite different from those in a mining community, where miners work as employees of a large bureaucratic organization." He also recognized that geographic isolation was a common factor for single-resource communities, almost guaranteeing that they remained with a single industry (p. 394). In the mountainous landscape of British Columbia, isolation has certainly been a factor.

The influence of large corporations and their control of company towns and employees eventually became the target of government policy. A major reassessment of urban and rural policy in British Columbia coincided with the significant expansion of the mining, energy, and forestry resource sectors from the 1950s to the early 1970s. This period is referred to as the megaproject era: a time of very large resource extraction projects with large and expensive technologies and, most often, foreign corporations in control of the investment. The provincial government, recognizing that its economic well-being came from the harvesting and processing of resources, saw new single-resource communities on the resource frontier as inevitable.

The Instant Towns Act of 1965 was enacted by the provincial government largely in response to the negative conditions of the company towns. Bradbury points out some differences between company towns and **instant towns**. Instant towns had "a much greater degree of advance planning than had been the case in earlier single enterprise communities ... [This] reflected an attempt to correct some of the systemic problems of the traditional company town such as high population turnover, unbalanced demographic profiles, settlement impermanence, social instability, isolation and corporate dominance and control; it was also an attempt to be rid of the stigma of the company town label" (Bradbury 1978, 118).

The company town reflected the times and conditions in the first half of the century, but times and conditions were quite different by the 1960s. Large multinational corporations were in control and investing in capital-intensive technologies, markets were much more global and uncertain, and governments at all levels played a greater role in everyday life. In many ways these were good times economically – and the instant town reflected these new opportunities. Yet the boom economy was not to last. The destabilization of currencies with the removal of the gold standard, the energy crisis, recessions, and the transition to post-Fordism resulted in the restructuring of resource- based economies (see Chapter 7). In this turbulent economic climate, instant towns and all resource-based communities were highly vulnerable, with their basic survival often in doubt.

This chapter compares life in and the conditions of the historical company town to the much more recent instant town. Were the instant towns much better communities in which to live and work? As well, it examines urban policy for the future. What are the options for living conditions and communities for workers as the resource frontier develops?

THE COMPANY TOWN

During the gold rush of 1858, 25,000 to 30,000 people came to British Columbia to find gold (Chapter 4). Numerous communities were rapidly constructed, the largest of which was Barkerville. As fewer and fewer gold discoveries were made, many became ghost towns. These were some of the first single-resource communities in British Columbia. They were hastily built affairs, mainly of frame structure, and most were at high risk of fire. They were not company towns, because they were settled by independent miners. The company town was defined by the fact that the company not only owned and operated the mine or mill but also built the town, owned the housing, and ran the company store. Even later, during the silver boom of the West Kootenays, newly established communities such as Silverton, Sandon, and Three Forks did not qualify as company towns, even though some rather large companies operated the mines adjacent to them.

The company town became very common on the landscape of British Columbia by the early 1900s. The early coal mining communities of Nanaimo and Cumberland were two of the first company towns. Coastal pulp mills soon followed, and later mines such as Britannia Beach and Cassiar became company towns. In the early days, almost all of these single-resource communities existed in remote settings. Few were linked by rail, and coastal communities were serviced only by slow-moving vessels. Even Britannia Beach, today only minutes away from Vancouver on the way to Squamish and Whistler, was hours away by boat in the 1920s, and Ocean Falls was days away. Movement was not easy.

Company towns did not have a normal distribution of population. There were few women and children, and no elderly people. Single men dominated the workforce and the community, although this changed as the corporations realized that married workers were more stable and less likely to move. Stability was of particular importance as jobs became more skilled and the turnover of labour caused a decrease in productivity. The trade-off for the company was to provide housing for married men and their families and to invest in one-room schools. Nevertheless, education was not a priority.

The work day and work week were much longer than today, with twelve-hour days six days a week and low wages. This left little time for social activities, except on Sundays. The long hours of work were often in hazardous conditions, especially in underground mines, where there was the danger of cave-ins, blowouts, and diseases related to the inhalation of dust particles. Even minor accidents were often fatal because of the lack of proper medical knowledge and facilities, and the inability to transport people to the few hospitals the province had (McGillivray 1980, 3).

These rather dismal working and social conditions describe a general way of life that was not confined to company towns. The struggle of the working class to improve its lot was not easy anywhere on the resource frontier, and in the company town the struggle for change was at its most difficult.

Distinguishing the company town from other single-resource communities was the control that the corporation had over its labour force. These were "closed" communities, exempt from the Municipal Act, and the company set the rules: "Most company towns existed outside the municipal laws applied to other urban settlements" (Bradbury 1979, 54). The provincial government did try to make them more open: "Attempts had been made, in 1919 and 1948, to penetrate the overall control maintained by the companies over their towns, but little was attempted in terms of changing the ownership of the settlement or altering the ownership of the means of production by large corporate bodies" (p. 54). Unions were seen by corporations as a threat to profits and management rights. They were banned, and workers agitating to form unions within company towns were quickly fired and sent packing (Walker 1953, 3). There was no room for organized labour in a company town.

Housing was another avenue of company control, and it reflected class, status, and ethnicity. Company towns were planned on the principle of residential segregation (Bradbury 1978; Porteous 1987). This form of spatial organization separated workers according to status, with dormitories for single men, small cottages for married labour, larger houses for foremen, and substantial housing (preferably on a hill overlooking the town site) for managers. Chinese, First Nations, and other visible minorities were housed in separate locations, such as across a stream. All rents were deducted from the paycheque.

The company store represented even greater control, as it was the only source of both basics and luxuries. Given a monopoly on retail goods, it was not uncommon for the company to charge exorbitant prices, and workers were encouraged to use a system of credit to pay for any purchases. A "bob-tailed" cheque was a pay slip without any funds, because the worker owed more to the company than he received in wages at the end of the pay period. The control over labour was powerful and multifaceted.

The struggle to change this system was not confined to British Columbia. It was much more universal. New political and social philosophies were tempered and tested by world war and depression. And conditions did change. Unions formed in British Columbia and became ever stronger and more militant in their demands, political parties advocated economic and social change, and technologies changed inside and outside the workplace. Better rail and road systems reduced isolation and placed even greater pressure on company towns to become more open. Slowly, the towns transformed as workers commuted from adjacent or nearby communities, breaking the company's control over housing and its monopoly on retail sales. Better working conditions, with improved safety, fewer hours, and improved pay, also became part of the transition as mines and mills reluctantly accepted a unionized workforce. Mergers and takeovers resulted in larger and larger corporate entities with greater global connections and investments.

By the end of the 1960s, British Columbia was in the process of an unprecedented expansion of resource development. Large multinational corporate investment was encouraged by the provincial government through tax incentives and provision of roads and energy supplies. As the resource frontier pushed forward into relatively isolated locations, new communities were needed to house workers. Not wanting to carry on the negative company town experience, the provincial government passed the Instant Towns Act in 1965.

INSTANT TOWNS

Figure 15.2 shows the main instant towns across the province. Initially eight instant towns were built between 1965 and 1971: Gold River (1965), Mackenzie (1966), Sparwood (1966), Fraser Lake (1966), Logan Lake (1970), Tahsis (1970), Elkford (1971), and Granisle (1971). Tumbler Ridge, in the Peace River region, was the last instant town to be created, in 1981. Table 15.1 traces the population of these towns to 2009. Other instant towns were planned, mainly around mineral extraction, but because of lower world market prices for metals, these communities never materialized. The instant towns were planned with the involvement of corporations and the provincial government.

Instant towns came under the Municipal Act and were therefore open communities with locally elected councils, unlike company towns. Curvilinear streets, smart three-bedroom houses, apartment complexes for single employees, and mobile home parks characterized their layout and housing options. Many workers owned their own homes. Retailing in the instant town differed greatly from the company town's monopoly via the company store. The instant town included a small shopping centre with independent retail franchises that suited small communities. Most instant towns did not have sufficient population to warrant a hospital, but clinics and dental offices were set up for regular visits by medical professionals. Modern elementary schools were constructed, and when a community became large enough, a high

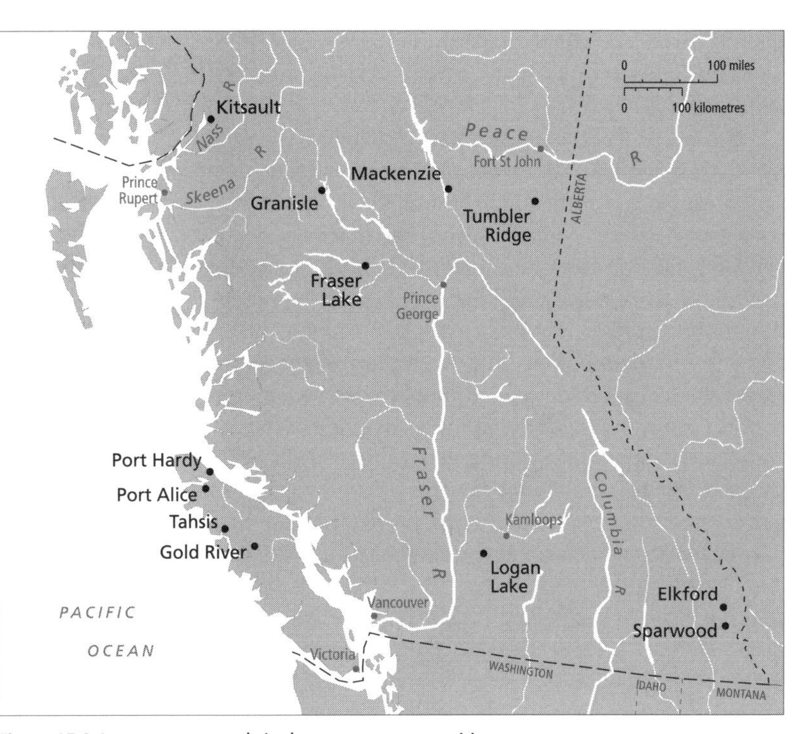

Figure 15.2 Instant towns and single-resource communities
Source: Modified from Porteous (1987, 385).

Table 15.1

Population of instant towns, 1971-2009

Town (year established)	1971	1981	1986	1991	1996	2001	2003	2006	2009
Gold River (1965)	1,896	2,225	1,879	2,130	2,041	1,359	1,340	1,362	1,425
Mackenzie (1966)	2,332	5,890	5,542	5,790	5,997	5,206	5,316	4,616	3,827
Sparwood (1966)	2,990	4,161	4,540	4,135	3,982	3,812	3,954	3,680	3,804
Fraser Lake (1966)	1,292	1,182	1,545	1,300	1,344	1,268	1,267	1,129	1,112
Logan Lake (1970)	3	2,637	2,001	2,380	2,492	2,185	2,287	2,198	2,189
Tahsis (1970)	1,351	1,739	1,445	1,090	940	626	559	366	381
Elkford (1971)	–	3,126	3,187	2,925	1,729	2,589	2,671	2,517	2,591
Granisle (1971)	–	1,430	646	805	446	353	349	365	396
Tumbler Ridge (1982)	–	–	4,566	4,794	3,775	1,851	2,459	2,491	2,189

Sources: BC Stats (2006, 2010).

school was added. A recreation centre with a curling rink, an arena, and a gymnasium was a standard facility in the instant town.

A comparison of the old company town with the new instant town reflects differences in time, structure, and attitudes. On the surface, the instant town appears to represent a radical departure from the company town era. Unionization guarantees safer work conditions, the forty-hour work week, and high hourly wages. Recreational opportunities abound, especially for those who enjoy the outdoors. Vastly improved road systems and satellite television receivers connect the most isolated communities. Yet, the fundamental existence of the instant town, dependent on the economic viability of a single resource along with heavy corporate investment in the main source of employment, has not changed.

The film *No Life for a Woman,* produced by the National Film Board in 1979, was based mainly on the instant town of Mackenzie. As its title implies, the film depicts a male-dominated town where the council prioritizes decisions that facilitate the industry, where there are no women's centres or daycare facilities, and where women have few employment opportunities outside the home. Living in a trailer park, where units are crammed together and offer little privacy, causes anxiety and tension, especially in the winter when heavy snowfall makes it impossible to get out. Indeed, these communities tend to have high rates of violence and suicide, and of drug and alcohol abuse. Many of these social problems have been recorded by Lucas (1971), Bradbury (1978), McGillivray (1980), and Porteous (1987). Patricia Marchak (1983, 303), in her examination of instant towns related to the forest industry, recognized similar problems: "By their nature, such towns are isolated, with shallow roots."

The instant town was created by corporate economic interests based on resource extraction. This is a critical point, because the world market prices for resource commodities are rarely stable and a corporation may decide to shut its plant down temporarily or permanently when market price drops. The instant town of Kitsault, northeast of Prince Rupert, is a case in point. No sooner had it been planned and the first buildings erected than it became an instant ghost town (it was put up for sale for $7 million in 2004 and sold in January 2005 for an undisclosed sum). The world market price of molybdenum, the impetus for building the town, was in decline. With little prospect of an increase in the short or long run, the corporation pulled out.

The landscape of uneven development results from the multiple influences of wars, recessions, depressions, new technologies that make resources obsolete, and decisions by multinational corporations about resource procurement. The impacts are both unpredictable and uneven. Some mines and mills close while others remain in operation. The inevitable vulnerability and uncertainty is the basis of many of the social concerns within resource-dependent communities.

Forestry and mining are capital-intensive industries, and employees sometimes face being replaced by machinery. The Sullivan Mine at Kimberley employed 2,122 people in 1949, for example, but this was reduced to 660 by 2000, and then the mine closed in 2001 (Ward 1998, C4). Figures for pulp-and-paper mills and sawmills show similarly shrinking employment numbers due to capital-intensive technologies (Marchak 1983; Barnes et al. 1992; Rees and Hayter 1996; Halseth and Sullivan 2002).

The new instant towns, sanctioned by public policy, were "sold" to the public of British Columbia as permanent communities, and workers were encouraged to take out mortgages to purchase housing. Writers such as Bradbury (1979) suggest that the provincial government used the Instant Towns Act to facilitate corporate investment, trapping the worker in the process by shifting the cost of housing from the company to the employee while considerably reducing the employee's option of leaving. For the worker with a mortgage, anxiety increased as British Columbia experienced restructuring, layoffs, and, in some cases, closure of the resource operation. With the serious economic recession of the 1980s, the federal government recognized the vulnerability of single-resource communities in a report entitled *Canada's Single-Industry Communities: A Proud Determination to Survive* (Canada, Ministry of Supply and Services 1987). The report suggested that planning for single-resource communities had been inadequate: "The absence of long term planning and involvement in the past, and to a lesser extent in the present, has resulted in a rather haphazard development of the country's resource frontier" (p. v).

Government policy in the 1960s did not address the fundamental vulnerability of single-resource communities. Rather, by encouraging the construction of instant towns, the provincial government gave these communities the illusion that they were permanent.

Randall and Ironside (1996) question the generalizations and stereotypes about single-resource communities and shed new light on their composition. They show that the communities are not homogenous, that they have far more women in the workforce than has been assumed, and that they do not depend totally on a single resource for employment. Still, an obvious strategy for single-resource communities is to develop a broader economic base, both for the existing resource and for other resources. The town of Chemainus, the "Little Town That Did," is often held up as a prime example of a successful response to the recession of the early 1980s (Rees and Hayter 1996; see also Chapter 7).

Four of the towns listed in Table 15.1 exemplify the challenges that instant towns face following the permanent closure of their single resource. The copper mine at Granisle closed in 1992; the pulp mill at Gold River, on the west coast of Vancouver Island, closed in 1999; the saw mill at Tahsis closed in 2001; and the coal mines at Tumbler Ridge in the Peace River region – Quintette and Bullmoose – closed in 1999 and 2003, respectively. Unlike Chemainus, all four of these communities are relatively isolated, thus making the transition to tourism, the retirement industry, or other economic activities a challenge. As Table 15.1 indicates, however, these towns are not ghost towns and continue to thrive principally on tourism and the retirement industry. In the case of Tumbler Ridge (as mentioned in Chapters 11 and 14), coal mining rebounded and two new coal mines opened, giving the community some stability, at least until the mines no longer function. Communities such as Kimberley were in a similar situation when the Sullivan Mine, the longest-operating mine in the province, closed in 2001; and they too have successfully diversified into tourism, recreation, retirement, and forestry-related employment. The video recording *Britannia: A Company Town* (2000) relates the history and decline of Britannia Beach and how, once the mining and smelting came to an end, its residents attempted to retain their community and deal with the legacy of environmental issues.

The world market price for metals, forest products, coal, oil, natural gas, or even new resources is likely to rise, at which point it will be necessary to build new communities on the resource frontier. What lessons have been learned from past experience and what form will new urban policy take? Porteous (1987) has proposed several alternatives to instant towns: communities deliberately structured for commuters to go to; expansion of existing centres; and new large, permanent cities in these isolated regions. Permanence is the key; if it had been applied as a decisive principle, government policy would not have permitted instant towns such as Logan Lake. In that particular example, the nearby existing community of Ashcroft would have been expanded. The federal government has made this transition with its policy of resource development in the Yukon, Northwest Territories, and Nunavut. Single-resource communities in these territories are dwindling, and new mining operations "now tend to rely on flying in the bulk of the labour force on a rotational basis, rather than establish[ing] permanent communities" (Usher 1998, 380). Alternatively, if the frontier has numerous resources to develop where no communities currently exist, then a relatively large, permanent instant city could be developed with a broad range of services based on several resources. Randall and Ironside (1996) suggest a similar strategy for existing single-resource communities, namely, the expansion of the economic base into two or more industries.

As discussed in Chapter 7, British Columbia has evolved a core-periphery relationship in which the periphery relies to a large degree on resource exploitation. The uncertain global economic climate, especially since the 1970s, has revealed the difficulties in mines or mills remaining economically viable. The problem has unhinged the economic stability of single-resource communities in particular. One lesson for the provincial government is to review its urban policies as new resources develop in isolated locations. For existing single-resource communities, it is essential to explore ways to diversify the economic base. Provincial policy could establish a reserve fund to assist in relocation, restructuring, and other hardships endured by members of single-resource communities when their mine or mill closes.

SUMMARY

Earlier chapters focused on individual resources such as forests, fish, and minerals, tracing the factors responsible for change and the impact these industries have on the landscape of British Columbia. An examination of single-resource communities shifts the focus from resource development to the human factor of workers and their living and working conditions.

The early single-resource communities, particularly mining towns, were by and large unplanned and hastily built efforts to house miners and provide services. Many became ghost towns. Company towns, with much greater investment in plant and equipment and a complete infrastructure – housing, company store, services, and so on – became popular by the early twentieth century. The provincial government encouraged both the investment and the employment. The first half of the century saw changes in work conditions, unionization, and expectations about working and living conditions. The company town, with its overwhelming control of workers and all who lived within its borders, became an anachronism and an embarrassment to the provincial government.

New single-resource communities were required for the megaproject era that began in the 1960s. Consequently, a new urban model was conceived and promoted by the provincial government: the instant town. These open communities began to appear by 1965. They were different from company towns in all respects except their dependence on a single resource, and in most cases, a single corporation. Changing economic conditions from the 1970s onward – an increasingly global economy, competition, and uncertainty – led to tough economic times and the realization that it was the corporations that benefited most from the Instant Towns Act.

Caution should be exercised in creating any single-resource communities in the future; they come with expectations of permanence, and dependence on a resource is anything but permanent. For existing single-resource communities, therefore, the struggle is to diversify their economic base. Not to do so could add more ghost towns to the map of British Columbia. Of course, the ability to diversify depends on many factors, such as geographic location, accessibility, and the availability of other resources. It is a challenge.

REFERENCES

Barnes, T.J., D.W. Edgington, K.G. Denike, and T.G. McGee. 1992. "Vancouver, the Province, and the Pacific Rim." In *Vancouver and Its Region,* ed. G. Wynn and T. Oke, 171-99. Vancouver: UBC Press.

Bradbury, J.H. 1978. "The Instant Towns of British Columbia: A Settlement Response to the Metropolitan Call on the Productive Base." In *Vancouver: Western Metropolis,* ed. L.J. Evenden, 116-29. Western Geographical Series vol. 16. Victoria: University of Victoria.

–. 1979. "Towards an Alternative Theory of Resource-Based Town Development in Canada." *Economic Geography* 55 (2): 47-66.

Canada, Ministry of Supply and Services. 1987. *Canada's Single-Industry Communities: A Proud Determination to Survive.* Ottawa: The Ministry.

Halseth, G., and L. Sullivan. 2002. *Building Community in an Instant Town: A Social Geography of Mackenzie and Tumbler Ridge, British Columbia.* Prince George: UNBC Press.

Lucas, R.A. 1971. *Minetown, Milltown, Railtown: Life in Canadian Cities of Single Industry.* Toronto: University of Toronto Press.

McGillivray, B. 1980. "Single Resource Communities in Canada: The Vulnerability of People and Communities, Past and Present." *Canadian Studies Bulletin* (November): 3-6.

Marchak, P. 1983. *Green Gold: The Forest Industry in British Columbia.* Vancouver: UBC Press.

Porteous, J.D. 1987. "Single Enterprise Communities." In *British Columbia: Its Resources and People,* ed. C.N. Forward, 382-99. Western Geographical Series vol. 22. Victoria: University of Victoria.

Randall, J.E., and R.G. Ironside. 1996. "Communities on the Edge: An Economic Geography of Resource-Dependent Communities in Canada." *Canadian Geographer* 40 (1): 17-35.

Rees, K., and R. Hayter. 1996. "Enterprise Strategies in Wood Manufacturing, Vancouver." *Canadian Geographer* 40 (3): 203-19.

Usher, P. 1998. "The North: One Land, Two Ways of Life." In *Heartland and Hinterland: A Regional Geography of Canada,* 3rd ed., ed. L. McCann and A. Gunn, 357-94. Toronto: Prentice-Hall.

Walker, H.W. 1953. *Single Enterprise Communities in Canada.* Report to Central Mortgage and Housing Corporation by the Institute of Local Government. Kingston: Queen's University.

Ward, D. 1998. "A Town That Wouldn't Die." *Vancouver Sun*, 18 October, C1, C4.

FILMS

Kreps, B., director. 1979. *No Life for a Woman*. Documentary, 26 mins. Serendipity Films and National Film Board of Canada.

Vaisbord, D., director. 2000. *Britannia: A Company Town*. Documentary, 46 mins. Screen Siren Pictures and National Film Board of Canada.

INTERNET

BC Stats. 2006. "British Columbia Municipal Census Populations, 1921-2006." www.bcstats.gov.bc.ca/data/pop/pop/mun/mun1921_2006.asp.

–. 2010. "Population Estimates: Total Populations Only." www.bcstats.gov.bc.ca/data/pop/pop/estspop.asp#totpop.

Urbanization

16

A Summary of People and Landscapes in Transition

British Columbia entered the new millennium as a highly urbanized society, and Vancouver stands out as a world city – a hub in the global network (Short and Kim 1999, 53) – with a population expected to exceed 3 million within the next two decades. No other urban centre in the province will come close in size, but it is increasingly recognized that the spatial order of employment and most of the decisions that affect the province will be made in cities generally. Vancouver, a major urban complex with links to the Pacific-based economies, will show the greatest growth. A hierarchy in which one large urban agglomeration dominates over many smaller communities is not random. The evolution of this urban system is best understood from a historical perspective.

The unique physical characteristics of the province, with its vertical and rugged landscape, have been significant in confining settlement and urbanization mainly to river valleys and coastal inlets. The various patterns of urbanization reflect 200 years of non-Native settlement and development. The antecedents of the current urban system are the fur trade forts erected to secure and control territory for British colonial interests, principally for material gain. Competition for territoriality was initially between the British and the Spanish and, to some degree, the Russians. Later, there was even more serious and sustained competition with the Americans. These struggles shaped and reshaped the political boundaries that eventually became British Columbia. They also refined the basis of its urbanization and the important connecting links between centres.

The early fur trade posts represented the first semi-permanent presence of non-Natives and the disintegration of the traditional way of life for First Nations. The forts increased dependence on European goods and were focal points for trade, missionary zeal, and the spread of diseases, although European diseases had reached the west coast of North America long before forts appeared. With the discovery of coal, gold, and silver, and the development of forests, fish, and other resources, the forts gave way to resource- and transportation-based communities. For First Nations, the arrival of Europeans caused an accelerating process of deterritorialization as well as cultural destruction through waves of disease, alcohol, a reserve system without treaties, and major pressures to assimilate (Table 16.1). First Nations were reterritorialized in a landscape shaped largely by British values of civilization that were well represented in the emerging urban centres, often with adjacent reserves and residential schools. First Nations were set aside on small parcels of land but became part of the urban fabric over time as they sought economic opportunity that was rarely available on the reserve. Asian populations were treated in a similarly discriminatory way, and relegated to racially segregated regions of the community (e.g., Chinatowns).

Table 16.1

Native and non-Native population, 1782-1870

	Native	Non-Native
1782	>200,000	n/a
1835	70,000	n/a
1854	n/a	450
1858-60	47,000	25,000-30,000
1870	25,661	10,586

Sources: Harris (1997, 28, 30); Muckle (1998, 37); Duff (1965, 2); Robin (1972, 14); Meen (1996, 97, 109).

Staples theory, discussed in Chapter 7, can assist in understanding the growth of communities as resource after resource was discovered throughout British Columbia and exploited for export. Political decisions about the geographic location of important backward linkages, such as the Canadian Pacific Railway (CPR) and other rail lines, were factors in time-space convergence, often giving economic growth to the communities created or connected by the railway. In some cases, growth was significant. Railway companies, and the CPR specifically, acquired enormous economic advantage by taking control of much of the best land and resources in the province. The urban system that evolved also reflected corporate land use decisions.

Urbanization is a dynamic process that concentrates both population and numerous functions, including the sale of goods and services, financial transactions, trade, manufacturing, and government and administrative services. In many ways, cities are the repository of societal values. As these values change, they modify the landscape. The growth, and sometimes decline, of communities, especially smaller ones, is tied mainly to the viability of a changing economic base. In British Columbia, a fragile economic base turned some communities into ghost towns. Others experienced rapid and often unplanned growth due to new industries, transportation developments, and an expanding economic base. Urban centres also grew because of amalgamation or annexation. The boundaries of many communities have been redrawn, in some cases many times, to incorporate more and more of the surrounding territory.

Population growth and urbanization accelerated after the Second World War. Economic conditions resulted in an unprecedented demand for the forests and minerals of British Columbia. New capital-intensive technology and control by large corporations significantly influenced migration from rural to urban settings. New resource towns were created, major hydroelectric dams were built, pipelines criss-crossed the province, and roads, railways, and airports expanded throughout the megaproject era. Most urban centres, new or old, expanded during this boom period.

The dynamics of post-Fordism – including increased international competition, corporate control and concentration of resources by multinationals, and a host of global economic crises – particularly challenged resource-dependent communities. Global conditions also provided opportunities for banking, finance, health, education, real estate, tourism, and many other urban services in the new knowledge- and information-based economy. A spatial pattern emerged that favoured large urban centres. New technologies of transportation and communication also developed at this time, radically changing the traditional form of urban centres and resulting in location decisions that blurred the distinction between urban and rural. Automobile suburbs, shopping centres, and industrial parks were the first developments to cause the relocation of traditionally centralized business functions. The new global communications technologies of fax, e-mail, Internet, e-commerce, and a host of information service providers, all available to the individual household, have only accelerated the move to often less expensive dwellings at ever greater distances from the metropolitan centre. There are opposite forces, such as agglomeration economies, that generate ever greater service employment in urban centres (e.g., banking, finance, administration, law, entertainment, and many other activities) and result in high-rise office complexes, entertainment facilities, and convention centres. These employment opportunities are a spur for condominium construction, thus increasing the density of urban cores.

The overview of British Columbia presented in this chapter, traced mainly through the development of communities and an evolving urban system, is developed within three periods. The first is the British control of the territory through the Hudson's Bay Company until the province joined Confederation in 1871. This period set one pattern for urbanization. The second, from Confederation to 1951, saw an explosion of urbanization influenced by railway development in combination with resource exploitation and land speculation. The third period spans the post-Second World War boom to the present. These three are tied to census data periods to show clearly the statistical basis of population change and urbanization.

The terms "community" and "urban centre" are used synonymously here, although the census defines urban centres more specifically. Up to 1951, incorporated communities were considered urban, and all other population was rural. An incorporated community has a geographically defined boundary that encompasses fairly dense population and provides a level of local autonomy with an elected mayor and council. Each province has its own way of defining incorporated urban areas. In British Columbia, the Municipal Clauses Act (1896) classified incorporated places into cities and district municipalities. As Donald Higgins (1977, 36) explains, "while city status was presumably intended for the more densely populated settlements and district municipalities for rural areas, such a distinction became lost over time. Some cities had as few as two hundred residents, and some districts close to forty thousand people." He also points out that the category of "village" was created in 1920 and "town" was added in 1958.

The first two incorporated municipalities in British Columbia pre-date Confederation: New Westminster (1860) and Victoria (1862). The Consolidated Municipal Act (1872) encouraged more incorporations. By the census definition, however, there were only two urban centres in British Columbia in 1871, three in 1881, and four by 1891. Nevertheless, many other unincorporated communities both prior to and during this time were very much a part of the evolving urban system. From 1961, the census definition of urban also included unincorporated areas if they had a high enough population density per square mile ratio (initially 1,000 per square mile, although this changed again in the 1981 census). Statistics used in this chapter are from the census and reflect the changing definitions of "urban."

SETTING AN URBAN PATTERN ON THE WESTERN FRONTIER

As observed in Chapter 4, the British succeeded in adding what is now called British Columbia to their colonial empire in pursuit of valuable sea otter pelts, but the nature of the sea otter trade did not require permanent forts on the coast. Permanent forts did appear in British Columbia by the early 1800s, however, as part of the network of fur trade forts that the North West Company and later the Hudson's Bay Company extended across the country. These forts served many functions, of which the entrepôt trade with First Nations was the most obvious. Year-round fortified communities of a few Europeans represented a military presence, colonization, and the presence of British values and institutions. This was the beginning of reterritorialization. Maps indicating the location of these forts and their transportation networks became a base of knowledge for others, whether missionaries or fortune seekers (see Figures 4.1 and 4.2, pp. 68 and 69).

Fort Victoria, constructed at the southern end of Vancouver Island, was much more than a fur trading post. A.H. Siemens (1972, 13) calls it the first permanent coastal settlement in British Columbia. Although it was established to secure British interests, the community also had land for farming and forests, and fish and coal were readily accessible for exploitation by the Hudson's Bay Company. Fort Langley, the Hudson's Bay fur trade fort at the western end of the Fraser Valley, turned to farming and the salmon trade as furs were exhausted. While the interior forts still relied on the fur trade, a number of forts appeared on the coast as a result of the increasing demand for coal. Fort Simpson, near present-day Prince Rupert, and Fort Rupert, at the north end of Vancouver Island, were producing coal by the 1840s. By the 1850s, the Hudson's Bay Company had developed the rich seams of coal on the southeastern coast of Vancouver Island and built the new town of Nanaimo.

The discovery of gold, first on the Queen Charlotte Islands, aroused economic and political interest in securing this territory for the British. These initial discoveries were neither large nor sustained. The major one was on the lower reaches of the Fraser River, and by 1858 some 25,000 to 30,000 gold seekers had arrived. Many came from the exhausted goldfields of the San Francisco Bay area, while others came from across Canada, Europe, and China. By the early 1860s, further discovery of significant amounts of gold up the Fraser and into the Cariboo was the main catalyst in opening up British Columbia and changing its landscape permanently.

Gold along the Fraser provided the most important single impetus for early rural and urban settlement in the province (Siemens 1972, 11). Frame-structured towns such as Emory Creek, Boston Bar, Richfield, and Barkerville were rapidly erected adjacent to the gold discoveries. Other communities, such as Port Douglas, Yale, Lillooet, and Quesnel (named Quesnelle Mouth then), acted as service centres on the various routes up the Fraser and into the Cariboo. Other, more minor discoveries of gold were made in the 1860s on the Stikine, Peace, and Columbia Rivers. They did not become large enough to warrant infrastructure investments such as the Cariboo Road, nor were communities developed, but by 1863 these finds had been instrumental in shaping the political boundaries of the territory: north to the sixtieth parallel and east to the 120th degree of longitude.

The lower Fraser River and Cariboo gold rush left a significant legacy for settlement and urbanization. This was a turning point for British Columbia. Before the gold strike, the area had been under the control of the Hudson's Bay Company, which was interested in extracting its resources with a minimum amount of permanent settlement. The construction of the Cariboo Road between 1862 and 1864 placed a permanent line of connected communities on the map of British Columbia, from Victoria by paddlewheeler to New Westminster and Yale, and then by road to Barkerville. The urban system was evolving with Victoria as its anchor.

Victoria was the largest centre and owed its growth to its location as the main entrance and exit to the territory. It became a service centre for travellers as well as the gateway for the import and export of goods. Victoria was also the main political, administrative, and financial centre, especially after 1866, when the colonies of Vancouver Island and British Columbia were joined. Growth also occurred because of manufacturing, "including tanneries, sash mills, a foundry, a soap factory, a ship yard and a gas works," and by 1870, "the brewing and distilling industries were concentrated around Victoria" (Meen 1996, 111). Nanaimo emerged as the main supplier of coal during the early 1860s. New Westminster, while losing a bid to become the capital city, was another important centre, providing services for a growing farming settlement in the adjacent Lower Fraser Valley.

British values incorporated a white supremacist attitude that saw all visible minorities as inferior and groups such as the Chinese, who represented a significant percentage of the non-Native population by the mid- to late 1860s, as undesirable. These racist views resulted in Chinatowns being constructed as separate enclaves in the urban landscape of British Columbia.

Women were scarce in the territory, although Sharon Meen (1996, 111) informs us that the few women who lived in the Cariboo "seized this unusual opportunity, running hotels, restaurants, saloons and laundries. In 1869, twelve were listed as business women in their own right." She also relates the arrival of two bride ships from Britain that delivered approximately a hundred women to Victoria in 1862 and 1863 (p. 116). The effort to correct the imbalance between the numbers of men and women was also an attempt to create a more stable, permanent colony and communities where couples would build homes and raise families.

The gold rush communities were structurally different from corporate-controlled communities such as Victoria and Nanaimo. Placer mining communities were largely haphazard, disorganized, frame-structured, and fire-prone settlements, rapidly constructed and just as rapidly abandoned when the gold ran out. After 1865, the flood of people into the region was rapidly reversed. Emory Creek, Richfield, and eventually Barkerville joined the list of resource-depleted ghost towns, establishing an enduring pattern. Even Victoria, whose population was estimated at 6,000 in 1863, had declined to one-half this size by 1870 (Meen 1996, 117).

By the time British Columbia entered Confederation in 1871, the landscape had undergone major changes. The political boundaries of the province had been shaped and reshaped by international challenges and the lure of gold. The economy had been transformed, as had the communities where most economic transactions occurred. The Hudson's Bay Company, the largest corporate monopoly in Canada, developed many resources other than fur, and its forts served to secure colonial control, but they engendered little permanent settlement outside of their enclosures. The Fraser River and Cariboo gold rush broke this corporate pattern of control, and a new system of urban centres emerged. Thousands of individuals settled the land, while thousands more came to make their fortune and leave. The resulting communities evolved with an extremely racist British value system, which was incorporated into the institutions of religion, justice, education, and land use.

The decision to join Confederation was an important political direction, particularly because provincial status came with the promise of a railway linking central and eastern Canada to the very distant Pacific. Many Americans had arrived with the gold rush, rekindling concerns of annexation, and the US purchase of Alaska in 1867 only heightened these fears. Confederation ended the anxiety.

THE SHAPING OF BRITISH COLUMBIA WITH RAILWAY AND RESOURCE TOWNS, 1871 TO 1951

Railway construction and the anticipated economic boom did not materialize immediately. Growth was slow from 1871 to 1881, but the influence of CPR construction can be seen in the figures for the 1881 census. The population nearly doubled over the next ten years, and by 1921 had increased tenfold, to over half a million people. By 1951, over a million people resided in British Columbia (Table 16.2). Another significant change during this time was the shift from a largely rural-based population in 1871 to a more urban-based one by 1951. Rapid population and urban growth was a response to new technologies of time-space convergence and of resource extraction, to

Table 16.2

Population and urbanization: British Columbia and Canada, 1871-1951

	British Columbia			Canada		
	Population	% rural	% urban	Population	% rural	% urban
1871	36,247	84.8	15.2	3,689,257	80.4	19.6
1881	50,387	72.0	18.0	4,324,810	74.3	25.7
1891	98,173	57.5	42.5	4,833,239	68.2	31.8
1901	178,657	49.5	50.5	5,371,315	62.5	37.5
1911	392,480	48.1	51.9	7,204,838	54.6	45.4
1921	524,582	52.8	47.2	8,787,949	50.5	49.5
1931	694,263	56.9	43.1	10,376,786	46.3	53.7
1941	817,861	45.8	54.2	11,506,655	45.7	54.3
1951	1,165,210	47.2	52.8	14,009,429	38.4	61.6

Sources: Statistics Canada (1983b, series A2-14); (1983c, series A67-69).

new market demands, and to major political and economic growth promotion schemes. New urban patterns also evolved in response.

The Pacific Scandal of the 1870s delayed the promised railway for nearly ten years, and population growth was therefore modest during these years. Coal-mining activity continued, but employment in placer mining diminished and gold-mining communities disappeared. The appearance of fish canneries, initially at the mouth of the Fraser River, provided limited seasonal employment, because only sockeye was considered suitable for canning at this time. The forest industry was gaining importance, with substantial mills on Burrard Inlet and at the south end of Vancouver Island. Pockets of fertile agricultural land, particularly in the southwestern corner of the province, attracted some settlers while ranching in the interior sustained others. Victoria was connected to the regular ocean traffic routes to San Francisco and London, but these were long and slow routes. Transportation within British Columbia consisted of river navigation, the Cariboo Road, and trails – all slow routes as well.

The CPR changed the face of the province in many ways. Initially, there were debates over the route through the province. A major political debate was whether the terminus should be on the mainland or on Vancouver Island and, if it was to be on Vancouver Island, how near to the existing capital city of Victoria. Figure 16.1 outlines the routes of the Canadian Pacific Railway Survey (CPRS), conducted by Sandford Fleming. The survey's importance is pointed out by Frank Leonard (2002, 166): "That two dominion governments made four 'final' decisions on the Pacific terminus certainly underlines the political import of the survey." Two of those "final" decisions confirmed Esquimalt as the terminus of the CPR. The CPR however, calculated the financial choices of

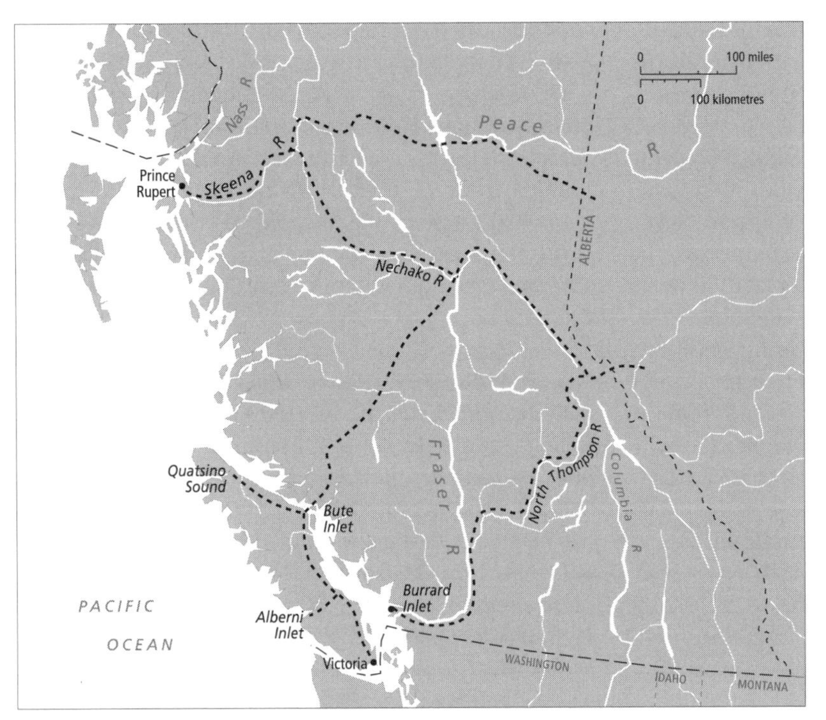

Figure 16.1 Canadian Pacific Railway Survey routes through British Columbia
Sources: Modified from Leonard (2002), 169, 176; Hayes (1999), 177.

each route and showed that the Homathko River, Bute Inlet to Esquimalt route "exceeded the Burrard Inlet by 287 miles and $20 million" (p. 173). A more northerly route to the mouth of the Skeena River (near Prince Rupert), was surveyed and discarded principally based on the survey recommendation: "The Skeena area [was found to be] unsuitable for any significant agriculture because of its poor soil, excessive rainfall and cloudy weather" (p. 177). In the end, the federal government selected the terminus of Port Moody on Burrard Inlet on the basis of financial considerations. The lure of 6,000 acres (2,428 hectares) of land convinced the CPR to extend the rail lines down the inlet to Coal Harbour, creating Vancouver.

Over 15,000 Chinese labourers were brought in to build the railway, and many remained in British Columbia because the promise of return passage to China was never fulfilled. First Nations were still in the majority in 1881 but, decimated by disease, they were outnumbered by 1891 and segregated on small reserves (Table 16.3). The majority of immigrants were from eastern and central Canada or Europe. They viewed this newly connected frontier as an economic opportunity.

The CPR initiated rapid growth, but this era also saw the development of international, continental, and regional linkages. The steam engine transformed the economy and the landscape, mitigating isolation and establishing better connections to national and world markets. Steam technology for moving trains and ships reduced the time-distance between locations and therefore the price of moving bulk goods of low value. The export market for fish, forest, and mineral commodities, all of which were abundant in the province, expanded greatly. Transportation developments facilitated agricultural production and farming settlements.

Technological progress was not restricted to transportation. All sectors of resource extraction underwent change, and in most cases, increased production. As a result, a great number of people were attracted to British Columbia, and land speculation became a major industry.

Railway companies acquired federal and provincial land grants and cash subsidies for the building of rail lines. Crown grants entitled them to the forest and mineral rights adjacent to the railway. John Belshaw (1996, 149) records 116 railway charters between 1883 and 1903, noting that many were for "bogus railways." By 1914 the number of charters had increased to 212 (Seager 1996, 209). Land and its resources had become the negotiating basis of all levels of government and were a strong motivating force in opening up Vancouver Island and the interior of British Columbia.

Many individuals and groups of people seeking a better life either developed or speculated in land. The north end of Vancouver Island and the mid-coast (e.g., Cape

Table 16.3

Native, non-Native, and Asian provincial population, 1871-1951

		Native		Non-Native		Asian	
	Total population	Population	%	Population	%	Population	%
1871	36,247	25,661	70.8	9,038	24.9	1,548[a]	4.3
1881	50,387	26,849	53.3	19,348	38.4	4,195[a]	8.3
1891	98,173	24,543	25.0	64,720	65.9	8,910[a]	9.1
1901	178,657	25,488	14.3	133,687	74.8	19,482[b]	10.9
1911	392,480	20,134	5.1	341,899	87.1	30,447[c]	7.8
1921	524,582	22,377	4.3	462,715	88.2	39,490[c]	7.5
1931	694,263	24,599	3.5	620,320	89.4	49,344[b]	7.1
1941	817,861	24,882	3.0	752,264	92.0	40,715[b]	5.0
1951	1,165,210	28,478	2.4	1,111,693	95.4	25,039[c]	2.2

a Chinese only.
b Chinese and Japanese origin.
c Chinese, Japanese, and East Indian origin.
Sources: Strong-Boag (1996, 276); Kerr (1975, 86); Johnston (1996, 191); Statistics Canada (1983c, series A125-163).

Table 16.4

Incorporated communities by region, 1871-1951

	Total	Vancouver Island/ central coast	Lower Mainland	Kootenay	Okanagan	South central interior	North coast/ northwest	North central interior	Peace River/ northeast
1871	2	1	1	–	–	–	–	–	–
1881	3	2	1	–	–	–	–	–	–
1891	4	2	2	–	–	–	–	–	–
1901	22	4	3	10	3	2	–	–	–
1911	26	4	4	10	4	3	1	–	–
1921	34	8	6	9	5	4	1	1	–
1931	48	8	10	12	5	4	3	6	–
1941	52	9	10	12	6	4	3	6	2
1951	73	15	12	18	8	7	3	7	3

Sources: Census of Canada (1881, table 1); (1951, 6-85); BC Stats (2004a).

Scott, Sointula, and Hagensborg) were populated by Scandinavians who had expectations of vastly improved accessibility to markets through railway access. Early in the 1900s, the residents of Walhachin, a British settlement on the Thompson River south of Kamloops, had expectations of turning the arid landscape into another Okanagan Valley with fruit farming. Later, settlers in the Peace River region had expectations of farming one of the last agricultural frontiers in North America. Twenty thousand homesteads were claimed between 1928 and 1931 in the area, though it should be noted that this figure includes the Alberta portion of the Peace River region (Strong-Boag 1996, 281).

The pace of development after 1881 spawned new spatial patterns of urbanization in this expanded resource frontier. Table 16.4 shows the number of incorporated communities in each decade from 1871 to 1951. Few communities were incorporated until 1901, but then there was steady growth until 1941, and twenty-one new communities were added between 1941 and 1951, signalling the beginning of the postwar boom. The table classifies the communities by region to give some sense of the evolution of economic development (see Figure 1.3, p. 9, for a map of the regions). The regions of Vancouver Island and the Lower Mainland were the first to have incorporated towns, with Victoria and New Westminster, respectively. The development of coal added Nanaimo by 1881, and Vancouver was a significant addition by 1891. The rapid increase in incorporated communities by 1901 shows the development of the Kootenays, Okanagan, and the railway communities in the south central interior. The north coast/northwest made the list of regions containing incorporated communities in 1911 with the completion of the Grand Trunk Pacific to Prince Rupert. The isolated Peace River/northeast region was the last to incorporate communities. By this measure, the Kootenay region dominated from 1901 to 1951.

Table 16.5 uses the 1951 census data to rank incorporated communities of 1,000 or more people. Population size tells us a lot more than simply the number of communities in an area and sheds a new light on economic development for the various regions of British Columbia. Comparing the totals at the end of Table 16.5 to those in Table 16.4 reveals that many of the communities incorporated after 1901 were relatively small. Table 16.5 also sections off cities attaining a population of 10,000 or more inhabitants. Three cities reached 10,000 by 1911 – Vancouver, Victoria, and New Westminster – and these were the only three for the next thirty years. By 1951, the number of such centres had doubled, with North Vancouver, Trail, and Penticton added to the list.

Although only three communities in British Columbia were incorporated by 1881, there were many unincorporated communities at this time, including those connected to the Cariboo Road, farming and ranching communities, mining communities, railway construction camps, and even old fur trade posts. All were small and many were tenuous.

Table 16.5

Incorporated communities over 1,000 population by rank, 1871-1951

1871	1881	1891	1901	1911	1921	1931	1941	1951	Community	Region
–	–	13,685	27,010	100,401	163,220	246,593	275,353	344,833	1 Vancouver	Lower Mainland
4,161	5,925	16,841	20,919	31,660	38,727	39,082	44,068	51,331	2 Victoria	Vancouver Island
1,356	1,500	6,641	6,499	13,199	14,495	17,524	21,967	28,639	3 New Westminster	Lower Mainland
–	–	–	–	8,196	7,652	8,510	8,914	15,687	4 North Vancouver	Lower Mainland
–	–	–	1,360	1,460	7,573	3,020	9,392	11,430	5 Trail	Kootenay
–	–	–	–	–	3,979	4,640	5,777	10,548	6 Penticton	Okanagan
–	–	–	–	4,184	6,393	6,350	6,714	8,546	7 Prince Rupert	North coast/northwest
–	–	–	261	1,663	2,520	4,655	5,118	8,517	8 Kelowna	Okanagan
–	–	–	1,594	3,772	4,501	6,167	5,959	8,099	9 Kamloops	South central interior
–	–	–	–	–	1,056	2,356	4,584	7,845	10 Port Alberni	Vancouver Island
–	–	–	802	2,671	3,685	3,927	5,209	7,822	11 Vernon	Okanagan
–	1,645	4,595	6,130	8,306	6,559	6,745	6,635	7,196	12 Nanaimo	Vancouver Island
–	–	–	5,273	4,476	5,230	5,992	5,912	6,772	13 Nelson	Kootenay
–	–	–	–	–	–	–	–	5,933	14 Kimberley	Kootenay
–	–	–	277	1,657	1,767	2,461	3,675	5,663	15 Chilliwack	Lower Mainland
–	–	–	–	–	2,053	2,479	2,027	4,703	16 Prince George	North central interior
–	–	–	6,156	2,826	2,848	2,097	3,657	4,604	17 Rossland	Kootenay
–	–	–	1,196	3,090	2,725	3,067	2,568	3,621	18 Cranbrook	Kootenay
–	–	–	–	–	–	–	518	3,589	19 Dawson Creek	Peace River/northeast
–	–	–	–	–	–	–	–	3,507	20 Westview	Okanagan
–	–	–	–	–	998	702	1,807	3,323	21 Alberni	Vancouver Island
–	–	–	–	–	1,178	1,312	1,539	3,232	22 Port Coquitlam	Lower Mainland
–	–	–	1,600	3,017	2,782	2,736	2,106	2,917	23 Revelstoke	South central interior
–	–	–	–	–	1,178	1,843	2,189	2,784	24 Duncan	Vancouver Island
–	–	–	–	–	–	1,314	1,957	2,668	25 Mission	Lower Mainland
–	–	–	–	–	810	1,219	1,737	2,553	26 Courtenay	Vancouver Island
–	–	–	1,640	3,146	2,802	2,732	2,545	2,551	27 Fernie	Kootenay
–	–	–	–	–	627	830	1,786	2,389	28 Salmon Arm	South central interior
–	–	–	–	–	1,030	1,260	1,512	2,246	29 Port Moody	Lower Mainland
–	–	–	746	3,295	1,151	1,443	1,706	2,094	30 Ladysmith	Vancouver Island
–	–	–	–	–	–	–	–	1,986	31 Campbell River	Vancouver Island
–	–	–	–	–	–	–	–	1,979	32 North Kamloops	South central interior
–	–	–	–	–	–	374	515	1,668	33 Hope	Lower Mainland
–	–	–	1,012	1,577	1,469	1,298	1,259	1,646	34 Grand Forks	Kootenay
–	–	–	–	–	–	–	–	1,628	35 Lake Cowichan	Vancouver Island
–	–	–	–	–	–	695	1,153	1,626	36 Creston	Kootenay
–	–	–	–	–	–	446	653	1,587	37 Quesnel	North central interior
–	–	–	–	–	–	–	–	1,350	38 Cranberry Lake	Kootenay
–	–	–	–	–	–	–	–	1,329	39 Castlegar	Kootenay
–	–	–	–	703	1,389	1,296	940	1,251	40 Merritt	South central interior
–	–	–	–	–	–	999	759	1,204	41 Smithers	North central interior
–	–	–	–	–	983	989	977	1,126	42 Armstrong	Okanagan
–	–	–	–	–	–	–	–	1,000	43 Oliver	Okanagan
									Other communities[a]	
–	–	–	732	1,657	2,161	2,371	885	971	Cumberland	Vancouver Island
–	–	–	1,359	778	371	171	363	809	Greenwood	Kootenay
2	3	4	13	19	26	28	29	43	Total[b]	

a Other communities that attained 1,000 population during 1871-1951.

b Includes only communities of 1,000 or over.

Sources: Farley (1979, 2); BC Stats (2004a).

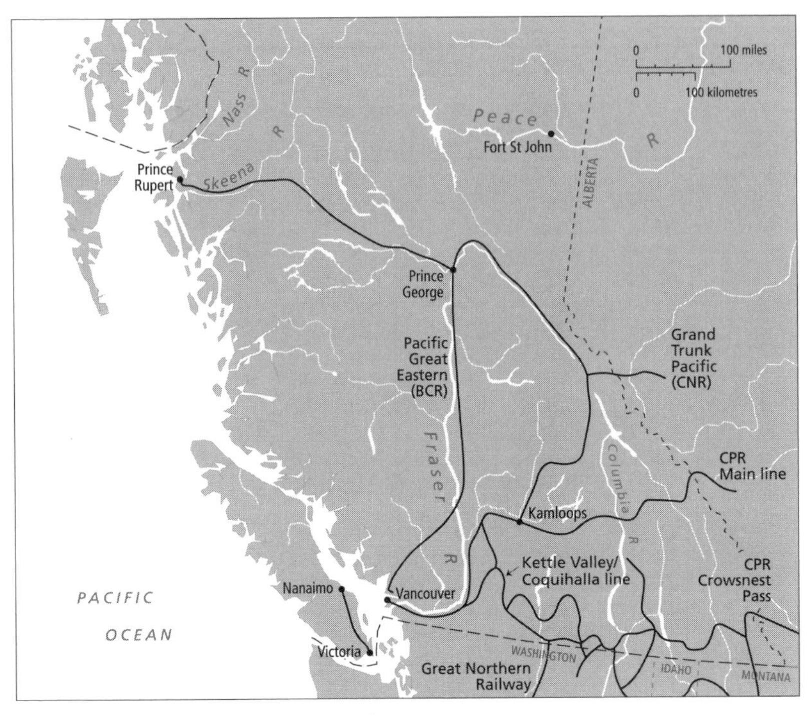

Figure 16.2 Main rail lines in British Columbia to 1952
Source: Modified from Forward (1987), 4; Galois (1990), Plate 21.

By the 1891 census, the most significant change was the creation of the city of Vancouver, which was incorporated in 1886. According to Robert Chodos (1973, 53), "The city of Vancouver, B.C.'s metropolis and the third largest city in Canada, is entirely a CPR creation, down to its name." The CPR was responsible for surveying the city into a grid pattern. It also owned much of the land, many of the buildings, and the major infrastructure of a national railway and international port. With these important transportation facilities in place, Vancouver attracted a variety of economic functions: warehousing for import export activities, sawmills, grain elevators, bulk storage facilities, financial institutions including a stock exchange, and manufacturing. The city became a central location for much of western Canada: "Vancouver's rise to prominence as a symbol of the new industrialism is undeniably related to the expansion of the provincial staple economy and to the intermediary role the city performed within the regional space economy" (McCann 1978, 34). By the 1901 census, the population of Vancouver had surpassed that of Victoria.

The 1901 census shows twenty-two incorporated communities. Four of them were on Vancouver Island: the provincial capital and the three coal mining communities of Nanaimo, Ladysmith, and Cumberland. Ten were in the Kootenays. This mineral-rich region experienced a mining boom in silver (e.g., Sandon), coal (e.g., Fernie), lead and zinc (e.g., Trail), and copper (e.g., Phoenix and Greenwood). Figure 16.2 shows some of the railways built to serve this region. Three supply and service centres appeared in the Okanagan – Vernon, Kelowna, and Enderby – while Kamloops and Revelstoke, both on the CPR main line, served the south central interior region. The Lower Fraser Valley, with its proximity to Vancouver, attracted a considerable number of settlers.

The urban pattern for 1911 was similar to that of 1901 in that it was dominated by communities in the Kootenays. It should be recognized, however, that the economic volatility of mining resulted in half of these communities losing population. Merritt appeared as one of the centres on the Kettle Valley/Coquihalla rail line, which was under construction in 1911. Vancouver's population exploded – experiencing a nearly 400 percent increase over 1901 – significantly influencing growth in the adjacent areas of New Westminster and North Vancouver. Vancouver Island's four communities also continued to grow. Construction of a second national railway, the Grand Trunk Pacific (absorbed into the CNR in 1922), with Prince Rupert as the northern terminus, opened up the north central interior region.

The First World War and enhanced transportation systems gave some impetus to growth for the next decade. Electricity was a new technology, and the increasing demand for it after 1910 resulted in a rudimentary provincial grid of electric lines connecting both communities and industries. Many remote communities and industries, not on the provincial grid, created their own electrical systems. The opening of the Panama Canal in 1914 was a significant factor in time-space convergence internationally and facilitated the movement of bulk goods to and from British Columbia. One of the consequences of new forms of energy and resource competition was a serious reduction in the demand for BC coal. Between 1913 and 1920, both output and employment dropped by 50 percent (Seager 1996, 217). The drop in population in the coal-mining communities of Nanaimo, Ladysmith, and Fernie reflects the impact of these dynamics. New technologies of lode mining and smelting developed and initiated smelter communities such as Anyox, in the north coast/northwest region, and Britannia Beach, just north of Vancouver. Neither was incorporated. In the Kootenays, both Phoenix and Sandon dropped off the map of incorporated communities as a result of the decline in their mining fortunes.

The years from 1921 to 1931 saw the postwar expansion of the forest, mining, and fishing industries. Fourteen incorporated communities, most of them small, were added during this time, nine of them along the CNR line and the newly completed (1921) Pacific Great Eastern Railway, which ran between Squamish and Quesnel. Prince George, with a population of nearly 2,500, was the dominant service centre for the north central interior region. Then the Great Depression began in 1929. These were "bust" times. British Columbia's resource-based economy was hit hard, and many urban centres had slow growth or lost population. Racial tensions during this period affected both the Chinese and Japanese communities as the Chinese Immigration Act of 1923 banned Chinese from entering Canada, and further immigration rules in 1923 and 1928 seriously restricted the Japanese.

The Depression of the 1930s took its toll on employment, and the tough times lasted until the end of the decade and the advent of the Second World War. The coal-mining community of Cumberland, for example, lost nearly two-thirds of its population, thus falling below 1,000. Overall population growth slowed down; only four small communities were incorporated during this period, two of which were in the Peace River/northeast region. Creston, in the Kootenays, was the only community newly crossing the 1,000 population threshold.

The decade from 1941 to 1951 saw a turnaround of the economy due to the demands of the Second World War and the boom following it. Ship building in the Burrard Inlet employed approximately 17,000 workers alone, and sent North Vancouver's population well over the 10,000 mark (Belshaw and Mitchell 1996, 320). Forest, mineral, and fish products were in great demand. Hydroelectric dams were built to tap the tremendous energy potential of the province, and oil and natural gas discoveries prompted growth in the Peace River/northeast region. The Alaska Highway, with Dawson Creek as Mile 0, was constructed by the Americans as an important part of the North American military infrastructure, and it made the whole of northern British Columbia more accessible. The road system throughout the province was expanded and upgraded – newly paved in many areas – as automobile and truck transportation provided competition for rail transportation and diesel fuel provided competition for coal. The addition of twenty-one incorporated communities, increasing the total to forty-three with a population over 1,000 and six over 10,000, reflected provincial growth as a whole. Some population shifts were related to the forced evacuation of Japanese Canadians in 1942 from coastal British Columbia to the interior, mainly in the Kootenays.

Examining incorporated communities from 1871 to 1951 tells only part of the urban story for British Columbia in this period. By the 1890s and early 1900s, company towns had appeared on the map. Corporations established mining towns (e.g., Michelle, Cumberland, Britannia Beach), pulpmill towns (e.g., Port Mellon, Woodfibre, Powell River), and cannery communities (e.g., Klemtu, Namu), most of them "closed" communities (see Chapter 15). Some grew and became incorporated, while others became ghost towns.

POSTWAR BOOM TO GLOBAL UNCERTAINTY: INSTANT TOWNS TO WORLD CITIES, 1961 TO THE PRESENT

Urbanization, settlement, resource development, transportation systems, and values – all transformed from the 1950s on. British Columbia experienced prosperous times until the 1970s, and then a series of unpredictable global economic events, most of them negative, restructured the economy of the province and of the world. The population quadrupled between 1951 and 2011 – one indicator of economic prosperity – but growth was not uniform throughout the province and there was a very distinctive shift to urban areas.

The 1950s to the 1970s are referred to as "the long boom." It was a wave of unprecedented growth in North America, and British Columbia was swept along with it. Higher disposable incomes, increased dependence on automobile use, new homes in the suburbs for the postwar baby boomers, rapid immigration, and new consumer goods for all were part of improving living conditions and a move toward greater urbanization. This higher standard of living throughout North America translated into a need for more and more of British Columbia's resources.

Oil, natural gas, and electricity became the most important forms of energy, and regions such as the Peace River/northeast had all three. Oil was discovered just north of Dawson Creek at Boundary Lake, while natural gas came mainly from the vicinity of Fort St. John and Fort Nelson. Employment and growth came not only from resource exploration and development but also from the construction of elaborate pipeline systems throughout the province. The enormous hydroelectric potential of the rivers of British Columbia was recognized by the provincial government in the 1950s and became an important component of its industrial strategy. Construction jobs contributed to the employment and population growth of the Kootenay and Peace River regions in response to the demand for dams and large transmission lines.

The interior had a number of wood processing plants but no pulp mills until 1961. The increased demand for pulp, paper, lumber, and other wood products made the largely untapped forests of the interior economical to exploit by the 1960s. Multinational corporations, most of them foreign, built new pulp mills throughout the province and through consolidation achieved vertical integration of all aspects of the forest industry. The urban growth related to this massive investment was felt in both existing and new towns: Castlegar and Skookumchuck in the Kootenays; Kamloops in the south central interior; and Quesnel, Prince George, and Mackenzie in the north central interior. Prince George alone had three mills. Although the most obvious impact was on the interior, many of the old pulp mills on the coast were either upgraded or phased out. The Ocean Falls mill was shut down and new mills were built at Gold River on Vancouver Island and Kitimat in the north coast/northwest region.

A major expansion into mining occurred as the world market prices of copper, molybdenum, silver, lead, and zinc increased. In the early 1970s, the price of gold was unpegged and allowed to find its own level; it rose dramatically (see Table 10.3, p. 182). The value of coal made a dramatic rebound as coking or metallurgical coal used in the steel industry and thermal coal for producing electricity came into demand, especially in Japan. Large, foreign-owned multinational corporations opened mines on Vancouver Island and throughout the interior, investing in capital-intensive technologies such as enormous earth-moving machinery and open-pit techniques. In some cases, new communities were built (e.g., Logan Lake, Fraser Lake, Sparwood, Elkford), while other mining operations enhanced existing communities.

The initial expansion and upgrading of road systems in the 1950s became "pavement politics" in the 1960s and 1970s. In fact, the whole transportation system was transformed as the provincial government became involved in airport expansions, ferry systems, and the extension of the Pacific Great Eastern Railway (renamed the British Columbia Railway, or BCR, in 1971 and leased to CN in 2004) from Prince George to Fort St. James in 1967 and to Fort Nelson by 1971. These developments provided accessibility and assisted in the growth of the expanded urban system.

The provincial and federal governments were also involved in education, health, and developing the social safety net, which not only meant new schools, hospitals, and other government institutions but also a community college system throughout British Columbia.

Most of this rapid growth affected the urban system. Suburbanization went hand in hand with highway and freeway expansion and a growing number of private automobiles, from 153,325 in 1951 to 544,310 by 1971 (Strong-Boag 1996, 288). By 2008, over 3 million vehicles had been registered in the province (Statistics Canada 2009). The move to the rural-urban fringe was prompted by the desire to own single-family dwellings, a prospect that was becoming more difficult as land prices climbed in the growing established communities. Residential and commercial sprawl moved farther and farther from the urban centres. Most urban functions – retail, wholesale, entertainment, banking, manufacturing – in the form of shopping centres and industrial parks were motivated to find new suburban, commutershed locations. They then became the catalyst for further growth. The process caused enormous problems for high growth regions such as the Lower Mainland, the Victoria-Saanich peninsula on Vancouver Island, and the Okanagan. Agricultural land, never in abundance in British Columbia, was consumed for urban-industrial uses at a rapid rate. Servicing low density regions with water, sewer, and solid waste facilities presented other challenges. The need for planning was obvious.

The provincial government responded in various ways to all of these developments. The Instant Towns Act of 1965 accommodated the construction of new resource towns for the booming resource industry. New instant towns such as Gold River, Mackenzie, Logan Lake, Elkford, and Sparwood were designed as "permanent" incorporated communities and deliberately structured differently from the old company town model. Nevertheless, they were single-resource communities and therefore vulnerable to world market conditions. (See Chapter 15 for a fuller discussion of instant towns.)

Regional districts were also created as part of the new legislation in 1965, creating a federation of incorporated and unincorporated electoral areas, all with political representation and municipal powers under the Municipal Act. This allowed the unincorporated areas of regional districts to become involved in planning, in public utilities for water, sewer, and solid waste, and in many other functions, from parks and recreation to dog control, that had previously been restricted to incorporated urban areas.

By 1972, the provincial government had passed the Agricultural Land Commission Act, which zoned all agricultural land in the province and prohibited much of its conversion to urban uses. Urban amalgamation or annexation was also encouraged by the provincial government, and many urban boundaries expanded (e.g., Kamloops, Prince George, Kelowna, Nanaimo). Vancouver city exceeded 400,000 people and therefore became a census metropolitan area, but it was more practical to view this centre within the Greater Vancouver Regional District (GVRD) – Vancouver and the rapidly growing adjacent communities of North Vancouver, Burnaby, Richmond, Surrey, and so on – as one large metropolitan area. Vancouver's role in the province was changing as it gained more and more control of resource development throughout the province (Robinson and Hardwick 1968; Denike and Leigh 1972). Victoria, the main urban centre for the Capital Regional District (CRD), had a population of nearly 200,000 by the 1971 census and became the second census metropolitan area in the province.

By the 1970s and '80s, traditional ways of making a living had been radically altered. Internationally, currencies were encouraged to float and find their own levels, and the energy crisis that began in 1973 caused the world to reassess its energy supply and demand. The 1970s were a period of high inflation, increased wages, and speculation in land. By contrast, the first half of the 1980s was a serious reminder of the 1930s Depression, marked by high unemployment and a major reduction in demand for resources.

Trade and investment had become much more global by the 1970s and 1980s, as technologies of time-space convergence facilitated the rapid movement of goods, ideas, services, capital, and people. International markets also shifted radically from Atlantic-based trade to the Pacific, placing BC, especially Vancouver, at Canada's front door. These technologies also aided new global organizational structures that allowed corporations to fragment the production process. Post-Fordism, with its flexible specialization, challenged the traditional, centralized industrial assembly line process of producing goods. (See Chapter 7 for the shift from Fordism to post-Fordism.)

British Columbia felt the impact of these changes. The employment of capital-intensive technologies in mines and mills began to reduce the number of unionized workers. Employment in the province fell by 79,000 from 1981 to 1984 with losses of, among others, 3,100 in forestry, 10,400 in mining, 23,600 in construction, and 37,700 in manufacturing (Belshaw and Mitchell 1996, 334). British Columbia suffered more than any other province during the recession of the 1980s. Restructuring, downsizing, and adjusting to the new world order of business meant uncertainty for not only the resource industries but also manufacturing and services. The economy partially recovered in the late 1980s, but the Canada-United States Free Trade Agreement (FTA) of 1989 created further uncertainty going into the 1990s.

Governments made substantial investments in an attempt to buoy the economy. One of the new communities to appear in British Columbia during this time was Tumbler Ridge, the centre for northeast coal. The provincial and federal governments spent considerable amounts of money extending the BCR line from Anzac to Tumbler Ridge (the only electric rail line in the province) and double tracking the CNR line to Prince Rupert, where the new coal port of Ridley Island was constructed. (For a fuller explanation of the controversial nature of this investment, see Chapter 12.) Other major government investments included the Coquihalla Highway and Expo 86. These two projects had profound impacts on tourism for the province. As the traditional industries were cutting back, the provincial government also became aware of the retirement industry. The Okanagan, the southeastern side of Vancouver Island, and the Lower Mainland became more popular destinations for retirees across Canada and helped these regions grow.

As seen earlier, the population of British Columbia doubled in the twenty years between 1951 and 1971. It took the next forty years for the population to double again because of the economic turbulence from the 1970s to the present. Most importantly, almost all of this increase of nearly 2.4 million people took place in incorporated communities (Table 16.6).

The 1990s saw the uncertain global economic climate continue. First the FTA and then the 1994 North American Free Trade Agreement (NAFTA) affected some industries more than others (e.g., see Chapter 13 on agriculture). Change in the territorial status of Hong Kong from a British colony to a part of China in 1997 increased immigration of Hong Kong Chinese to British Columbia, along with substantial transfers of capital and investment, mainly to the Lower Mainland region. By the end of the 1990s, the "Asian economic flu" had affected the strong economic links between British Columbia and the Asia-Pacific countries and resulted in another recession for the province.

The economic polarization between the highly urban and integrated core of the Lower Mainland plus the southern end of Vancouver Island and the rest of the province intensified. As Lewis Robinson (1998, 342) remarks, "Georgia Strait links the urban centres rather than separating them; this southwestern region may accurately be described as the heartland of British Columbia." Characterized by knowledge- and information-based employment, higher value-added manufacturing, and administrative functions, this core has also gained an ever greater share of the incoming population (p. 344). No longer just a port city with traditionally strong links to the rest of the province and western Canada, by the 1990s, Vancouver had become an integrated part of the Asia-Pacific region and a major city (Davis and Hutton 1994; Hutton 1997).

The hinterland of the province, with its reliance on export of resources, struggled through the 1990s. The forest industry had to face a reduction in the amount of wood available and some of the highest production costs in the world for products such as pulp. Nor has mining been spared. Low world market prices for metals such as copper and energy minerals such as coal have reduced production, threatening the closure of communities such as Logan Lake and Tumbler Ridge.

The new millennium did not usher in any optimism for economic or political stability. At the North American and global level, the dot-com crash beginning in 2000 put into question the long-term value of the Internet in terms of buying and selling goods and services, and increased the skepticism of investors. The terrorist attack

Table 16.6

Population and urbanization, British Columbia and Canada, 1951-2011

Year	Population of British Columbia	Percentage		Population of Canada	Percentage	
		Rural	Urban		Rural	Urban
1951	1,165,210	47.2	52.8	14,009,429	38.4	61.6
1961	1,629,082	27.4	72.6	18,200,600	30.4	69.6
1971	2,184,621	24.3	75.7	21,962,100	23.9	76.1
1981	2,744,465	22.0	78.0	24,342,600	24.3	75.7
1991	3,282,061	19.6	80.4	27,296,900	23.4	76.6
2001	4,078,447	15.3	84.7	31,110,600	20.3	79.7
2011[a]	4,572,617	13.8[b]	86.2[b]	34,018,957[c]	19.2[b]	80.8[b]

a Estimated from BC Stats (2009b).
b Estimated from Statistics Canada (2010b).
c Estimated from Statistics Canada (2010a).
Sources: McGillivray (2010, 27); BC Stats (2009b); Statistics Canada (2010a, 2010b).

of 11 September 2001 left us with images of the New York World Trade Centres collapsing and led to increased security measures, hampering the movement of people and goods crossing international borders. Global conflict also increased with the US wars in Afghanistan and Iraq.

By 2002, investor confidence in the stock market and mutual funds was further eroded by the bankruptcies of Enron (2001) and Worldcom, Inc. (2002) in the United States. The Worldcom bankruptcy, which had potential links to government complicity, was the largest in US history at the time, but Lehman Brothers Holding, Inc. (2008) now holds the record. The Softwood Lumber Agreement between Canada and the United States came to an end in 2001, but a new agreement could not be negotiated. The United States unilaterally imposed tariffs and duties amounting to 32 percent. This had a profound impact on sawmills within British Columbia; many reduced production and others closed. Rulings by the World Trade Organization state that the movement of Canadian lumber into the American market has not harmed US producers, but US appeals have stalled the reopening of the border to "free trade." An agreement was finally signed in 2006, but the already depressed industry was hit even harder with the recession in the last quarter of 2008, when over seventy plants shut down. By 2008, communities such as Mackenzie had experienced all of their forest mills shutting down and an 80 percent unemployment rate. The world market price for softwood lumber and pulp has only recently (spring 2010) made a recovery.

Trade, investment, and travel were also influenced by pandemics. Severe acute respiratory syndrome broke out in Asia in late 2002. SARS was then transported to Canada, first to Toronto and eventually to Vancouver. An avian flu outbreak in the Fraser Valley in 2004 brought renewed fears of a deadly virus, and that fear was enhanced by the swine flu (H1N1) scare of 2009-10.

Following the war in Afghanistan (2001), the war in Iraq, begun in 2003, furthered the United States's war on terrorism following 9/11. One consequence of the war was a restricted supply of oil and a record-breaking world market price per barrel (US$140 in August 2008) that pushed up prices at the gas pumps and heavily influenced the cost of transportation.

At the national and provincial levels, other events in the 2000s both stimulated and hindered the economy. Canada's agreement with the United States over salmon in 1999 ended the fish war between the two countries; however, many salmon stocks have collapsed and salmon farming, which brings in higher revenue, has become prominent, although it is often blamed for reduced runs of wild salmon. The McRae-Pearse Commission of 2004 recommended that all commercial fishing be privatized (individual transferable quota) with twenty-five-year licences, and many commercial species are managed through ITQs (see Chapter 9). The price of natural gas escalated in the early 2000s and has continued to spur further exploration and development, particularly in the Peace River/northeast region. As well, there has been increasing political pressure to lift the moratorium on offshore drilling. The unstable price of metals and coal, which caused temporary closures at Myra Falls (coal) and Highland Valley Copper (Logan Lake) and a permanent closure at Tumbler Ridge in the early 2000s recovered and reached record prices by 2008. New mines opened, and more exploration ensued; however, the recession of 2008 led the price of all minerals except gold to collapse. The poor economic development of the late 1990s and early 2000s caused outmigration to exceed inmigration (see Table 7.9, p.122). The recovery of metal and energy prices by 2004 reversed the interprovincial migration pattern as new mines and exploration attracted many to British Columbia.

Part of the heartland/hinterland polarization is the perception that the highly urbanized core extends its values and political influence into the resource-oriented remainder of the province. Legislated increases in parks, preservation of wildlife areas, and protection of watersheds are often viewed as driven by "urban environmentalists." Many British Columbians, especially those in resource-based industries, are also concerned about the provincial government's recognition of Aboriginal title and the negotiation of the Nisga'a, the Tsawwassen, and other modern treaties. (For more details see Chapter 5.)

Table 16.7 shows the number of incorporated communities of 10,000 or more people, ranked by their 2009 population. This list stands in marked contrast to the list of incorporated communities for 1951, when only six had exceeded this number (Table 16.5). In the over fifty years since 1951, the urban system has evolved into 153 incorporated communities in twenty-seven regional districts.

In the same period, the population has shifted from 52 percent to over 86 percent urban. The urban hierarchy of four census metropolitan areas (CMAs, population over 100,000) and twenty-three census agglomerations (CAs, with between 10,000 and 100,000) in the 2006 census reflects new urban relationships from global to local.

More specifically, this pattern reflects the regionalization of British Columbia into a core/periphery distribution of population in which the core is highly urbanized and integrated with diversified employment frequently tied to global economic transactions (e.g., information, banking and financial services, and high-tech industries). The core, or heartland, area includes three CMAs and six other urban centres listed in Table 16.7: North Cowichan, Courtney, Nanaimo, Chilliwack, Whistler, and Squamish. This relatively small geographic area contains approximately two-thirds of the population of the province, and is likely to retain an even greater share in the future.

Outside the core is the periphery, or hinterland, of British Columbia, which largely relies on resource extraction and processing. Within this large geographic area, three urban centres have emerged to perform important regional service roles: Kelowna (another CMA) for the Okanagan, Kamloops for the south central interior, and Prince George for the north central interior. Table 16.7 provides a comparison between the populations of 1996, 2001, and 2009; Prince George has lost population during this time, a reflection of how this important service centre remains tied to a forest industry that has not been stable. Kelowna's population has increased greatly, a development that can be attributed to its international airport and favourable climate, which is a magnet for the retirement industry. The city is also an important administrative centre. The regional centres below 20,000, such as Port Alberni, Prince Rupert, Terrace, Powell River, Dawson Creek, Kitimat, and Quesnel, also perform important regional services but they depend much more on primary resources; these centres have lost population since 1996. As well, many of the communities below 10,000 depend on a single resource. These hinterland communities are the most vulnerable of all in the urban hierarchy.

Table 16.7

Municipalities over 10,000 population, 2009, and change from 1996

Municipality	2009	2001	1996	% change
1. Vancouver, CMA	2,318,526	2,092,902	1,891,465	18.4
2. Victoria, CMA	351,314	326,753	304,287	13.4
3. Kelowna, CMA	184,375	154,241	89,442	51.5
4. Abbotsford/ Mission, CMA	173,133	156,073	105,403	39.1
5. Kamloops	87,017	81,153	76,394	12.2
6. Nanaimo	84,228	75,106	70,130	16.7
7. Chilliwack	76,106	65,228	60,186	20.9
8. Prince George	74,547	75,206	75,150	-0.8
9. Vernon	38,968	34,548	31,817	18.4
10. Penticton	33,250	31,642	32,097	3.5
11. Campbell River	31,328	29,181	29,982	4.3
12. North Cowichan	29,493	27,158	26,222	11.1
13. Courtney	24,216	18,865	18,016	25.6
14. Fort St. John	19,457	16,437	15,683	19.4
15. Cranbrook	19,161	18,721	18,831	1.7
16. Port Alberni	17,741	17,795	19,163	-8.0
17. Salmon Arm	17,220	15,540	15,181	11.8
18. Squamish	17,181	14,867	14,549	15.3
19. Comox	13,444	11,316	11,460	14.8
20. Powell River	13,338	13,085	13,610	-2.0
21. Prince Rupert	12,846	15,032	16,714	-30.1
22. Parksville	11,783	10,451	9,796	16.9
23. Terrace	11,675	12,376	13,298	-13.9
24. Dawson Creek	11,514	10,788	11,579	-0.6
25. Lake Country	11,409	9,518	9,330	18.2
26. Summerland	11,243	10,824	10,927	2.8
27. Williams Lake	11,090	11,596	10,895	1.8
28. Coldstream	10,388	9,316	9,289	10.6
29. Whistler	10,228	9,461	7,592	25.8
30. Kitimat	9,226	10,775	11,136	-20.7
31. Quesnel	9,710	10,271	10,532	-8.5

Sources: BC Stats (2006, 2010).

SUMMARY

Since British Columbia joined Confederation, profound changes have occurred within a relatively short period. In 1871, the population was barely over 36,000, there were only two incorporated communities, and the province was only 15 percent urban. In 2009, British Columbia had over 4.5 million inhabitants and was over 86 percent urban. The increase in urbanization and population can be followed regionally, as it has been tied to the discovery and exploitation of resources and development of transportation systems that went hand in hand with corporate and political decisions.

Initially, coastal water routes and interior trails, roads, and river systems linked resource regions. The coming of the CPR was the beginning of the railway era, which brought many new technologies for exploiting resources throughout the mountainous landscape of British Columbia. Railways, steam engines, and ships also reduced British Columbia's isolation globally. Urbanization increased dramatically, reflecting the spatial patterns of transportation and resource penetration. From the 1890s to the 1950s, world events such as war and depression helped determine the economic viability of resource extraction and thus of many communities.

World events played an even greater role in shaping the geography of British Columbia from the 1960s on, as transportation technologies shrank time and space, and business transactions and commerce began to be conducted on an international scale. The way people made a living changed radically, forcing a reorganization of British Columbia into a new regional geography of heartland and hinterland. From an economic and urban perspective, two economies evolved: the peripheral resource-based economy, relying on primary industry processing and oriented to the export market; and the service-based economy of the core, where knowledge and information are the important commodities.

In the future, as the world shrinks and greater pressures are exerted toward freer trade, the new economy will have the advantage. This translates into a disproportionate amount of urban and economic growth, mainly for the southwestern area of the province. The problem in this region will be to manage this growth in order to maintain the quality of life. This does not bode well for the hinterland because many of its communities remain attached to the resource-based economy. Signs of change were evident when primary-industry (e.g., agriculture, forestry, fishing, or mining) employment in 1996 amounted to 85,800 jobs; by 2009, this number had been reduced to 74,600 (BC Stats 2009a). A number of the more isolated regions – the north coast/northwest, the Peace River/northeast, and the Kootenays – may actually decline in population. For individual communities in this vast and differentiated geographic hinterland, economic diversification will continue to be the major challenge.

REFERENCES

Belshaw, J.D. 1996. "Provincial Politics 1871-1916." In *The Pacific Province: A History of British Columbia,* ed. H.J.M. Johnston, 134-64. Vancouver: Douglas and McIntyre.

Belshaw, J.D., and D.J. Mitchell. 1996. "The Economy since the Great War." In *The Pacific Province: A History of British Columbia,* ed. H.J.M. Johnston, 313-42. Vancouver: Douglas and McIntyre.

Chodos, R. 1973. *The CPR: A Century of Corporate Welfare.* Toronto: James Lorimer.

Davis, H.C., and T.A. Hutton. 1994. "Marketing Vancouver's Services to the Asia Pacific." *Canadian Geographer* 38 (1): 18-27.

Denike, K.G., and R. Leigh. 1972. "Economic Geography 1960-1970." In *British Columbia,* ed. J.L. Robinson, 139-45. Toronto: University of Toronto Press.

Duff, W. 1965. *The Indian History of British Columbia.* Victoria: Royal British Columbia Museum.

Farley, A.L. 1979. *Atlas of British Columbia: People, Environment, and Resource Use.* Vancouver: UBC Press.

Forward, C.N. 1987. "Evolution of Regional Character." In *British Columbia: Its Resources and People,* ed. C.N. Forward, 1-24. Western Geographical Series vol. 22. Victoria: University of Victoria.

Galois, R. 1990. "British Columbia Resources." In *Historical Atlas of Canada.* Vol. 3: *Addressing the Twentieth Century, 1891-1961,* ed. D. Kerr and D. Holdsworth. Toronto: University of Toronto Press.

Harris, C. 1997. *The Resettlement of British Columbia: Essays on Colonialism and Geographical Change.* Vancouver: UBC Press.

Hayes, D. 1999. *Historical Atlas of British Columbia and the Pacific Northwest.* Vancouver: Cavendish.

Higgins, D.J.H. 1977. *Urban Canada: Its Government and Politics.* Toronto: Macmillan.

Hutton, T.A. 1997. "The Innisian Core-Periphery Revisited: Vancouver's Changing Relationships with British Columbia's Staple Economy." *BC Studies* 113 (Spring): 69-98.

Johnston, H.J.M. "Native People, Settlers, and Sojourners, 1871-1916." In *The Pacific Province: A History of British Columbia,* ed. H.J.M. Johnston, 165-204. Vancouver: Douglas and McIntyre.

Kerr, D.G.G., ed. 1975. *Historical Atlas of Canada.* 3rd ed. Toronto: Nelson.

Leonard, F. 2002. "'A Closed Book': The Canadian Pacific Railway Survey and North Central British Columbia." In *Western Geography* (Canadian Association of Geographers) 12: 163-84.

McCann, L.D. 1978. "Urban Growth in a Staple Economy: The Emergence of Vancouver as a Regional Metropolis, 1886-1914." In *Vancouver: Western Metropolis,* ed. L.J. Evenden, 17-42. Western Geographical Series vol. 16. Victoria: University of Victoria.

McGillivray, B.P. 2010. *Canada: A Nation of Regions.* 2nd ed. Toronto: Oxford University Press.

Meen, S. 1996. "Colonial Society and Economy." In *The Pacific Province: A History of British Columbia,* ed. H.J.M. Johnston, 97-132. Vancouver: Douglas and McIntyre.

Muckle, R.J. 1998. *The First Nations of British Columbia.* Vancouver: UBC Press.

Robin, M. 1972. *The Rush for the Spoils: The Company Province 1871-1933.* Toronto: McClelland and Stewart.

Robinson, J.L. 1998. "British Columbia: Canada's Pacific Province." In *Heartland and Hinterland: A Regional Geography of Canada,* 3rd ed., ed. L. McCann and A. Gunn, 321-55. Toronto: Prentice- Hall.

Robinson, J.L., and W.G. Hardwick. 1968. "The Canadian Cordillera." In *Canada: A Geographical Interpretation,* ed. J. Warkentin, 438-72. Toronto: Methuen.

Seager, A. 1996. "The Resource Economy 1871-1921." In *The Pacific Province: A History of British Columbia,* ed. H.J.M. Johnston, 205-52. Vancouver: Douglas and McIntyre.

Short, J.R., and Y. Kim. 1999. *Globalization and the City.* New York: Addison Wesley Longman.

Siemens, A.H. 1972. "Settlement." In *Studies in Canadian Geography: British Columbia,* ed. J.L. Robinson, 9-31. Toronto: University of Toronto Press.

Strong-Boag, V. 1996. "Society in the Twentieth Century." In *The Pacific Province: A History of British Columbia,* ed. H.J.M. Johnston, 273-312. Vancouver: Douglas and McIntyre.

INTERNET

BC Stats 2006. "British Columbia Municipal Census Population, 1921-2006." www.bcstats.gov.bc.ca/data/pop/pop/mun/mun1921_2006.asp.

–. 2009a. "Employment and Unemployment Rate by Detailed Industry, 15 Years and Over, Annual." www.bcstats.gov.bc.ca/data/lss/rate_ind/empurindbc.pdf.

–. 2009b. "Population Projections." www.bcstats.gov.bc.ca/data/pop/pop/dynamic/PopulationStatistics/index.asp.

–. 2010. "Population Estimates: 2006-2009". www.bcstats.gov.bc.ca/data/pop/pop/estspop.asp#totpop.

Census of Canada. 1881. "Table 1: Areas, Dwellings, Families, Population, Sexes, Conjugal Condition." 1: 6-8. www.collectionscanada.gc.ca/databases/census-1881/index-e.html.

–. 1951. "Table 6: Population by Census Subdivision, 1871-1951." 4: 63. www.statcan.gc.ca/.

Statistics Canada. 1983a. *Historical Statistics of Canada.* Catalogue 11-516-X1E. Series A125-63. Origins of the population, census dates, 1871-1971. www.statcan.gc.ca/pub/11-516-x/sectiona/4147436-eng.htm.

–. 1983b. *Historical Statistics of Canada.* Catalogue 11-516-X1E. Series A2-14. Population of Canada, by province, census dates 1851-1976. www.statcan.gc.ca/pub/11-516-x/sectiona/4147436-eng.htm.

–. 1983c. *Historical Statistics of Canada.* Catalogue 11-516-X1E. Series A67-69. Population, rural and urban, census dates, 1851-1976. www.statcan.gc.ca/pub/11-516-x/sectiona/4147436-eng.htm.

–. 2009. "Road Motor Vehicles, Registrations, Annual." CANSIM Table 405-0004. www.statcan.gc.ca/.

–. 2010a. "Estimates of Population: Canada, Provinces and Territories." CANSIM Table 051-0005. www.statcan.gc.ca/.

–. 2010b. "Population and Dwelling Counts, for Canada, Provinces and Territories by the Statistical Area Classification, 2006 and 2001 Censuses – 100% Data." www.statcan.gc.ca/.

Glossary

Aboriginal rights. Rights derived from the historical use of land that include the right to use the land and the right to self-government. These rights are constitutionally recognized by the Constitution Act of 1982, which protects them but fails to describe their nature.

Aboriginal title. Ownership or control of the historical territory of various tribes, clans, or bands of First Nations and Inuit. The geographic boundaries of such territory were not static, and overlaps in territorial use for resource procurement during the year were common among First Nations. Identification of historical boundaries that define Aboriginal title is important in establishing resource rights and compensation in the modern treaty process.

allowable annual cut. The volume of wood allowed by the provincial Ministry of Forests and Range to be harvested each year from a forest region, or timber supply area. The calculation takes into account the density, species, ages, and growth rates of merchantable timber. The AAC also has to be adjusted over time in relation to changing forest conditions, such as reduction in old-growth volumes, environmental restrictions, removal of forest land for other land uses, fires, and disease.

biomagnification. Process of the food chain whereby low levels of toxic wastes (e.g., parts per billion of mercury) are consumed by one- and two-celled animals and passed up the chain. These tiny animals are consumed by those on the next level, which are in turn consumed. In the process, the level of toxic materials magnifies, or accumulates in ever increasing amounts. Because humans are at the top of the food chain and consume food such as fish, contamination levels have in some cases been high enough to cause serious illnesses and death. These consequences to human health are often referred to as the environmental backlash and raise serious questions about so-called safe levels of toxic wastes allowed into the environment.

clastics. Small fragments of rock created by the weathering process and frequently carried by water and deposited as sediments. The buildup of these sediments results in both pressure and heat, and the process eventually forms **sedimentary rock.**

common property resource. A resource owned in common by the public and managed by government bodies (e.g., salmon, the foreshore, water).

company town. A **single-resource community** created and controlled by a corporation. This was common historically in British Columbia, particularly with mining/smelting communities and coastal pulp mills. The company was the only employer and controlled housing and the company store. Most of the resulting communities were "closed," meaning that they were not regulated under the province's Municipal Act.

comprehensive claims. Claims for compensation made by First Nations for Aboriginal title lands that have not been covered by treaties or were never ceded by First Nations. The federal government recognized the need to negotiate comprehensive claims as a result of the *Calder* case (1973), although it was not until 1981 that it was willing to exchange the undefined Aboriginal land right for concrete rights and benefits. The provincial government of British Columbia did not recognize these claims until 1992. *Contrast* **specific claims.**

convection precipitation. Process by which incoming solar radiation heats the earth and air in proximity to the earth, causing the warm air, which contains moisture from evaporation and transpiration, to rise in the atmosphere, where it cools, condenses, forms clouds, and often results in thundershowers. In British Columbia this is mainly a summer phenomenon, when the solar radiation is its most intense.

cultural region. An area of the populated world defined by having common cultural characteristics. These characteristics can be further subdivided into formal (e.g., common language areas), functional (e.g., timber supply areas), and vernacular/perceptual cultural regions (e.g., sense of place).

deformation. General term to describe processes that produce folding, faulting, and changes to the surface of the earth.

deterritorialization. The process by which a group of people lose their territory and their traditional ways of living. In British Columbia, First Nations have undergone deterritorialization as a result of European colonization and settlement.

earthquake. A tectonic vibration resulting from the clash of two crustal plates. These movements may be a result of volcanic eruption; a **transform fault,** in which two plates slide past one another; or **subduction,** in which one plate slides under another. The vibrations caused by this movement radiate from the epicentre of the earthquake, and the magnitude of the waves is calculated logarithmically on the Richter scale. A quake reading 3.0 produces little shaking, whereas one reading 8.0 produces major movement, especially at the epicentre or its vicinity.

erosion. The transportation of rock sediments and weathering of rock through the action of water, wind, glaciation, and mass wasting (gravity).

Eurocentrism. The imposition of the value systems of a European culture in order to judge other cultures. The term is derived from *ethnocentrism,* the judgment (often negative) of other cultures in terms of one's own.

fiduciary trust. The legal obligation, under the Indian Act, for the federal government to act in the best interests of First Nations.

flash flooding. Flash flooding is a result of intense and sustained rainfall, and occurs in coastal watersheds, including Vancouver Island and Haida Gwaii, mainly in the winter months. Flooding is the natural hazard responsible for the greatest costs of destruction in British Columbia.

Fordism. The assembly line process of manufacturing standardized products, usually at centrally located plants. The term is derived from the process used by Henry Ford to assemble automobiles at the turn of the twentieth century.

frontier mentality. The mentality of people arriving in British Columbia during the Fraser River and Cariboo gold rush period with the intention of getting rich and leaving. This attitude led to much damage to the environment, some irreversible.

geomorphology. The study of the processes that change the surface of the earth.

geophysical hazard. The assessment of risk from the earth's forces. These forces can be categorized in relation to tectonic, climatic, and gravitational forces. Geophysical hazards are also referred to as natural hazards. They are assessed in terms of the threat or risk to human property and/or life.

ghost town. A community, usually developed because of a single resource, in which the main employment base has terminated and most or all the residents have left.

hard energy path. A term used by A. Lovins (1977) to describe ever increasing dependence on nonrenewable energy sources, a focus on the supply side of energy, future dependence on coal and nuclear energy as oil and gas run out, and enormous costs for energy. In this model, individuals are alienated from energy decisions because large corporations and/or large government bodies control both the energy and the planning for future energy. *Contrast* **soft energy path.**

head tax. A tax on individuals coming into a country. Within the context of British Columbia's history, the term applies specifically to the levy of a series of head taxes by the federal government in an attempt to reduce Chinese immigration. This became a barrier to the spatial diffusion of the Chinese to British Columbia.

ice jam flooding. This can occur on any river that freezes over in winter. The ice is frequently in layers, and where the stream slows, or where there are river obstructions such as bridge piers, the ice is more stable, or stuck firmly to the shore. When ice breaks up – usually in the spring, but unusual winter warming can also result in ice flows – the ice fragments jam where the ice remains stable and thus form a dam where the water backs up, flooding upstream areas. The pressure of the water behind the ice jam can result in a sudden release of the jam, thus causing rapid flooding of downstream areas.

igneous rock. Rock formed from the molten state, either rapidly through exposure to cooler surface environments such as air or water, or much more slowly if the material does not reach the surface of the earth. Extrusive igneous rock, such as basalt, has undergone the rapid cooling process and has high density and weight. The slow cooling process results in intrusive igneous, which has much larger-grained rock structures with lower density and weight, such as granite.

industrial economy. Employment from the production of goods (secondary industry). A distinction is made in **staples theory,** as well as in core-periphery (heartland-hinterland) analysis, between industry based on the manufacturing of consumer goods (e.g., automobiles), which is largely done by core regions, and the relatively few forward and backward linkages in processing (or manufacturing) raw materials (e.g., logs to lumber). Gaining industry, and its associated occupations, was a major impetus to urban growth until the 1970s.

instant town. A single-resource community created in British Columbia under the Instant Towns Act (1965). Major new investment in the resource frontier in the 1960s and '70s (especially forestry and mining) created a need for new communities. These single-resource communities were a response to the negative conditions of the **company town;** they were regulated under the Municipal Act and operated in a similar fashion to other villages and towns in the province.

isostasy. The balance between the weight of continental crusts pushing down into the mantle and the uplifting forces of the mantle itself. Weathering and erosion both wear down and reduce the weight of physiographic features such as the Rockies. This loss of weight, and height, results in an uplifting of the Rockies. The fairly rapid removal of the vast sheets of ice covering much of the northern portion of North America only 10,000 years ago resulted in isostatic rebound, or a similar uplifting of the earth with the removal of this enormous weight.

lode mining. The process of crushing rock and extracting the minerals of value. Most metals occur in "bound" form, found along with other minerals within rock. Lode mining technologies were employed in British Columbia in the late 1800s and early 1900s, mainly to extract silver, gold, and copper. Although some placer mining for gold continues today, nearly all metal mining production in the province is through lode mining. The technologies of lode mining are sophisticated, costly, and usually require large corporate financing.

magma. The liquid interior portion of the earth where extreme temperatures melt rock. Magma makes its way to the surface of the earth through plate tectonics and becomes igneous rock.

manifest destiny. A mid-nineteenth-century American belief that the United States was destined to expand through all of western North America and eventually, according to some advocates, to cover the entire continent.

metamorphic rock. Rock formed through the intense pressure and heat from, or chemical infusion from, new molten material intruding into existing rock structures. Rocks undergoing this process can develop entirely new physical properties and chemical structure.

modern treaties. Contemporary treaties negotiated as compensation – in the form of land, resources and resource management, or money – for the extinguishment of Aboriginal title. These treaties also include options for self-government and greater autonomy for First Nations. Treaty negotiations in British Columbia did not occur until 1992, at which point agreement between the provincial and federal governments resulted in a six-stage process.

New Caledonia. The name given to the central Interior Plateau region of British Columbia by Simon Fraser in 1806. The Americans referred to this territory, and the area south of it, as the Oregon Territory. Fraser's name remained after the **Oregon Treaty** (1846) extended the US-Canada border west along the forty-ninth parallel from the Rockies to the coast, but in 1858 Queen Victoria named the area British Columbia to avoid confusion with a French colony in the south Pacific also named New Caledonia.

Oregon Treaty. An 1846 boundary agreement between the United States and Britain to continue the border along the forty-ninth parallel from the Rockies to the Pacific coast. Vancouver Island, which extends south of the fortyninth parallel, was allowed to remain part of the British North American territory. British sovereignty over the area, then known as the Oregon Territory, began with this agreement.

orogeny. The process of mountain building caused by tectonic forces that have folded and faulted land masses through compression.

orographic effect. Precipitation resulting from relatively warm, moist air being forced up mountain barriers, where it cools and condenses. It is analogous to a saturated sponge being squeezed.

placer mining. The process of mining stream beds for gold. Since gold can be found in a pure state as dust or nuggets and is one of the heaviest elements, it tends to settle in stream beds. There it can be recovered by some of the simplest and least expensive technologies: a shovel and gold pan. The discovery of gold in British Columbia, and consequent gold rush in the mid-1800s, was the result of placer mining. Many other technologies were also employed, including dams and sluices, hydraulic systems, and dredges, but these are more costly.

plate tectonics. A combination of two older hypotheses, continental drift and sea-floor spreading, plate tectonics theory is essential to the understanding of geomorphology and geology. It asserts that the earth's crust is made up of large and small plates, which move in a manner somewhat analogous to a conveyor belt, through magma being forced to the surface of the earth in some geographic locations and destroyed in others. Regions where molten material comes to the surface and pushes plates apart are known as **rift zones,** and regions where plates are pushed under or over one another are known as **subduction** zones. Regions where plates simply push past each other in a parallel manner are referred to as **transform faults.** Plate movement results in **earthquakes, volcanic activity,** and mountain building.

post-Fordism. The new technologies and economic conditions that apply to the period after Fordism. Since the mid-1960s, multinational corporations have employed techniques of flexible specialization (short-run production through contracting out) to manufacturing goods at a global scale. The result has been the bankruptcy, merger, or restructuring of many corporations and uncertainty for all.

private property resource. A resource that can be held, or controlled, by private interests, such as private property.

rain shadow effect. Relatively low precipitation in a given region because of mountain barriers. In southwestern British Columbia, westerly winds are forced to rise over the Insular Mountains of Vancouver Island. This **orographic effect** wrings out much of the moisture from the air mass before it passes over the mountains. As it descends on the lee (east) side of the mountains, the air mass expands and absorbs moisture, leaving the region from Victoria to Vancouver considerably drier than the west side of Vancouver Island.

region. An area of the surface of the earth that can be distinguished through physical and/or human characteristics.

regional geography. A subfield of the discipline of geography in which spatial phenomena are described and studied by dividing the world into areas having common physical and/or human characteristics.

resource. Any naturally occurring substance of value to a society. This definition implies that resources are culturally defined.

reterritorialization. The new set of values, institutions, and "rules" established as one cultural group gains control over the territory of another group. In British Columbia a primarily British value system was imposed on the First Nations inhabiting the territory.

rift zone. A region of earth where magma reaches the crust's surface and splits crustal plates apart. The ocean floors of the world are where the earth's crust is thinnest and where most rift zones occur, resulting in sea-floor spreading.

rock cycle. The process of rocks constantly being recycled as part of the earth's physical process. Igneous rocks created from the molten state are subject to weathering and erosion and may become sedimentary rocks; heat, pressure, and chemical action produce metamorphic rock; tectonic processes may cause these rocks to go back to the molten state.

sedimentary rock. A relatively soft rock formed through the bonding, or cementing together, of sedimentary materials (e.g., limestone, shale).

service economy. Employment from the provision of services, including knowledge and information, as opposed to manufacturing. Services were categorized as tertiary industry until the complexity of service occupations required a division into tertiary and quaternary industries by 1961. Through this division, service workers dealing directly and face to face with the public, such as sales across the counter in retail and fast food outlets, remain categorized as tertiary workers. Quaternary workers carry out transactional services (and are sometimes referred to as transactional workers). Lawyers, accountants, government employees, and those in tourism, real estate, and the expanding information and education sectors are examples. Through Census Canada, quaternary employment can also be divided into producer services (or government and corporate services) and consumer services (or services to individuals). The major shift to quaternary industry employment is the basis of the service-based economy, which favours large urban centres because of their global communications and many economic, political, and cultural functions.

single-resource community. A community in which the main employment is derived from one resource. In British Columbia single-resource communities have developed mainly through the harvest of fish, forests, and minerals, although some agricultural communities have occurred, as have tourist communities such as Whistler.

snow-melt flooding (spring run-off flooding). This type of flooding is associated with the interrelated factors of drainage basin size, snowpack over the winter season, and the spring weather conditions responsible for the rate of snowpack melt. Spring run-off flooding affects communities and built environments adjacent to the many rivers draining the interior of the province.

soft energy path. A term used by A. Lovins (1977) to describe an alternative to the **hard energy path.** The approach involves a focus on the demand side of energy, recognition that energy influences lifestyle, and the necessity for individuals to become responsible for their own energy requirements. Lovins advocates technology changes that both develop renewable energy sources and conserve energy.

spatial diffusion. A concept of movement through time and over space employed to trace the spread (adoption) of ideas or innovations, people, and goods from one geographic location to another. Also known as the spread effect, spatial diffusion identifies the "barriers" (forces that prevent movement) and "carriers" (factors assisting movement) that produce the spatial distribution of phenomena.

specific claims. Claims for compensation by individual bands and tribal councils based on an alleged breach of fiduciary duty or responsibility on the part of Canada. These claims are often for reserve lands that have been taken without compensation. Seizure of such land occurred in many ways in British Columbia, from the outright annexation of reserve land to the construction of roads, railways, hydroelectric lines, and pipelines through reserve lands. *Contrast* **comprehensive claims.**

staples theory. A theory of Canada's economic development based on the exploitation of five resources: fish, furs, timber, wheat, and minerals. Regional economic growth occurs through the discovery, development, and export of these resources, and some regions undertake resource manufacturing as well. Economic historian Harold Innis suggested this theory in the 1930s to account for the regional development of Canada.

subduction. Plate tectonic activity that occurs where plates collide and one plate overrides the other. The overridden plate – usually the heavier, oceanic plate – bends downward and descends, or subducts, into the mantle. Mountain building occurs as a result of compression along the boundary where two plates collide. Subduction zones also result in deep oceanic troughs and continental volcanic activity because of the friction of subduction.

sustainability. Also referred to as sustainable development. Development implies the use of resources, but sustainable development recognizes that renewable resources can be exploited beyond their ability to reproduce and that nonrenewable resource use often has major negative effects on the ecosystem. This concept, which emerged in the early 1980s, has important implications for appropriate technologies in resource harvesting and use.

tenure. A system of allowing private corporations access to publicly held land and resources. Tenure involves various types of arrangements: licences, leases, and grants. The main issue of tenure in British Columbia is allowing private corporations timber rights to provincially controlled forest land.

terranes. Fragments of oceanic or continental plates. When plates collide these fragments attach themselves to the adjacent continental plate. Much of British Columbia is made up of attached, or accreted, terranes, making the geology of this province very complex.

territory. The boundaries of a geographic area within which political control is exerted.

time-space convergence. Change in transportation technologies that reduces the time required to move or communicate between geographic locations. Also referred to as time-space compression or collapse. Expressions such as "the world is shrinking" recognize that modern satellite communications, airline flights, and expressway systems allow communication and movement on many geographic scales. It is important to recognize that changes to movement are not equally distributed, however, and some geographic locations are therefore more isolated and remote than others.

transform fault. An area of the earth's crust where one crustal plate pushes past another crustal plate in a parallel manner. Both the Pacific Plate and North American Plate are moving north, for example, but the Pacific Plate is moving more rapidly. The two plates are often in a "stuck" position along the transform fault from Haida Gwaii north to Alaska, and a major earthquake occurs when they move.

volcanic activity. The result of the eruption of molten material, or **magma,** that has come to the surface of the earth's crust. This activity is frequent in **rift zones,** where magma comes to the surface under the ocean. Volcanoes are also associated with **subduction** zones, such as the one off the west coast of British Columbia. Many parts of the interior and west to the coast have been active volcanic areas throughout the past 150 million years.

weathering. Breaking down. Two broad divisions, chemical weathering and mechanical weathering, categorize the agents involved in this process. Chemical weathering is the decomposition of minerals through agents such as water, carbon dioxide, and oxygen, which form acids. These agents sometimes combine with organic materials during this process. Mechanical weathering is the breaking down of rocks, mainly through water running into rock cracks or fissures and then freezing, expanding, and breaking the rock into fragments.

Index

Page references where glossary terms appear in **bold** are also marked in **bold.** "f" after a page number indicates a figure, and "t," a table.